QUANTITATIVE ANALYSES OF BEHAVIOR

Standing, left to right: 1. Michael L. Commons 2. Leonard Green 3. John E.R. Staddon 4. Barbara Burkhard 5. Richard J. Herrnstein 6. B.F. Skinner 7. R. Duncan Luce 8. Joseph J. Pear 9. Brian L. Rector 10. Peter A. de Villiers 11. William Timberlake 12. Charles P. Shimp 13. John W. Donahoe 14. Marvin Z. Deluty 15. William Vaughan, Jr. 16. David S. Olton 17. Alan C. Kamil

Seated, top step, left to right: 1. James Smith 2. John H. Kagel

Seated, bottom step, left to right: 1. James Allison 2. John Gibbon 3. William M. Baum 4. Howard Rachlin 5. John A. Nevin 6. Stephen E.G. Lea 7. Ben A. Williams 8. Gene M. Heyman 9. Alan Silberberg 10. James E. Mazur

QUANTITATIVE ANALYSES OF BEHAVIOR

Matching and Maximizing Accounts

Volume II
of
Quantitative Analyses of Behavior
A Five-Volume Series

Edited by

MICHAEL L. COMMONS
Harvard University

RICHARD J. HERRNSTEIN
Harvard University

HOWARD RACHLIN
State University of New York
Stony Brook

BALLINGER PUBLISHING COMPANY
Cambridge, Massachusetts
A Subsidiary of Harper & Row, Publishers, Inc.

International Standard Book Number: 0-88410-739-6

Library of Congress Catalog Card Number: 81-2654

Printed in the United States of America

Library of Congress Cataloging in Publication Data (Revised)

Main entry under title:

Quantitative analyses of behavior.

Based on the annual Harvard Symposium on Quantitative Analysis of Behavior, held in April 1978.
Includes bibliographies and indexes.
Contents: v. 1. Discriminative properties of reinforcement schedules / edited by Michael L. Commons, John A. Nevin – v. 2. Matching and maximizing accounts / edited by Michael L. Commons, Richard J. Herrnstein, and Howard Rachlin.
1. Conditional response – Congresses. I. Commons, Michael L. II. Nevin, John A. III. Harvard Symposium on Quantitative Analyses of Behavior. [DNLM: 1. Behavior – Congresses. 2. Reinforcement schedule – Congresses. BF 319.5.R4 D611 1978]
BF319.Q36 150.19′43 81-2654
ISBN 0-88410-377-3 (v. 1)
ISBN 0-88410-739-6 (v. 2)

Volumes in the **QUANTITATIVE ANALYSES OF BEHAVIOR** series:

- Volume I: DISCRIMINATIVE PROPERTIES OF REINFORCEMENT SCHEDULES. Edited by Michael L. Commons, *Harvard University* and John A. Nevin, *University of New Hampshire*
- Volume II: MATCHING AND MAXIMIZING ACCOUNTS. Edited by Michael L. Commons, *Harvard University*, Richard J. Herrnstein, *Harvard University*, and Howard Rachlin, *State University of New York, Stony Brook*
- Volume III: ACQUISITION. Edited by Michael L. Commons, *Harvard University*, Richard J. Herrnstein, *Harvard University*, and Allan R. Wagner, *Yale University*
- Volume IV: DISCRIMINATION PROCESSES. Edited by Michael L. Commons, *Harvard University*, Richard J. Herrnstein, *Harvard University*, and Allan R. Wagner, *Yale University*
- Volume V: REINFORCER VALUE – THE EFFECT OF DELAY AND INTERVENING EVENTS. Edited by Michael L. Commons, *Harvard University*, John A. Nevin, *University of New Hampshire*, and Howard Rachlin, *State University of New York, Stony Brook*

HARPER & ROW, PUBLISHERS
NEW YORK
Cambridge
Hagerstown
Philadelphia
San Francisco
1817
London
Mexico City
Sao Paolo
Sydney

CONTENTS

List of Figures xi

List of Tables xxi

Preface—*Michael L. Commons* xxiii

Introduction—*Richard J. Herrnstein* xxvii

PART I FINE STRUCTURE IN INSTRUMENTAL RESPONDING

Chapter 1
Toward Analyzing the Continuity of Behavior
—*Joseph J. Pear, Brian L. Rector, and Joseph A. Legris* 3

Chapter 2
How Reinforcers Are Aggregated in Reinforcement-Density Discrimination and Preference Experiments
—*Michael L. Commons, Michael Woodford, and John R. Ducheny* 25

Chapter 3
A Molecular Approach to Ratio Schedule Performance
—*James E. Mazur* 79

Chapter 4
Reinforcement and the Local Organization of Behavior
—*Charles P. Shimp* 111

Chapter 5
The Interchangeover Time as a Molecular Dependent Variable in Concurrent Schedules
—*Alan Silberberg and John M. Ziriax* 131

Chapter 6
Some Persistent Issues in the Study of Matching and Maximizing—*John A. Nevin* 153

PART II ECOLOGY OF INSTRUMENTAL RESPONDING

Chapter 7
The Mechanism of Optimality in Foraging
—*Stephen E. G. Lea* 169

Chapter 8
An Ecological Perspective on the Study of the Allocation of Behavior—*Alan C. Kamil, Jerrilynn Peters, and Frederick J. Lindstrom* 189

Chapter 9
Staying and Shifting: Their Effects on Discrimination Learning—*David S. Olton* 205

Chapter 10
Instrumental Behavior and Foraging in the Wild
—*William M. Baum* 227

PART III MULTIPLE AND CONCURRENT SCHEDULES

Chapter 11
Behavioral Competition, Contrast and Matching
—*John E. R. Staddon* 243

Chapter 12
Choice and the Rescorla-Wagner Model
—*William Vaughan, Jr.* 263

Chapter 13
Do Interactions in Multiple and Concurrent Schedules
Have a Common Basis?—*Ben A. Williams* 281

PART IV PUNISHMENT

Chapter 14
Maximizing, Minimizing, and Matching between
Reinforcing and Punishing Situations
—*Marvin Z. Deluty* 305

Chapter 15
Toward a Quantitative Theory of Punishment
—*Peter A. de Villiers* 327

PART V ECONOMIC MAXIMIZATION THEORY

Chapter 16
Economics of the Matching Law
—*Howard Rachlin* 347

Chapter 17
Preference and Response Substitutability
in the Maximization of Behavioral Value
—*Barbara Burkhard* 375

Chapter 18
Ratio Schedules of Reinforcement and their Relation
to Economic Theories of Labor Supply
—*Leonard Green, John H. Kagel,
and Raymond C. Battalio* 395

PART VI MATCHING THEORY

Chapter 19
Melioration as Behavioral Dynamism
—Richard J. Herrnstein 433

Chapter 20
Is Time Allocation Unconditioned Behavior?
—Gene M. Heyman 459

PART VII BEYOND MATCHING OR MAXIMIZING

Chapter 21
A Unified Principle of Reinforcement:
Some Implications for Matching
—John W. Donahoe, Michael A. Crowley,
Will J. Mallard, and Kenneth A. Stickney 493

Chapter 22
Constraints on Performance in Two Elementary
Paradigms—*James Allison* 523

Chapter 23
The Emperor's Clothes: Assumptions of the Matching
Theory—*William Timberlake* 549

Name Index 569

Subject Index 575

About the Editors 583

List of Contributors 585

LIST OF FIGURES

1-1	Overall session-to-session propensity, speed, and response-rate data from several variable-interval schedules and extinction in Experiment 1	6
1-2	Diagram of the apparatus used in Experiment 2	10
1-3	Selected segments of head-movement data, shown as distance from the key over time, from session 1 of VI 1 minute for bird P-21	12
1-4	Selected segments of head-movement data from sessions 2-5, 10, and 15 of VI 1 minute for bird P-21	14
1-5	Selected segments of head-movement data from sessions 20 and 0, 20 and 91 of VI 1 minute for bird P-21	18
1-6	Selected segments of head-movement data from sessions 1-3, 10, 16, 20, and 60 of VI 1 minute for bird P-22	20
2-1	A stimulus example from a discrimination trial and contingencies during an entire trial; the second-link reinforcing stimulus from a preference trial and contingencies during an entire trial	29
2-2	The distribution of reinforcement for the first center-key peck for 4-cycle and 6-cycle substimuli	34
2-3	Left-key-peck probability for the 4-cycle discrimination data for each individual substimulus	44
2-4	The extent to which the curves generated from the forgetting models fit values of substimuli derived from the multivariate regression estimates	54

2-5 Titrated effective value plotted against the perceived value 58
2-6 Theoretical plots of the $V_{S_{N_d}}$ for the logit and for the probit transformations 70

3-1 The general form of the relation between the duration of the remaining instrumental requirement and the value of instrumental behavior predicted by Eq. (3.6) 84
3-2 Mean postreinforcement pause durations in Powell's (1968) experiment compared to the results of the computer simulation 86
3-3 For Powell's 1968 experiment, approximate proportions of time spent eating and key pecking compared to the results of the computer simulation 88
3-4 Representative cumulative records generated in the computer simulation of Powell's (1968) experiment 89
3-5 Relative frequencies of postreinforcement pauses in consecutive 27-second bins for several FR sizes in Powell's (1968) experiment 90
3-6 Predicted and obtained postreinforcement pause durations in the experiments of Winograd (1965) and Kaplan (1956) 91
3-7 Predicted and obtained postreinforcement pause durations in the experiments of Powell's (1969) experiment, where both ratio size and reinforcer duration were varied 92
3-8 Relative durations of lever pressing and combined schedule behaviors for each subject in the study of the response effort 93
3-9 Sizes of FR schedules that were about equally preferred to the VR 120 schedule plus the switching requirement for subjects in the experiment of Sherman and Thomas (1968: experiment 1) 98
3-10 The relative durations of lever pressing and the combined schedule behaviors for Mazur's (1977) study of VR schedules 100
3-11 For the study of satiation in FR schedules, the relative durations of the schedule behaviors and the conditional probabilities of drinking, shown for successive tenths of a session in each condition 103
3-12 The results of a computer simulation for the study of FR schedules and satiation 105

4-1 The relative frequency of occurrence of sequences as a function of their lengths (the number of responses in them) 121
4-2 The relative frequency of the shorter of two reinforced sequences of 2 and 4 responses, for single subjects 123
4-3 The relative frequency of the shorter of two reinforced sequences of 2 and 8 responses, for single subjects 124

4-4 The relative frequency of the shorter of two reinforced sequences, each curve an average for four subjects 125

5-1 Matching data, changeover-probability curves, and sequential statistics from OMNI 1 simulation 138
5-2 Conditional probability of a changeover as a function of time between changeovers for concurrent VI 60-second VI 180-second schedules as a function of time between changeovers from a computer simulation 141
5-3 Probability of reinforcement on concurrent VI 60-second VI 180-second schedules as a function of time between changeovers from a computer simulation 142
5-4 Conditional probability of a changeover as a function of time between changeovers for concurrent chain VI 120-second VI 40-second versus chain VI 120-second VI 80-second schedules from Silberberg et al., 1978: experiment 4 144
5-5 Probability-of-reinforcement curves from a computer simulation where preference is varied 146
5-6 Probability of reinforcement and the conditional probability of a changeover as a function of time between changeovers, from Nevin (1979) replication and a computer simulation 148

6-1 Conditional probability of pecking green given a prior peck to red, as functions of the total number of switches 159
6-2 Shimp and Hawkes (1974) data replotted to show the ratio of short to long interresponse times as a function of the ratio of reinforcers received for these interresponse times 162

7-1 Flowchart of the schedule of reinforcement used to simulate a foraging situation in Lea, 1979 175
7-2 Simulated foraging (Lea, 1979), showing the effect of varying the search-state fixed-interval schedule parameter on the probability of accepting the "long-prey" schedule when it was offered 177
7-3 Effect of varying the rate of occurrence of the short prey (per unit time spent for aging) on probability of accepting the long prey 178
7-4 Effect of varying the rate of occurrence of the long prey (per unit time spent for aging) on probability of accepting it when encountered 179
7-5 Effect of observed ratio of reinforcement immediacy on observed response-rate ratio in the choice state 180

8-1 A *Catocala relicta* on a white birch tree trunk 193

8-2	The mean giving-up and attack times across the last two sessions at each travel time for jays' search for crytic moths	197
8-3	The mean giving-up and attack times across the last two sessions at each travel time for jays on a red/green discrimination	199
9-1	Results of the radial-arm maze test	213
9-2	Summary of the procedure for the discrimination test	216
9-3	Results of the three-table maze test	217
10-1	The organism and environment represented as a feedback system	229
10-2	The dependence of rate of reinforcement on response rate imposed by several conventional variable-interval schedules (feedback functions)	230
10-3	Feedback functions for foraging: showing predation (reinforcement) as a function of rate of foraging	234
10-4	Feedback function for foraging when averaging interval T is relatively long	236
10-5	Feedback functions for foraging in the same patch when various averaging intervals T are assumed	237
10-6	Feedback function for foraging when averaging interval is relatively short	238
11-1	Pressure model for equilibria in concurrent and multiple VI VI schedules	248
11-2	Contrast as a function of the reinforcement for the interim activity or for the instrumental response in the unchanged component, deduced from the pressure model	250
11-3	Matching relations on multiple VI VI schedules, deduced from the pressure model	251
11-4	Ratio of response rates in a multiple schedule as a function of reinforcement rate on a concurrent VI schedule, deduced from the pressure model	254
11-5	A competition model for local contrast effects	257
12-1	Value of key A as a function of time at delivery of reinforcement	265
12-2	The equilibrium values of keys A and B as a monotonically strictly increasing function of the local reinforcement rates on those keys	266
12-3	Flow diagram of the basic variables assumed	268
12-4	Nonlinear and linear variable-interval 2-minute schedules	269
12-5	Log ratio of behavior on left and right keys as a function of log ratio of obtained reinforcements	270

12-6 Function corresponding to the sum of a linear interval and a ratio schedule 271
12-7 Schedule produced by subtracting a ratio from a linear interval schedule and schedule produced by subtracting a linear interval from a ratio schedule 272
12-8 Response rate and reinforcement rate on schedules with negative slopes as well as on schedules with zero slopes 273
12-9 Relative time on right and rate of reinforcement for three birds on an interdependent schedule 275
12-10 Relative time spent in blue for birds in the fourth experiment, following the first and second training conditions 278

13-1 Responses per minute in the constant VI 3-minute component of a multiple schedule, averaged over four subjects, as a function of component duration 284
13-2 Hypothetical pattern of local response rate postulated by additivity theory for the VI component of a multiple VI EXT 286
13-3 Local response rates for individual subjects trained on a multiple VI 1-minute EXT schedule 288
13-4 Local rate of responding within the VI component of a multiple VI 1-minute EXT schedule, averaged over subjects 289
13-5 The sequences of component presentation used in Williams (1979) 292
13-6 Response rates, averaged over subjects, during the circle and lines components, which had identical schedules 293

14-1 For Experiment 1, the proportion of time spent in component 1 as a function of the proportion of obtained punishers in component 2 and the proportion of time allocated to component 1 as a function of the square roots of the relative delays until shock 307
14-2 For Experiment 2, the proportion of time spent in component 1 as a function of the proportion of obtained punishers in component 2 and the proportion of time allocated to component 1 as a function of the square roots of the relative delays until shock 312
14-3 For Experiment 3, the proportion of responses and time allocated to lever 1 as a function of the scheduled proportion of reinforcements to lever 1 315
14-4 For Experiment 3, the proportion of responses and time allocated to lever 1 as a function of the obtained proportion of reinforcements to lever 1 318

14-5	For Experiment 4, the proportion of responses and time allocated to lever 1 as a function of the obtained proportion of reinforcements to lever 1	320
14-6	For Experiment 5, the proportion of responses and time allocated to lever 1 as a function of the obtained proportion of reinforcements to lever 1	321
15-1	Deviation of relative response rates and relative time distribution from the obtained relative reinforcement frequencies as a function of punishment intensity; suppression ratios	335
15-2	Analysis of baseline and punishment data from Farley (1980: experiment 1) in terms of least-squares fit linear functions relating log response ratios to log reinforcement ratios	336
16-1	Flow diagram of a molar choice situation	349
16-2	The feedback function, $C = aT^m$, where $a = 1$ and $m = 0$ (variable-time schedule), $m = 0.2$ (variable-interval schedule), and $m = 1.0$ (variable-ratio schedule)	351
16-3	Amount of C_1 versus amount of C_2	354
16-4	Amount of rootbeer versus amount of tom collins for a rat in the experiment of Rachlin, Green, Kagel, and Battalio, 1976	363
16-5	The slope of the budget line of Figure 16-4 (a_1/a_2) versus the ratio of consumption (C_1/C_2) by the rat with that budget	364
16-6	A plot of C_1 versus C_2 for completely substitutable commodities	367
17-1	Indifference curves for combinations of two responses	378
17-2	Indifference surfaces shown in three-dimensional space	380
17-3	The locus of distributions of three responses contained in a constant time interval	381
17-4	Intersection of indifference surfaces with the triangular plane of Figure 17-3 to show indifference contours on the triangle	382
17-5	Indifference contour that bounds a convex set and one that does not bound a convex set	383
17-6	Indifference contour showing nearly complete substitutability between two responses	384
17-7	Values of distributions for two subjects	385
17-8	Geometrical representation of a ratio-like schedule	386
17-9	Representation of the toy-snack experiment in triangular coordinates	387
17-10	Gross-sections of three-dimensional indifference surfaces whose sides are parallel planes	389

18-1 Indifference curves for consumption and leisure (not working) and the constraint on choices imposed by a ratio schedule of reinforcement 398
18-2 Effects of income-compensated wage changes 402
18-3 Mean rates of responding for individual birds as a function of the different random-ratio schedules of reinforcement with no free consumption 411
18-4 Expansion paths implied by the minimum-distance hypothesis under two different ratio schedule requirements 416
18-5 Expansion paths obtained under RR 25 and RR 100 schedules of reinforcement for each bird as the number of free reinforcements delivered was varied 417
18-6 Feedback function for concurrent VI VR schedule and hypothetical indifference curves for consumption and leisure 421

19-1 Concurrent VR VR in which alternative 1 reinforces with a higher probability than alternative 2 438
19-2 The logged ratio of times spent on the components of the concurrent VI VR as a function of the logged ratio reinforcements received from them, for four pigeons on six pairs of schedule values 441
19-3 R_D and R_T for a hypothetical concurrent VI VR; t_1 gives the proportion of time allocated to the VI 442
19-4 For four pigeons, the four transitions where matching predicts a shift away from equal responding to the green and red alternatives 445
19-5 Relative responding to the green alternative as a function of relative reinforcement received therefrom, for each of four pigeons, averaging over the final six sessions of eight experimental conditions 446
19-6 Programmed local rates of reinforcement on left and right alternatives as a function of time on right, measured as a proportion of unit time 447
19-7 Overall reinforcements as a function of relative time on the right alternative, for two experimental conditions (adapted from Vaughan, 1981) 448
19-8 Local rate of reinforcement on the right minus local rate of reinforcement on the left as a function of relative time on the right alternative (adapted from Vaughan, 1981) 449
19-9 Maximization and melioration predictions for conditions in Vaughan's experiment (1981) 450

20-1 Sequence of events in a concurrent VT VT schedule 462
20-2 Three different switching patterns in a concurrent schedule 463

20-3	The expected reinforcement rate in a concurrent VT 60-second VT 180-second schedule as a function of time spent at the VT 60-second schedule	466
20-4	Expected reinforcement rate in a concurrent VT 60-second VT 180-second schedule with a fixed-time switching process	467
20-5	The relation between changeover delay duration and changeover rate	470
20-6	The relation between time allocation and reinforcement rate in a concurrent VT 60-second VT 180-second schedule that uses a changeover delay	471
20-7	A comparison of typical performance and reinforcement rate maximization predictions in concurrent VT VT schedules	473
20-8	A comparison of typical performance and reinforcement rate maximization predictions (Poisson switching) in concurrent VT VT schedules	474
20-9	Performance and maximization theory predictions in procedures that use a changeover delay	476
20-10	Performance and maximization theory in concurrent schedules that use a changeover delay and a response requirement	477
20-11	Performance and maximization theory in interdependent interval schedules	479
20-12	Performance and maximization theory predictions in a concurrent VI VR experiment (Herrnstein and Heyman, 1979)	480
20-13	Switching in three different concurrent schedule experiments	483
21-1	A space displaying the time allocated among three hypothetical behaviors	504
21-2	The relative amount of time allocation and the relative rate of wheel running during the reinforced component of a concurrent VI 20-second, EXT schedule	512
22-1	Hypothetical example of experimental constraints in the controlled-time paradigm	524
22-2	Theoretical labor supply curves	526
22-3	Total drinking time as a function of total lever-holding time in baseline and on nine cumulative time schedules	529
22-4	Total food consumed as a function of total lever presses in baseline and on twelve fixed-ratio schedules	530
22-5	Total licks as a function of total lever presses in baseline and on variable-interval schedules	531
22-6	Total licks as a function of total lever presses in baseline and on concurrent fixed-ratio schedules	533
22-7	Total 45-milligram food pellets as a function of total lever presses on four concurrent fixed-ratio schedules	535

22-8 Total 45-milligram food pellets as a function of total lever presses on four concurrent fixed-ratio schedules 536
22-9 Hypothetical example of experimental and theoretical constraints in the controlled-amount paradigm 537
22-10 Rate of 25 water licks as a function of the one lever press 539
22-11 Rate of 10 and or 100 saccharin licks or water licks as a function of the rate of licking the dry spout 540

LIST OF TABLES

2-1	Description of stimuli	32
2-2	Simple concurrent schedule compared with concurrent chain schedule	36
2-3	Plan for the analysis, with data from the preference study compared with data from the discrimination study	38
2-3a	Organization of the presentation of the data, analysis, and corresponding theory	41
2-4	Three transformations of $p(L)$ fit to the multivariant model of the contribution of each reinforcer to the aggregate perceived value	50
2-5	Models for the discrimination study and the preference study	52
4-1	Experimental conditions	119
5-1	Simulated schedule contingencies and simulated relative rate	145
6-1	First-order sequential statistics for responding and reinforcement in a simulation of Nevin's (1969) study	156
6-2	Matrix of first-order dependencies	157
6-3	Matrix of first-order dependencies assuming 500 switches in 1,000 trials and matching to relative reinforcement at 0.75	158
13-1	Sequence of conditions for subjects trained on multiple VI 3 VI 1, multiple VI 3 EXT, and multiple VI 3 signaled VI 1	285

13-2 Rates of responding during components with VI 3-minute schedules as a function of the schedule during Blue 295
13-3 Following-schedule effects on behavior in a concurrent-chain schedule 298

14-1 Sequence of conditions in Experiments 1 and 2 and the results 308
14-2 Sequence of RI x-minutes schedules for lever 2 for each subject in Experiment 3 314
14-3 Responses per minute and obtained reinforcements per minute on lever 1 and lever 2 at each of the scheduled proportion of reinforcers in Experiments 3-5 316

15-1 Parameter values and percentage of the data variance accounted for by least-squares fits of Eqs. (15.3) and (15.9) to absolute response rates from Farley (1980: experiment 1) 339

17A-1 Observed mean distance ($n = 4$) between Y and Y' on all restriction lines 393

18-1 Summary of results: mean of last five sessions 408
18-2 Substitution effects (percentage decrease in responding) as a function of baseline random-ratio requirement 410
18-3 Income effects (percentage decrease in responding as a function of free food) as a function of random-ratio requirement 410

19-1 Probability of food reinforcement in a dark-key period 444

20-1 Overall expected reinforcement rate in a concurrent VT 60-second VT 180-second schedule as a function of time allocation and changeover delay duration 471

PREFACE

The study of behavior has consisted of a number of somewhat separate traditions. One tradition, starting with Thorndike and then Skinner, has experimentally analyzed the control of behavior by events subsequent to it. A second tradition, starting with Bechterev and Pavlov and coming down to the present through Watson, Spence, and others, has analyzed the control and transfer of control by events antecedent to behavior.

After the 1920s both traditions have become more quantitative. In the experimental analysis of behavior, quantifiable variables, such as the rate of responding, were used to represent the behavioral outcomes. At the same time, more elaborate quantitative studies were carried out in the Hullian tradition. Quantifiable measures such as response probability and latency were introduced. In that period, and extending through the 1950s, mathematical models were developed by Hull, Spence, Estes, Bush, and Mosteller and Logan, among others. Both groups carried out some parametric studies in the tradition of psychophysics. By the early 1960s mathematical psychology had developed to deal with problems from a number of domains. In each domain explicit mathematical models were proposed for the processes by which performances were acquired and maintained within that domain. Although the models generated a number of experiments, they were of limited generality.

"Quantitative analysis" now generally refers to the fact that theoretical issues are represented in quantitative form. An analysis is not a matter of fitting arbitrary functions to data points. Rather, each parameter and variable represents part of a process that has both a theoretical and an empirical interpretation. The matching law, a model of maintained performance, is one example from the analysis of behavior tradition. The Rescorla–Wagner model of acquisition processes is a second example. These models represent effects of interactions of environmental and behavioral events. Since neither model requires otherwise, there is the possibility that both the organism and the environment modify each other. A quantitative analysis has forced an explicit representation of a number of variables within the phenomenon and formulation of the relations between those variables. The models also have explicit representational notions and economical rules about parameter estimation, ratio of number of data points to number of parameters, and cross-situation parameter invariance. The rules of interaction may be represented by an arithmetic that accounts for the results of a large class of studies. The models are designed to account for the maximum amount of variance found in a number of experimental situations to which the processes described by the model apply. The adequacy of quantitative models is tested by comparing obtained data to theoretically predicted data. This is to be contrasted with the testing of relatively simple hypotheses. As in other areas of science, looking for the generality of a formulation has made the models more testable. Independent routes of verification are possible because of the increased scope of the models.

Each volume of the series examines a particular topic that has been discussed at the annual Harvard Symposium on Quantitative Analyses of Behavior. The topic for Volume I is the discrimination of schedules of reinforcement. It was chosen because it represents an area that has been highly quantified through the application of psychophysical methods and analyses to maintained performances. Volume II, the present volume, explores matching and maximizing accounts of the allocation of behavior, another area that has been highly quantified. It explores the generality of such formulations and how they apply to the behavior of animals in both the field and the laboratory and to humans in choice situations in economics. Acquisition data and models are considered in Volume III, and Volume IV reports studies of discrimination processes. Volume V will address the topic of how reinforcement value is affected by delay and intervening events.

The contents of this volume were first prepared for and presented at the second Symposium on Quantitative Analyses of Behavior at Harvard University, in June 1979, on maximization and matching. Subsequently, the chapters were revised, updated, and rearranged into the seven topical parts found herein.

The symposium out of which this second volume grew was organized by Michael L. Commons, Richard J. Herrnstein, John A. Nevin, and Howard Rachlin. For the third, fourth, and fifth symposia, Allan Wagner also served on the program committee. The symposia were supported in part by the Department of Psychology and Social Relations at Harvard University and by the Dare Association. Local arrangements were provided by Patrice M. Miller and Julie Bisbee, with assistance from John R. Ducheny, Stuart L. Mattson, Susan Herrnstein, and Donald Loveland.

We wish to thank R. Duncan Luce for serving as discussant and Alexandra W. Logue, John A. Nevin, Herbert L. Roitblat, and Michael Woodford for reviewing several chapters for which they had special competence. Patrice M. Miller informally reviewed many of the chapters, helping to improve their style and organization. It was a pleasure to work with Carol Franco (editor), Michael Connolly (president of Ballinger), Steven A. Cramer (project editor), Cynthia Insolio Benn (copy editor), Gerry Galvin (production manager), Barbara Roth (production assistant), Michael Quinlin (advertising and promotion coordinator), Leslie Zheutlin (senior marketing assistant), and Thomas G. Greene (compositor). We would like to thank them for their help in producing and marketing the book.

Michael L. Commons
Harvard University

INTRODUCTION

Richard J. Herrnstein

The authors of this volume address the general area of maximization and matching with varying degrees of literal-mindedness, as the range of contributions testifies. It may be helpful to state here the editors' conception of the unifying topic and its relation to the twenty-three chapters that follow.

Maximization theory formalizes the intuition that behavior is, or at least tends to be, rational. Analogously, it extends to behavior the evolutionary presumption that the laws of behavior should yield the best possible outcome in some sense, or else they would not have survived natural selection. The mathematical structures of neoclassical economic analysis express the maximization intuition most elegantly and with little if any extra conceptual baggage, so it is not surprising that they should infiltrate other behavioral sciences, such as psychology, behavioral biology, or ethology. Part V, Rachlin (Chapter 16), Burkhard (Chapter 17), and Green Kagel, and Battalio (Chapter 18) set forth the economic conception of maximization as applied to the behavior of individual organisms under experimental conditions. These authors argue that the mathematically crystallized intuition of economic theory, conceived for the aggregated human behavior of the marketplace, also describes the behavior of any organism being controlled by reinforcement.

A quite different conception of maximization is offered by Shimp (Chapter 4) and Silberberg and Ziriax (Chapter 5), in Part I, The Fine Structure of Instrumental Responding. These two chapters develop possibly the simplest formulation of maximization, which is, in effect, that an organism looks only one or a few responses ahead, choosing to perform whichever has the highest momentary probability of reinforcement or some approximation to it. The result should be sequential patterns of responding that track the transiently shifting probabilities of reinforcement. These authors suggest that the predicted sequences can be found in data properly scrutinized, but Nevin (Chapter 6) is doubtful.

The burden of proof must be on any theory that argues against maximization, just because it is arguing against everyday as well as biological intuition and the formal structures of economic analysis. The matching law is such a nonmaximizing account of behavior, as elaborated, though differently, in Chapters 19 and 20, by Herrnstein and by Heyman. According to matching, the distribution of behavior across sources of reinforcement equals, or matches, the relative frequencies of obtained reinforcement, which may or may not result in maximization depending on whether, in a given environment, matching and maximizing do or do not converge. Herrnstein, Heyman, and Nevin claim in their chapters that distributions of choice responses and sequential response patterns both support the matching law rather than either momentary or economic maximization. Commons, Woodford, and Ducheny (Chapter 2) provide further evidence for the adequacy of the matching law in accounting for undermatching, as does Mazur (Chapter 3) for response sequences under schedules of intermittent reinforcement. An even finer scale look at behavioral sequences is described by Pear, Rector, and Legris (Chapter 1), who have breached the integrity of the pigeon's key peck and have found unanticipated orderliness in the usually unrecorded fragments of behavior leading to the reinforced operant. Although it is not yet possible to relate the findings of Pear et al. to the issue of maximization versus matching, the question of response topography is central to the issue, as detailed in Part II, The Ecology of Instrumental Responding.

Lea (Chapter 7) ties ecological optimal foraging theory to the concepts and findings of laboratory research on instrumental conditioning. He considers, but does not try to answer here, the question of whether behavior in natural settings is literally optimal, as called for by maximization theory, or, as called for by matching theory, only approximately and contingently so, depending on environmental feedback functions. Baum (Chapter 10) deals with a similar question, but he also tries to estimate the long-term effects on the replenishing of environments, if behavior obeys either maximizing or matching laws. Kamil, Peters, and Lindstrom (Chapter 8) and Olton (Chapter 9), like Lea, construct the parallels between the variables governing naturally occurring foraging and discrimination learning as studied in the laboratory. Both chapters note that optimal behavior in the wild may take on a different aspect in the laboratory, suggesting that the underlying behavioral law may be a more circumscribed source of optimization than a literal maximization process would be.

Parts III and IV deal, respectively, with multiple schedules and punishment, two procedures that have been much examined in the laboratory in relation to reinforcement theory. In multiple schedules, the schedule of reinforcement shifts from time to time, correlated with a discriminative stimulus. The resulting changes in behavior bear a strong though unaccounted for similarity to the interactions among the alternatives in a choice situation. Staddon (Chapter 11) derives some of the properties of responding on multiple schedules from a model in which reinforced behavior is pictured as competing for available time with background activities. Different features of multiple-schedule responding are dealt with by Vaughan (Chapter 12), who adapts the Rescorla–Wagner theory of associative conditioning for the control of response strength. Staddon's and Vaughan's theories both touch peripherally on the issue of maximization versus matching in the distribution of choice, with Staddon shading marginally toward the former and Vaughan toward the latter. Williams' Chapter 13 reviews the evidence showing that the control of responding in each component of a multiple schedule is more affected by the component to come than the component just passed, a finding that challenges both maximizing and matching theories.

Deluty (Chapter 14) and de Villiers (Chapter 15) present results of punishment procedures using aversive electric shock. De Villiers concludes that the evidence, his own results and others', supports the matching theory, with aversive stimuli having a dual effect, as a negative quantity of reinforcement and as a source of positive reinforcement for competing activities. Deluty's results, in contrast, may be more easily accounted for by a straightforward maximization account.

The final part offers three views not readily aligned either with matching or maximizing theory, though obviously related to them. Donahoe, Crowley, Millard, and Stickney (Chapter 21) attempt to derive both operant and respondent conditioning from a model whose lineage can be traced to Premack's relational theory of reinforcement. Among its other features, their model encompasses both matching and the Rescorla-Wagner theory of association, as does Vaughan's (Chapter 12). Allison (Chapter 22) also presents a model with clear Premackian ancestry, although it differs in focus from that of Donahoe et al. Allison employs some of the methods of economic analysis while reserving judgment about its central assumption, maximization. Timberlake, in the book's final chapter, expresses analogous reservations about matching theory. He highlights the implicit assumptions that lead from direct laboratory observations of matching to a matching theory that claims ecological generality. Since the test of ecological generality has yet to be passed by any quantitative theory of behavior, Timberlake's cautionary chapter seems the proper place to end a book whose subject is far from closed.

I FINE STRUCTURE IN INSTRUMENTAL RESPONDING

1 TOWARD ANALYZING THE CONTINUITY OF BEHAVIOR

Joseph J. Pear, Brian L. Rector,
and Joseph A. Legris

This chapter does not deal with either matching or maximizing per se but, rather, with the problem of conceptualizing and studying behavior as a continuous process in time. The issue of measuring behavioral continuity is relevant to theories of matching and maximizing, however, because both types of theories attempt to describe how behavior adjusts to reinforcement variables operating over continuous time periods. Refinements in techniques for measuring behavioral continuity should be useful in examining the continuous interaction between behavior and these variables and therefore in clarifying the processes studied in experiments on matching and maximizing.

Two experiments arising from attempts at such refinements in our laboratory will be described here. The first experiment concerns the partitioning of response rate into two constituents: (1) the proportion of session time the subject allocates to the experimentally reinforced activity and (2) the rate at which responses occur during the time the subject allocates to the reinforced activity. The second experiment provides a more detailed examination of the experimentally reinforced activity.

EXPERIMENT 1: PARTITIONING RESPONSE RATE

It seems reasonable to think of a single reinforcement schedule as a concurrent schedule composed of an arranged contingency for some particular activity

This research was supported by Grant No. A7461 from the Natural Sciences and Engineering Research Council of Canada.

under experimental study and unspecified contingencies for other activities. If it is assumed that at any given moment only one activity can occur and that changeover time between activities is zero, it follows that the total time for any given period is the sum of the times allocated to the experimental activity and to the other activities. Given these assumptions, let *propensity* be defined as

$$p = t/T \quad , \tag{1.1}$$

and *speed* be defined as

$$s = n/t \quad , \tag{1.2}$$

where T is the total time during some given period in the experiment (a session, say) and t and n are the time allocated to the experimentally reinforced activity and the number of responses emitted during that period. Then response rate r during the period is

$$r - sp \quad , \tag{1.3}$$

since

$$sp = (n/t)\ (t/T) = n/T \quad , \tag{1.4}$$

which is the conventional measure of response rate.

Equation (1.3), first formulated by Gilbert (1958), partitions response rate into two constituents. If speed is constant over the range of some reinforcement variable (such as reinforcement rate), then variations in response rate as a function of this variable can be expressed simply as variations in propensity. Thus, for two concurrently reinforced activities of identical topography, the matching relation

$$r_A/r_B = x_A/x_B \tag{1.5}$$

reduces to

$$p_A/p_B = x_A/x_B \quad , \tag{1.6}$$

where r_A and r_B are the response rates of the two activities A and B, p_A and p_B are their propensities, and x_A and x_B are their reinforcement rates (cf. Baum and Rachlin, 1969). Also note that in this formulation, as Rachlin (1973) has pointed out, speed is equivalent to the constant representing asymptototic response rate in Herrnstein's (1970) formulation for single and concurrent schedules.

One purpose of Experiment 1 was to test directly the assumption of the constancy of speed. Another was to determine how propensity varies as a function of reinforcement rate. In order to conduct these tests under the simplest possible conditions, only single schedules of reinforcement were studied. Since the experiment has been reported in detail elsewhere (Pear and Rector, 1979), only the points most relevant to the present purposes are described here.

Method and Results

In a standard operant chamber for pigeons, a thin metal plate was placed on the floor under the response key. Pressure on the plate operated microswitches underneath it. When the bird stood on the plate the key light was red; when the bird was off the plate the key light was green. Grain reinforcement was used to maintain key pecking on various reinforcement schedules. Key pecking was reinforced only when the bird stood on the plate. In addition, key pecks were never reinforced less than 8 seconds after the bird had just stepped onto the plate, so as to reduce the likelihood that activities immediately preceding stepping onto the plate would be adventitiously reinforced. Thus, the relative amount of session time spent standing on the plate constituted a measure of propensity. (In making this calculation, feeder-operation time was subtracted from total session time.)

The behavior of eight pigeons, which were deprived to 80 percent free-feeding weight, was examined under several values of several basic single reinforcement schedules. Cumulative records indicated that, in general, key pecking occurred only when the birds stood on the plate and substantial pausing occurred only when the birds were off the plate. Thus, the relative time on the plate appeared to be a reasonably accurate measure of propensity, and the rate of responding during that time appeared to be a reasonably accurate measure of speed, under the reinforcement schedules examined. These were fixed ratio, fixed interval, variable ratio, and variable interval; however, discussion is limited here to the findings concerning variable-interval schedules.

Figure 1-1 shows propensity, speed, and response rate for three pigeons exposed to variable-interval (VI) schedules. Note that the propensity data were quite consistent in several respects across birds. First, changing to a lower reinforcement rate (that is, changing from VI 1 minute to VI 5 minutes for P-111 and P-112, and from VI 5 minutes to VI 10 minutes for all three birds) generally produced an immediate decrease in propensity, sometimes followed by a complete or partial recovery to the propensity level prior to the decrease in reinforcement rate. Second, with the reintroduction of VI 1 minute propensity increased to or above its baseline level (except in the case of P-113, which had already recovered its baseline propensity level prior to the reintroduction of VI 1 minute). Third, during extinction propensity decreased to zero for all three birds.

In contrast to propensity, speed and response rate showed little consistency within or across birds. Clearly the hypothesis that speed is constant was not confirmed. Instead, speed might be best characterized as gradually changing, or drifting, over sessions in a manner little related to changes in reinforcement rate. Consequently, response rate also tended to be less strongly related to reinforcement rate than was propensity.

Figure 1-1. Overall session-to-session propensity, speed, and response-rate data from several variable-interval schedules (VI 1 minute, VI 5 minutes, VI 10 minutes) and extinction for birds P-111–P-113 in Experiment 1. Some sessions have been omitted to conserve space. (From Pear and Rector, 1979. Copyright 1979, Society for the Experimental Analysis of Behavior.)

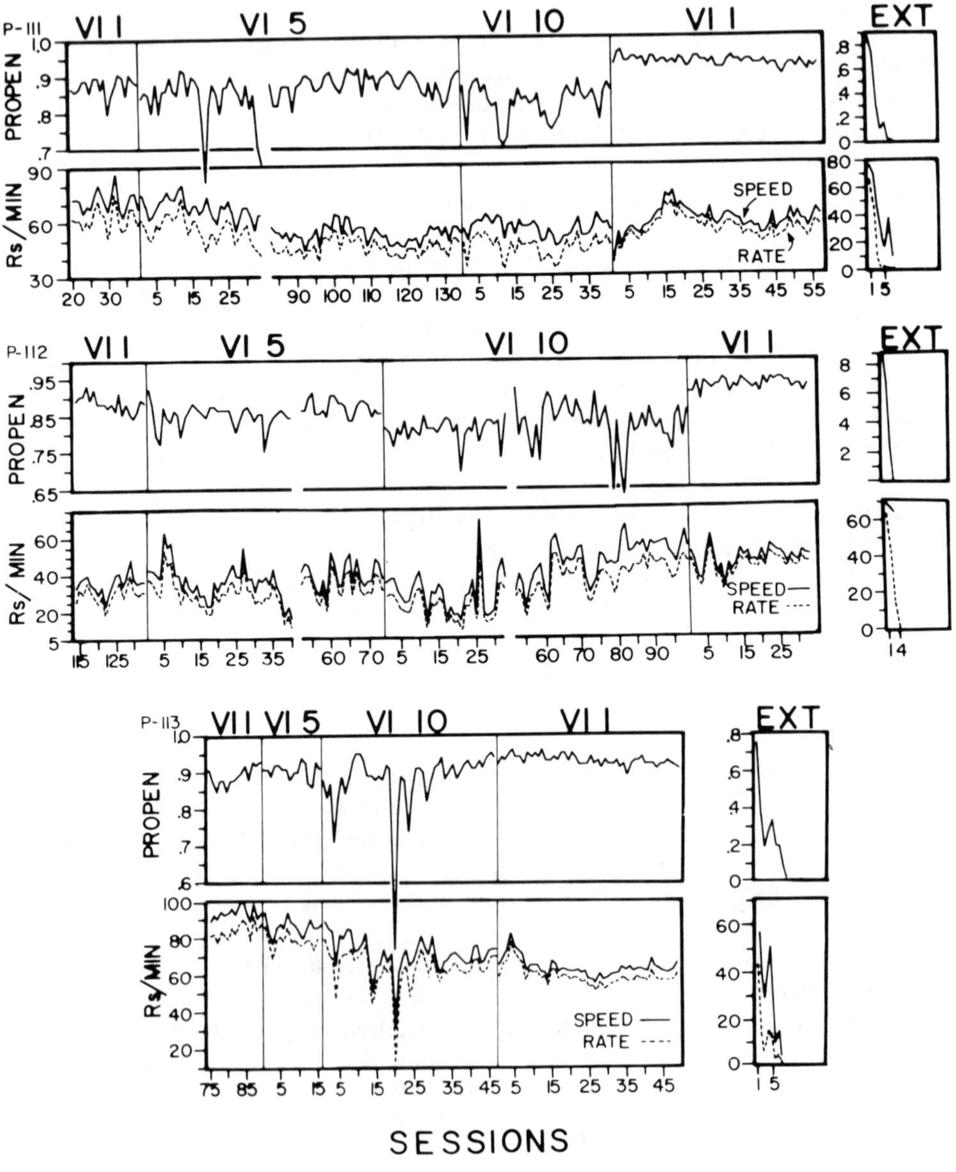

Discussion

The data appear to be incompatible with any simple application of the matching relation between response rate and reinforcement rate to single schedules of reinforcement. On the other hand, Figure 1-1 indicates that propensity varied directly with reinforcement rate, although the changes occurring in propensity were much smaller than would be predicted by a straightforward substitution of propensity for response rate in Herrnstein's (1970) extension of the matching law to single schedules. This point will not be developed here, however. For the present purpose the point to be noted is that the data show a more reliable relation between propensity and reinforcement rate than between response rate and reinforcement rate.

This result suggests that propensity is a more reliable measure of the probability of the experimentally reinforced activity than is response rate. A possible reason may be that response rate is distorted by changes in speed caused by uncontrolled variations in the structure of the activity, that is, the sequence of movements made by the bird while it is engaging in the experimentally reinforced activity. According to this interpretation, which is similar to the positions taken by Shimp (1975 and Chapter 4 in this volume) and by Commons, Woodford, and Ducheny (Chapter 2), the drifting of speed over sessions is explained by the coincidence between reinforcement and specific structural variations in behavior. Any particular structural variation that happens to be followed by reinforcement tends to be repeated. Subsequent reinforcements then tend to maintain it until some other variations happens to occur and be reinforced (cf. Skinner, 1948). It seems reasonable to assume that successive structural variations would generally be quite similar, although sudden large changes probably also occur occasionally. This would explain the gradual drifting in speed and consequently response rate in the present data as well as the occasional abrupt changes that also occurred with no evident relation to changes in reinforcement rate. (Similar effects in variable-interval schedules have been observed by others, such as Blough and Blough [1968].)

In this view the concept of an experimentally reinforced activity constantly in flux replaces the more familiar picture of the stereotypical operant response depicted or implied in many accounts of operant behavior. Although the former view is consistent with the data obtained in the present experiment, it remains to be explained why other studies of variable-interval schedules have typically obtained a direct functional relation between response rate and reinforcement rate (e.g., Catania and Reynolds, 1968). The reason may have to do with the fact that those studies did not use a procedure that reinforced only behavior occurring close to the response key, as was the case in the present experiment. Shimp (1974) and Platt (1979) have shown that the functional relation typically obtained between response rate and reinforcement rate on variable-interval schedules is likely due in part to a greater tendency for long interresponse times to be selectively reinforced by lean than by rich variable-interval schedules. It may be that when the experimentally reinforced activity is restricted to a small region of the chamber, as was the case in the present experiment, this tendency for long interresponse times to be selectively reinforced by lean variable-interval schedules is less because structural variations covering a large area of the chamber are unlikely to be reinforced.

The present finding of less consistency with speed and response rate than with propensity seems therefore to be due to the fact that reinforcement was preceded by a number of topographically different key pecks with different interresponse times, whereas standing on the plate consistently preceded reinforcement. Although reinforcement may also have been preceded by several different ways of standing on the plate, all of these affected the continuous time measure of propensity in the same way. These considerations suggest that propensity is a good variable for use in molar studies of behavior. They also indicate that neither propensity nor response rate can provide a comprehensive account of the complex interplay between behavior and reinforcement. What seems to be needed for a detailed and comprehensive analysis of operant behavior is a method for recording the relevant behavior throughout the intervals between the discrete points in time that are typically recorded as responses. The development of such a method was the main objective of the second experiment.

EXPERIMENT 2: A STRUCTURAL ANALYSIS OF OPERANT BEHAVIOR

Motion picture film or video tape could be used to obtain a detailed record of the behavior occurring between discretely recorded operant responses, but performing any sort of meaningful quantitative analysis on the resulting data would be extremely cumbersome and difficult. Some means must be obtained for abstracting essential quantitative aspects from continuous samples of behavior.

One way of approximating this goal is to put a fixed dark point on a pigeon's head and to record the motion of this point through space while the pigeon engages in key pecking under the control of some schedule of reinforcement. The absolute distance of this point from the key as the pigeon engages in key pecking and other activities would describe a wave function over time. Many different types of precise quantitative analyses could be performed on the resulting data, especially since there exists a well-developed mathematics of wave functions. With this eventual objective in mind, we designed and constructed an apparatus to collect such data.

Experiment 2 was carried out using this apparatus to obtain some preliminary data on the "behavior waves" generated by a variable-interval schedule. The procedure used in the previous experiment to restrict the experimentally reinforced activity to the region close to the key was not used in this experiment, so that the behavior studied would be that generated by a standard variable-interval schedule.

Method

The main components of the apparatus are illustrated in Figure 1-2. Two perpendicularly oriented television cameras are directed toward two transparent walls of an operant chamber for pigeons. The walls through which the cameras view are made of clear glass; the other two walls are made of white opaque Plexiglas. On one of the opaque walls is mounted an aluminum panel, also painted white, containing a response key and a food-hopper aperture. The response key is 2.8 centimeters in diameter, and its center is located 19.8 centimeters from the floor and 27.5 centimeters from the closer side wall; the food-hopper aperture is centered near the base of the wall. Each wall is 58.0 centimeters wide and 38.0 centimeters high.

A white pigeon with its head painted with black dye is placed in the chamber. The TV cameras are connected to an electronic logic system programmed to compute at 1/30-second intervals the spatial (x, y, z) coordinates of the center of the highest dark area (the pigeon's head) scanned by the cameras. This logic system in turn is connected to a microprocessor that can be programmed to collect and analyze the data in any desired manner, store the data on magnetic tape, and display the data on the output of a chart recorder.

In the present experiment, a force of 0.11 newton on the response key closed a switch for electronically detecting key pecks. No special feedback sound was provided for key pecks that closed the switch. Reinforcement consisted of 3 seconds of access to the food hopper, correlated with a tone and an increase in the intensity of the feeder light. The chamber was continuously illuminated, including the key and the food-hopper aperture, so that shadows and other dark areas were minimized.

Two experimentally naïve White Carneaux pigeons, deprived to 80 percent free-feeding weight, served in this experiment. Each bird's head was painted black so as to be the only dark area in the chamber. Thus, the apparatus recorded the motion of a hypothetical point on the pigeon's head in three-dimensional space as the pigeon engaged in key pecking and other activities in the chamber. From these data the apparatus computed the absolute distance of the head from the key at 1/30-second intervals.

Key pecking was first shaped and then maintained on continuous reinforcement for several sessions. Then the birds were given a number of 1/2-hour daily sessions of VI 1 minute. Samples of the head-movement data were examined over somewhat randomly selected continuous time periods during many of these sessions. Data occurring during and just after a food presentation were not recorded, mainly because the bird's head was in the hopper aperture during a food presentation and therefore not visible to the TV cameras. The aim was to deter-

Figure 1-2. Diagram of the apparatus used in Experiment 2.

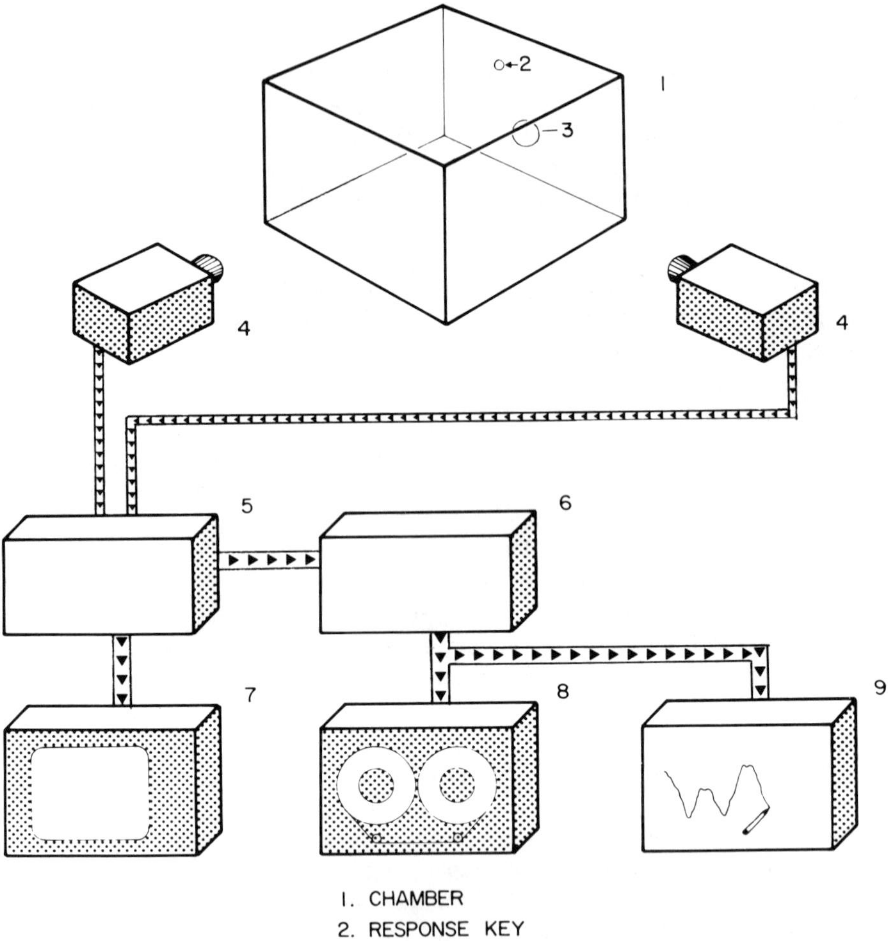

1. CHAMBER
2. RESPONSE KEY
3. FOOD APERTURE
4. TV CAMERA
5. LOGIC SYSTEM
6. MICROCOMPUTER
7. TV MONITOR
8. TAPE RECORDER
9. PLOTTER

mine the changes that occurred in the behavior waves as a function of exposure to a VI 1-minute schedule over a large number of sessions.

Results and Discussion

Figure 1-3 shows representative segments of data from session 1 of VI 1 minute for bird P-21. The segments are labeled alphabetically in chronological order. Segment A is a record of the head movements resulting in the first five key pecks of the session. Note that the wave pattern is fairly regular, probably as a carryover from the preceding sessions of continuous reinforcement and that the troughs of each wave corresponded to a key peck (operationally defined, of course, as the closure of the switch behind the key). As seen from segments B and C, however, the wave pattern then changed to one of lower amplitude and higher frequency in which there were long series of waves whose troughs did not correspond to key pecks. Such troughs are *nonterminal*, because they are associated with the absence of the terminal member of the key-pecking sequence (that is, the operationally defined key peck). Segments D and E show the emergence of lower frequency waves, with somewhat higher amplitudes, superimposed on the higher frequency nonterminal wave pattern. Segment F depicts a type of wave that occurred infrequently in this and subsequent sessions but is shown here for the sake of completeness. Its atypical high amplitude and low frequency (long period) suggest that it is the product of some activity other than key pecking (the type of activities occurring when the bird was not standing on the plate in Experiment 1). The remaining three segments, G, H, and I, show low-amplitude, low-frequency waves, with superimposed high-frequency waves that are not as regular or as clearly defined as the high-frequency waves in the earlier segments. This development foreshadowed a decline in the predominance of the high-frequency waves.

Figure 1-4 shows representative segments from sessions 2-5, 10, and 15 of VI 1 minute for P-21. Segment A of session 2, taken from near the beginning of the session, shows a low-amplitude, high-frequency nonterminal wave pattern similar to the pattern occurring in the early part of session 1 (see segments B and C of Figure 1-3). This tendency for the behavior at the beginning of sessions to "regress" to the wave patterns of earlier sessions was typical, as will be seen. Segments B-E of session 2 show the emergence of high-amplitude, low-frequency waves, similar to those that occurred in the latter part of session 1 except that the amplitude tended to be larger in session 2. These high-amplitude waves often had nonterminal troughs, and sometimes the nonterminal troughs between two such waves was itself quite high, as in segment D. Note also that in these segments the high-amplitude waves are asymmetrical, the highest crest occurring left of center with respect to the corresponding interresponse time, hence giving a positively skewed shape to the wave pattern between key pecks.

Figure 1–3. Selected segments of head-movement data, shown as distance from the key over time, taken from session 1 of VI 1 minute for bird P–21. The data in this and subsequent figures were collected every 1/30 second and smoothed by averaging each point with the one immediately preceding it and the one immediately following it. The vertical marks on the lines under the curves indicate key pecks.

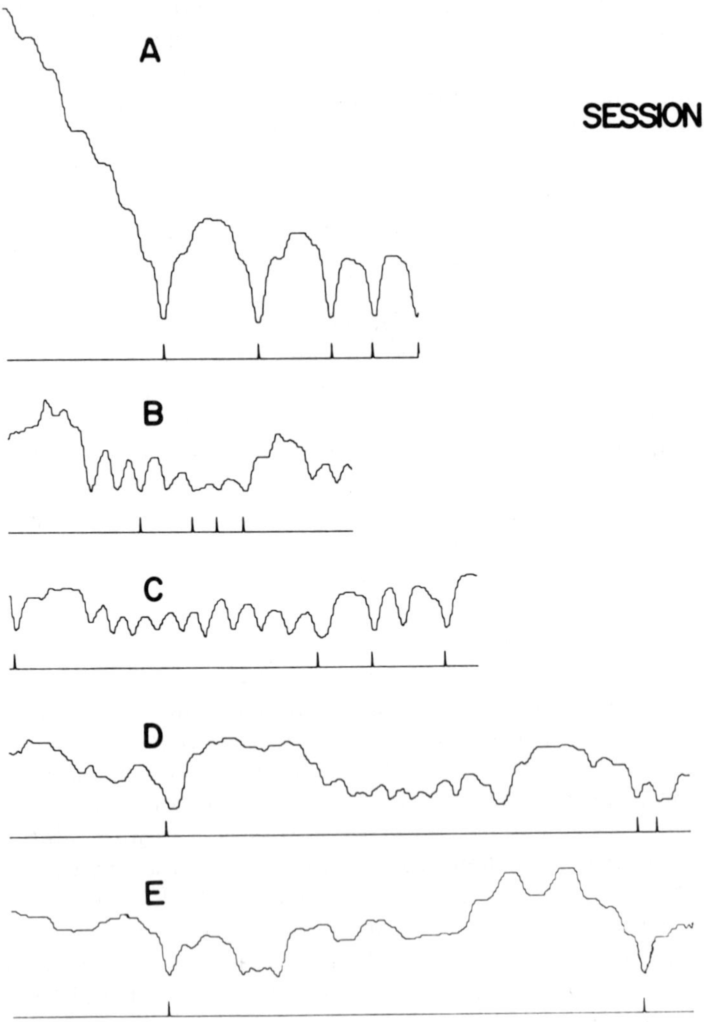

Figure 1–3. continued

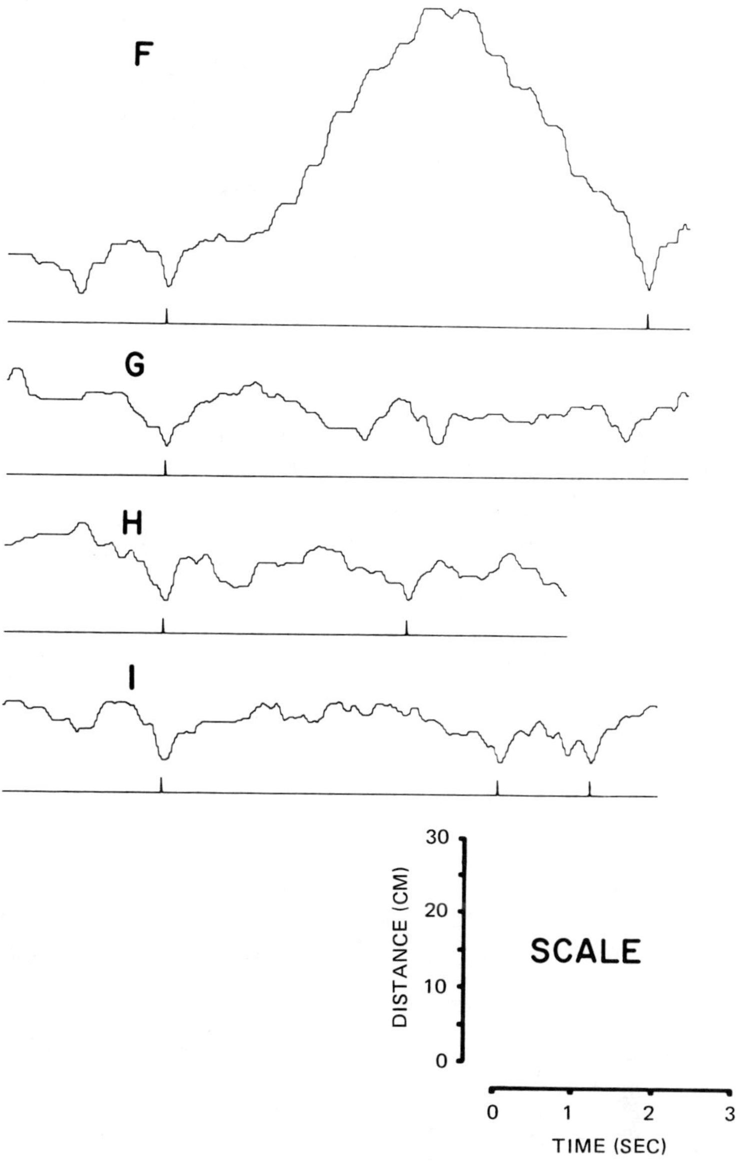

Figure 1–4. Selected segments of head-movement data, shown as distance from the key over time, taken from sessions 2–5, 10, and 15 of VI 1 minute for bird P–21. (The data were collected as described in Figure 1–3, and the vertical marks indicate key pecks.)

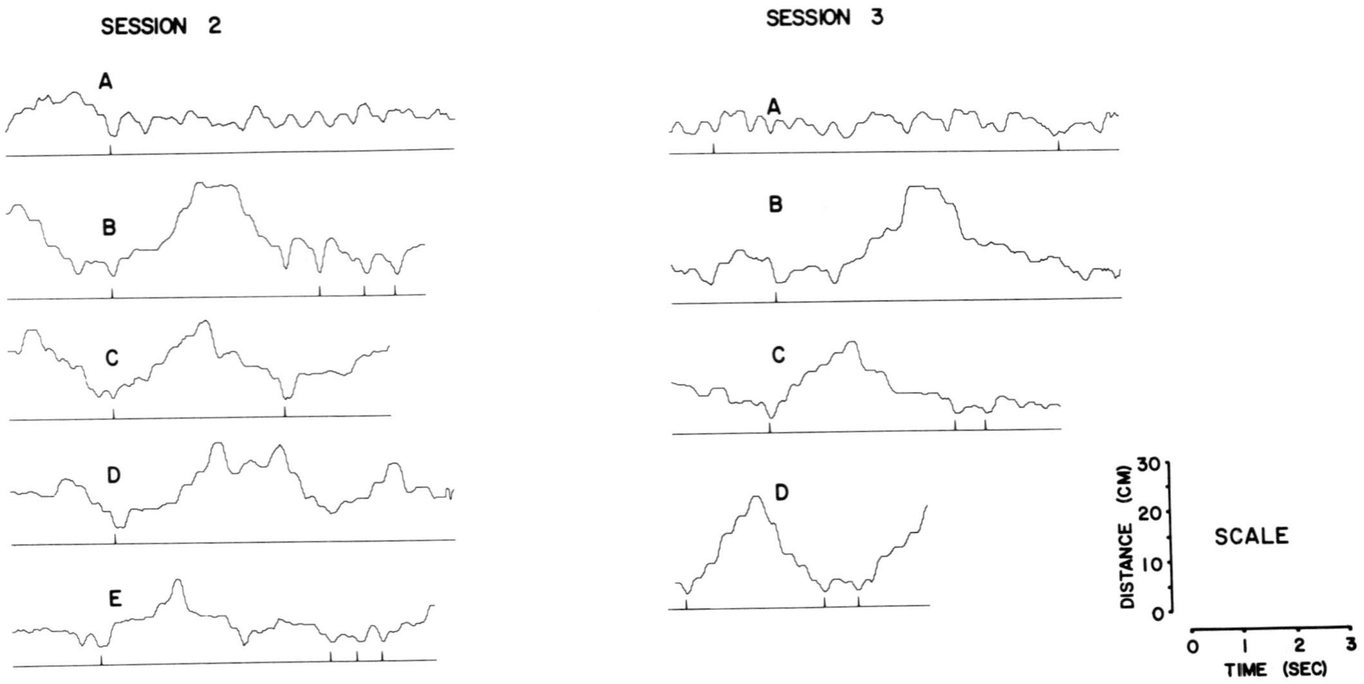

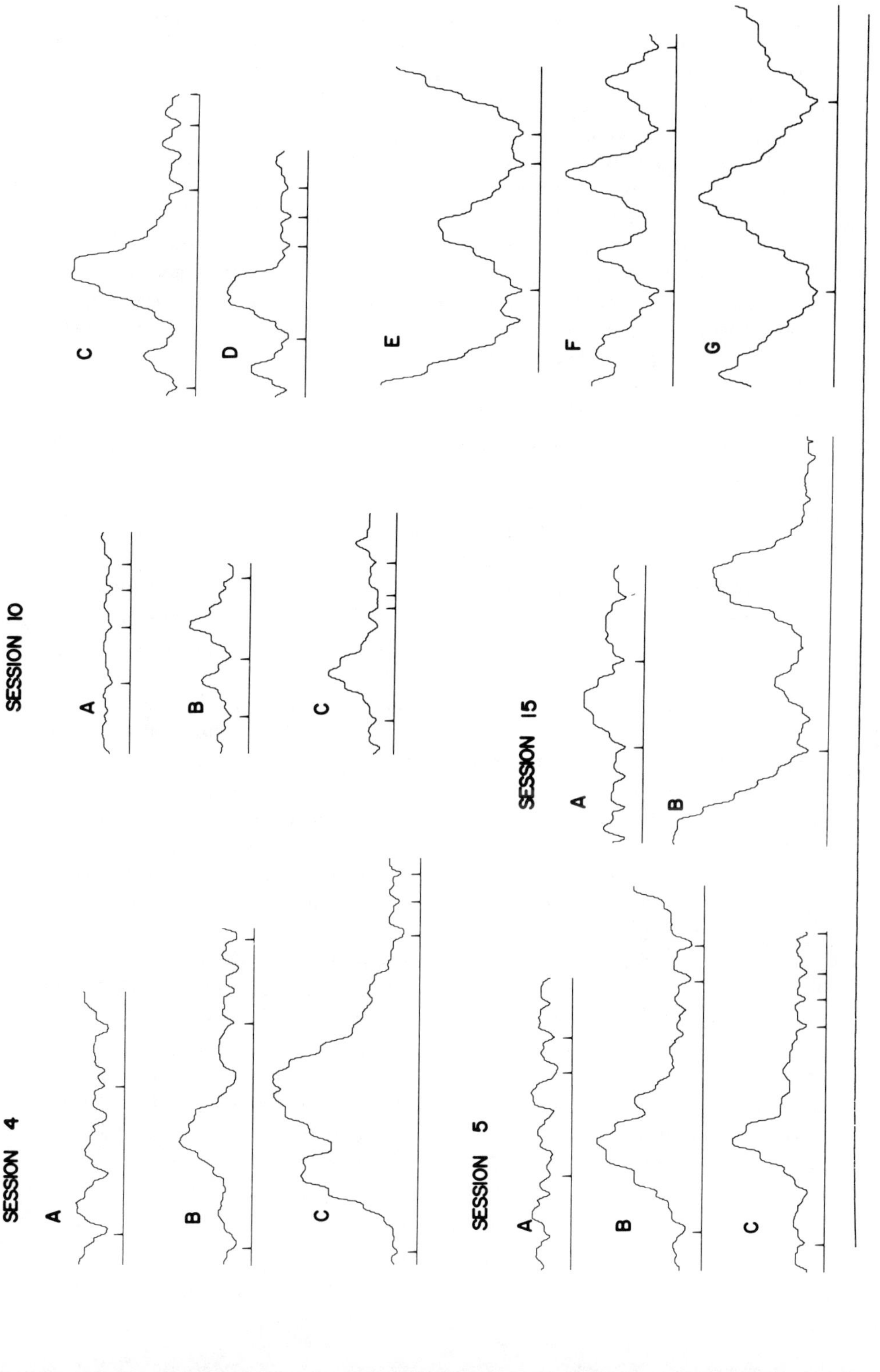
SESSION 4
A
B
C
SESSION 5
A
B
C
SESSION 10
A
B
C
D
E
F
G
SESSION 15
A
B

Near the beginning of session 3, as seen in segment A of that session, a low-amplitude, high-frequency nonterminal pattern appeared, showing again the regression characteristic of early parts of sessions. This was followed again by high-amplitude, low-frequency waves as indicated in segments B–D. Note that as in the previous session, the wave pattern between responses tended to assume a positively skewed shape. Some fairly symmetrical high-amplitude waves also occurred, however, as seen in segment D.

The data from sessions 4 and 5 are similar to the data from sessions 2 and 3. A low-amplitude, high-frequency nonterminal pattern occurred near the beginning of each session and was followed by the emergence of higher amplitude waves as the session progressed. The main difference from the high-amplitude waves of the earlier sessions seems to be a tendency for the highest crest between two responses to occur closer to the center of the interresponse time. This tendency apparently foreshadowed the highly symmetrical pattern that emerged in later sessions.

Near the beginning of session 10, as seen in segment A of that session, a very low amplitude prevailed, giving a rather flat appearance to the wave pattern. The high-frequency nonterminal pattern that occurred near the beginning of earlier sessions is therefore difficult to detect, if indeed it is present. Another difference from the earlier sessions is that the higher amplitude waves that emerged were more symmetrical than those in the earlier sessions and not as high in amplitude.

Segment A of session 15 again shows a low-amplitude, high-frequency nonterminal pattern similar to that occurring in the early part of previous sessions. The high-amplitude waves that then emerged were quite high in comparison to the high-amplitude waves of the sessions shown previously in the figure. Sometimes the troughs on one side or the other of these waves were nonterminal, as in segments B, C, and F. At other times neither of these troughs was nonterminal, as in segments D, E, and G. In the latter cases the high-amplitude waves were fairly symmetrical within their respective interresponse times. Whether bounded on either side by a nonterminal trough, however, the high-amplitude waves in this session appear to have more regular shapes than those in the previous sessions shown, perhaps indicating that a repetitive or stereotypical pattern was beginning to emerge.

Figure 1–5 shows representative segments from sessions 20 and 90 for P–21. The beginning of session 20, seen in segment A, shows again the low-amplitude nonterminal pattern characteristic of the early parts of previous sessions. As before, these waves were replaced by ones of higher amplitude, but the amplitude was lower than had been characteristic of most of the high-amplitude waves of the sessions seen previously (note the scale change in the figures). In addition, these high-amplitude waves were much more regular and symmetrical and more seldom nonterminal than were the high-amplitude waves of the previously shown sessions.

Low-amplitude, high-frequency waves still occurred at the beginning of sessions even as late as session 90, as seen in segment A of that session, although there seem to have been fewer nonterminal troughs at the beginning of these later sessions than had occurred at the beginning of earlier sessions. The most striking difference from the earlier sessions is the predominance of very symmetrical high-amplitude waves in the latter part of the session. These waves rarely had nonterminal troughs.

Figure 1-6 illustrates the head-movement data over sessions for P-22. The data for P-22 are very similar to the data for P-21 and are therefore presented in a more abbreviated form. The figure shows representative wave segments from the latter portions of sessions 1-3, 10, 16, 20, and 60. Only one segment per session is shown. During sessions 1 and 2 the waves were quite low in amplitude, as had been the case for P-21 (see Figure 1-3). The wave exhibited an almost flat appearance, however, and the high-frequency, nonterminal pattern cannot be as clearly seen as it can for P-21. Nevertheless, the fact that the waves were not completely flat can be taken as indicating the presence of a nonterminal pattern to at least some degree. The early portions of subsequent sessions were characterized by similar low-amplitude waves, whereas the later portions showed high-amplitude waves—as was the case for P-21.

P-22's high-amplitude waves followed an interesting course of development. As indicated in the figure, they first appeared in session 3 as a bulge arising from the low-amplitude flat pattern. At first these bulges were asymmetrically located within their corresponding interresponse times, at right of center. By session 10 their heights had increased and shallow, nonterminal troughs had emerged to separate the main portion of these bulges from most of the low-amplitude region to their left. Over sessions these troughs on the left gradually deepened and became more frequently terminal, until by session 60 the high-amplitude waves tended to be centrally located within their interresponse times. Along with this development, the low-amplitude waves assumed higher frequencies (their periods became shorter) and took on more rounded shapes. A very consistent wave pattern eventually emerged in which fairly symmetrical high-amplitude waves were preceded and followed by trains of fairly symmetrical low-amplitude waves. As with P-21 after many sessions of VI 1 minute, nonterminal troughs were now extremely rare.

Although fairly stereotypical (but different) wave patterns eventually emerged for both birds after prolonged exposure to VI 1 minute, it should be noted that a large number of sessions were required for this to occur. Moreover, we cannot say at this time whether additional sessions would have produced further changes in the wave patterns, as implied by the discussion of Experiment 1. It seems fair to say, however, that the systematic development in the behavior waves gives ground for caution in quantitatively equating operant responses in the absence of knowledge about their structural characteristics.

Figure 1-5. Selected segments of head-movement data, shown as distance from the key over time, taken from sessions 20 and 90 of VI 1 minute for bird P-21. (The data were collected as described in Figure 1-3, and the vertical marks indicate key pecks.)

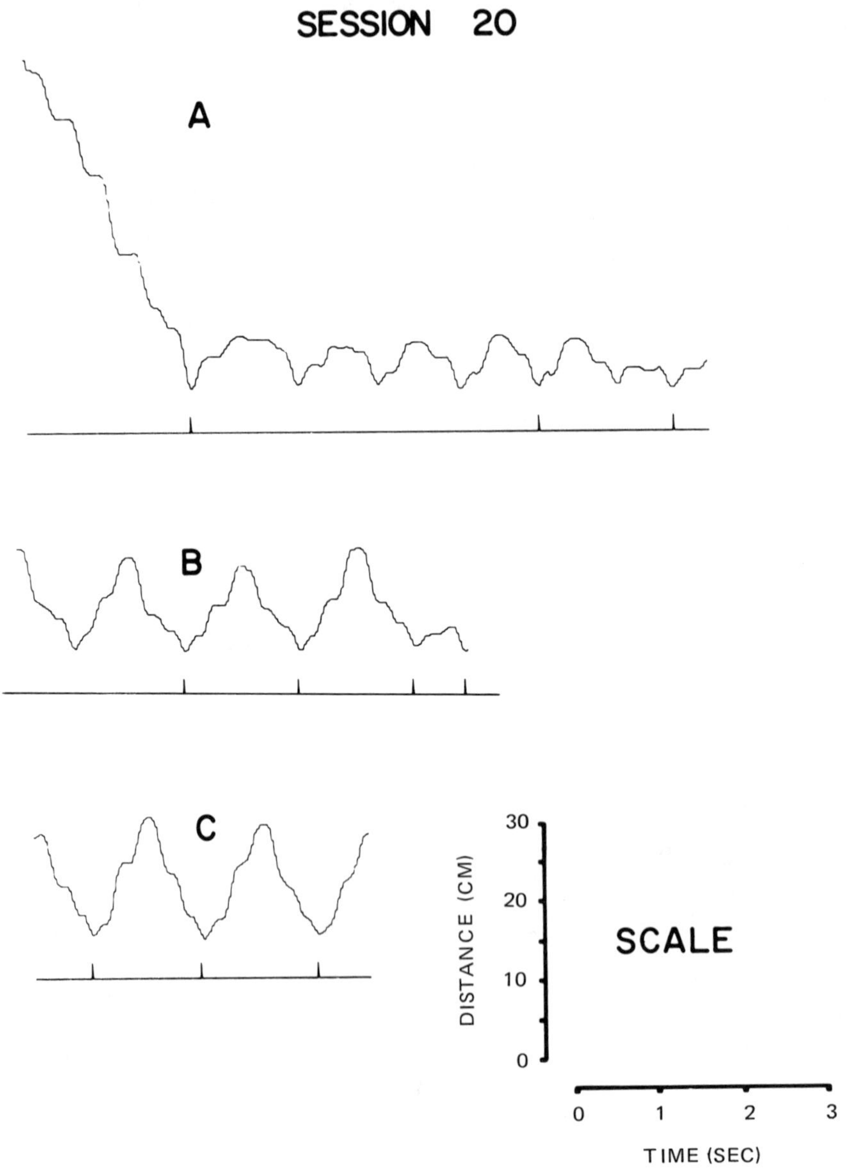

Figure 1-5. continued

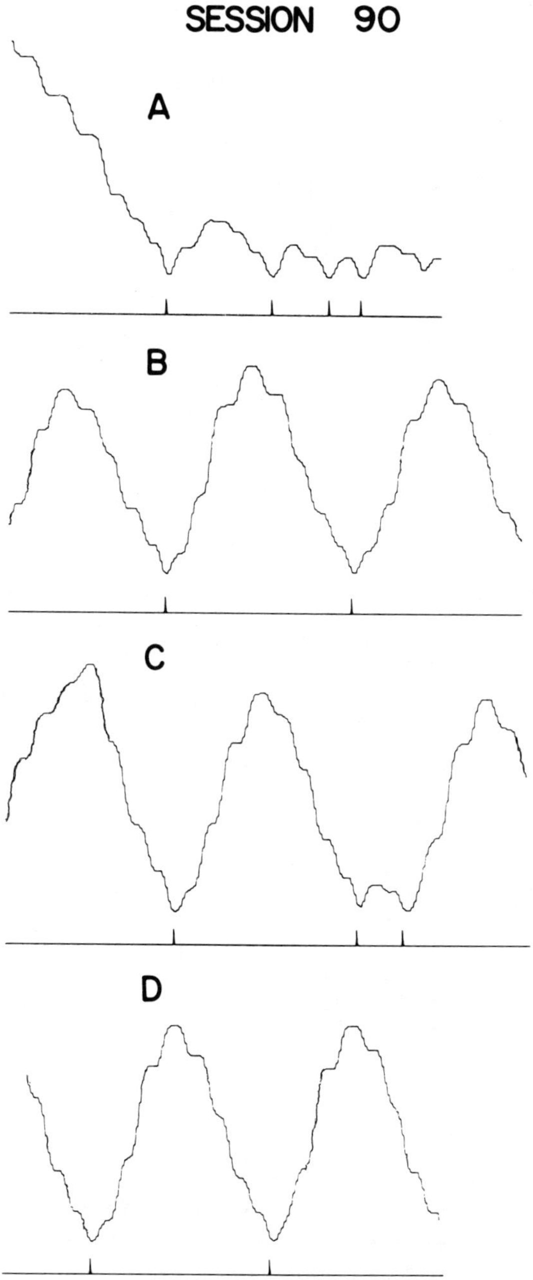

Figure 1-6. Selected segments of head-movement data, shown as distance from the key over time, taken from sessions 1-3, 10, 16, 20, and 60 of VI 1 minute for bird P-22. (The data were collected as described in Figure 1-3, and the vertical marks indicate key pecks.)

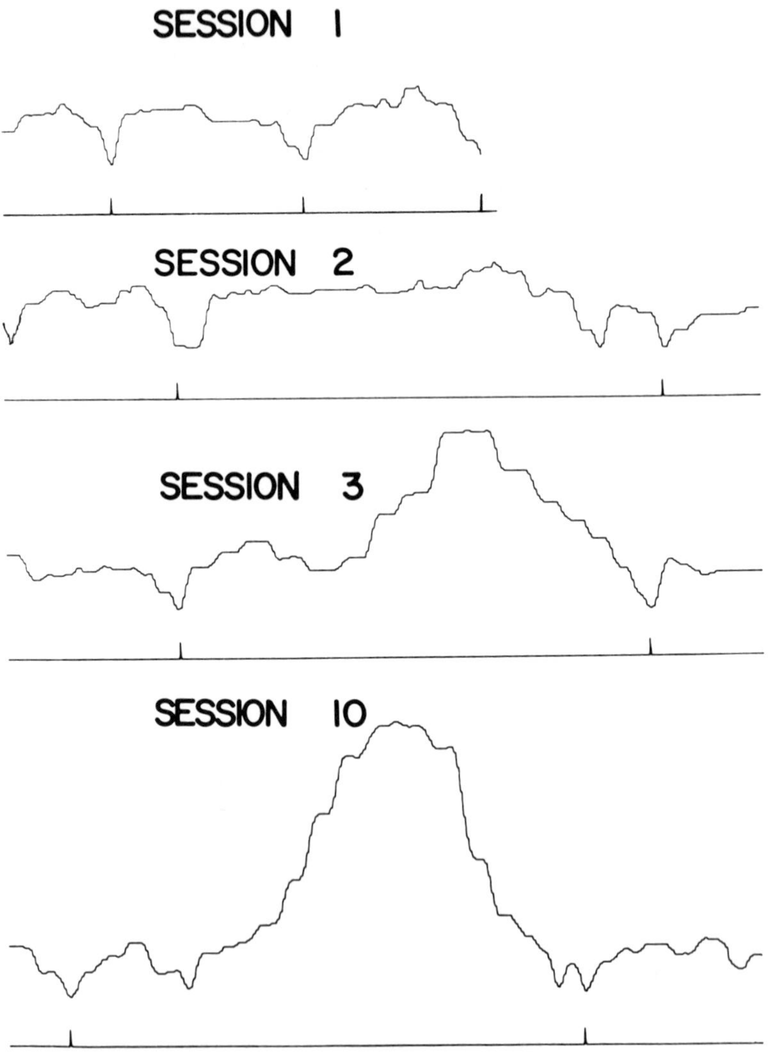

Figure 1-6. continued

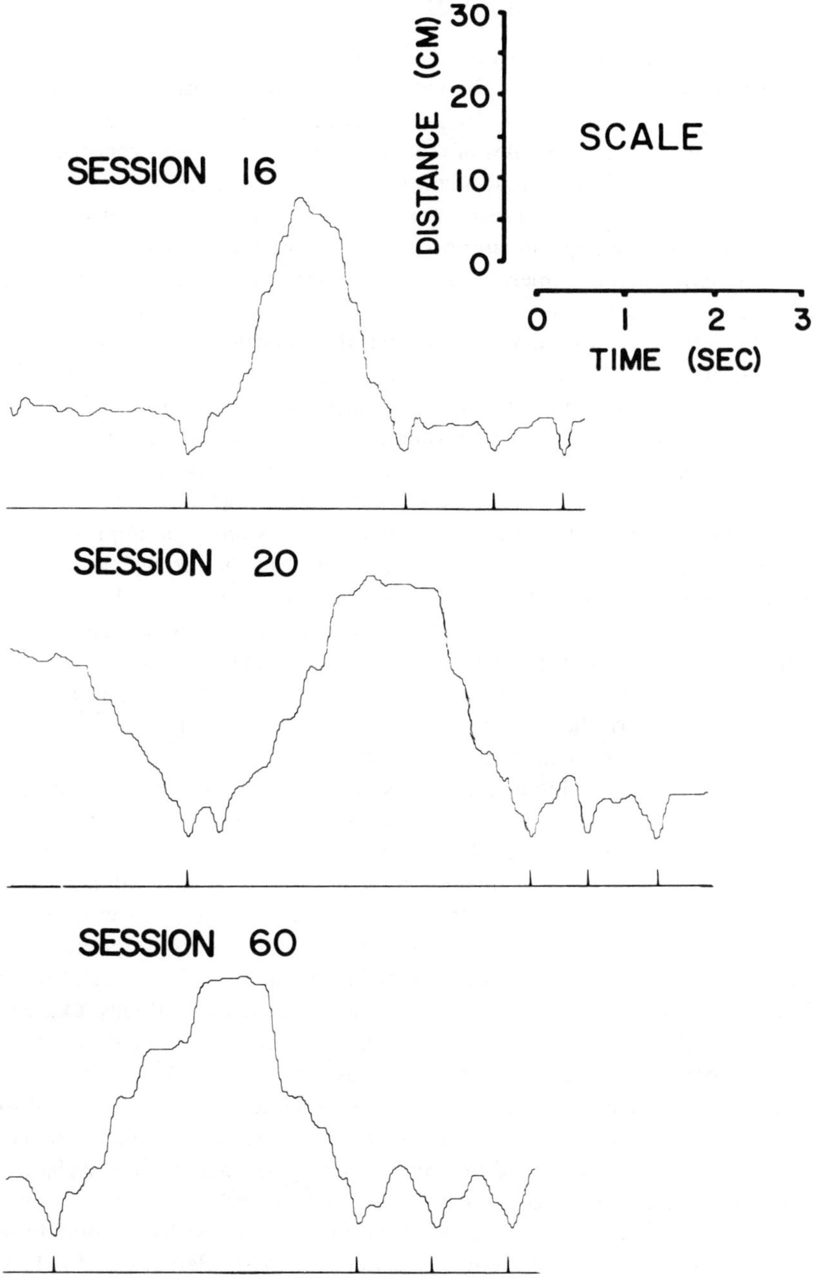

CONCLUSIONS AND DIRECTIONS FOR FUTURE RESEARCH

The first experiment described in this chapter indicated that a detailed study of the effects of reinforcement on the specific continuous motions involved in responding may be necessary to obtain a satisfactory theoretical account of responding under various schedules of reinforcement. The second experiment illustrated the use of a technique that might prove useful in this task.

With only two birds in the second experiment, generalizations must be tentative. The results, however, do support the major inference made from the data of the first experiment, namely, that the form or structure of an experimentally reinforced activity can vary *systematically* (as opposed to randomly—cf. Skinner, 1938: 37) over a wide range, even when the conditions of reinforcement are held constant. In addition, there were several interesting similarities in the behavior of the two birds in the second experiment that seem worthy of emphasis: (1) Upon the introduction of VI 1 minute after continuous reinforcement, a low-amplitude, high-frequency nonterminal wave pattern occurred (that is, the birds remained near the key and frequent movements toward the key were made that did not result in key-pecks); (2) after several sessions high-amplitude waves of lower frequency emerged, which at first generally had irregular shapes and often were bounded on one or both sides by nonterminal troughs; (3) over a large number of sessions high-amplitude waves began to predominate and the waves became much more regular with considerably fewer nonterminal troughs; (4) the later sessions typically began with a low-amplitude wave pattern resembling the wave pattern that predominated during the very early sessions.

These findings raise a number of interesting questions. For example, what process accounts for the increase over sessions in the amplitude of the key-pecking movements (the increase in the distance the bird moved between contacts with the key)? Is the increase due to an "adventitious shaping" process (cf. Skinner, 1948); or can it be explained by inhibitory tendencies interacting with excitatory tendencies (cf. Wilkie, 1974), thus causing the bird to move increasingly further from the key after approaching it and emitting a nonreinforced peck; or again, might it reflect some "schedule-induced" behavior (cf. Staddon, 1977)? (Of course, these possibilities are not necessarily mutually exclusive.) Does amplitude vary as a function of reinforcement rate, and if so can this account in part for the functional relation that has been found between response rate and reinforcement rate on variable-interval schedules (e.g., Catania and Reynolds, 1968)? Along similar lines, what are the characteristics of the behavior waves generated by other schedules, and how are they related to the characteristic response-rate patterns generated by those schedules?

Many previous studies have examined the role of response topography in reinforcement processes (e.g., Hefferline and Keenan, 1961; Herrnstein, 1961; Notterman, 1959; Notterman and Mintz, 1962). The present study's method of examining topography continuously in time is another approach. The main advantage of this new approach seems to be in the type of information it provides concerning the effects of reinforcement contingencies on response rate. Such information cannot be obtained by any approach concentrating solely on response rate or even interresponse times, no matter how detailed the analysis. For example, perusal of Figures 1-3 to 1-6 reveals that interresponse times of the same length can be produced by quite different head-movement patterns, and these differences may reflect different controlling variables. Consequently, all interresponse times of the same length should not necessarily be treated as equal. Conversely, the data also show that similar wave patterns can result in interresponse times of quite different lengths due to the presence of nonterminal troughs. In this connection it might be noted that by providing a record of nonterminal troughs, the method described here raises the possibility of treating the occurrence of a response as a continuous variable with values ranging from 0 to 1. Such a treatment might have important mathematical and hence theoretical advantages over the usual treatment of a response as a discrete event.

It might be objected that the complexities introduced by this new approach unnecessarily undermine the simplicity and generality of the concept of the operant, which has proved to be extremely useful because of these features. Being sympathetic to the concept of the operant, we are aware of this danger. The comprehensiveness and precision to be gained by this approach, however, may more than offset the loss in simplicity it entails. With regard to generality, we shall merely mention that continuity is widely regarded to be an important property of behavior. Any method that can deal successfully with this property under a specific set of circumstances may well yield results that will generalize in unexpected ways to other, quite different circumstances. For the present, the primary focus must be directed to the orderliness of the data.

REFERENCES

Baum, W.M., and H.C. Rachlin. 1969. Choice as time allocation. *Journal of the Experimental Analysis of Behavior 12*:861–74.

Blough, P.M., and D.S. Blough. 1968. The distribution of interresponse times in the pigeon during variable-interval reinforcement. *Journal of the Experimental Analysis of Behavior 11*:23–27.

Catania, A.C., and G.S. Reynolds. 1968. A quantitative analysis of the responding maintained by interval schedules of reinforcement. *Journal of the Experimental Analysis of Behavior 11*:327–83.

Gilbert, T.F. 1958. Fundamental dimensional properties of the operant. *Psychological Review 65*:272–82.

Hefferline, P.F., and B. Keenan. 1961. Amplitude-induction gradient of a small human operant in an escape-avoidance situation. *Journal of the Experimental Analysis of Behavior 4*:41–43.

Herrnstein, R.J. 1961. Stereotypy and intermittent reinforcement. *Science 133*:2067–69.

____. 1970. On the law of effect. *Journal of the Experimental Analysis of Behavior 13*:243–66.

Notterman, J.M. 1959. Force emission during bar pressing. *Journal of Experimental Psychology 58*:341–47.

Notterman, J.M., and D.E. Mintz. 1962. Exteroceptive cueing of response force. *Science 135*:1070–71.

Pear, J.J., and B.L. Rector. 1979. Constituents of response rate. *Journal of the Experimental Analysis of Behavior 32*:341–62.

Platt, J.R. 1979. Interresponse-time shaping by variable-interval-like interresponse-time reinforcement contingencies. *Journal of the Experimental Analysis of Behavior 31*:3–14.

Rachlin, H. 1973. Contrast and matching. *Psychological Review 80*:217–34.

Shimp, C.P. 1974. Time allocation and response rate. *Journal of the Experimental Analysis of Behavior 21*:491–99.

____. 1975. Perspectives on the behavioral unit: Choice behavior in animals. In W.K. Estes, ed., *Handbook of Learning and Cognitive Processes*, Vol. 2, *Conditioning and Behavior Theory*, Hillsdale, N.J.: Lawrence Erlbaum.

Skinner, B.F. 1938. *The Behavior of Organisms: An Experimental Analysis.* New York: Appleton-Century-Crofts.

____. 1948. "Superstition" in the pigeon. *Journal of Experimental Psychology 38*:168–72.

Staddon, J.E.R. 1977. Schedule-induced behavior. In W.K. Honig and J.E.R. Staddon, eds., *Handbook of Operant Behavior*, Englewood Cliffs, N.J.: Prentice-Hall.

Wilkie, D.M. 1974. Stimulus control of responding during a fixed-interval reinforcement schedule. *Journal of the Experimental Analysis of Behavior 21*: 425–32.

2 HOW REINFORCERS ARE AGGREGATED IN REINFORCEMENT-DENSITY DISCRIMINATION AND PREFERENCE EXPERIMENTS

Michael L. Commons, Michael Woodford, and John R. Ducheny

In the controversy over matching versus maximizing there are two traditions. According to the first, free operant concurrent choice procedures are used; what is thought to be matched or maximized is the amount of reinforcement obtained for responding over a period. Previous illustrations of matching (e.g., Herrnstein, 1970, Chapter 19 of this volume, and Herrnstein and Vaughan, 1980) and maximizing (Rachlin, Kagel and Battalio, 1981), which are accounts of individual choice behavior in operant conditioning experiments, have looked only at the aggregate amount of reinforcement received on a particular schedule. In many situations, such as the concurrent variable-interval, variable-ratio (conc VI VR) schedules described by Herrnstein and Heyman (1979), aggregate data are sufficient to indicate that the experimental subjects do not allocate behavior in a way that would maximize obtained aggregate reinforcement. As Herrnstein has mentioned (personal communication 1981), it is always possible to add constraints that "save" maximizing in this situation. One of these could be that the subjects were not fully sensitive to the contingencies applied to their behavior. Then, the deviation from obtained aggregate maximization would often be roughly consistent with a hypothesis that aggregate relative rate or probability of behavior is "matched" to aggregate obtained relative rate or probability of reinforcement.

Portions of this chapter were reported at the Harvard Symposium on Quantitative Analyses of Behavior, 1979, and the Midwestern Association for Behavior Analysis, Chicago, 1976, by A.A. Thompson, M.P. Krupa, J.J. Andersen, and M.L. Commons, and at the Eastern Psychological Association, Boston, 1977, by M.P. Krupa, M. Hirvonen, J. Pohl, M. Lamb, and A.A. Thompson. These authors collected the 4-cycle discrimination data, wrote the computer programs, and prepared the preliminary reports. The line of research

In the second tradition discrete-trial noncorrection choice procedures are used; what is thought to be matched or maximized is the amount of reinforcement programmed (e.g., Estes, 1957; Neimark and Estes, 1967). However, in a discrete-trial choice situation, the four accounts only make two distinguishable predictions. Matching of obtained reinforcement and maximizing of either obtained or programmed reinforcement all predict that one alternative comes to be chosen to the exclusion of others. The fourth possibility, that of matching of programmed or expected payoff, appears as undermatching of obtained payoff. Estes (1957) proposed a matching law for such expected payoff in a probability learning situation. With pigeons, matching to expected payoff has been found in a discrete-choice reinforcement density discrimination (Commons, 1979, 1981). Here, unless otherwise specified, matching or maximizing refer to programmed reinforcement or expected payoff rather than obtained reinforcement.

The aggregate data in both traditions, however, are ill suited to examination of the possibly more sophisticated local maximizing model proposed here. One of the purposes then of this chapter is to account for undermatching of obtained reinforcement and matching of expected payoff with such a model. It is hypothesized that at the time of each individual choice, the subject acts so as to maximize the expected utility of reinforcement resulting from that particular choice. However it has an incomplete awareness of the effects of the choice that are available to it.[a] More specifically, this model supposes imperfect discrimination of stimuli, including reinforcers, and even more imperfect memory for past

---

leading up to the discrimination experiment was developed under the guidance of John Anthony Nevin. John R. Ducheny collected the 6-cycle discrimination data and he and Joel R. Peck supervised the data collection for the preference experiment. The data acquisition on the preference experiment was carried out with the support of Richard J. Herrnstein. The computer programming was done by Rose P. Meegan and Gilbert Yee with assistance from Wei-o Huang and Katrine Norris. The computer analysis was performed at the Northeastern University Computer Center. The graphs were made by Rose P. Meegan, Brian D. Cabral, Martin N. Davidson, and Rodney Wong. Drafts were read and edited by Patrice M. Miller, Sherry Luna, James E. Mazur, Margaret Vaughan, Theodore L. Allen, and Philip F. Kelleher. The recommendations made by R. Duncan Luce and Richard J. Herrnstein after hearing presentations of the material were invaluable. Howard Rachlin and Richard J. Herrnstein, the coeditors of this volume, gave considerable time and effort toward making the manuscript more readable. Research was supported by grants to Richard J. Herrnstein, by grants from the Dare Association, Inc., to the first author and a National Science Foundation Fellowship to the second author.

a. A simpler hypothesis would be that the subject maximizes the expected value of reinforcement, measured, say, in milligrams of grain received. However, the reinforcing value to the subject of a food pellet need not be proportional to its weight, and so a rationally maximizing subject need not regard the prospect of a 50 percent chance of receiving 60 milligrams and a 50 percent chance of receiving 20 milligrams as equal in value to a certainty of receiving 40 milligrams of grain. In order to take this into account, it is assumed instead that the subject maximizes the expected value of the utility of the reinforcement, where the utility is taken to be some monotonically increasing (but not necessarily linear) function of the amount of grain received.

events. These two phenomena can explain the observed deviations from performance values predicted by full-information aggregate maximization. Whether or not the local maximizing model that contains these two hypotheses is good, however, it is easy to support and difficult to reject with aggregate data. This is because the values of the model's parameters relating to discriminability and forgetting cannot be directly estimated from aggregate level observation. They appear as free parameters that may always be adjusted to make the theory's predictions roughly consistent with the aggregate data. The predictions generated in this fashion are not easily disproved; accordingly, confirmation of the theory is trivial.

The studies reported in this chapter seek to determine the reinforcement value of sequences of reinforcers at the local, nonaggregated level. Value is measured using two kinds of choice procedures. In the first procedure, called discrimination (Experiments 1 and 2), the subject's choice is reinforced, S^{R+}, for correctly determining from which of two reinforcement schedules, one rich and one lean in reinforcement, a given reinforcement sequence has been taken. In Experiment 1, one of sixteen different sequences of reinforcers was presented on a trial; in Experiment 2, one of 64. For each different sequence of reinforcers the proportion of trials on which the subject indicates that the richer density schedule has been sampled is separately recorded. This proportion reflects the richness or *perceived value* of that sequence. This procedure makes it possible to examine how sequences of reinforcers are valued at the local level.

In the second procedure, called preference (Experiment 3), the subject allocates its time and responses between two keys in the first link of a concurrent chain schedule. One of the second links of this schedule consists of a *comparison* sequence of reinforcers and the other a *standard* sequence of reinforcers; the sequences used here are the same 16 sequences used in Experiment 1. In the comparison link, the sequence of reinforcers is changed to a new sequence only after its value has been determined. In the standard link, the sequence of reinforcers is always the same. The number of milligrams of food reinforcement delivered within the standard link is varied until the amount required to make the subject indifferent between the two second links is found. This value that reflects *effective value* is recorded for different sequences of the comparison schedule.

Both the discrimination and the preference procedures allow sequences of reinforcers to be scaled as to relative aggregate reinforcement provided, with the discrimination experiments indicating the perceived value and the preference experiment indicating the effective value. It will be shown that these two measures of aggregate reinforcement value coincide closely.

In addition, because the value of each sequence of reinforcers is separately measured, it becomes possible to determine the disaggregated effect of each individual reinforcer according to its position in the sequence. The more sophisticated local maximizing hypothesis just described yields quite specific predictions as to what the relative reinforcement values of different sequences should be. The assumption of a forgetting process is no longer a rationale for the introduction of ad hoc parameters, as in the case of aggregate data, since the precise nature of the forgetting process may be determined by comparing the aggregate values of different sequences of reinforcers. Consistency of the data with the sophisticated local maximizing hypothesis is no longer trivial. This hypothesis will be shown to account in detail for the relative perceived values, and, since the effective values have the same structure as do the perceived values, it is believed that a similar maximizing model can account for the effective values. Therefore, apparent matching in aggregated data may be accounted for by a local maximizing model allowing for imperfect discrimination and memory loss.

EXPERIMENT 1: DISCRIMINATION PROCEDURE

Each of four White Carneaux pigeons was run in one 256-trial session per day. Trials consisted of a stimulus period followed by a choice period, as shown at the bottom left of Figure 2-1. The task was similar to one in which the subject identifies whether a randomly chosen urn is the rich one with three reinforcer balls and one nonreinforcer ball, or whether the urn is the lean one with one reinforcer ball and three nonreinforcer balls. On each trial there are four draws from that trial's urn; these draws take place in the stimulus period. After a ball is drawn it is put back into the urn. After the four draws from the urn presented on that trial, the subject indicates whether the sample came from the rich or the lean urn; this takes place during the choice period.

The stimuli presented during the stimulus period are to be described *macroscopically* (the whole stimulus) and *microscopically* (the parts of the stimulus). First there will be a general description of the procedure, including the stimulus and choice periods, and then the contingencies in these periods will be described more specifically.

The stimuli to be discriminated in the present study were samples from one of two modified *T* schedules (Commons, 1979; Schoenfeld and Cole, 1972; Schoenfeld, Cumming, and Hearst, 1956; Weissman, 1961). The rich schedule had the reinforcement S^{R+} probability for a lighted center-key peck, C, of $p(S^{R+} | C) = 0.75$. The lean schedule had a $p(S^{R+} | C) = 0.25$. A sample from a schedule consisted of 4 cycles, which in the standard case were each 3 seconds in length. At the beginning of each cycle, the center key was illuminated. The cycles, c_i or just i, were numbered so that c_4, occurring at the beginning of the stimulus period, is the furthest from choice, and c_1, occurring at the end of the stimulus period, occurs just before the choice period. The first center-key peck in each of the four cycles darkened the center key and was reinforced with a probability of 0.75 (if the rich schedule was sampled) or 0.25 (if the lean schedule was sampled). Although the reinforcement probability was neither 0 nor 1

Figure 2-1. A stimulus example from a discrimination trial shown (*top left*) by means of a state diagram for a sample from a 3-second cycle schedule to illustrate what may happen if center-key pecks occur (and do not occur) when reinforcement has been programmed (or not). In reality it would be very unusual because there is a cycle without a peck occurring in it. The contingencies during an entire trial are shown (*bottom left*). The stimulus period contains substimuli of the form shown in the top left portion. The choice period contain a single reinforcer for correct choice or extinction for incorrect ones. The choice period immediately follows the stimulus period. There is no intertrial interval.

The second-link reinforcing stimulus from a preference trial (*top right*) is the mirror opposite of the stimulus example shown on the left. It illustrates what may happen if side-key pecks occur (and do not) occur when reinforcement has been programmed (or not) on cycles during the second link. In reality it would be very unusual because of the cycle with no peck occurring in it. The contingencies during an entire trial are shown (*bottom right*). The choice period consists of a concurrent VI 12-second, VI 12-second schedule. The first response after the interval completes the VI requirement. When the left requirement is met, the left-key changes color and has the comparison sample substimulus programmed on it. Likewise, when the right requirement is met, the second-link standard substimulus is presented on the right key, which also changes color. There is no intertrial interval.

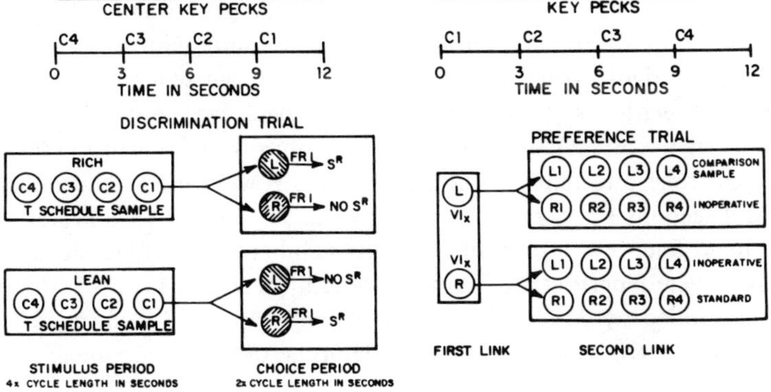

on a cycle i, a center-key peck was ($v_i = 1$) or was not ($v_i = 0$) reinforced on each of these 3-second cycles. Only the first center-key peck in any one cycle was ever reinforced.

An example of what happens in a reinforcement schedule sample presentation, during the stimulus period of a trial, is shown in the upper left of Figure 2-1 for the substimulus 1110 (reinforcement programmed for c_4, c_3, and c_2, but not c_1). Each cycle began with the illumination of the center key. Again, this happens in all the samples. In the fourth cycle before choice (c_4) the first center-key peck, labeled A, darkened the key and was reinforced. Likewise at the beginning of the third cycle before choice (c_3) the center-key light came on and the first center-key peck (labeled B) darkened the key and was reinforced. The other center-key peck during the same cycle, labeled C, was not reinforced. At the beginning of the second cycle before choice (c_2) the center-key light came on again. There was no center-key peck during that cycle (an unlikely occurrence). Therefore the center-key light remained illuminated into the first cycle before choice (c_1), and the programmed reinforcer was not picked up. At the beginning of the first cycle before choice (c_1), the center-key light just remained on. The first peck in that cycle, labeled D, darkened the key. There was no reinforcement programmed for this trial, so none was obtained. The next center-key peck, E, had no effect since it was the second one in this cycle. For the trial as a whole, only two out of three reinforcers were obtained. Although this sample could have come from either the rich or the lean schedule, it was more probable that it came from the rich schedule because it had three reinforcements programmed.

At the onset of the choice period the side keys were illuminated; the center key stayed dark or was darkened in those rare cases where no key peck occurred in the last cycle of the substimulus. Duration of the choice period was always twice the standard or base length cycle, as shown in the lower left of Figure 2-1. The first side-key peck, whether correct or not, darkened both keys, and no further pecks were counted. If a substimulus sample from the rich schedule had been presented on the center key, the first left-key peck was reinforced (fixed ratio-1, FR-1, a hit or left correct); a right was not reinforced (extinction, EXT, a miss or right error). If a sample from the lean schedule had been presented on the center key, the first right-key peck was reinforced (a correct rejection or right correct); a left was not reinforced (a false alarm or left error).

Although three different standard cycle lengths were used—2 seconds, 3 seconds, and 4 seconds ($T = 2, 3, 4$)—only the data for the 3-second standard are reported here. Each was run until the birds stabilized. In daily sessions the standard (base) cycle length was of standard length (T seconds) on 224 trials, doubled (T multiplied by 2) on 16 probe trials, and tripled (T multiplied by 3) on another 16 probe trials. The position of the probe trials within a session was randomly distributed.

Stimulus Description

Division of the stimuli into four levels of analysis is crucial for understanding the relation between stimuli at the aggregated level and stimuli at the more disaggregated levels. There are four levels of description of the *stimuli*:

1. On a *macro* level each probabilistic reinforcement schedule, either the rich (p = 0.75) or the lean (p = 0.25), is viewed as a single *stimulus*, here called S_{rich} (S_R) and S_{lean} (S_L), respectively; sampling considerations are of no concern. These two *stimuli* are equivalent to cued random-interval (RI) or variable-interval (VI) schedules with a T-second limited hold and average intervals equal to $4/3T$ second (rich) and $4T$ seconds (lean), where T = 3 seconds for the standard case being considered. On a variable schedule (VI) after a certain amount of time (interval) the first response is reinforced.

2. At the *molar* level all samples from the two reinforcement schedules (called at this level *substimulus samples*), with the same number of reinforcers are viewed as the same *stimulus.* Each substimulus S_n has a reinforcement density $D(S_n)$ equal to N_d, the number of center-key-peck reinforcement opportunities over the four cycles in the substimulus. Density, the defining characteristic of a stimulus at this level of analysis, ranges from 0 to 4 reinforcers per substimulus, $D(S_n) = N_d = 0, 1, 2, 3, 4$, giving rise to 5 such *stimuli*, designated S_{N_d}, as shown in Table 2-1. This level of stimulus description distinguishes among the number of reinforcements within a substimulus; it does not distinguish among substimuli with the same number of reinforcers but with those reinforcers programmed on different cycles within the $4T$-second stimulus period. The frequency of the occurrence of the possible samples with a given number of reinforcers is shown in the left panel of Figure 2-2 and in Table 2-1. There are 16 possible sequences of reinforcement or nonreinforcement in a 4-cycle sample. The probability that a particular set of 4 events will occur was obtained by expanding the binomial $(p + q)^4$ with p = 0.75 or 0.25 and $q = 1 - p$. Actual presentation frequencies deviated somewhat from the expected frequencies because of sampling, as shown in Table 2-1. The modal and mean density of the rich schedule is 3; the modal and mean density of the lean schedule is 1.

3. At the *molecular* level the definition of a *stimulus* involves the sequence or pattern of reinforcement in a sample. At this level substimuli are described not with a one-digit density number but with a four-digit binary number. A number like 0001 indicates a sample with 3 cycles without reinforcement opportunities followed by 1 cycle with a reinforcement opportunity. The substimuli S_n are numbered from 0000 to 1111. The leftmost digit represents the cycle furthest in time from choice c_4, and the rightmost represents the cycle immediately before choice c_1.

Table 2-1. Description of stimuli. For each of the five possible molar classifications of substimuli with number of reinforced cycles, N_d, 0–4, and each of the 16 possible molecular substimuli, S_n, 0000–1111, the theoretical and sampled probabilities of reinforcement are given for the rich (LC), $p = 0.75$, and the lean (RC), $p = 0.25$, schedules. The sum of the molecular frequencies equals the molar frequencies. On a molecular level 0 indicates a nonreinforced cycle and 1 a reinforced cycle, the digit on the right being closest to choice. The theoretical and sampled numbers of occurrences of molar classified substimuli and molecular substimuli in the 256 trials from a session are also given. The last column shows the probability of a left-key peck or a right-key peck being correct. This is the same as the probability of reinforcement for a left-key peck and a right-key peck.

Molar Level: Number of Reinforcers in a Substimulus N_d	Molecular Level: Substimulus Number in Binary Notation	Gross Molar: Rich Frequency (LC)		Gross Molar: Lean Frequency (RC)		Gross Molar: Frequency of Corrects (LC + RC)		Probability of a L or R Being Correct $p(LC\|S_{N_d})$ $(p(RC\|S_{N_d}))$
		Theoretical	Sampled	Theoretical	Sampled	Theoretical	Sampled	
0	0000	0.5	1	40.5	38	41	39	.0256
		0.5	1	40.5	38	41	39	(.9744)
	1000	1.5	1	13.5	17	15	18	
	0100	1.5	1	13.5	16	15	17	
	0010	1.5	3	13.5	14	15	17	
	0001	1.5	2	13.5	10	15	12	
1		6.0	7	54.0	57	60	64	.2093
								(.8907)
	1100	4.5	3	4.5	4	9	7	
	1010	4.5	5	4.5	6	9	11	
	1001	4.5	6	4.5	4	9	10	
	0110	4.5	3	4.5	5	9	8	
	0101	4.5	6	4.5	6	9	12	
	0011	4.5	4	4.5	4	9	8	
2		27.0	27	27.0	29	54	56	.4821
								(.5179)

	1110	13.5	17	1.5	2	15	19	
	1101	13.5	14	1.5	1	15	15	
	1011	13.5	11	1.5	0	15	11	
	0111	13.5	12	1.5	2	15	14	
3		54.0	54	6.0	5	60	59	.9152 (.0847)
	1111	40.5	38	0.5	0	41	38	
4		40.5	38	0.5	0	41	38	
		128	127	128	129	256	256	1.0000 (.0000)

Figure 2-2. The distribution of reinforcement for the first center-key peck ($S^{R+}|C$) for 4-cycle (*left*) and 6-cycle substimuli (*right*) are shown. The number of center-key pecks and proportion of trials on which reinforcement occurs in a sample is shown for the rich (positive-slope hatching) and lean (negative-slope hatching) schedules. The respective $p(S^{R+}/C)$ are 0.75 and 0.25 for the 4-cycle substimuli and 0.767 and 0.233 for the 6-cycle substimuli.

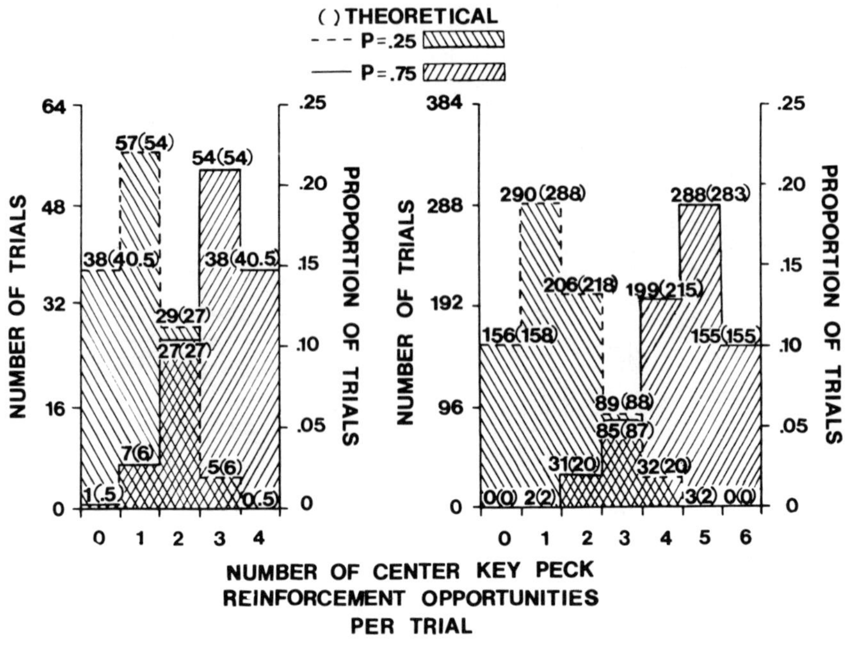

4. At the *micro* level the definition of a *stimulus* depends on whether or not there is a reinforcement opportunity on a particular cycle c_i, i cycles before choice, irrespective of what is programmed for its neighbors. There are 8 such *micro* stimuli (4 with reinforcers on each of the 4 cycles and 4 without reinforcers on each of the 4 cycles).

A molecular substimulus, even 0000 or 1111, could be sampled from either schedule stimulus. What differed therefore was not the substimulus that could be sampled from the schedule stimuli but the frequency with which a substimulus came from the rich schedule or the lean schedule. These frequencies are presented in Table 2-1. A comparison of columns 1 and 2 with the last column in the table shows that the greater the number of cycles in a substimulus having a reinforcement opportunity, the greater the likelihood that the rich schedule was in effect. Therefore, the rich schedule was more likely to be in effect when 3 or 4 reinforcement opportunities were programmed, and the lean schedule was more likely to be in effect when 0 or 1 reinforcement opportunities were presented on a trial. Substimuli with 2 reinforcement opportunities occurred about equally often given either schedule. Again, when the rich schedule was in effect $p(S^{R+} | L)$ increased. Birds were run until their performances stabilized. The data reported here were collected in the 5 sessions following stabilization.

EXPERIMENT 2: SIX 3-SECOND CYCLE SAMPLES

Method and Procedure

The subjects and apparatus for this experiment were identical to those described in Experiment 1, except that one less subject was used. The discrimination task was similar to that used in Experiment 1, except that the number of cycles was increased from 4 to 6. The two schedules to be discriminated were, again, two overlapping distributions (stimuli) of reinforcement opportunities, one with $p(S^{R+} | C) = 0.767$ (the rich schedule) and one with $p(S^{R+} | C) = 0.233$ (the lean schedule). These probabilities will be treated as approximately the same as for the 4-cycle experiment for purposes of later analysis. The stimulus on each trial was one of the 64 substimuli that could be generated by taking samples of size 6 from one of the two T schedule stimuli. The frequency of occurrence of the possible samples with a given number of reinforcers for the 6-cycle case is shown in the right panel of Figure 2-2. The modal density of the rich schedule is 5 and the lean schedule is 1; mean value can be obtained by multiplying the probability of occurrence of a reinforcer times the number of cycles (e.g., $0.767 \times 6 = 4.60$ for the rich and 1.40 for the lean). The value of T was again equal to 3 seconds, and there were no probe trials.

EXPERIMENT 3: PREFERENCE PROCEDURE

Each of ten White Carneaux pigeons was run in one 128-trial session per day. A trial consisted of the first and second links of a concurrent chain schedule as shown in the bottom right-hand portion of Figure 2-1. In the top half of Table 2-2, a simple version of a concurrent chain schedule is shown. In a concurrent schedule two schedules are in effect at the same time, one on each key. The center key is inoperative. The bird pecks either the right or the left key. The second link on either key is activated by the completion of the contingency on the first link. Both keys are freely available during the first link, but once the bird has entered the second link on either key the other key is deactivated. In this simple version the first link on both the right and the left keys is a 12-second variable-

Table 2-2. Simple concurrent schedule compared with concurrent chain schedule used here.

Simple Concurrent Schedule

VI 12 seconds ⟶ Reinforcer

VI 12 seconds ⟶ Reinforcer

Concurrent Chain Schedule

VI 12 seconds ⟶ v_1 v_2 v_3 v_4 on left comparison key

VI 12 seconds ⟶ 0 V_{mg} 0 0 on right standard key

$$v_i = \begin{cases} 0 \text{ milligrams} \\ 20 \text{ milligrams} \end{cases}$$

$$V_{mg} = \text{varies}$$

interval schedule. The second link is a single reinforcer. Completing the VI 12-second requirement on either of the keys would lead to the obtaining of the reinforcer programmed for that key. In the more complex case shown both at the bottom right of Figure 2-1 and bottom of Table 2-2 the single reinforcer of the second link is replaced by one of two samples from a *T* schedule. Therefore the second link's *T* schedule sample on a given key serves as the reinforcer for completing the requirement on the same key during the first link. Presumably for this example different values of the reinforcers in the second link would lead to differential responding on the left and right keys. The samples in the second link are the same as those used in the 4-cycle discrimination experiment (Experiment 1), so that the aggregation of sequences of reinforcers in the two kinds of experiments may be directly compared. Again, a schematic diagram of the first and second links is shown in the bottom half of Table 2-2.

In order to understand the contingencies in the second link let us return to a short description of the reinforcement samples. Recall that the reinforcement samples in Experiment 1 consisted of four 3-second cycles. A reinforcer is either programmed or not on each cycle. A total of 16 such samples, called substimuli, are possible. The 16 substimulus samples each consisted of four 3-second cycles. Each cycle began with the illumination of the left key, and the first left-key peck was ($v_i = 1$) or was not ($v_i = 0$) reinforced. Subsequent pecks during a cycle were not reinforced. The cycles c_i were numbered so that the c_1 was the cycle at the beginning of the second link right after choice, and c_4 is the cycle furthest from choice occurring at the end of the second link. Notice that this is exactly the same as was done in Experiment 1. In both cases c_1 is the cycle closest to choice and c_4 the cycle furthest away; what changes is where the choice occurs relative to the four cycles. For the preference experiment, however, the binary-numeral notation for substimuli will be reversed. That is, 0111 means $\nu_1 = 0, \nu_2 = 1, \nu_3 = 1, \nu_4 = 1$. This is so that left to right will continue to indicate temporal ordering. An example of an unlikely response pattern to the 0111 substimulus sample is shown in the upper right of Figure 2-1. At the beginning of the first cycle, c_1, the key was illuminated. The first left-key peck darkened the left key and was not reinforced. Although the reinforcement was set up during the second cycle and in the actual experiment the bird would always peck and obtain it, for the sake of illustration, a cycle without a peck is shown here. Because there was no peck, no reinforcer was delivered during that cycle and the light remained on. At the end of that cycle, the reinforcer was set up again. In the third cycle after choice the bird did peck and obtained the reinforcer; the key immediately darkened. The same thing happened during the fourth cycle, c_4.

Two aspects of these reinforcement samples serve as independent variables: (1) the pattern of reinforcement, that is, on which of the four cycles reinforcers were delivered, and (2) the amount, in milligrams, of the reinforcer delivered on a particular cycle. On the left key, the *comparison* key, the pattern of reinforcement varied while the amount delivered on a cycle stayed constant. On the right key, the *standard* key, the reinforcement amount was varied but the pattern was always the same one. The conditions determining when these dimensions of the reinforcement samples were varied will be described next.

On the left, comparison key, the peck that completed the VI led to the comparison substimulus. The same comparison substimulus was present for a minimum of 30 sessions. The maximum number of sessions it was run was determined by conditions discussed later. Throughout the study all reinforcers obtained by comparison-key-pecks in a cycle were 20 milligrams.

On the right, standard key, the peck that completed the VI led to the standard substimulus, $0V_{\text{mg}}00$. In this substimulus no reinforcers were programmed on the first, third, and fourth cycles. Reinforcement, consisting of varying number of pellets, was always programmed on the second cycle. The amount of reinforcement, V_{mg}, during the second cycle was varied by the experimenter. The number of sessions that had a given value V_{mg} programmed depended on conditions to be outlined. To get values such as 30 milligrams using just 20-milligram pellets, 0.5 of the c_2 cycles would have two pellets and 0.5 would have one pellet. A similar probabilistic delivery of reinforcement was used to generate values smaller than 20 milligrams. In this second case, the hopper light still went on when a reinforcer was to be earned, but the pellet was delivered probabilistically to generate approximately any required value between 0 and 20 milligrams.

The treatment of probabilistic reinforcement as equivalent to certain delivery of the mean reinforcement amount implicitly assumes that the utility obtained from reinforcement is a linear function of amount. In fact, as will be shown, utility is a slightly concave function of the number of pellets. But since the exponent obtained for the power function is not too much less than one (approxi-

Table 2-3. Plan for the analysis, with data from the preference study compared with data from the discrimination study for both the perceived value and the effective value (there is no model in the effective value and the models in the elemental data). Column 1 shows the substimuli in preference data in digital form and Column 2 shows the same substimuli in binary form detailing the reinforcement pattern. The equivalent substimuli in the discrimination case run just the reverse (0001 in the preference case is equivalent to 1000 in the discrimination case). The cycles are read from left to right; a 0001, having the reinforcer, is the fourth cycle after choice in the preference case. The titrated values in milligrams are shown for each bird that was tested on a given comparative substimulus; brackets indicate that the bird eventually died after the particular data were being collected. The mean value in milligrams is shown (Column 4) along with its standard deviation.

Preference				*Discrimination*				
Substimulus Number								
Digital	*Cycles*	*Titrated* V_{mg}	$\overline{V}_{mg}$ *(S.D.)*	$\overline{Z}_{p(L)}$ *(S.D.)*	$Z_{p(L)1}$	$Z_{p(L)2}$	$Z_{p(L)3}$	$Z_{p(L)4}$
00	0000		0	-2.420 (.30)	-2.035	-2.750	-2.371	-2.517
01	0001	12_{30} [8_{85}] 4_{346} 7_{58} 9_{58}	8.000 (2.92)	-2.026 (.53)	-1.492	-2.086	-2.739	-1.787
02	0010	10_{346} 14_{1} 10_{1} 12_{30}	11.500 (1.91)	-1.741 (.78)	-1.025	-2.043	-2.704	-1.193
03	0011	29_{346} 30_{346} [30_{85}]	29.667 (.577)	-0.874 (.84)	-0.262	-0.529	-2.108	-0.595
04	0100	20_{58} 18_{59} 20_{59} 24_{62} 18_{161} 20_{161} [20_{85}] 20_{62}	20.000 (1.85)	-1.255 (.45)	-1.025	-1.155	-1.918	-0.923
05	0101			-0.596 (.61)	-0.449	-0.449	-1.460	-0.026
06	0110	40_{1}	40.000 —	-0.326 (.85)	0.131	-0.169	-1.565	0.299
07	0111	94_{62}	94.000	0.658 (.84)	1.396	0.804	-0.553	0.983

08	1000	44_{59}	44.000 —	-0.783 (.44)	-0.470	-0.430	-1.383	-0.849
09	1001	[80_{85}]	80.000 —	0.05 (.77)	0.288	0.537	-1.097	0.472
10	1010			0.290 (.64)	0.272	0.734	-0.609	0.764
11	1011	[70_{85}] [50_{84}]	60.000 (14.14)	1.216 (.79)	1.684	1.534	0.038	1.609
12	1100	[70_{84}]	70.000 —	0.900 (.65)	0.863	1.698	0.111	0.929
13	1101			1.794 (.84)	1.696	2.536	0.657	2.287
14	1110	[70_{84}] [82_{84}]	76.000 (8.48)	1.820 (.50)	1.732	1.960	1.192	2.394
15	1111	144_{30} 153_{346} 160_{161} [80_{102}]	134.250 (36.75)	2.386 (.51)	2.983	2.276	1.754	2.532

mately 0.7), this method of interpolating values is reasonably accurate over the range of values in question. An extrapolation of Mazur's (1981) results shows that probabilistic delivery of reinforcement is a reasonable procedure for obtaining intermediate amounts.

Mazur, using a Stubbs–Pliskoff procedure (a concurrent interdependent procedure), showed that probabilistically delivered reinforcement, $p(S^{R+}) = 0.50$ was valued as 0.50 and $p(S^{R+}) = 0.33$ was valued as 0.33 (in his procedure relative value was measured by relative rate). In Mazur's study, relative left-key pecking rate, Rel B_L (his measure of relative value) is equal to the rate of left-key pecking, B_L, divided by the sum of the left-peck rate and the right-peck rate, B_R, Rel $B_L = B_L/(B_L + B_R)$. This same relation, between probabilistically delivered reinforcement and relative rate (as defined for this situation) should hold in the present situation, which uses a concurrent chain schedule (instead of a simple concurrent schedule) where one of the pellets delivered in the second link's standard substimulus was probabilistically delivered as described before. The probabilistic reinforcement p can be used to obtain a reinforcement amount reasonably equivalent to the actual milligram value desired, as follows: For m = number of pellets,

$$V_{\text{mg}} = m(20 \text{ mg}) + p(\text{reinforcement}) \times 20 \text{ mg} = 20(p + m) \text{ mg}$$

For example, to obtain a reinforcement amount equal to 10 mg, m would equal 0 and p would equal 0.5, and therefore $V_{\text{mg}} = 0.5 \times 20 \text{ mg} = 10 \text{ mg}$.

Three kinds of information were recorded: first, rates of left-key and right-key pecking during the initial links; second, which comparison substimulus was programmed on the left key during the second link, and third, the amount of food, V_{mg}, delivered for the reinforced right-key peck in the second link standard substimulus $0V_{\text{mg}}00$. The amount of food delivered in the second link standard substimulus was varied in order to get the titrated amount of food. The titrated amount of food was defined as the number of milligrams of food necessary to produce indifference between the comparison and the standard. Such indifference was defined as a relative rate of left-key pecking equal to 0.5 ± 0.1 for 6 consecutive sessions. There also had to be 6 consecutive sessions for food amounts that produced relative rates above the 0.5 indifference value and 6 consecutive sessions below.

The titrated amount of food V_{mg} was not the only parameter that was systematically varied. After V_{mg} had been found for a particular substimulus, the substimulus that served as the comparison substimulus was changed and a new V_{mg} was obtained for the new substimulus. A given substimulus for which V_{mg} is found is referred to as a point. This procedure was not carried out using all substimuli with all birds. Which birds were run on which substimuli is shown on the left side of Table 2-3. Next to each substimulus number (listed in the column entitled cycles) is a list of the titrated amounts obtained for that substimulus by each bird. Entries are in the following form: Titrated amount$_{\text{bird number}}$. Thus,

Table 2-3a. Organization of the presentation of the data, analysis, and corresponding theory.

	Elemental Data	*Transformation*		*Addivity*		*Decrementing*	
Perceived Value		Exp.	Formal Theory	Exp.	Formal Theory	Exp.	Formal Theory
		X	X	X	X	X	X
Effective Value		X	X	X	X	X	

the entry 12_{30} under substimulus 0001 means that bird 30 is indifferent between V_{mg} = 12 mg and substimulus 0001. Numbers in brackets are for birds that died in the course of the study; three of the ten birds died during the study, which took over three years to run. Some of the points were replicated by the same bird.

Results of the Three Experiments

The analysis is presented in four parts: (1) elemental data for perceived value and effective value of a substimulus; (2) transformations that make the value of the molar level *stimuli* a linear function of the reinforcement density; (3) the degree to which the aggregation of reinforcers within a substimulus is a linear function of the contribution provided by each reinforcer when the transformations from (2) are applied; and (4) the decrementing functions that best account for the decrease in value as a reinforcer occurs further from choice. (See Table 2-3a.) The analysis will be presented first with the corresponding model following.

ELEMENTAL DATA

The Discrimination Situation

The most elemental form of the discrimination data examines the choices that the pigeons make following a substimulus presentation. A left peck to a substimulus, $L|S_n$, is the decision that indicates that the substimulus is rich. The probability of a left-key peck to the given substimulus is $p(L|S_n) = (L|S_n)/(L|S_n + R|S_n)$. This value is called bias in signal detection theory (Nevin, 1969; Green and Swets, 1966), the model used here and in previous presentations of similar data (see Commons, 1979; 1981). This represents the tendency to indicate that the substimulus comes from the rich schedule. Although only single left-key peck or right-key peck dichotomous choice can occur on each trial,

$p(L \mid S_n)$ indicates the proportion of times the pigeon indicates that the sample was from the rich schedule, or in the following sense, the perceived richness of the sample.

It is assumed that probabilistic responding is a result of the combination of signal and random noise in the perceptual and memory processes for reinforcers in the schedule sample. For a substimulus whose perceived density is very high when measured over a number of trials, the perceived density on a trial will be high no matter whether the noise processes increase it or decrease it, unless the noise is extremely large, and so the subject will almost always peck left (high $p(L \mid S_n)$). Similarly, for a substimulus whose perceived density is very low when measured over a number of trials, the perceived density on a trial will be low no matter what the sign of the noise terms, and so the subject will almost always peck right (low $p(L \mid S_n)$). When the substimulus has an intermediate mean perceived density, it will sometimes be perceived as high density and sometimes low density, depending on the noise terms, and so $p(L \mid S_n)$ will have an intermediate value. Furthermore, $p(L \mid S_n)$ will be higher the higher the mean perceived density, and hence $p(L \mid S_n)$ can be used as an indicator of the relative perceived density of different substimuli.

Decision Rules

The quantity $p(L)$ versus some stimulus dimension, $p(L \mid \text{“}S\text{”})$ is a decision rule. A decision rule states that the conditional probability of a particular choice given that certain situations or parameters of situations have occurred. In the discrimination studies it describes the control of choice by various parameters of a substimulus, such as reinforcement amount and proximity of reinforcement opportunity to choice.

Decision rules express a functional relation similar to the one between relative response rate and relative reinforcement rate in concurrent schedule research. They will be used to provide a framework in terms of which various versions of maximization laws and matching laws may be examined. Whether maximizing or matching best accounts for the decision rules shown here has implications for which is a better account in general. The results here should apply to other cases because the situation is just a parametric variation of concurrent schedules.

Nevin (1969) has pointed out that experiments similar to this one can be characterized as multiple schedules. In this case the multiple schedule has two components that are concurrent schedules of the following form: conc FR-1, EXT and conc EXT, FR-1. The stimuli marking the presence of a particular multiple component, rather than being a red or green illuminated key, are different density reinforcement schedules. Since the schedules are not perfectly discriminable, the multiple schedule is to some extent a mixed schedule. This situation is similar to illuminating two yellow lights, one signaling that a left-red is reinforced on FR-1, with the right-green key on extinction, and the other signaling that the red and green keys have the schedules reversed on them. In either case, the relative payoff probability (rate) for making a choice in the presence of a given substimulus is known and presumably the birds' choices are controlled by this probability in a similar fashion to the way choice is controlled in the more usual concurrent schedule studies. However, relative obtained reinforcement rate for each choice is *not* the same as the probability of a choice being correct.

Matching of obtained outcome is the same as maximizing of preprogrammed or obtained reinforcement in a discrete choice situation. The more the animal goes to the more highly paid-off side, the more reinforcement is picked up on that side relative to the other. A maximizing account at the molar level would predict that a decision rule that maximizes reinforcement for choice would be the following: The optimizing ideal observer always pecks right for density 0 and 1 substimuli and always pecks left for density 3 and 4 substimuli. For density 2 substimuli, a maximizing account would make no prediction since reinforcement for making either choice was essentially equal. For the 6-cycle case, the optimizing ideal observer would always peck right for densities 0, 1, and 2, always peck left for densities 4, 5, 6, and follow any rule for density 3.

The Empirical Decision Rules at the Molar and Molecular Level: A Matching of Programmed Reinforcement Account Works at the Molar and Fails at the Molecular Level. Recall that substimuli may be described on the molar and molecular levels. On the molar level only reinforcement density is considered and substimuli take on reinforcement density values from 0 to 4 (or from 0 to 6). On a molecular level, there are 16 substimuli, each indicating a particular reinforcement pattern that may occur over the 4 cycles, or 64 substimuli, each indicating a particular reinforcement pattern that may occur over the 6 cycles. The substimuli are described in binary notation. For example, substimuli from the 4-cycle case with a density value of 1 at the molar level can have molecular values of 0001, 0010, 0100, and 1000.

Figure 2-3 shows both the molecular and molar level data. The panels in the left column of the figure show data for the 4-cycle discrimination experiment (Experiment 1). The bottom three panels of the right column show data for the 6-cycle discrimination experiment (Experiment 2). The top right panel shows 4-cycle preference data. The discrimination data will be discussed first.

At the molecular level each panel (whether for the 4-cycle or the 6-cycle data) shows one bird's $p(L)$ for each substimulus density during the test sessions on that condition. Each point indicates the probability of a left-key-peck given a specific substimulus, $p(L \mid S_n)$.

At the molar level the decision rules are indicated by the average left-peck probability for substimuli at a given density $p(L \mid S_{N_d})$. These points are the

Figure 2-3. Left-key peck probability $p(L)$ for the 4-cycle discrimination data for each individual substimulus, arranged so that all same-reinforcement density substimuli are grouped above that reinforcement density number, N_d. The solid line shows the average $p(L)$ for each substimulus group. At the top of the right column is a graph of the values of V_{mg}, the amount of food in milligrams necessary to produce indifference between the standard and comparison substimuli, plotted against number of reinforcers in a substimulus. This plot, which is from the preference data, is similar to that seen for the discrimination data. The bottom three panels in the right column show $p(L)$ values for each individual substimulus for the 6-cycle discrimination data. Since there are 64 such substimuli, they are not identified individually here. The average $p(L)$ for each substimulus group is shown by the solid line.

---

filled-in circles. The curve through these points is close to one where the probability of left is equal to the probability of programmed payoff for that choice. The shape of the curves appears to be ogival.

The panels of this figure show clearly that choice probabilities at the molecular level, reflecting the value of substimuli, do not generally fall close to the programmed reinforcement matching curve, $p(L \mid S_{N_d})$. This suggests that same reinforcement density substimuli are not given equivalent value on a molecular level, although they are equivalent on a molar level by definition. That is, at the molecular level, substimuli with the same number of reinforcers in them act differently in determining the left-key-peck probability within the discrimination procedure.

A Simple Maximizing Account and Matching of Obtained Reinforcement Do Not Account for Empirical Decision Rules at the Molar and Molecular Level. Neither molar nor molecular maximizing of payoff or matching of obtained payoff are supported by the discrimination data in Figure 2-3. Remember that the assigned values for substimuli should be the minimum for substimuli with densities below the mean (0 and 1 in the 4-cycle case and 0, 1, and 2 in the 6-cycle case and should be maximum for values above the mean (3 and 4 in the 4-cycle case and 4, 5, and 6 in the 6-cycle case). In none of the panels of Figure 2-4 is this maximization nor the equivalent obtained payoff matching indicated. Neither the average nor the individual values for substimuli (for densities other than 2) fall on the discontinuous maximizing function, $p(L \mid S_{N_d} = 0 \text{ or } 1) = 0$, $p(L \mid S_{N_d}) = 2 =$ any value, and $p(L \mid S_{N_d} = 3 \text{ or } 4) = 1$. With respect to matching to obtained payoff, there is strong undermatching. This would be consistent with the data of Chung and Herrnstein (1967) since reinforcers do not immediately follow the end of the first link. They found that different delays produce different values. A matching of obtained reinforcement account is consistent with the data but does not account for it.

Figure 2-3

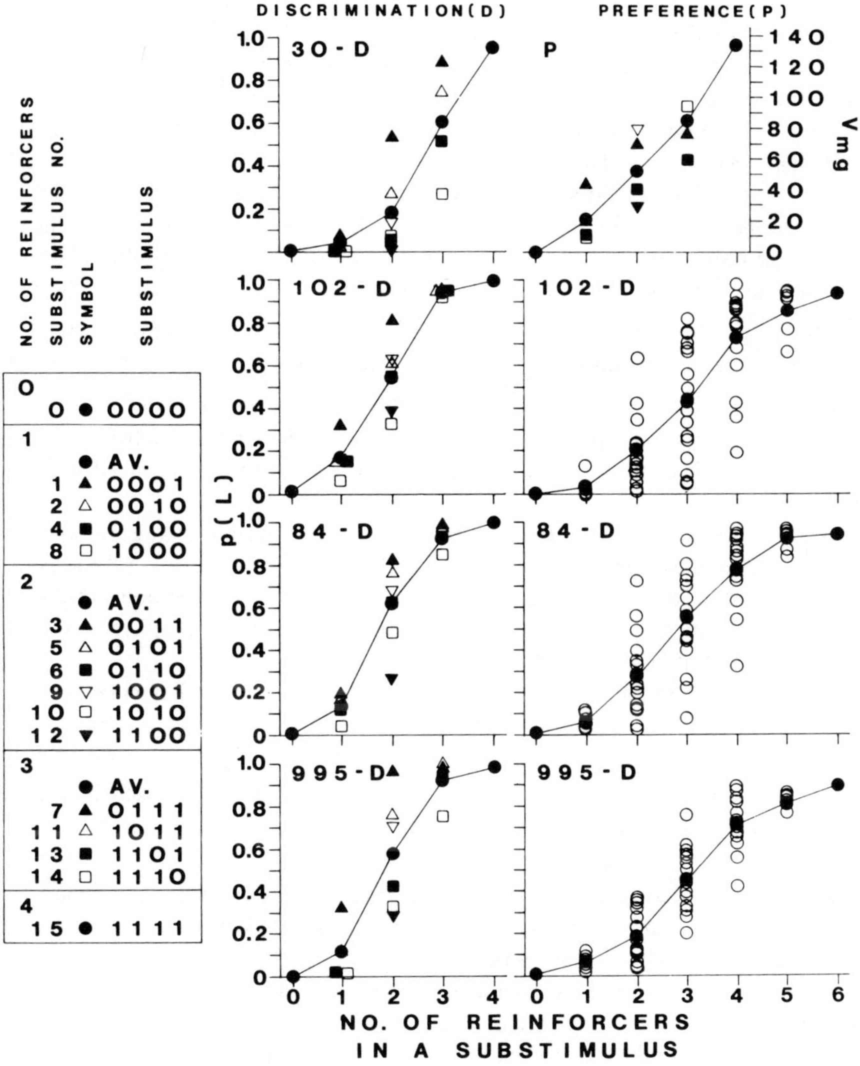
DISCRIMINATION(D)
PREFERENCE(P)
30-D
P
102-D
84-D
995-D
p(L)
V mg
NO. OF REINFORCERS
SUBSTIMULUS NO.
SYMBOL
SUBSTIMULUS
0 0 ● 0000
1 ● AV. 1 ▲ 0001 2 △ 0010 4 ■ 0100 8 □ 1000
2 ● AV. 3 ▲ 0011 5 △ 0101 6 ■ 0110 9 ▽ 1001 10 □ 1010 12 ▼ 1100
3 ● AV. 7 ▲ 0111 11 △ 1011 13 ■ 1101 14 □ 1110
4 15 ● 1111
NO. OF REINFORCERS IN A SUBSTIMULUS

ELEMENTAL DATA

The Preference Situation Obtaining the Titrated Value

In the preference case the value V_{mg} of a comparison sample is equal to the amount of food needed to titrate the value of the substimulus $0V_{mg}00$ in which the reinforcers are delivered the second cycle into the second link. Remember that the value of $0V_{mg}00$ is titrated when the relative rate Rel B_L is approximately 0.5, showing near indifference between the standard $0V_{mg}00$ sample and the comparison. However, the amount of food that produced exactly Rel $B_L = 0.5$ was not possible to find through an adjustment procedure. To obtain a more precise value in milligrams V_{mg} that would produce indifference, a regression equation was used to interpolate. Along the way to finding the titration amount, relative rates Rel B_{L_k} were found for different amounts in the standard, V_{mg_k}, on a session k. The independent variable in the regression equation was reinforcement amount V_{mg_k}, and the dependent variable was the relative rate Rel B_{L_k}. The number of milligrams necessary to produce exactly a 0.5 relative rate for both the comparison substimulus and the standard substimulus was then found. These values were very close to the values that produced near 0.5 relative rates.

Data analyzed in the preference situation included each V_{mg} at each point for each bird, as shown on the left of Table 2-3. There were 33 cases. All the points except 0000, 0101, 1010, and 1101 were run at least once. The point 0000 produces a value of $V_{mg} = 0$ but cannot be run because all the reinforcer amounts in the second link would be zero.

Another way to think about the data is to keep in mind that bias, or $p(L)$, and V_{mg} both reflect value. One can thereby see that a somewhat similar set of results should be found for $V_{mg} \mid S_{N_d}$ with the preference procedure substimulus S_{N_d}, that have the same number of reinforcers as for $p(L \mid S_{N_d})$, which is reflective of perceived value.

One Aspect of a Generalized Programmed Reinforcement Matching Account for the Preference Data Fails at Both the Molar and the Molecular Level. The value V_{mg} did not linearly increase with N_d (density) but was more of a power function, as shown in the top right panel of Figure 2-4. Recall that the average value for same density substimuli (molar measure) was ogival within the discrimination procedure. A suitable transformation would make each of these linear. Again, in the preference case the V_{mg} value of each comparison substimulus with the same number of reinforcers but different pattern $V_{mg} \mid S_N$ (molecular level) deviated from the mean effective value for that group $V_{mg} \mid S_{N_d}$ (molar). The order of deviations in value looks similar to that found in the discrimination procedure.

In order to evaluate matching in the preference case, we must use what is called the generalized matching account for obtained reinforcement (Baum and Rachlin, 1969). There is no equivalent way to test matching of programmed reinforcement in the preference experiment to that discussed for the discrimination case since relative rate was always set at 0.5. The preference procedure results do not support one aspect of the generalized matching of obtained reinforcement law of Baum and Rachlin (1969) if it is extrapolated to programmed reinforcement. Aggregate molar theories that average the effects of reinforcement across patterns of reinforcement and only worry about the total values V^t have suggested that total value is equal to the amount of reinforcement delivered on a given instance a, times frequency of reinforcement r (Baum and Rachlin, 1969), so that $V^t = a \cdot r$. However, the upward curvature shows that increasing amounts of reinforcement in the standard are necessary to balance increases of frequency of reinforcement in the comparison. As shown in the top right of Figure 2-3, it took more than twice as much reinforcement to balance twice as frequent reinforcement as shown by comparing the V_{mg} for density 4 and for density 2 substimuli. The molar values are indicated by closed circles and conform to a power function, which is inconsistent with the linear system required by the generalized matching law account.

At the molecular level, a generalized matching law account fails because in addition to what was just stated, the position of the reinforcer, with respect to the decision that evaluates value, matters. This, of course, would be expected (see Rachlin and Green, 1972).

Generalized Maximizing Fails to Account for Preference Data at the Molar and Molecular Level. Since value again is treated as the product of amount and frequency (Rachlin, Battalio, Kagel, and Green, 1981), a maximizing account will fail on the same grounds as just previously stated for the generalized matching of obtained reinforcement account. Neither of these findings is terribly surprising because, as Rachlin himself points out, increasing reinforcement might have marginal decreasing values.

ANALYSIS OF RESULTS

Plan of the Analysis

Since previous accounts do not account for either the molar or the molecular data by definitively showing either matching or maximizing, a different approach to explaining the data is taken here. Also accounts of undermatching of response probability to obtained payoff used unmeasured parameters. The samples are clearly being evaluated in a way different than suggested by these models. Next, models that might better approximate this evaluation process by using molecular level information in the substimuli will be examined.

For the 4-cycle discrimination procedure, Commons (1981) showed that there was a transform that would yield a measure of perceived value that equaled the sum of the contributions of each reinforcer occurring in the substimulus. This means that in a reinforcement schedule sample the value of the sample is the sum of the contributions of each reinforcer. However, a number of transforms were not tried in this previous work. The contribution of each reinforcer decreases the further away it occurs from the choice used to determine value. Only one decrementing function was tried; this left a number of questions to be answered about the way reinforcers are aggregated.

There are a number of steps in describing the processes by which reinforcing events in reinforcement schedule samples are aggregated. A description of part of the processes of how a series of reinforcers are valued, when they are to be perceived as stimuli, and when they are to be preferred as reinforcers, has four parts. The first step is to see what transformation makes additive the effect of each reinforcer on perceived value. Second, data on how the impact of reinforcers decreased as a function of how many cycles of possible reinforcement intervened between it and a choice that was used to measure value, are presented. Models for weighting the contribution of each reinforcer in a schedule substimulus is found. Fourth, for the preference data, the decrementing function that accounts for the decrease in the contribution of each reinforcer the further it occurs from choice is found. In general, all the models suggest that the value of samples are approximately additive after being appropriately transformed and weighted. The differences between the models, for both the discrimination and the preference case, are the transformation and weighting schemes proposed. After these steps have been completed for both procedures, values of substimuli obtained with the discrimination procedure are directly compared to values obtained with the preference procedure.

A Single Model Predicting Both Molar and Molecular Values

Since the average of same density substimuli was not linear in either the discrimination or the preference data, transformations that would make them linear would give information about the aggregate or molar functions. However, if the averages that appear ogival in the discrimination case and as a power function in the preference case can be explained by predicting each disaggregated, individual substimulus value, a much stronger explanation of how decisions are made will be at hand. As was shown in Commons (1981) additivity of the contributions of each reinforcer at the molecular level ensures that the most linear form of the molar relations will be obtained; that is, the relation between the average of values of substimuli with the same number of reinforcers and that number is linear. This also explains matching of programmed reinforcement probability at the molar level.

Transformations That Allow Additivity

Additivity is best shown by seeing how variance can be accounted for by a multivariate linear regression with the contribution of each reinforcer being represented by the coefficient for that position. Let the predicted value V^* equal the sum of each coefficient a_i times v_i, which indicates whether a reinforcer is present or not. A star after a V means that this represents a predicted value. The predicted value of a substimulus will be called $V^*_{S_n}$, whether this value is derived from the preference case or the discrimination case. The S_n refers to the perceived or obtained value of a substimulus number. The transformed value allowing additivity will be referred to as *perceived value* (discrimination case) and as *effective value* (preference case). If V_{S_n} is scaled simply as $p(L)$, it is highly nonlinear. In the preference case, if $V_{\mathrm{mg}} | S_n$ is simply scaled as V_{mg}, it is also nonlinear. The first thing is to see how transformation of V_{S_n} increases the amount accounted for by a linear multiple regression. Again, this can be done for both the perceived value and the effective value.

Discrimination Experiments

Transformation Needed to Get Perceived Value. Recall that bias $p(L)$ is expected to be a monotonic function of perceived value. We find that as the number of reinforcers in the substimulus increases, $p(L)$ does not linearly increase. A good measure of perceived value for molar stimuli should linearly increase as the density (number of reinforcers in a substimulus) increases. Perceived value V_{S_n} should be the sum of the contribution of each reinforcer, as was previously discussed:

$$V_{S_n} = T(p(L)) = a_0 + \sum_{i=1}^{n} a_i v_i \quad .$$

The letter T stands for the transformation that makes the multiple regression the most linear. The a_i are the weights applied to v_i, where $v_i = 1$ is when a reinforcer did and $v_i = 0$ is when a reinforcer did not occur on the ith cycle away from choice. From the shape of the curve relating $p(L)$ to density N_d in Figure 2-4 the transformations that might work are the probit (inverse probit transformation), the logit, and the arcsine, which is written $\sin^{-1}$. The probit transformation maps cumulative probabilities, the area under the normal curve $p(L)$ from $-\infty$ to z, into their respective z scores. This transformation is just the opposite of one in which the z score is converted into a cumulative probability, as in a z test. The logit transformation is

$$\mathrm{logit}\,(p(L)) = \log \frac{p(L)}{1-p(L)} = \log \frac{p(L)}{p(R)} \quad ,$$

and the $\sin^{-1}(p(L))$ is the value whose sine is $p(L)$. All these transformations yield values of $p(L|S_n)$ ranging from approximately −2.5 to 2.5. A scale translation constant a_0, which is estimated in the multiple regression, is also included in the transformation, to get the values to go down to a negative value. This is done because the sum $a_i v_i$ is always positive with a reinforcement instance arbitrarily assigned a value of 1 and an instance of nonreinforcement arbitrarily assigned a value of 0.

These three transformations of $p(L)$ were tested to see which produced the best-fit regression line to the points produced by the discrimination procedure shown in Figure 2-3. The results are summarized in Table 2-4. The probit transformation did the best, as indicated by an $r^2 = 0.8175$, with the logit very close behind with an $r^2 = 0.8097$. The $\sin^{-1}$ transformation did significantly worse, with $r^2 = 0.2707$.

The sum of $a_i v_i$ was assumed to have an exponent equal to 1 after $p(L)$ was transformed. This, along with the specific function describing the decrease in weights a_i, the further a reinforcer occurs from choice, are discussed in the next section.

The Models for the Decremental Contributions of Reinforcers to Perceived Value as They Recede from Choice. Four models for weighting the values of reinforcers in the sample substimuli were tested. The four models tested for the 4- and the 6-cycle discrimination were: the multivariate linear, the exponential, the hyperbolic, and the linear. Each of the latter models may be considered a special case of the multivariate linear:

$$T(p(L)) = (a_0 + \sum_{i=1}^{4} a_i v_i)^n \quad .$$

The sum was raised to some power n to see how close it would come out to 1 and to see if there was a multiplicative effect between terms. The value of a schedule sample V is equal to the sum of the effects of a reinforcer on each cycle, all raised to some power.

Table 2-4. Three transformations of $p(L)$ fit to the multivariant model of the contribution of each reinforcer to the aggregate perceived value.

Cycles	*Discrimination Form*	*Transformation*	r^2	*df*	*MSRes*	*F*
6	Multivariate $\Sigma a_i v_i$	Probit, $z^{-1}(p(L))$	.8175	186	.213	138.9
		Logit, $\log \frac{p(L)}{1-p(L)}$	.8097	186	.234	131.9
		$\text{Sin}^{-1}(p(L))$	.2707	186	.336	11.5

The multivariate model estimates each weight a_i as a separate parameter. The other models each involve only two free parameters (in addition to a_0 and n). Muth (1960) in forecasting theory and Wickelgren (1974) have argued for exponentially weighted models, while Mazur (1981) has suggested a hyperbolic weighting. No one supports the fourth kind of model, the linear model, but since it is another simple two parameter family, it was included for purposes of comparison. The hyperbolic hypothesis has the most support in the operant literature.

As Mazur points out in the next chapter of this volume, a shift from preferring one immediate small reinforcer over a slightly larger but slightly delayed one, to preferring the reverse after an initial delay is added to both, could not occur with exponentially decreasing value. This shift means the curve for the smaller reinforcer is above the larger for short common initial delays. With longer common initial delays, the curves then cross. With an exponential decay function, the heights of the curves always have the same ratio one to another and could not cross. A hyperbolic curve, however, would cross.

In Table 2-5, the models for 4-cycle and 6-cycle cases are compared. The exponent for all the models in the 4-cycle samples is very close to 1.08 with a standard deviation of 0.14. The exponent for the 6-cycle models is 0.85 with a standard deviation of 0.82. These are not significantly different from 1.0, satisfying the use of that exponent elsewhere for discrimination data. The residual squared errors for the exponential and hyperbolic models are not very different from that found for the multivariate, showing that they well represent the weighting functions, the exponential doing slightly better in the 4-cycle case and the hyperbolic doing much better in the 6-cycle case. The linear models do not do as well; however, the differences among all the models are not significant in the range that had data. The last three models require four parameters to be estimated.

Figure 2-4 shows the degree to which the various models fit the points predicted by the multivariate regression. This simply is a way of illustrating graphically the data statistically examined in Table 2-5. Two ways that the models could be visually compared within an experiment as well as across experiments are either to compare the weights generated by the various models or to compare how well they predict substimuli that have only a single reinforcer, as that reinforcer moves further and further from choice. The latter method was used because it included the effects of the power to which the sums were raised as well as a_0, the minimum value in the discrimination cases. Instead of using actual data points, which would not reflect the entire data set, from which it would be hard to visualize the weighting functions, values for substimuli generated by the multivariate fit were used. Since each reinforcer occurred only once in any of these substimuli, the degree to which a model predicts the value of a multivariate substimulus reflects how well it estimates each independent weight for the contribution of a reinforcer on a cycle.

Table 2-5. Models for the discrimination study and the preference study. The second column gives the model's name and the third column shows the right hand side of the equations. The coefficients for these models are listed in order (Columns a_0 - a_6) with standard deviations beneath. Column n lists the exponent calculated for each model and standard deviation. Column *RS* lists the residual sum of squares. The rightmost Columns *df*, *MS*, and number

	Discrimination				
Cycles	*Form*	*Equation*	a_0	a_1	a_2
4	Multivariate	$a_0 + (\Sigma a_i v_i)^n$	-2.542	1.776	1.279
	S.D.		.27	.41	.30
	Exponential	$a_0 + (\Sigma a_1 v_i e^{-a_2 c_i})^n$	-2.544	2.502	.3397
	S.D.		.27	.62	.058
	Hyperbolic	$a_0 + (\Sigma a_1 v_i / (a_2 + c_i))^n$	-2.550	3.547	.9743
	S.D.		.27	1.00	.47
	Linear	$a_0 + (\Sigma v_i (a_1 + a_2 c_i))^n$	-2.548	2.090	-.3751
	S.D.		.27	.48	.10
6	Multivariate	$a_0 + (\Sigma a_i v_i)^n$	-2.599	1.696	1.339
	S.D.		.20	.36	.29
	Exponential	$a_0 + (\Sigma a_1 v_i e^{-a_2 c_i})^n$	-2.598	2.138	.2423
	S.D.		.20	.46	.024
	Hyperbolic	$a_0 + (\Sigma a_1 v_i / (c_i + a_2))^n$	-2.536	3.122	.3783
	S.D.		.19	.31	.22
	Linear	$a_0 + (\Sigma v_i (a_1 + a_2 c_i))^n$	-2.599	1.757	-.2192
	S.D.		.21	.37	.050

	Preference				
Cycles	*Form*	*Equation*	a_0	a_1	a_2
	Combined Preference				
4	Multivariate	$(\Sigma a_i v_i)^n$	—	7.290	7.394
	S.D.		—	2.60	2.08
	Exponential	$(\Sigma a_1 v_i e^{-a_2 c_i})^n$	—	11.505	.2087
	S.D.		—	5.00	.096
	Hyperbolic	$(\Sigma a_1 v_i / (c_i + a_2))^n$	—	33.300	2.407
	S.D.		—	16.25	2.04
	Refeed	$(a_1 \Sigma a_i v_i)^n$	—	7.173	—
	S.D.		—	2.44	—
	Linear	$(\Sigma v_i (a_1 + a_2 c_i))^n$	—	9.992	-1.281
	S.D.		—	3.80	.79

Table 2-5. continued
of parameters estimated. These values are the results of nonlinear regression where the independent variable predicted is the $p(L)$ of a substimulus in the preference case and the effective Value V_{mg} of the discrimination variables. The other independent variable is whether or not a reinforcer occurred on a given cycle.

a_3	a_4	a_5	a_6	n	RS	df	MS	*Number of Parameters Estimated*
.8518	.6763	—	—	1.081	22.312	58	.385	6
.22	.20	—	—	.14				
—	—	—	—	1.079	22.392	60	.373	4
—	—	—	—	.14				
—	—	—	—	1.081	22.541	60	.376	4
—	—	—	—	.14				
—	—	—	—	1.077	22.893	60	.382	4
—	—	—	—	.14				
.9000	.7261	.7393	.5442	.8490	36.387	184	.198	8
..19	.17	.17	.14	.082				
—	—	—	—	.8448	37.371	188	.199	4
—	—	—	—	.081				
—	—	—	—	.8578	37.061	188	.197	4
—	—	—	—	.081				
—	—	—	—	.8493	39.110	188	.208	4
—	—	—	—	.083				

a_3	a_4	a_5	a_6	n	RS	df	MS	*Number of Parameters Estimated*
2.395	5.945	—	—	1.561	6976.8	28	249.2	5
1.41	1.80	—	—	.15				
—	—	—	—	1.453	8324.6	30	277.5	3
—	—	—	—	.15				
—	—	—	—	1.443	8295.7	30	276.5	3
—	—	—	—	.15				
—	—	—	—	1.382	8510.2	31	274.5	2
—	—	—	—	.14				
—	—	—	—	1.468	8401.1	30	280.0	3
—	—	—	—	.15				

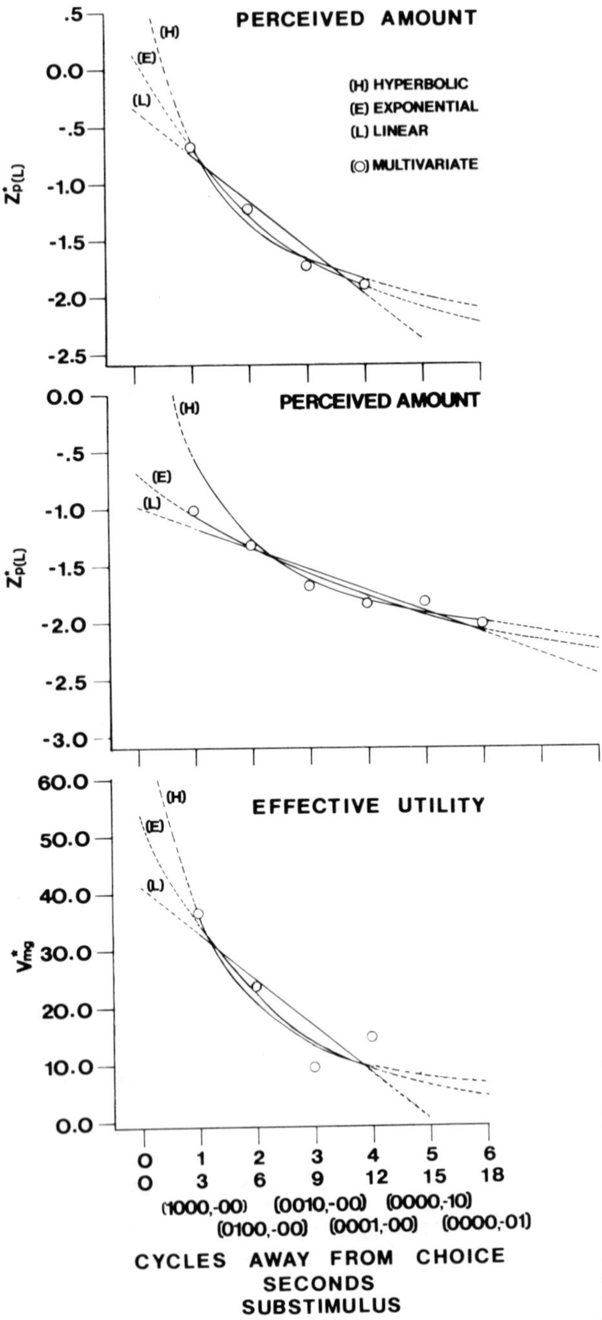

Figure 2-4. The extent to which the curves generated from the forgetting models fit values of substimuli derived from the multivariate regression estimates is shown. The curves represent the three forgetting models, hyperbolic, exponential, and linear. The points represent the predicted substimulus values. Substimuli are arranged so the one reinforcer in them recedes further from choice as one goes further to the right. The top two panels show perceived values for substimuli found from 4- and 6-cycle discrimination studies; the bottom panel shows effective values for substimuli found from the 4-cycle preference study.

A number of things are clear from an examination of this figure. First, the curves are always in the same order outside the region where the fitted points lie. The hyperbolic is the most curved and falls above the other curves, the exponential is next, and finally the linear lies below. Within the range of N_d values of the set of substimuli being predicted, the order is reversed. Second, the hyperbolic curve misses the first point in the 6-cycle case but was closer to the other points than the other curves. This suggests that with more points the hyperbolic may be even more favored. Third, the fits would seem to be better differentiated if there were more cycles. Mandell (1981) has pigeons discriminate longer schedule samples successfully, but because she used no samples of shorter length, her data could not be analyzed as was done here. The curves suggest that the different models diverge most clearly in their predictions for the value of reinforcers very close to choice. But in order to program reinforcers closer to choice, cycle length would have to be shortened. There are problems in doing that. When the cycle length was shortened to 2 seconds (Commons, 1981), there was a loss of discriminability because the cycles were occurring so close together that the birds missed picking up reinforcers.

Preference Experiment (Effective Value)

Transformation Needed to Get Effective Value. Recall that V_{mg} is expected to be a monotonic function of effective value. We find that as the number of reinforcers in the substimulus increases, V_{mg} did not linearly increase (as was shown in the top right panel of Figure 2-3). Effective value $T(V_{\text{mg}})$ should be a transformation of V_{mg} that is equal to the sum of the contribution of each reinforcer, if we follow a process that is parallel to that used for perceived value. Recall that additivity is best shown by a multivariate linear regression where the predicted variable is the transformation of V_{mg}:

$$T(V_{\text{mg}}) = V_{\text{mg}}^{1/n} = \sum_{i=1}^{4} a_i v_i \;.$$

Only a power function was tried here. The value for $1/n$ found in the multivariate regression was 0.7. Its inverse was 1.41 with a standard deviation of 0.14, which was significantly different from 1 ($p < 0.01$). This result is reported in the bottom of Table 2-5. That exponent $1/n$ explains the curvature seen in the top right panel of Figure 2-3, which was described in the section on the elemental preference data.

The transformation T is interpreted as indicating that the effective value of a reinforcer is not proportional to the amount of food received but, rather, is related to that amount via a concave utility function $U(x) = x^{1/n}$. If the effective value of a food pellet of size x (in mg) is $x^{1/n}$, and the effective value, EV, of a sequence of reinforcers is a weighted sum of their individual effective values

$$\sum_{i=1}^{4} a_i \, (x_i)^{1/n} \quad ,$$

then the effective value of a substimulus (v_1, v_2, v_3, v_4) will be

$$EV = \sum_{i=1}^{4} a_i \, (v_i \cdot 20 \text{ mg})^{1/n}$$

$$= (20 \text{ mg})^{1/n} \sum_{i=1}^{4} a_i \, v_i \quad ,$$

whereas the effective value of the comparison schedule $(0V_{\text{mg}}00)$ will be

$$EV = 0 + a_2 \, (V_{\text{mg}})^{1/n} + 0 + 0$$

$$= a_2 \, (V_{\text{mg}})^{1/n} \quad .$$

The value V_{mg} at which indifference will be observed must be when

$$a_2 \, (V_{\text{mg}})^{1/n} = (20 \text{ mg})^{1/n} \sum_{i=1}^{4} a_i \, v_i \quad .$$

So that

$$V_{\text{mg}} = 20 \text{ mg} \cdot \left(\frac{\sum_{i=1}^{4} a_i \, v_i}{a_2} \right)^n \quad .$$

This is the regression equation, to be estimated next.

THE PREFERENCE DATA

The Models for the Decremental Contributions of Reinforcers to Effective Value

The models tested were the same as those in the perceived value case: multivariate, hyperbolic, exponential, and linear. (Note, however, one important difference: The value of n should be different from 1 in these regressions, because $T(V_{\text{mg}}) = V_{\text{mg}}{}^{1/n}$ is the transformation needed to get effective value, whereas n should equal 1 in the discrimination regressions, because the correct nonlinear transformation $z_{p(L)}$ has already been applied to the left side variable. An exponent greater than 1 has an interpretation in terms of a concave utility function for consumption of grain pellets in the case of the preference data, whereas it would have no such interpretation in the case of the discrimination data.) The outcomes are shown in the bottom part of Table 2-5. The plots of the models for the derived points are shown in Figure 2-4. The derived points were found in the manner described for the discrimination data. Both the table and the figure show that the multivariate model accounts for the most variance in the data, the residual being the smallest. (This is necessarily so, as it is a more general model than the other three.) The hyperbolic accounts for the next most variance, doing slightly better than the exponential, with the linear doing quite a bit worse. Still the mean square errors were not different enough to produce significant F statistics between models. Table 2-5 lists one extra model tested for this procedure; it will be discussed in the next section.

The "forgetting" curves in Figure 2-4 were in the same order, and with similar spacing to the ones seen for the discrimination data. The variability of the preference data was much greater since the number of cases was smaller, so deviations might not be as meaningful.

Perceived Value and Effective Value of a Substimulus Are a Linear Function of One Another: The Aggregation Process Must be Isomorphic. If control by reinforcers decreases as they recede from choice in the same manner in both the perceived value and effective value cases, then the function $a_i = f(c_i)$ should be the same in both cases. Not only the functional form—hyperbolic, exponential, or linear—should be the same, but the coefficients should be related.

The next-to-last model in Table 2-5 for the preference case used the values of a_i found in the 4-cycle discriminative multivariate model as estimates of the coefficients in the preference multivariate case. This procedure did quite well in accounting for variance and used only two parameters. The residual sum of squares was still less than that found for the linear model though more than the hyperbolic and exponential models. However, the mean square error (MS) was the smallest because the number of parameters estimated was only two. This shows that the coefficients from the discrimination case adequately account for the decremental weighting in the preference case.

This concludes the first way of showing that the perceived value of a sample from a schedule is directly related to the effective value of that same sample.

The second way of showing this is to plot directly the relation between the V_{mg} obtained in the preference experiment for a schedule sample and its perceived value $z_{p(L)}$ from the discrimination experiment. This is shown in Figure 2-5. The regression equation relating V_{mg} to $z_{p(L)}$ should yield a comparable exponent of 0.7 to the previous regression since $z_{p(L)}$ is a linear function of perceived value, as shown previously. Analytically this method of comparison is the same as the previous method but it is more traditional within psychology. The means of the values across birds for both the discrimination procedure and for the preference procedure at each of the 12 points were used in this second regression:

Figure 2-5. Titrated effective value V_{mg} is plotted against the perceived value $Z_{p(L)}$. The numbers refer to the substimulus numbers for either case. The lines indicate one standard deviation. Absence of a line in the vertical director means there is only one bird.

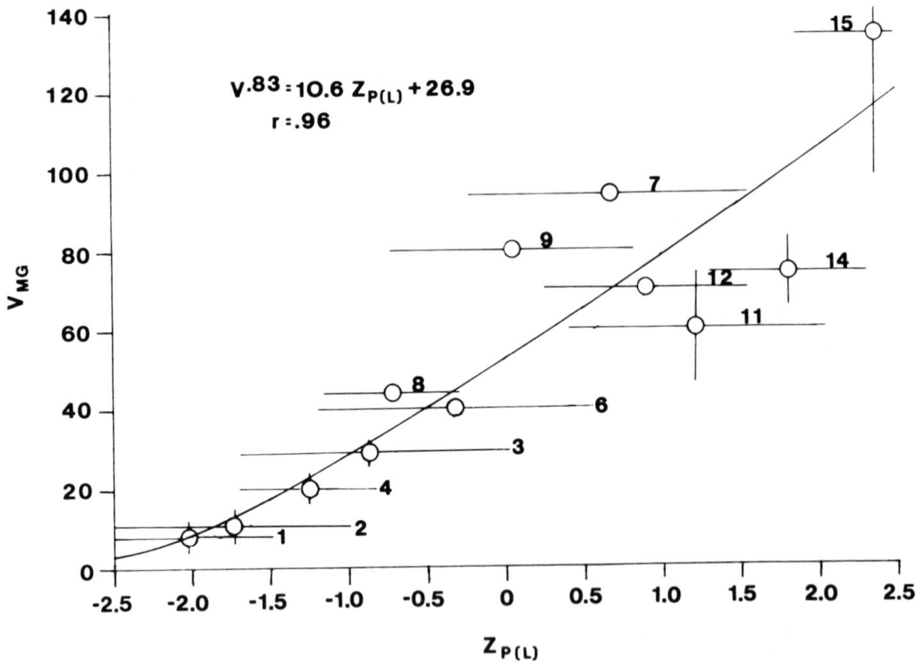

Transformed effective value = Linearly transformed perceived value

$$V_{\mathrm{mg}}^{1/n} = \left(\sum_{i=1}^{4} b_i v_i \right)^m = a_0 + \sum_{i=1}^{4} a_i v_i = c_0 + c_1 z_{p(L)} \quad .$$

Logging both sides and performing the linear regression, the values obtained are

$$V_{\mathrm{mg}}^{0.83} = 10.6 \, z_{p(L)} + 26.9 \quad ,$$

or in a slightly different form:

$$V_{\mathrm{mg}} = (10.6 \, z_{p(L)} + 26.9)^{1.2} \quad .$$

The correlation $r = 0.9632$ between mean effective value ($V_{\mathrm{mg}}^{.83}$) and mean perceived value $z_{p(L)}$ for each substimulus was reasonable.

Signal Detection Analysis of Molecular Matching and Maximizing: Is Molar Matching a Consequence of Molecular Processes Given Forgetting? The results of the discrimination experiment described in Commons (1979 and 1981) are presented as evidence of matching to relative expected payoff (programmed reinforcement) at the molar level. The relative $EP(L \mid S_{N_d})$ is the ratio of the expected payoff for a left-key-peck, $EP(L \mid S_{N_d})$, to the sum of the expected payoff for both key pecks:

$$\mathrm{Rel}\; EP(L \mid S_{N_d}) = \frac{EP(L \mid S_{N_d})}{EP(L \mid S_{N_d}) + EP(R \mid S_{N_d})} \quad .$$

The expected payoff for a left-key peck in a given stimulus condition, $EP(L \mid S_{N_d})$, may be interpreted as the product of the conditional probability of reinforcement for a left-key peck in the given stimulus condition, $p(LC \mid S_{N_d})$ and its reinforcement utility U_L :

$$EP(L \mid S_{N_d}) = p(LC \mid S_{N_d}) \cdot U_L$$

$$EP(R \mid S_{N_d}) = p(RC \mid S_{N_d}) \cdot U_R$$

$$= [1 - p(LC \mid S_{N_d})] \cdot U_R \quad .$$

(We will allow for the possibility that the utility of reinforcement is not a linear function of the amount of food received.) Relative expected payoff retains the common notion of programmed relative amount of reinforcement (Catania, 1963). According to the matching law (Estes, 1957; Commons, 1979), the probability of pecking left for a substimulus of density N_d should be

$$p(L|S_{N_d}) = \text{Rel}\,EP(L|S_{N_d}) = \frac{EP(L|S_{N_d})}{EP(L|S_{N_d}) + EP(R|S_{N_d})} \;.$$

Now, for the 4-cycle T schedule used in Experiment 1, using Bayes' law:

$$EP(L|S_{N_d}) = p(LC|S_{N_d}) \cdot U_L$$

$$= \frac{p(S_{N_d}|LC)}{p(S_{N_d}|LC) + p(S_{N_d}|RC)} \cdot U_L$$

$$= \frac{\binom{4}{N_d}\frac{3}{4}^{N_d}\;\frac{1}{4}^{(4-N_d)}}{\binom{4}{N_d}\frac{3}{4}^{N_d}\;\frac{1}{4}^{(4-N_d)} + \binom{4}{N_d}\frac{1}{4}^{N_d}\;\frac{3}{4}^{(4-N_d)}} \cdot U_L$$

$$= \frac{\binom{4}{N_d}\;3^{N_d}\left(\frac{1}{4}\right)^4}{\binom{4}{N_d}\;3^{N_d}\left(\frac{1}{4}\right)^4 + \binom{4}{N_d}\;3^{(4-N_d)}\left(\frac{1}{4}\right)^4} \cdot U_L$$

$$= \frac{3^{N_d}}{3^{N_d} + 3^{(4-N_d)}} \cdot U_L$$

Likewise,

$$EP(R|S_{N_d}) = p(RC|S_{N_d}) \cdot U_R$$

$$= \frac{p(S_{N_d}|RC)}{p(S_{N_d}|LC) + p(S_{N_d}|RC)} \cdot U_R$$

$$= \frac{3^{(4-N_d)}}{3^{N_d} + 3^{(4-N_d)}} \cdot U_R \;.$$

Hence, the matching law for programmed reinforcers would predict that

$$P(L \mid S_{N_d}) = \text{Rel } EP(L \mid S_{N_d})$$

$$= \frac{\dfrac{3^{N_d}}{3^{N_d} + 3^{(4-N_d)}} \cdot U_L}{\dfrac{3^{N_d}}{3^{N_d} + 3^{(4-N_d)}} \cdot U_L + \dfrac{3^{(4-N_d)}}{3^{N_d} + 3^{(4-N_d)}} \cdot U_R}$$

$$= \frac{3^{N_d} \cdot U_L}{3^{N_d} \cdot U_L + 3^{(4-N_d)} \cdot U_R} .$$

The term U_L is the utility value of the reinforcer received for a correct left-key peck, and U_R is the utility value of the reinforcer received for a correct right-key peck. The following discussion uses $p(L)$ instead of $p(L \mid S_{N_d})$. The prediction of matching of expected payoff (also referred to as programmed reinforcement) can be simplified using a logistic transformation:

$$\frac{p(L)}{1 - p(L)} = \frac{p(L)}{P(R)} = \frac{3^{N_d} \cdot U_L}{3^{(4-N_d)} \cdot U_R}$$

$$\log \frac{p(L)}{1 - p(L)} = N_d \log 3 + \log U_L - (4 - N_d) \log 3 - \log U_R$$

$$= \log \frac{U_L}{U_R} + (2N_d - 4) \log 3 \quad . \tag{2.1}$$

Hence molar matching predicts that the logistic transformation of $p(L)$ should be a linear function of the density N_d.

Figure 2-3 showed that if the average $p(L)$ value for all substimuli of a given density is plotted against N_d, the plot is of approximately the right shape. This suggests that the logistic transform of $p(L)$ is approximately a linear function of density of a sample at the molar level.

Unfortunately this theory does not have a systematic account of the great differences observed in the values of $p(L)$ for different substimuli of the same density. If the birds were really matching to relative expected payoff, given the information available to them, only the density of a given substimulus should matter. Instead, for substimuli of a given density, the birds are observed to peck left more frequently when the reinforcers occur later in the substimulus. In fact, we have shown above that *molecular* level data for the zero-bias experiment ($U_L = U_R$) are well accounted for by a relation of the form:

$$z_{p(L)} = a_0 + \sum_{i=1}^{4} a_i v_i \ , \qquad (2.2)$$

where $v_i = 1$ for a substimulus with a reinforcer in cycle c_i, $v_i = 0$ for a substimulus with no reinforcer in that cycle, and the a_i are a decreasing sequence of positive weights given in Table 2-5. This account of the *molecular* level data implies that suitably transformed *molar* data should yield a linear plot, specifically:

$$z^*_{p\,(L\,|\,S_{N_d})} = \frac{N_d - 2}{4} \sum_{i=1}^{4} a_i \ , \qquad (2.3)$$

where $z^*_{p}\,(L|S_{N_d})$ represents the mean of the $z_{p(L)}$ values for all substimuli of a given density. (For derivation see Commons, 1981: 78-80.) The z value of $p(L)$ is found from the inverse probability of $p(L)$, also known as the *probit* transformation. Since this probit transformation of the data is not too different from the *logit* transformation used in Eq (2.1) for the range of values in question, a theory capable of predicting Equation (2.2) would also explain the general characteristic of the *molar* data that was interpreted as evidence of matching of expected value at the molar level.

Such a theory can in fact be constructed, on principles strictly compatible with utility maximization. The difference between this theory and the simple maximizing accounts rejected in Commons (1979) is that we assume that at the time of choice the bird can only remember the substimuli with error, the error being greater in the case of reinforcers more distant in time. (This was a maximization model suggested at the time but not developed.) The formal theory of utility maximizing choice under such conditions is well known as *signal detection theory* (Green and Swets, 1966; Egan, 1975). Since psychophysics experiments generally involve single stimuli, rather than sequences of stimuli of the sort we use, a full derivation of the implications of signal detection theory for an experiment of this sort is presented next.

Let y_i be the amount of reinforcement in a cycle c_i, as remembered by the bird at the time of choice. Assume that y_i is a normally distributed random variable with variance σ_i^2 and mean μ_i in the case that $v_i = 1$, and mean $-\mu_i$ and variance σ_i^2 in the case that $v_i = 0$. Assuming the two means to be additive inverses involves no loss of generality; only the ratio $d' = 2\mu_i/\sigma_i$ (to use the standard psychophysical index of discriminability of the reinforcer on cycle c_i) turns out to have observable significance. Four times the squared inverse of d', which is $\delta_i = \sigma_i^2/\mu_i^2 = 4/(d')^2$, will be used instead of d' because it simplifies certain formulas. Then the distribution of y_i, in the case of the rich schedule, $S_R : v_i = 1$

with probability 3/4, $v_i = 0$ with probability 1/4, will have probability density function (PDF):

$$f(y_i) = \frac{3}{4}\,\frac{1}{\sqrt{2\pi}\,\sigma_i}\cdot e^{\frac{-(y_i-\mu_i)^2}{2\sigma_i^2}} + \frac{1}{4}\,\frac{1}{\sqrt{2\pi}\,\sigma_i}\cdot e^{\frac{-(y_i+\mu_i)^2}{2\sigma_i^2}}\,.$$

In the case of the lean schedule, $S_L : v_i = 1$ with probability 1/4, $v_i = 0$ with probability 3/4, it will have density function:

$$f(y_i) = \frac{1}{4}\,\frac{1}{\sqrt{2\pi}\,\sigma_i}\cdot e^{\frac{-(y_i-\mu_i)^2}{2\sigma_i^2}} + \frac{3}{4}\,\frac{1}{\sqrt{2\pi}\,\sigma_i}\cdot e^{\frac{-(y_i+\mu_i)^2}{2\sigma_i^2}}\,.$$

If a bird chooses left or right so as to maximize the expected payoff, given the particular remembered values y_1, y_2, y_3, y_4, of the sequence of events v_1, v_2, v_3, v_4 in the schedule sample, it will employ a likelihood ratio criterion. That is, it will peck left only when the ratio

$$\frac{EP(L \mid y_1, y_2, y_3, y_4)}{EP(R \mid y_1, y_2, y_3, y_4)} > 1$$

Now,

$$\frac{EP(L \mid y_1, y_2, y_3, y_4)}{EP(R \mid y_1, y_2, y_3, y_4)} = \frac{U_L\ p(y_1, y_2, y_3, y_4 \mid S_R)}{U_R\ p(y_1, y_2, y_3, y_4 \mid S_L)} = \frac{U_L\ \Pi_i\, f(y_i \mid S_R)}{U_R\ \Pi_i\, f(y_i \mid S_L)}$$

$$= \frac{U_L}{U_R}\,\frac{\prod_{i=1}^{4}\left(\frac{3}{4}\,e^{\frac{-(y_i-\mu_i)^2}{2\sigma_i^2}} + \frac{1}{4}\,e^{\frac{-(y_i+\mu_i)^2}{2\sigma_i^2}}\right)}{\prod_{i=1}^{4}\left(\frac{1}{4}\,e^{\frac{-(y_i-\mu_i)^2}{2\sigma_i^2}} + \frac{3}{4}\,e^{\frac{-(y_i+\mu_i)^2}{2\sigma_i^2}}\right)}$$

Taking out the $\Pi\,(1/4)$ from top and bottom yields

$$= \frac{U_L}{U_R}\prod_{i=1}^{4}\frac{3\,e^{\frac{-(y_i-\mu_i)^2}{2\sigma_i^2}} + e^{\frac{-(y_i+\mu_i)^2}{2\sigma_i^2}}}{e^{\frac{-(y_i-\mu_i)^2}{2\sigma_i^2}} + 3\,e^{\frac{-(y_i+\mu_i)^2}{2\sigma_i^2}}}$$

and then expanding, yields

$$= \frac{U_L}{U_R} \prod_{i=1}^{4} \frac{3\,e^{\frac{-(y_i^2 - 2y_i\mu_i + \mu_i^2)}{2\sigma_i^2}} + e^{\frac{-(y_i^2 + 2y_i\mu_i + \mu_i^2)}{2\sigma_i^2}}}{e^{\frac{-(y_i^2 - 2y_i\mu_i + \mu_i^2)}{2\sigma_i^2}} + 3e^{\frac{-(y_i^2 + 2y_i\mu_i + \mu_i^2)}{2\sigma_i^2}}}$$

Then the term, $e^{\frac{-(y_i^2 + \mu_i^2)}{2\sigma_i^2}}$ may be factored out and cancelled top and bottom:

$$= \frac{U_L}{U_R} \prod_{i=1}^{4} \frac{3\,e^{\frac{(\cancel{2}y_i\mu_i)}{\cancel{2}\sigma_i^2}} + e^{\frac{-(\cancel{2}y_i\mu_i)}{\cancel{2}\sigma_i^2}}}{e^{\frac{(\cancel{2}y_i\mu_i)}{\cancel{2}\sigma_i^2}} + 3\,e^{\frac{-(\cancel{2}y_i\mu_i)}{\cancel{2}\sigma_i^2}}} \;;$$

dividing top and bottom by $e^{\frac{-(y_i\mu_i)}{\sigma_i^2}}$ yields

$$= \frac{U_L}{U_R} \prod_{i=1}^{4} \frac{3\,e^{\frac{(2y_i\mu_i)}{\sigma_i^2}} + 1}{3 + e^{\frac{(2y_i\mu_i)}{\sigma_i^2}}} \,.$$

The bird will peck left when this quantity is greater than 1. This is a nonlinear condition on the y_i values. But in the case that discrimination of the reinforcers is difficult (that is, all the δ_i are relatively large), and therefore $\mu_i << \sigma_i^2$ which would appear to be the case in this experiment, the following approximations may be used:

$$\frac{3\,e^{\frac{(2y_i\mu_i)}{\sigma_i^2}} + 1}{3 + e^{\frac{(2y_i\mu_i)}{\sigma_i^2}}} \simeq \frac{3\left(1 + \frac{2y_i\mu_i}{\sigma_i^2} + \frac{2y_i^2\mu_i^2}{\sigma_i^4}\right) + 1}{3 + \left(1 + \frac{2y_i\mu_i}{\sigma_i^2} + \frac{2y_i^2\mu_i^2}{\sigma_i^4}\right)}$$

$$\simeq \frac{1 + \frac{3}{2}\frac{y_i\mu_i}{\sigma_i^2} + \frac{3}{2}\frac{y_i^2\mu_i^2}{\sigma_i^4}}{1 + \frac{1}{2}\frac{y_i\mu_i}{\sigma_i^2} + \frac{1}{2}\frac{y_i^2\mu_i^2}{\sigma_i^2}}$$

$$\simeq \left(1 + \frac{3}{2}\frac{y_i \mu_i}{\sigma_i^2} + \frac{3}{2}\frac{y_i^2 \mu_i^2}{\sigma_i^4}\right)\left(1 - \frac{1}{2}\frac{y_i \mu_i}{\sigma_i^2} - \frac{1}{4}\frac{y_i^2 \mu_i^2}{\sigma_i^4}\right)$$

$$\simeq 1 + \frac{y_i \mu_i}{\sigma_i^2} + \frac{1}{2}\frac{y_i^2 \mu_i^2}{\sigma_i^4} \simeq e^{y_i \mu_i / \sigma_i^2} .$$

So the condition for the bird to peck left is

$$\frac{U_L}{U_R} \prod_{i=1}^{4} e^{\frac{y_i \mu_i}{\sigma_i^2}} > 1$$

$$\prod_{i=1}^{4} e^{\frac{y_i \mu_i}{\sigma_i^2}} > \frac{U_R}{U_L} .$$

Taking the log of both sides gives

$$\sum_{i=1}^{4} \frac{y_i \mu_i}{\sigma_i^2} > \ln \frac{U_R}{U_L} .$$

The approximations are quite robust. Note that both linear and quadratic terms in each of the four Taylor expansions are kept, so that the result is correct up to terms of order $(1/\delta_i)^3$; when δ_i is very large, $(1/\delta_i)^3$ is very small. Furthermore, we only care about the validity of the approximation in the region where $EP(L)/EP(R) \simeq 1$; that is, where

$$\sum_{i=1}^{4} (y_i \mu_i)/\sigma_i^2 \simeq \ln (U_R / U_L) ,$$

which in the case of only moderate bias, will mean y_i values such that none of the $(y_i \mu_i)/\sigma_i^2 = y_i/\delta \mu_i$ are very large.

We have found that the bird will peck left when

$$\sum_{i=1}^{4} \frac{y_i \mu_i}{\sigma_i^2} > \ln \frac{U_R}{U_L} .$$

But the expression on the left, as a sum of normally distributed multivariate random variables, is itself normally distributed, with mean

$$M = \sum_{i=1}^{4} \frac{(2v_i - 1)}{\delta_i}$$

and variance

$$V = \sum_{i=1}^{4} \frac{1}{\delta_i}$$

conditional on the occurrence of the substimulus described by the ν_i, since the conditional distribution of y_i is $N(\mu_i(2\nu_i - 1), \sigma_i^2)$. Therefore the probability of its being greater than ln (U_R/U_L), given the occurrence of that substimulus is given by

$$z_{p(L)} = \frac{M - \ln(U_R/U_L)}{V^{1/2}} \tag{2.4}$$

so that

$$z_{p(L)} = \sum_{i=1}^{4} a_i \nu_i + b \quad , \tag{2.5}$$

where

$$a_i = \frac{2}{\delta_i \left[\sum_{j=1}^{4} \frac{1}{\delta_j} \right]^{1/2}} \tag{2.6}$$

and

$$b = \frac{\ln(U_L/U_R) - \left[\sum_{j=1}^{4} 1/\delta_j \right]}{\left[\sum_{j=1}^{4} 1/\delta_j \right]^{1/2}} \quad .$$

We thus obtain the linear form of Eq. (2.2) for the zero bias case ($U_L = U_R$). In addition to predicting the linear form (2.2), the theory predicts that one should find

$$\sum_{j=1}^{4} a_j = 2\left[\sum_{j=1}^{4} \frac{1}{\delta_j} \right]^{1/2} = -2a_0 \quad .$$

This relation is approximately satisfied by the values given in Table 2-4,

$$-2a_0 = 5.084 \quad , \quad \sum_{j=1}^{4} a_j = 4.583 \quad .$$

The slight deviation probably indicates that the variances of the two normal distributions are not equal, as assumed previously, when $\nu_i = 1$ and 0, respectively.

The model can easily be generalized although the generalization will not be discussed here. We also obtain the result, verified by Commons (1979), that the effect of bias should be a vertical shift of the plot of suitably transformed response probability against density. (This molar result, of course, is also a prediction of the matching eq. (2.1).)

The Decrementing of Value: The Additive Noise Model of Forgetting Functions

The a_i coefficients can now be interpreted as indicating the rate at which discriminability of reinforcement declines with time. The value of δ_i (indistinguishability) was defined as $4/(d')^2$. The set of δ_i implied by a given set of a_i coefficients are easily reconstructed by inversion of Eq. (2.6):

$$\delta_i = \frac{4}{a_i \left(\sum_{j=1}^{4} a_j \right)} \quad . \tag{2.7}$$

For the data to be consistent with a model of maximizing with imperfect discrimination, then, the linear relation of Eq. (2.2) must be observed, and the a_i obtained must be such that

$$a_1 > a_2 > a_3 > a_4 > 0 \ .$$

The values reported in Table 2-5 generally satisfy this condition. More precise predictions as to the relation that should exist among the a_i depend upon a particular model of forgetting. It makes more sense to formulate a forgetting model in terms of the δ_i rather than the a_i since it is the δ_i that have a simple interpretation in terms of discriminability.

The simplest model of forgetting would be to assume that the *signal* stays constant while the amount of *noise* increases linearly with time. If it is assumed that the perceived value Y at the time of reinforcement is a random variable Y_o (random because of imperfect discrimination of reinforcement even at that time), and that with each cycle that passes thereafter there is added another random noise component e (with the random noise terms independent of Y_o and of each other, and all identically distributed), then the random variable Y_i, representing remembered values of reinforcement after i cycles have passed (in the case of the 4-cycle experiment), can be written as

$$Y_i = Y_o + \sum_{j=1}^{4} e_j \ .$$

Therefore,

$$E(Y_i) = E(Y_o)$$

$$\text{var}(Y_i) = \text{var}(Y_o) + i \cdot \text{var}(e)$$

and

$$\delta_i = \frac{\text{var}(Y_i)}{E(Y_i)^2} = \frac{\text{var}(Y_o)}{E(Y_o)^2} + i \cdot \frac{\text{var}(e)}{E(Y_o)^2} \quad .$$

The discriminability measure δ_i, as a function of cycle number i, is therefore predicted to be a linear function:

$$\delta_i = \delta_o + i\gamma \quad .$$

Substituting this expression into (2.6) yields

$$a_i = \frac{c}{d+i} \tag{2.8}$$

where

$$c = \frac{2}{\gamma}\left[\sum_{j=1}^{4} (\delta_o + j\gamma)^{-1}\right]^{-\frac{1}{2}}$$

and

$$d = \frac{\delta_o}{\gamma} \quad .$$

This is a hyperbolic forgetting function of the sort discussed previously and estimated as line 3 of Table 2–4. As we have seen, the data are consistent with a forgetting function of this form; the 6-cycle data in particular indicate a function of this form. It can be shown that Eq. (2.8) holds for the general N-cycle case, so that the same hyperbolic function is observed regardless of the number of cycles.

Furthermore, quite general predictions are generated regarding the relation that should exist between the a_i values found for experiments involving different numbers of cycles, independent of the particular forgetting model used. If it is assumed that the discriminability of a reinforcer occurring, say, 2 cycles before choice should remain the same no matter how many cycles preceded it, then precise predictions are obtained regarding the a_i values that should be found for, say a 6-cycle experiment, and those found for the 4-cycle experiment. It is not that the last 4 a_i should be exactly the same in both cases; rather, the a_i should be such that when inserted into Eq. (2.7), the last 4 δ_i obtained are exactly the same in both cases.

In particular this theory predicts that the linear plots of suitably transformed response probability against density will have less steep a slope the greater the number of cycles in the experiment. The molar matching Eq. (2.1) generalizes to

$$\log \frac{p(L)}{1-p(L)} = \log \frac{U_L}{U_R} + (2N_d - N)\log 3 \ , \qquad (2.1')$$

where N is the number of cycles. This implies that the plot should have the same slope no matter what the number of cycles, as shown in the top left panel of Figure 2-6. The linear plot predicted by the maximizing theory on the other hand, generalizes to

$$z^*_{p(L\,|\,S_{N_d})} = \frac{2N_d - N}{2N} \sum_{i=1}^{N} a_i \ , \qquad (2.3')$$

where Eqs. (2-4)-(2-7) remain the same except that all summations should be from 1 to N. From (2.6) we see that the slope of (2.3′) is

$$\frac{1}{N}\sum_{i=1}^{N} a_i = \frac{2}{N}\left[\sum_{i=1}^{N} \delta_i^{-1}\right]^{1/2} = \frac{2}{N^{1/2}}\left[\frac{1}{N}\sum_{i=1}^{N}\right]\delta_i^{-1\ 1/2} \ .$$

When another cycle is added to the experiment, a new term is added to the sum. Since the δ_i for the new cycle must be larger than any of the others, the average value

$$\frac{1}{N}\sum_{i=1}^{N} \delta_i^{-1}$$

decreases. Since this term in the numerator decreases and the N in the denominator increases, the slope of (2.3′) decreases, as shown in the top right panel of Figure 2-6 for the 4- and 6-cycle cases. Hence the predictions of the maximizing theory and the matching theory can be distinguished even for *molar* data. The *molar* data plots presented in Figure 2-3 support the maximizing theory, since the 6-cycle plot has less steep a slope. Equation (2.1′) also indicates that the matching of expected payoff theory predicts that the degree of vertical shift associated with a given bias condition should be independent of the number of cycles. Equation (2.6) on the other hand, indicates that the maximizing of expected payoff theory predicts that the degree of vertical shift associated with a given bias condition should be similar the larger the number of cycles. This is another difference at the level of *molar* data alone.

In the case of the additive noise model (hyperbolic forgetting function), the slope of (2.3′) is

$$\frac{2}{N}\left(\sum_{i=1}^{N} \delta_i^{-1}\right)^{1/2} = \frac{2}{N}\left(\sum_{j=1}^{N} (\delta_0 + j\gamma)^{-1}\right)^{1/2} .$$

Figure 2-6. Theoretical plots (*top right*) of the value $V_{S_{N_d}}$ (on the y axis) that is the logit

$$p(L|S_{N_d}) = \log \frac{p(L|S_{N_d})}{1 - p(L|S_{N_d})} = \log\ p(L|S_{N_d})/p(R|S_{N_d})$$

on the top left panel and probit $p(L) = z_{p(L|S_{N_d})}$. There are two x axes; the one on the top is for the number of reinforcers in a substimulus in the 4-cycle experiment and the one on the bottom is for the 6-cycle experiment. The middle panel is also a theoretical plot of $p(L)$ versus either the z transform of $p(L)$ or the phi transform of $p(L)$. Note that the slope in the center portion is the same but that the knee at A is sharper than at B. Also the same is true for the ceiling knee although this is not marked. In the theoretical plot of $p(L)$ versus the predicted transformed value, $\hat{z}_{p(L)}$ (*bottom left*), if the straight line is more descriptive, local maximizing is supported, whereas if the curved line is more descriptive, local matching is supported. In two theoretical plots of the mean of the residuals $z_{p(L)} - \hat{z}_{p(L)}$ versus $\hat{z}_{p(L)}$ (*bottom middle*), the dashed line is for local matching and the x axis represents local maximizing. The plot of the actual residuals against $\hat{z}_{p(L)}$ is shown bottom right.

This implies that given the slopes for any two values of N, the values of δ_0 and γ may be calculated, and from them predictions obtained for the slopes for all other values of N. Thus it is possible to test the additive noise model at the level of *molar* data alone, even though there is no obvious forgetting function at that level.

The most interesting test of the additive noise model, of course, compares the values for δ_0 and γ implied by *molecular* level data (for any given number of cycles), with the values implied by comparative *molar* level data between experiments with different numbers of cycles. The 6-cycle data, as analyzed, indicates a value of $d \simeq 7.4$ (where d is presented in equation (2.8)), which would imply $\delta_0 \simeq 0.48$, and hence that:

$$\frac{\text{slope } (N = 6)}{\text{slope } (N = 4)} \simeq \frac{4}{6} \cdot \left[\frac{\frac{1}{1.4} + \frac{1}{2.4} + \frac{1}{3.4} + \frac{1}{4.4} + \frac{1}{5.4} + \frac{1}{6.4}}{\frac{1}{1.4} + \frac{1}{2.4} + \frac{1}{3.4} + \frac{1}{4.4}} \right]^{1/2} \simeq 0.7\ ,$$

which is close to what is observed.

Figure 2-6

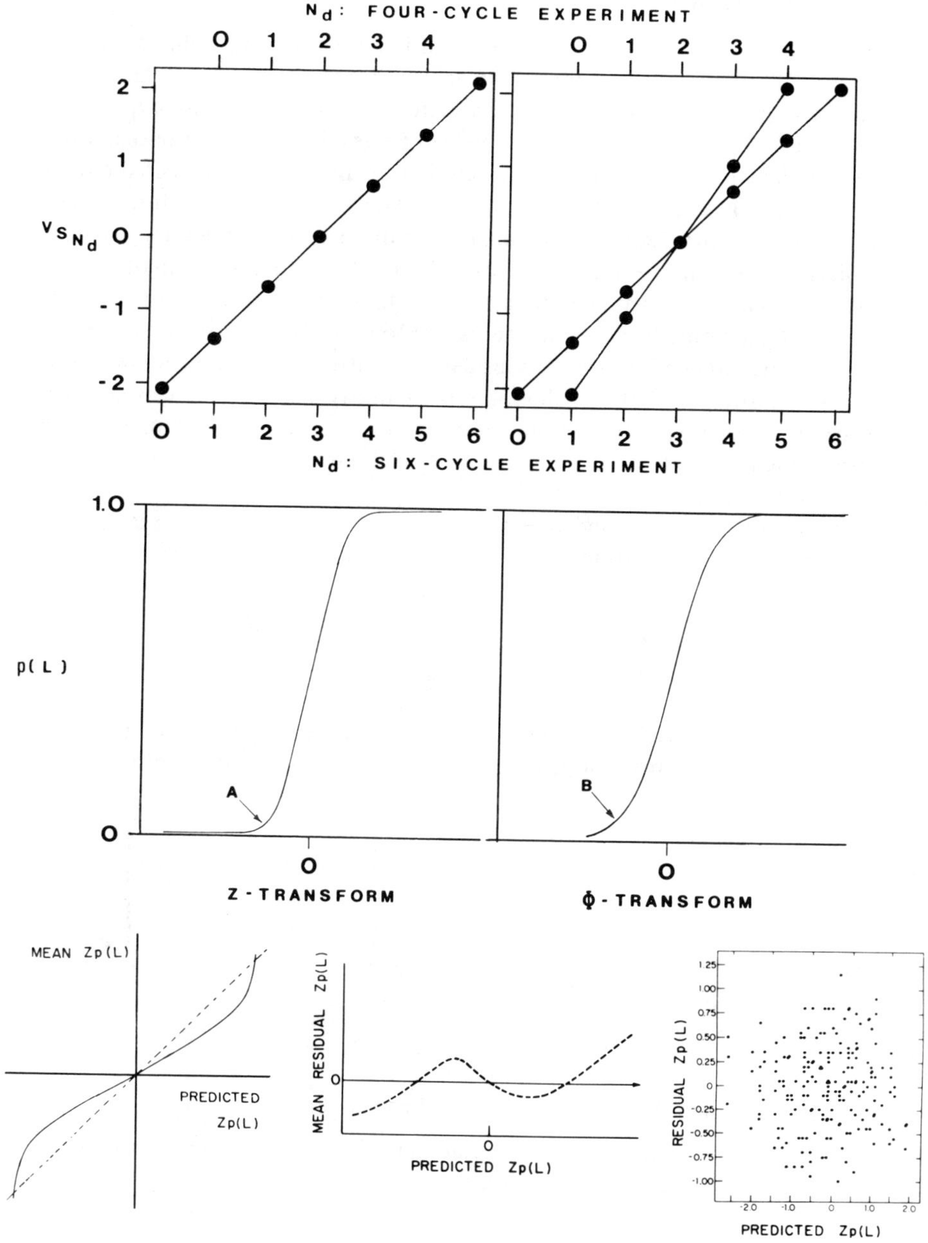

N_d: FOUR-CYCLE EXPERIMENT
0 1 2 3 4
0 1 2 3 4
2
1
$v_{S_{N_d}}$ 0
-1
-2
0 1 2 3 4 5 6 0 1 2 3 4 5 6
N_d: SIX-CYCLE EXPERIMENT
1.0
p(L)
A
B
0
0
Z-TRANSFORM
0
Φ-TRANSFORM
MEAN Zp(L)
PREDICTED
Zp(L)
MEAN RESIDUAL Zp(L)
0
0
PREDICTED Zp(L)
RESIDUAL Zp(L)
1.25
1.00
0.75
0.50
0.25
0
-0.25
-0.50
-0.75
-1.00
-2.0
-1.0
0
1.0
2.0
PREDICTED Zp(L)

Predictions of Local Matching Contrasted with Predictions of Local Maximizing

To the claim that a maximizing model with imperfect discriminability can explain both the *molecular* and the *molar* data better than does the molar matching law it might be objected that such a matching model could also explain the fact that different substimuli of the same density result in different probabilities of pecking left. In reply it can be observed that the matching of expected payoff hypothesis was introduced in Commons (1979) only because it was thought necessary to explain the shape of the *molar* plots of $p(L)$ against density; we have now shown that imperfect discrimination is sufficient to explain that, and no deviations from rationality need be assumed. Many other examples of apparent matching might similarly be explained as due to imperfect discrimination. However, if the matching hypothesis is considered of sufficient theoretical interest to justify contrasting it with the hypothesis of utility maximization even when thre are no gross qualitative results found to be inconsistent with maximizing behavior, then predictions of a matching model with imperfect discrimination may be easily derived.

It would then be predicted that the bird's probability of pecking left, given its recollection of the substimulus would be given by

$$\frac{p(L)}{1-p(L)} = \frac{EP(L\,|\,y_1, y_2, y_3, y_4)}{EP(R\,|\,y_1, y_2, y_3, y_4)} = \frac{U_L}{U_R} \cdot e^{\sum_{i=1}^{4} \frac{\mu_i y_i}{\sigma_i^2}} ,$$

using the same approximation as before. The total proportion of left pecks, when the actually occurring substimulus is ν_1, ν_2, ν_3, ν_4, will therefore be

$$p(L) = \frac{1}{\left[2\pi \sum_{i=1}^{4} \delta_i^{-1}\right]^{1/2}} \int_{-\infty}^{\infty} \frac{e^z}{\frac{U_R}{U_L} + e^z} \, e^{\left\{ \frac{\left[z - \sum_{i=1}^{4} \frac{(2\nu_i - 1)}{\delta_i}\right]^2}{2 \sum_{i=1}^{4} \delta_i^{-1}} \right\}} dz .$$

This expression depends on the ν_i only through the sum $\sum_{i=1}^{4} \nu_i/\delta_i$ and is a monotonically increasing function of that sum. Hence there will be some monotonic transformation Φ of $p(L)$ such that

$$\Phi\, p(L) = \sum_{i=1}^{4} a_i \nu_i , \tag{2.9}$$

where the relative sizes of the a_i will again have the interpretation,

$$a_i / a_j = \delta_j / \delta_i \ .$$

In particular, this theory also predicts that if $\delta_4 > \delta_3 > \delta_2 > \delta_1$, then $a_1 > a_2 > a_3 > a_4$. The definitive comparison of the local matching to the local maximizing hypothesis, then, will require careful examination of the transformation of the $p(L)$ data required to get a linear relation of the form (2.9). If a z transformation is correct, then local maximizing is indicated. Local maximizing theory would predict the relation of $z_{p(L)}$ to $a_0 + \sum_{i=1}^{4} a_i v_i = (\hat{z}_{p(L)})$ to be the straight line shown in the bottom left panel of Figure 2-6. Whereas if the correct transformation is a significantly gentler one (like the logit or arcsine or the one that is shown by the curve in the bottom left of Figure 2-6, deviation from maximizing behavior in the direction of local matching is indicated. Therefore local matching will be indicated by consistently positive residuals for the data points with large positive values for $\hat{z}_{p(L)}$, and consistently ngative residuals for the data points with large negative values for $\hat{z}_{p(L)}$. The bottom middle panel of Figure 2-6 shows another version of the same theoretical plot. If local matching were true, the mean of the residuals at each predicted $\hat{z}_{p(L)}$ value would be the curve. If the means of the residuals fell on the x axis, local maximizing would be true, since the x axis represents the fact that the mean of the residuals should not deviate from 0 as a function of $\hat{z}_{p(L)}$. The actual residuals, plotted against the predicted $\hat{z}_{p(L)}$, are shown in the bottom right panel of this figure. There is no evidence that the mean of the residuals follows the local matching curve. The data are consistent with local maximizing where the variance changes as a function of $\hat{z}_{p(L)}$.

This test could have been complicated by the fact that the derivation of the z transform in the case of maximizing theory relies upon the assumption that the remembered values of reinforcement y_i are normally distributed. This is the most common assumption in the applications of signal detection theory; the analysis could equally well be carried out for many other families of distribution functions (Egan, 1975). If the proper distribution is not normal, then the probit transformation should not yield a perfect linear regression even in the case of strict local maximizing behavior. Our derivation of Eq. (2.5) also depends on a Taylor series approximation, so that even if the assumption of normality is correct, one should not expect a perfect linear regression. Hence distinguishing between the fully general maximizing model and the fully general matching model is difficult. Still it is important to note that this analysis has revealed *no* regularities in the data that are not consistent with a generalized maximizing model.

Ethological Generalizations

The present situations bear on ethological interpretations of choice. Different patches in the environment may have changing amounts of reinforcement. The spatial distribution of reinforcers (prey) is converted into a temporal distribution of response-produced reintorcement (a reinforcement schedule) as the organism forages. The amount of prey in a patch may vary systematically from one patch to another by being richer or leaner so that the rate of obtaining reinforcement is either higher or lower. The prey density may also vary randomly so that one subpatch might be leaner than the average of another patch but the patch from which it comes may be richer overall. How does a forager in the nonhuman case and a decisionmaker in the human case combine those reinforcers within a subpatch (schedule sample)?

In the foraging situation the uneven distribution of prey may be decomposed into nonsystematic variation (noise) and systematic changes in prey density (signal) over an area that takes a period of time to obtain. Two sets of decisions might be made. First, the organism, from examining the sample, may forecast how dense the patch is. This is the discriminated or perceived value of the sample. This discriminative information can be used to control or shift behavior by serving as a cue to shift or stay. Second, the organism's decision to continue foraging may be reinforced with respect to the other experience yields. This is the effective value of the sample from the patch.

Hence there are two sets of behavior. One set consists of the choice to stay or to leave (shift) and may be controlled by the perceived value of the reinforcement sample. The second set obtains the reinforcers and is under the direct control of the bundle of reinforcers obtained in the schedule sample. Both economic (e.g. Pareto, 1906) and ethological (e.g., Krebs, Kacelnik, and Taylor, 1978; Lea, Chapter 7; and Kamil, Peters and Lindstrom, Chapter 8); theories have generally assumed that organisms maximize expected payoff in a choice situation. However, in psychology (e.g., Estes, 1957), interpretations of data on choice have tended to support the notion that organisms match their choice probabilities to the relative rate of the programmed outcome. At first glance these two sets of theories seem to be incompatible. In the type of choice situation studied here, however, momentary maximizing at the molecular level may result in overall molar matching. The underlying processes that establish the utility of the schedule sample may involve momentary maximizing when forgetting and imperfect discriminability of the sample exist.

CONCLUSION

Since the aggregation functions in the discrimination case and the preference case have been shown to be equivalent, a similar linear noise model should account for both. Models of the preference case and how local maximizing and an additive noise model might work there are left for a later exposition for a number of reasons.

First, it is necessary to have an analysis of acquisition data that reflects what is going on as response strength on each preference key shifts. This will be discussed by Commons, Woodford, Ducheny, and Peck (forthcoming). Second, it would be useful to have a better idea what the a_i weights look like for reinforcers occurring further from choice. A discrimination experiment with 6 to 18 cycles (Cabral, Woodford, and Commons, 1982) will allow for a more accurate determination of which model of perceived value produces the correct weighting scheme.

To determine better the correct weighting scheme in the preference situation, a new procedure must be used. Since the data for each point in the preference experiment (a concurrent chain titration schedule) take almost a year to collect, obtaining a sufficient number of points using that procedure with twelve cycles would be prohibitive. A new discrimination paradigm with the schedule samples as a consequence for correctly discriminating two similar stimuli would more rapidly establish the samples' values. By using signal detection analysis those values can be calculated from the bias the samples create.

It is likely that a similar principle to that of momentary maximizing of choice payoff will be true for effective utility of a schedule sample. The macro, molar, and molecular results for,the preference and discrimination experiments were homologous. The process that creates perceived value and effective value of a schedule sample should be very similar. If this is true, however, it would mean that behavior in the preference experiment is largely determined by discriminative properties of the substimuli. Problems in the discriminability of substimuli should manifest themselves most clearly in the rate at which behavior adjusts when the substimulus is changed. Hence in the case of the preference experiment, the best way to determine the type of noise process involved is by looking at acquisition, which will be the subject of a future analysis. Presently, undermatching of obtained payoff has been shown to be consequence of forgetting and of imperfect sensitivity to schedule differences. The degree of undermatching is a function of the amount of forgetting and of imperfect sensitivity to schedule differences. Local maximizing therefore will account for undermatching and obtained reinforcement that varies in degree from none to great. Obtained reinforcement matching accounts for obtained reinforcement maximizing; local maximizing and the additive noise processes should account for the traditional matching and maximizing of obtained payoff phenomena and predict when these accounts will hold.

Since aggregate obtained matching and aggregate obtained maximizing accounts predict the same local behavior in a concurrent VR VR situation, it would be good to have an experimental situation in which obtained matching and maximizing make different predictions. Local maximizing theory should be able to predict that, in a VR VI concurrent situation, matching to obtained reinforcement at the molar level is supported. The way to go about this is to modify the discrimination procedure described previously in this chapter. Remember that the stimuli to be discriminated overlap so that the concurrent FR-1, extinction programmed under any given stimulus condition appears as a rich VR versus a very lean VR for making a left versus right choice. The discriminability, or the lack thereof, of the stimuli indicating which side-key peck has the FR-1 program on it, controls the degree to which that FR-1 schedule appears as a VR schedule. With perfect discriminability one key would appear to have a FR-1 and the other key extinction. With no discriminability both keys would appear to have equal VR schedules programmed on them.

In a proposed discrimination experiment with imperfectly discriminable stimuli, each correct left choice is considered one behavioral unit. Left correct choices would be reinforced on a VI schedule. The other behavioral unit, a right correct choice, would be reinforced on a VR schedule that would yield approximately the same obtained rate of reinforcement as the left key. Suppose that each trial is 18 seconds long, 12 seconds for the stimulus period and 6 seconds for the choice period. The average of the VI intervals would be 72 seconds. The first left correct after the interval period (which are multiples of 18 seconds timed-out would be reinforced. The right corrects would be paid off on a VR schedule. For instance, the fifth right correct would be reinforced. The maximum programmed probability of reinforcement per trial for making a left choice on the VI would be 0.2, that is, the first left correct after four complete stimulus periods. For the VR it would be 0.2 also. The difference between the two is the VI does not require on the average four previous correct right choices, only that they had been programmed, whereas the VR requires four previous correct right choices. This would favor choosing the left VI key if the obtained reinforcement matching law described by Herrnstein (1970) is more correct than obtained reinforcement maximizing. The undermatching predicted by our local reinforcement maximization would produce undermatching, more indifference between the two alternatives than even predicted by obtained reinforcement matching. Obtained reinforcement maximization would favor almost exclusive preference for the VR key since only one choice out of five on the average has to go to the VI key to peck up at least half the reinforcers programmed and saved on it. It is hoped that the complexities of the situation will not make it impossible to extend the local programmed reinforcement maximizing model to this case.

REFERENCES

Baum, W.M., and H.C. Rachlin. 1969. Choice as time allocation. *Journal of the Experimental Analysis of Behavior 12*: 861–74.

Cabral, B.D.; M. Woodford; and M.L. Commons. 1982. Local maximizing versus local matching; How perceived values of reinforcement schedule samples that vary in length bear on those proposed mechanisms underlying matching found at the aggregate level. Presented at the Eastern Psychological Association.

Catania, A.C. 1963. Concurrent performances: A baseline for the study of reinforcement magnitude. *Journal of the Experimental Analysis of Behavior 6*: 299-301.

Chung S., and R.J. Herrnstein. 1967. Choice and delay of reinforcement. *Journal of the Experimental Analysis of Behavior 10*: 67-74.

Commons, M.L. 1979. Decision rules and signal detectability in a reinforcement density discrimination. *Journal of the Experimental Analysis of Behavior 32*: 101-20.

____. 1981. How reinforcement density is discriminated and scaled. In M.L. Commons and J.A. Nevin, eds., *Quantitative Analyses of Behavior*, vol. 1, *Discriminative Properties of Reinforcement Schedules*, Cambridge, Mass.: Ballinger.

Commons, M.L.; M. Woodford; J.R. Ducheny; and J.R. Peck. Forthcoming. The acquisition of performance during shifts between concurrent schedule terminal links. In M.L. Commons, R.J. Herrnstein, and A.R. Wagner, eds., *Quantitative Analyses of Behavior*, vol. 3, *Acquisition*, Cambridge, Mass.: Ballinger.

Egan, J.P. 1975. *Signal Detection Theory and ROC Analysis.* New York: Academic Press.

Estes, W.K. 1957. Theory of learning with constant, variable, or contingent probabilities of reinforcement. *Psychometrika 22*: 113-32.

Green, D.M., and J.A. Swets. 1966. *Signal Detection Theory and Psychophysics.* New York: John Wiley.

Herrnstein, R.J. 1970. On the law of effect. *Journal of the Experimental Analysis of Behavior 13*: 243-66.

Herrnstein, R.J., and G.M. Heyman. 1979. Is matching compatible with reinforcement maximization on concurrent variable interval variable ratio? *Journal of the Experimental Analysis of Behavior 31*: 209-24.

Herrnstein, R.J., and W. Vaughan. 1980. Melioration and behavioral allocation. In J.E.R. Staddon, ed., *Limits to Action: The Allocation of Individual Behavior.* New York: Academic Press.

Krebs, J.R.; A. Kacelnik; and P.J. Taylor. 1978. Test of optimal sampling by foraging great tits. In *Nature 275*: 27-31.

Mandell, C. 1981. A psychophysical analysis of time-based schedules of reinforcement. In M.L. Commons and J.A. Nevins, eds., *Quantitative Analyses of Behavior*, vol. 1, *Discriminative Properties of Reinforcement Schedules*, Cambridge, Mass.: Ballinger.

Mazur, J.E. 1981. Optimization theory fails to predict performance of pigeons in a two-response situation. *Science 214*: 823-25.

Muth, J.F. 1960. Optimal properties of exponentially weighted forecasts. *Journal of the American Statistical Association 55*: 299-306.

Neimark, E.D., and W.K. Estes. 1967. *Stimulus Sampling Theory.* San Francisco: Holden–Day.

Nevin, J.A. 1969. Signal detection theory and operant behavior. Review of D.M. Green and J.A. Swets, *Signal Detection Theory and Psychophysics. Journal of the Experimental Analysis of Behavior 12*: 475–80.

Pareto, V. 1906. Ophelimity in nonclosed cycles. *Giornale degli economisti* (2) 33 November. Also translated by A.P. Kirman in J.S. Chipman, L. Hurwiez, M.L. Richter, and Sonnenschein, eds., *Preferences, Utility, and Demand*, Harcourt, Brace and Jovanovich, 1971.

Rachlin, H.C., and L. Green. 1972. Commitment, choice and self-control. *Journal of the Experimental Analysis of Behavior 17*: 15–22.

Rachlin, H.; Battalio, R.; Kagel, J.; and L. Green. 1981. Maximization theory in behavioral psychology. *Behavioral and Brain Science 4*: 371–417.

Schoenfeld, W.N., and B.K. Cole. 1972. *Stimulus Schedules: The T–V Systems.* New York: Harper & Row.

Schoenfeld, W.N.; W.W. Cummings; and E. Hearst. 1956. On the classification of reinforcement schedules. *Proceedings of the National Academy of Sciences 42*: 563–70.

Weissman, A. 1961. Impairment of performance when a discriminative stimulus is correlated with a reinforcement contingency. *Journal of the Experimental Analysis of Behavior 4*: 365–69.

Wickelgren, W.A. 1974. Strength/resistance theory of the dynamics of memory atorage. In D.H. Krantz, R.C. Atkinson, R.D. Luce, and P. Suppes, eds., *Contemporary Developments in Mathematical Psychology: Learning, Memory, and Thinking.* San Francisco: Freeman.

3 A MOLECULAR APPROACH TO RATIO SCHEDULE PERFORMANCE

James E. Mazur

This chapter describes a quantitative molecular model of steady-state ratio schedule performance. The term *molecular* will be used to describe any model that accounts for moment-to-moment patterns in behavior. The reason for developing such a model is simple: The moment-to-moment behavior of subjects on ratio schedules exhibits many regularities, and a theory accounting for these regularities is needed.

The advantages of the quantitative molecular approach can be illustrated by contrasting this approach with what might be called the quantitative molar approach and the qualitative molecular approach. Theories such as those of Allison (1976 and Chapter 22 of this volume), Mazur (1975), Rachlin (1978 and Chapter 16), and Staddon (1979 and Chapter 11) are examples of the quantitative molar approach, for the purpose of each theory is to predict the overall durations or amounts of instrumental behavior (such as key pecking or lever pressing) and contingent behavior (such as eating or drinking) in entire experimental sessions. Like these molar theories, the molecular model to be described is designed to account for whole-session behavior durations but also for small-scale regularities such as the duration of the postreinforcement pause and the probability of pausing at other points between the presentation of two reinforcers. The potential advantage of a molecular theory is that such a theory could provide a more complete description of behavior by accounting for both momentary and overall patterns of behavior on ratio schedules.

All the previously unpublished experiments described in this chapter were supported by Grant MH–15494 from the National Institute of Mental Health to Harvard University. This chapter was written while the author was a NIMH postdoctoral fellow at Yale University.

The qualitative molecular approach is exemplified by the work of Ferster and Skinner (1957), who suggested many factors that may control an organism's moment-to-moment behavior on different schedules, including ratio schedules. For example, they proposed that the postreinforcement pause commonly observed on fixed-ratio (FR) schedules might occur because the presentation of a reinforcer acts as a discriminative stimulus for the absence of further reinforcement in the near future. The advantage of a quantitative theory over this sort of qualitative hypothesis is simply one of precision: A hypothesis about ratio schedule performance can be subjected to a more rigorous test when it is presented in the form of a mathematical model.

It should be emphasized that the molecular model presented here represents a working model, not a completed theory. New experimental results may require changes in the model. Some alternative theory built on very different basic assumptions (say, a momentary maximizing rule rather than the probabilistic choice rule used in this model) might be equally successful or more successful in accounting for the experimental results. Nevertheless, the present model integrates a large body of findings about performance on ratio schedules, and it makes many concrete predictions. I know of no data that are clearly inconsistent with the predictions of this model. Some of the model's predictions have not yet been tested, and the model suggests a number of areas where further research would be useful.

THE CONTROL OF BEHAVIOR BY MOMENTARY SCHEDULE REQUIREMENTS

Perhaps the single most important statement that can be made about the molecular control of behavior on ratio schedules is that the probability of instrumental behavior is highly dependent on the subject's proximity in time to the next reinforcement. Several lines of evidence indicate that behavior on ratio schedules is controlled by the amount of instrumental behavior that currently separates the subject from the next reinforcement. One type of evidence comes from multiple (mult) FR schedules, in which a subject is exposed to two or more different FR schedules and each schedule is signaled by its own discriminative stimulus. For example, Crossman (1968) presented pigeons with a mult FR 10 FR 100 schedule in which the components changed after each reinforcement. Crossman found that postreinforcement pause duration depended on the size of the upcoming ratio requirement, not on the size of the preceding ratio requirement. Several other researchers have obtained similar results (Davison and Over, 1966; Mintz, Mourer, and Gofseyeff, 1967; Schuster, 1959).

Another source of evidence is the behavior patterns observed on mixed (mix) FR schedules, in which no external discriminative stimulus indicates which FR schedule is currently in effect. If the sequence of components is regular, behavior is similar to that on multiple FR schedules—longer pauses precede the onset of responding when a longer FR schedule is currently in effect (Crossman, 1971; Findley, 1962; Keehn, 1971). However, if the components alternate randomly, after each reinforcement the subject cannot know which schedule is presently in effect without completing the ratio. In such schedules Ferster and Skinner (1957) and Crossman and Silverman (1973) observed a phenomenon called *priming*: After a fairly short postreinforcement pause, the organism emits a burst of responses (a prime) approximately equal in number to the size of the smaller FR; if no reinforcer is delivered, a pause occurs whose length is determined by the size of the longer FR schedule. Notice that the absence of a reinforcer after the smaller FR requirement has been met indicates that the longer FR schedule is in effect. This pattern of behavior suggests that whenever it is possible for the organism to predict the number of responses that must occur before the next reinforcement, pause length is determined by the size of that instrumental requirement.

A final bit of evidence comes from an examination of what will be called *within-ratio* pauses. These are interruptions in responding within a single ratio run, pauses other than postreinforcement pauses. Powell (1970) tallied the number of within-ratio pauses of greater than 5 seconds duration in successive fifths of each ratio run. He found the largest number of pauses in the first fifth, and in general successive fifths had fewer and fewer pauses.

All of these findings show that the probability of instrumental responding depends on the number of responses still to be completed before the next reinforcement is delivered. This information is not relevant to a molar theory, because by definition this type of theory does not consider momentary behavior probabilities. However, a molecular model can make use of this information, and a critical component of the present model is the assumption that the likelihood of instrumental behavior is inversely related to the size of the remaining instrumental requirement. The next section describes the basic features of this model.

AN OUTLINE OF THE MODEL

In most cases the predictions of this model were tested through the method of computer simulation: A computer program incorporated the model's equations, used a random number generator to introduce variability and attempted to simulate the moment-to-moment behavior of an organism on a ratio schedule. For each second of simulated session time the computer chose a behavior from one of three behavior categories—instrumental, contingent, or other behavior. The success or failure of the model could then be assessed by comparing the molar and molecular patterns of the simulated behavior with those of the behavior of actual subjects.

Unlike molecular models based on a *momentary maximizing* assumption (see Chapter 5 by Silberberg and Ziriax and Chapter 4 by Shimp), this model uses a *momentary matching* principle as its basic choice rule. The model assumes that for every second t, each behavior category i can be assigned a nonnegative *effective value* $v_{i,t}$. The model states that the probability $P(i, t)$ that behavior i will occur at time t *matches* or equals the relative effective value of that behavior, that is, its value divided by the total effective value of all behaviors possible at time t:

$$P(i, t) = \frac{v_{i,t}}{\sum_{i=1}^{n} v_{i,t}}, \tag{3.1}$$

for all possible behaviors i through n. This sort of probabilistic decision rule has many precedents (as in Herrnstein, 1961; Luce, 1959; Tversky, 1972), but the present model emphasizes the moment-to-moment changes that can occur in behavior values.

In a reinforcement schedule, as long as the reinforcer is not currently available, the contingent behavior cannot be performed, so Eq. (3.1) reduces to

$$P(I, t) = \frac{v_{I,t}}{v_{I,t} + v_{O,t}}, \tag{3.2}$$

where I refers to the instrumental behavior and O refers to other behavior. To complete the model, the procedures for calculating the effective values of instrumental and other behavior must be specified.

In this model a behavior's effective value at time t depends in part on which behavior was chosen at time $t - 1$, because the computer is biased by an *inertia parameter* toward a choice of whatever behavior was chosen in the previous second of the simulation. Without such a parameter the model would predict rapid oscillations among behaviors, since the computer makes a new choice of behavior each second. Obtaining bursts of behavior of appropriate durations is a problem that must be confronted by any molecular model (see Atkinson and Birch, 1970). The present model adopts a simple solution to this problem. The model assumes that each behavior i has an *absolute value* V_i, which does not depend on which behavior occurred last, whereas the effective value does depend on which behavior occurred last. For all behaviors except the last-occurring one, the effective value is equal to the absolute value. However, the inertia parameter S amplifies the effective value of the last-occurring behavior:

$$v_{i,t} = V_i \text{ if behavior } i \text{ did not occur at time } t - 1, \tag{3.3}$$

$$v_{i,t} = SV_i \text{ if behavior } i \text{ occurred at time } t - 1. \tag{3.4}$$

On the basis of some preliminary simulations, a value of 25 for S was found to provide bursts of behavior of reasonable length; this value was used for all the simulations reported in this chapter. The use of the inertia parameter allows us to proceed with the development of this model without a full understanding of the complex problem of burst durations. Although the inertia parameter is surely an oversimplification, the following argument can be made in its behalf. Inasmuch as switching to a new behavior involves the expenditure of time and effort, continuing an ongoing behavior has a special advantage, since no switching is involved. This advantage is represented in the model as a higher momentary effective value for the ongoing behavior.

Since the scale of values has no absolute units, V_O was arbitrarily set equal to 1. In calculating V_I, the model assumes that the typical instrumental behavior such as key pecking or lever pressing has a value of zero in a baseline situation and that it derives all of its value when placed in a contingency from the fact that it leads to the contingent behavior. For this reason the absolute value of instrumental behavior can be no greater than the value of contingent behavior, but it can approach the value of contingent behavior with very small ratio requirements. The equation for V_I takes into account four factors known to affect instrumental responding: (1) contingent behavior value (which reflects the subject's level of deprivation with respect to that particular reinforcer), (2) reinforcer magnitude, (3) the size of the remaining ratio requirement, and (4) instrumental response effort. Because this model measures all behaviors by their durations, reinforcer magnitude and ratio size are measured in units of time. Each of these requirements restricts the admissible forms for an equation of instrumental behavior value; the following equation was chosen because it is one of the simplest equations meeting all the requirements:

$$V_I = V_C \left(\frac{T_C}{T_C + (T_{I,t})^R} \right) , \tag{3.5}$$

where V_C is the absolute value of contingent behavior, T_C is the reinforcer duration, and R is a measure of response effort. $T_{I,t}$ is the amount of instrumental responding (measured in seconds) that at time t still must be performed before the next reinforcer is delivered. Notice that $T_{I,t}$ does not reflect the overall schedule requirements but the requirement that is in effect at a particular moment.

Before it was used in the simulations, one small modification was made in the equation for V_I. In Eq. (3.5), V_I is not sensitive to changes in R if the instrumental behavior has a duration of one second, since $(T_{I,t})^R = T_{I,t}$ for all values of R if $T_{I,t} = 1$. This problem was corrected by substituting the expression $(T_{I,t} + 1)^R - 1$ for $(T_{I,t})^R$, yielding the following:

$$V_I = V_C \left(\frac{T_C}{T_C + (T_{I,t} + 1)^R - 1} \right) . \tag{3.6}$$

Figure 3-1. The general form of the relation between $T_{I,t}$ (the duration of the remaining instrumental requirement) and V_I (the value of instrumental behavior) predicted by Eq. (3.6).

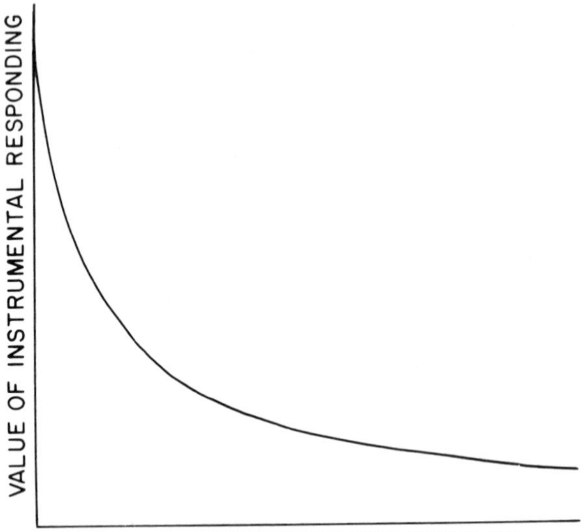

For all values of $T_{I,t}$, $(T_{I,t} + 1)^R - 1 > T_{I,t}$ if $R > 1$, and $(T_{I,t} + 1)^R - 1 < T_{I,t}$ if $R < 1$. Thus regardless of the size of $T_{I,t}$, its detrimental effect on V_I is increased if $R > 1$ and decreased if $R < 1$.

Figure 3-1 shows the form of the relation between $T_{I,t}$ and V_I that results from Eq. (3.6). The hypothetical value of instrumental behavior declines sharply as a function of increasing instrumental requirements. Many previous writers have proposed functions of similar shape to describe the relation between the strength or value of an instrumental behavior and the amount of time that separates the subject from the next reinforcement (e.g., Ainslie, 1975; Chung and Herrnstein, 1967; Hull, 1943; 135-64; Logan, 1965).

The basic assumptions of the model can now be summarized. In each second the probability of a behavior depends on its effective value relative to the effective values of all other possible behaviors, Eqs. (3.1) and (3.2). The inertia parameter S increases the effective value of the behavior that occurred in the previous second, Eqs. (3.3) and (3.4). The absolute value of the instrumental behavior depends on the absolute value of the contingent behavior, the reinforcer duration, the duration of instrumental behavior remaining before reinforcement, and the effort involved in the instrumental response, Eq. (3.6). The following sections describe how this model was used to simulate the results of several studies of FR and VR schedules.

THE SIMULATION OF FR PERFORMANCE

The operation of the model when simulating simple FR schedules is as follows. At the end of each reinforcement period, because contingent behavior becomes impossible, the simulation is momentarily returned to the other category of behavior, and the inertia parameter favors the continuation of other behavior ($\nu_{O,t} = SV_O$, and $\nu_{I,t} = V_I$). As long as other behavior is chosen by the simulation, the values of $\nu_{O,t}$ and $\nu_{I,t}$ do not change because the inertia parameter continues to favor other behavior, and $T_{I,t}$ is not reduced. In other words, the probability of switching to instrumental behavior is constant during each second of the postreinforcement pause. When the first instrumental behavior occurs, the inertia parameter then favors the continuance of instrumental responding (now $\nu_{O,t} = V_O$ and $\nu_{I,t} = SV_I$). The probability of switching back to other behavior declines each time the instrumental behavior occurs because the decreasing size of $T_{I,t}$ increases the value of instrumental responding V_I. In practice, once the simulated postreinforcement pause is ended by a choice of instrumental behavior, this behavior frequently continues until the reinforcer is presented. If other behavior does occur within a ratio run, the inertia parameter again favors the continuance of other behavior, and a pause of several seconds usually results.

When the reinforcer is presented, Eq. (3.1) is used to simulate a choice between contingent behavior and other behavior. In the simulation, V_I equals zero whenever the contingent behavior is available, and the inertia parameter favors contingent behavior over other behavior at the start of each reinforcement period ($\nu_{C,t} = SV_C$ and $\nu_{O,t} = V_O$). The inertia parameter initially favors contingent behavior for the following reason. The instrumental behavior must always occur just before a reinforcement period. Since the instrumental behavior presumably occurs only because it makes contingent behavior possible, it can be considered as part of a larger category of reinforcer-related behaviors, and the simulation favors the continuation of this category (via the contingent behavior). In any case the model's predictions for behavior during reinforcement periods cannot be tested on the basis of the published studies that are described later, because behavior during reinforcement periods has not generally been recorded. The tacit belief of most researchers seems to be that except for the time needed to reach the reinforcer the subject performs only the contingent behavior during a reinforcement period. Later some data will be presented showing that, as predicted by the model, the probability of contingent behavior is usually very high during reinforcement periods.

The Effects of Ratio Size. Perhaps the most extensive and systematic study varying the size of FR requirements was conducted by Powell (1968). Three pigeons were maintained at 80 percent of their free-feeding weights, and they were exposed to schedules ranging from FR 10 to FR 160 (for one subject, the largest ratio was FR 120). For each subject and each condition, Powell recorded the mean postreinforcement pause and the mean *running response rate*—the number of responses divided by the total session time excluding postreinforcement pauses and reinforcement periods. To simulate Powell's results T_C was set equal to 2 seconds (Powell stated that reinforcer durations varied between 2 and 3 seconds, but he presented no precise figures). To translate FR requirements into seconds of instrumental responding, it was necessary to estimate the average *local rate* of responding, or the rate of responding when both postreinforcement pauses and within-ratio pauses are excluded. Powell (1970) presented evidence that when within-ratio pauses are subtracted as well as postreinforcement pauses, the resultant response rates are virtually constant at all ratio sizes. The assumption that local rates of responding are invariant was adopted for all the simulations reported in this chapter. Based on the running response rate in the FR 10 conditions (where within-ratio pausing was presumably minimal), a local rate of 3.4 responses per second (0.29 seconds per response) was adopted for the simulation.

The points in Figure 3-2 are the mean postreinforcement pause durations for each ratio size that was tested for all three subjects. In order to approximate

Figure 3-2. Mean postreinforcement pause durations in Powell's (1968) experiment compared to the results of the computer simulation.

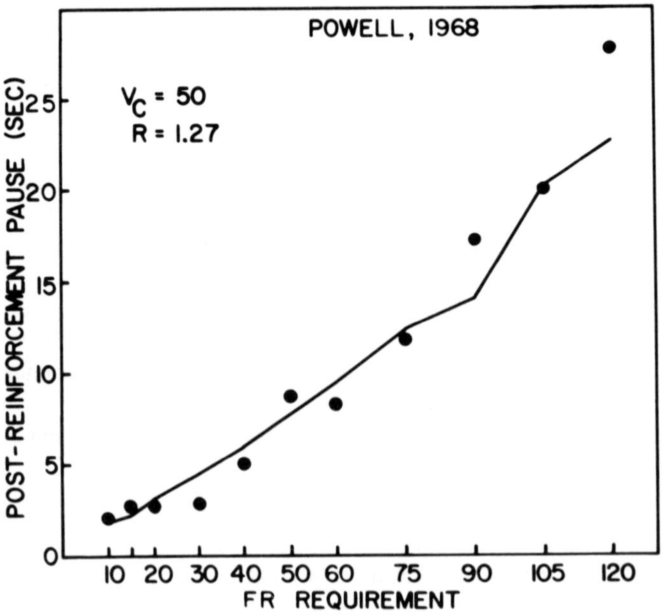

these results with the computer simulation, two parameters in Eq. (3.6), V_C and R, had to be treated as free parameters. Although several values of each parameter were tested, the parameter search was not exhaustive, so the results in Figure 3-2 may not represent an optimal correspondence between simulation and data. Nevertheless, values of 50 for V_C and 1.27 for R resulted in a reasonable approximation to the experimental results. The line in Figure 3-2 shows the results of 6,000 seconds of simulated session time for each ratio requirement. The value of 50 for V_C, the value of eating grain, suggests that the contingent behavior had a high value relative to that other behavior category (recall the V_O was arbitrarily set equal to 1). The fact that R had a value greater than 1 means that the detrimental effect of instrumental requirements more than doubled with each doubling of their duration. As will be seen, R was greater than 1 in every simulation reported in this chapter.

It was necessary to use two free parameters because both the value of the contingent behavior and response effort had to be taken into account, and there was no way to estimate these parameters in advance. However, a number of criteria can be used to evaluate a model that includes free parameters. One criterion is consistency from experiment to experiment. Once a value of 1.27 was obtained for R in this study, the same value was used for all other simulations of studies involving pigeons and key pecking. This (now constant) value of R proved to be adequate, as will be shown. Thus in many instances only one parameter had to be estimated from the data. As a second criterion, when an independent variable is changed from study to study or from condition to condition, a corresponding change should be observed in the appropriate free parameter. In the simulations to be described, variations in V_C seemed to be consistent with differences in subjects' motivational states and changes in R reflected changes in response effort.

A third criterion for a model with free parameters is how many different features of the data the model can account for once the two free parameters are estimated. The Powell (1968) study can be used to illustrate the number of predictions that can be generated by the model for a single experiment. Some of these predictions are at molar level: The lines in Figure 3-3 represent the predictions for the relative durations of eating and key pecking (from the same simulation that generated the predictions in Figure 3-2). Powell did not present behavior durations, but he presented enough information to estimate them. The data points in Figure 3-3 show the approximate proportions of time spent eating and key pecking in each condition; the correspondence between the simulated and experimental results is fairly close.

To provide a more complete picture of the model's predictions on the molecular level, Figure 3-4 shows some representative cumulative records generated by the simulation for three ratio sizes. This figure illustrates at least three predictions of the model that are supported by Powell's 1968 and 1970 findings: With larger ratio sizes, there are more within-ratio pauses. Within-ratio pauses

Figure 3-3. For Powell's (1968) experiment, approximate proportions of time spent eating (filled circles) and key pecking (open circles) compared to the results of the computer simulation.

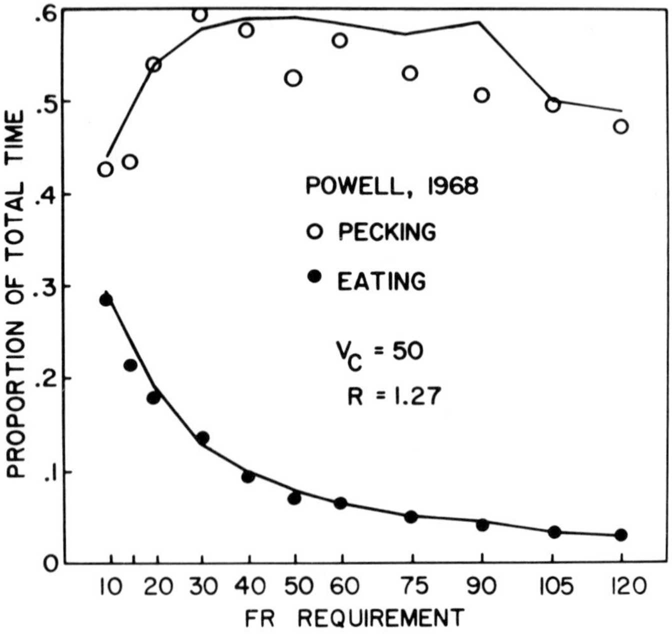

are more numerous and of longer duration near the beginning of a ratio run, and there are few pauses near the end of a ratio run. And finally, there is a great deal of variability in postreinforcement pause durations from one reinforcer to the next. The model's predictions about the frequency distribution of postreinforcement pauses can also be tested, because Powell presented this information for five ratio sizes. In Figure 3-5, that points represent the relative frequency of postreinforcement pauses in consecutive 27-second bins, averaged across all the subjects for which this information was presented. Since the model's predictions about the postreinforcement pause do not involve changes in the inertia parameter or in $T_{I,t}$, they can be calculated without the use of the computer program. Because of the model's probabilistic choice rule, an exponential distribution of postreinforcement pause durations is predicted. Figure 3-5 shows that the results closely approximated the predicted distributions.

This detailed examination of Powell's 1968 study shows that the model can account for many features of the results, on both the molecular and molar levels. The model's ability to account for empirical results is by no means confined to this one study. Figure 3-6 shows the results of two studies of postreinforcement

Figure 3-4. Representative cumulative records generated in the computer simulation of Powell's (1968) experiment.

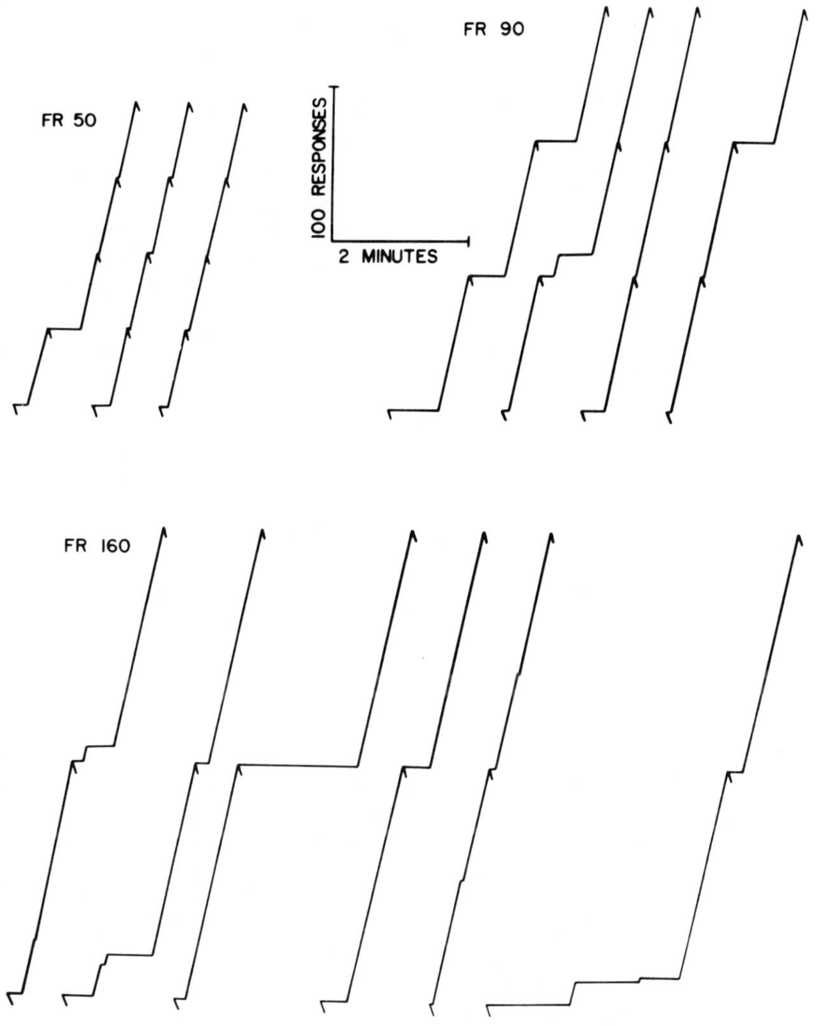

Figure 3-5. The relative frequencies of postreinforcement pauses in consecutive 27-second bins for several FR sizes in Powell's (1968) experiment. The lines are the predictions of the model.

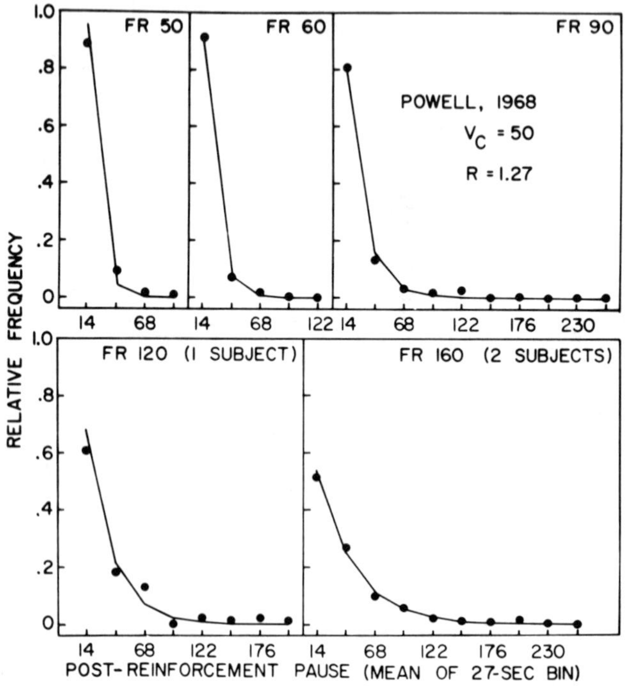

pauses in FR schedules where the subjects were rats. The reinforcer durations were very long in both cases: a 2-minute period of escape from a 0.8-milliampere shock in Winograd's (1965) study and a 1-minute period of escape from bright light in Kaplan's (1956) study. The model accounts for the general pattern of results in both experiments. The much lower value of V_C in Kaplan's study probably reflects the fact that escape from light was a much weaker reinforcer than escape from shock. (Notice that the postreinforcement pauses were much longer in Kaplan's study). The instrumental response involved a 0.60-newton force in Winograd's study and only a 0.13-newton force in Kaplan's study, and the different estimates of R between these two studies are consistent with this difference in response effort.

The Effects of Reinforcer Magnitude. Only a few studies (Inman and Cheney, 1974; Powell, 1969; Schwartz, 1969) have examined the effects of reinforcer magnitude on FR schedules. Powell's 1969 study is the most suitable for quanti-

Figure 3-6. Predicted and obtained postreinforcement pause durations in the experiments of Winograd (1965) and Kaplan (1956).

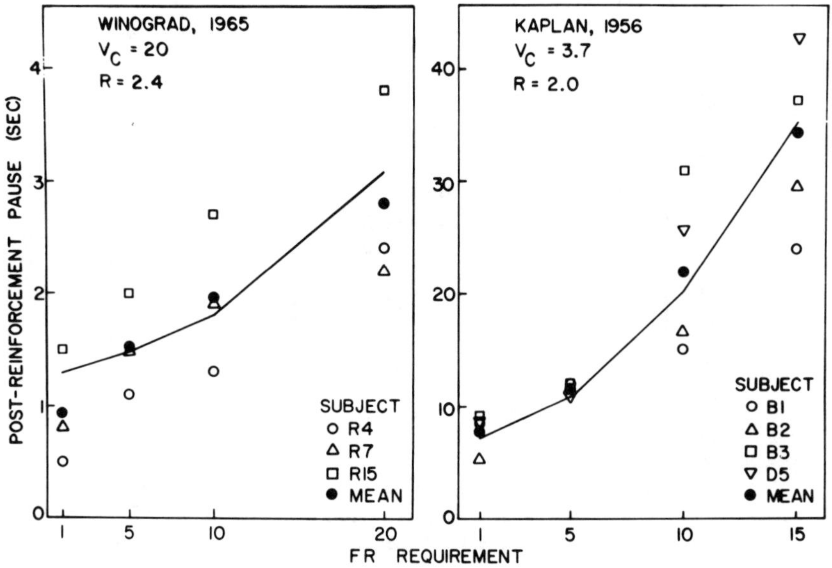

tative analysis. Powell examined a number of different FR requirements with reinforcer durations of both 2.5 seconds and 4 seconds. None of Powell's four subjects had experience with all the ratio sizes shown in Figure 3-7, but the data points are the average postreinforcement pause durations of all subjects tested at each ratio size. Although the results show a fair amount of variability, the effects of both ratio size and reinforcer duration are apparent. To obtain the predictions depicted by the two lines in Figure 3-7, R was set equal to 1.27, as in Powell's 1968 study. The higher value of V_C needed to approximate the results in Figure 3-7 (90 instead of 50) could be due to the somewhat higher level of food deprivation in the present study (75 percent of subjects' free-feeding body weights rather than 80 percent deprivation level in Powell's 1968 study). The important point of this analysis of Powell's 1969 study is that the model can account for the different postreinforcement pause durations that result from different reinforcer durations by inserting the appropriate values of T_C in Eq. 3.6 and without changing any other parameters.

The Effects of Response Effort. In an unpublished experiment I recorded the durations of instrumental, contingent, and other behavior when ratio size and response force were varied. The subjects were three male Lashley rats maintained

Figure 3-7. Predicted and obtained postreinforcement pause durations in Powell's (1969) experiment, where both ratio size and reinforcer duration were varied.

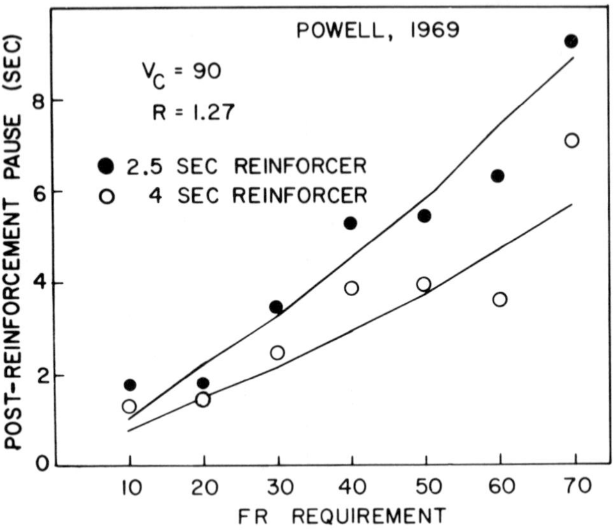

at 90 percent of their free-feeding weights, the instrumental response was pressing a lever, and the contingent behavior was drinking milk. Fixed-ratio schedules with ratio sizes from 1 to 60 were used, and in different conditions the force required to press the lever was either 0.18, 0.72, or 1.44 newtons. Each time a subject completed the ratio requirement, the drinking tube was presented, and it remained available until the subject drank for a cumulative total of 10 seconds. Each combination of ratio requirement and lever-force requirement remained in effect for nine sessions; the results of the last four sessions of each condition were used in the analyses. Short sessions were used so as to minimize satiation of drinking across a session. Each session lasted for 1,000 seconds or until the subject drank for a cumulative total of 240 seconds, which came first. A drinking timer started when a lick occurred and stopped when no licks had occurred in the last second. The lever-pressing timer operated in an analogous fashion. Unfortunately, postreinforcement pauses were not measured in this experiment, but the total behavior durations provide useful information about the model's ability to deal with the effects of effort.

The top six panels of Figure 3-8 present the results for each of the three subjects. In the left column, the proportion of time spent lever pressing is plotted for each combination of ratio size and response effort. For each subject and each force requirement, the relation between FR size and the overall probability of

Figure 3-8. From the study of response effort, the relative durations of lever pressing (*left*) and the combined schedule behaviors (*right*) are shown for each subject. The lines are the results of the simulation.

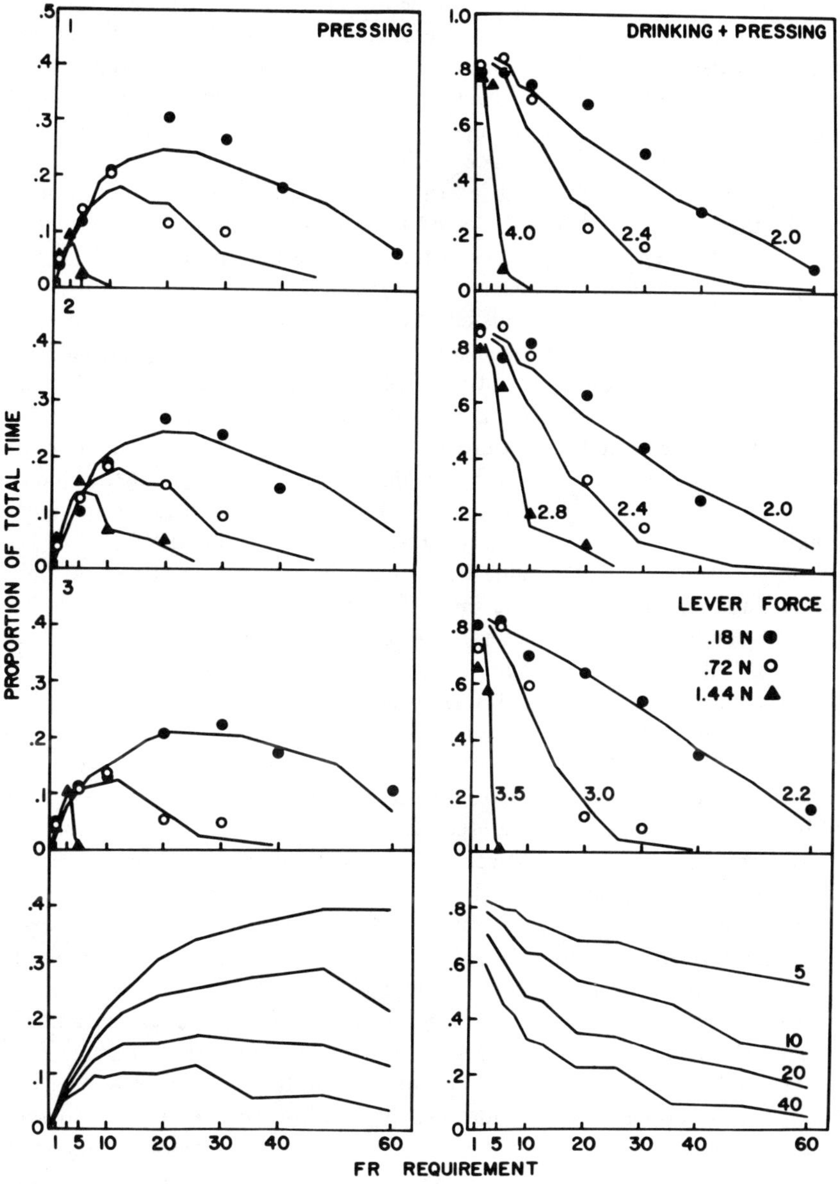

lever pressing was similar in form: The probability of lever pressing rose to a maximum at some intermediate ratio size and then declined. Increased force requirements affected these functions in two ways. First, the maximum probability of lever pressing occurred with a smaller ratio size. Second, the maximum probability was lower. In contrast to these bitonic functions, the right column of Figure 3-8 shows that the combined probabilities of the instrumental and contingent behaviors were monotonically related to FR size. The instrumental and contingent behaviors will be collectively referred to as the *schedule behaviors.* With greater lever-force requirements, the probability of schedule behaviors declined more rapidly with increasing ratio size.

To simulate the results of this experiment, V_C was set equal to 16. For each subject, three values of R were selected to approximate the results of the three lever-force requirements. The lines in the top six panels in Figure 3-8 show the simulated behavior probabilities, and the three values of R used to approximate the results for each subject are listed. As can be seen, the simulations mimicked the basic relations between behavior durations and FR size observed in this study (bitonic functions for instrumental behavior and monotonic functions for the schedule behaviors). Furthermore, notice that it was only necessary to increase the value of R to account for the effects of increasing response effort. It is important that the results from all three lever-force requirements could be simulated without changing the value of V_C, because as a measure of value of drinking milk V_C should not be affected by changes in response effort.

The results show a reasonable consistency between the value of R and response force requirements in the three studies where the subjects were rats. For the 0.18-newton force requirement, the results of the three subjects were simulated with values of 2.0, 2.0, and 2.2 for R. In comparison, R was set equal to 2.0 to simulate the results of Kaplan's 1956 experiment, where the response force requirement was 0.13 newton. The value of 2.4 used for the 0.60-newton force requirement in Winograd's 1965 study is the same as the value of R used for the 0.72-newton lever for two subjects in the present experiment. These relatively consistent values of R are especially encouraging in light of the many differences in procedure that separate these three studies, including differences in age and strain of subjects, session duration, and type of reinforcer.

This study of response effort also provided some information about behavior during reinforcement periods, because separate timers recorded the duration of drinking and the total duration the drinking tube was available. The probability of drinking when milk is available will be called the *conditional probability* of drinking. Because of the high value of drinking and the inertia parameter, the model predicts that the conditional probability of drinking will be very high. No systematic variations in the conditional probability of drinking were observed among subjects or across conditions. On the average, a subject accumulated the required 10 seconds of drinking in about 11 seconds. My casual observations indicated that it usually took a subject about 1 second to reach the drinking tube when it was presented, and this approach was followed by steady drinking until the tube was retracted. These findings therefore support the prediction that contingent behavior will have a very high probability during reinforcement periods.

A different way to incorporate a parameter for response effort in the model would be to use R as a multiplicative constant for $T_{I,t}$ rather than an exponent. However, the following analysis shows that a simple multiplicative parameter for response effort has difficulty accounting for the results of this study. A number of simulations were run with the following equation replacing Eq. (3.6):

$$V_I = V_C \left(\frac{T_C}{T_C + RT_{I,t}} \right) . \qquad (3.7)$$

The bottom row of panels in Figure 3-8 shows the results of simulations with $V_C = 16$ and $R = 5$, 10, 20, or 40. The problems with Eq. (3.7) are easiest to recognize in the bottom left panel. Regardless of the value of R, Eq. (3.7) predicts a relatively flat relation between ratio size and the probability of instrumental responding. The actual results showed more pronounced increases and decreased in instrumental behavior as ratio size increased, and Eq. (3.6) was much more successful in simulating this pattern.

Summary of the FR Simulations. The proposed model successfully predicts many characteristics of behavior on FR schedules, including mean postreinforcement durations, the shape of the postreinforcement pause frequency distributions, the probability of within-ratio pauses, and the overall durations of instrumental and contingent behavior. To account for changes in an independent variable, it was only necessary to change the value of the appropriate parameter: T_C for reinforcer duration, R for response effort, $T_{I,t}$ for ratio size. To simulate the results of several experiments, V_C and R were treated as free parameters. Estimates of V_C seemed reasonable considering the subjects' deprivation levels and the types of reinforcers used in the different studies. Similarly, estimates of R showed consistency from experiment to experiment.

APPLICATION TO OTHER RATIO SCHEDULES

What is unique about FR schedules is that the required number of responses is constant from one reinforcer to the next. But when the ratio requirement varies unpredictably from reinforcer to reinforcer, a subject cannot predict the current ratio requirement, so from the point of view of the model no single value of $T_{I,t}$ in Eq. (3.6) is entirely appropriate. Consider a simple case of a two-component mixed schedule, where one component is FR 25 and the other component is FR 75, and the two components alternate randomly, with equal probability. To generate predictions for this type of schedule, Eq. (3.6) is used to obtain two values of V_I, one for the FR 25 component and one for the FR 75 component. Then each value of V_I is multiplied by the probability that that component is in effect (in this case, the probability is 0.5 for each component), and these products are summed to obtain an average value of V_I. With Eq. (3.6) and the parameters used for the Powell 1968 study (local rate = 3.4 responses/second, $T_C = 2$ seconds, $V_C = 50$, $R = 1.27$), immediately after a reinforcement period V_I equals 6.32 for the FR 25 component and V_I equals 1.82 for the FR 75 component. The overall value of V_I for this mixed schedule is the average of V_I for the two components, or 4.07. Of course, this procedure for calculating V_I can be used for mixed schedules with any number of components. The procedure is formalized in the following equation:

$$V_I = \sum_{k=1}^{m} p_k V_C \left(\frac{T_C}{T_C + (T_{I,t,k} + 1)^R - 1} \right), \tag{3.8}$$

where m is the number of alternative ratio requirements that could be in effect, p_k is the probability that alternative k is currently in effect, and $T_{I,t,k}$ is the duration of the remaining instrumental requirement at time t if alternative k is in effect.

Mix FR FR Schedules. Fantino (1967) used a concurrent chains procedure to test pigeons' preferences for mix FR FR schedules compared to simple FR schedules with equal average response requirements. For example, in one condition pigeons chose between a mix FR 25 FR 75 schedule and a FR 50 schedule. To gain access to either schedule, the pigeons responded on two separate (concurrent) variable-interval (VI) schedules, and Fantino used relative response rates on the two VI schedules as measures of preference. Although the average response requirements were equal, subjects showed a preference for the mixed schedule. Eqs. (3.6) and (3.8) are not designed to predict behavior on concurrent chains schedules, but the postreinforcement size of V_I can be used as an estimate of the value of a ratio schedule (since Fantino's measures of preference were taken at times when the entire response requirement still remained for each ratio schedule). With a specific set of parameters, we saw earlier that V_I equals 4.07 for the mix FR 25 FR 75 schedule. With these same parameters V_I equals 2.94 for a FR 50 schedule. Thus, in agreement with Fantino's results, Eqs. (3.6) and (3.8) predict that the mix FR 25 FR 75 schedule will have a higher value than the FR 50 schedule. Fantino also found that preference for the mixed schedule increased as the difference between the lengths of the two components grew larger. For instance, a mix FR 1 FR 99 schedule was most highly preferred, followed by mix FR 10 FR 90, mix FR 25 FR 75, and finally FR 50. This trend is predicted by the model, which generates the highest value of V_I for the mix FR 1 FR 99 schedule, and the lowest value for the FR 50 schedule. The increased estimates of V_I that occur as the differences between schedule components increase are due to the high values of the short components. The rank orderings of V_I for these schedules are the same for any values of V_C, T_C, and the local response rate, as long as R is greater than 1.0.

Besides predicting a preference for mix FR FR schedules over simple FR schedules, the model predicts the priming pattern frequently observed on mix FR FR schedules. The simulation routine for these schedules proceeds as follows. After each reinforcement period, $T_{I,t,k}$ is at a maximum for both components. As instrumental behavior is chosen, $T_{I,t,k}$ decreases and so V_I increases. However, once the number of simulated responses equals the number specified by the shorter component, one of two changes must occur. If the short FR component is in effect, a reinforcement period will begin. If a reinforcement period does not begin, this is conclusive evidence that the longer component is in effect, so p_k now equals 0 for the short component and 1.0 for the long component. In this case a decrease in V_I occurs, because instead of a 0.5 probability of an imminent reinforcement from the short component the simulation must now take into account only the more distant reinforcement that will occur with the long component. This decrease in V_I results in the simulation of priming—an initial burst of responding is often followed by a fairly long pause before instrumental responding resumes.

A simulation of a mix FR 10 FR 100 schedule was conducted with the parameters used for Powell's 1968 study. Several features of this simulation were consistent with results from studies of mix FR FR schedules:

- Crossman and Silverman (1973) had found that the number of response composing a prime was not always close to the number required by the short component. Primes were often two or three times as long as the short FR component. Such longer primes also occurred in the simulation because the inertia parameter favored the continuation of instrumental behavior even when V_I decreased at the completion of the short FR requirement.
- Crossman and Silverman observed that primes did not occur every time the long FR component was in effect. Frequently a continuous burst of responses occurred until the reinforcer was delivered. This was also true in the simulation, again because of the inertia parameter.
- Crossman and Silverman noted that the presence or absence of a prime was not correlated with the duration of the post-reinforcement pause. In the simulation, this was true by definition, since the length of the postreinforcement pause has no effect on subsequent choices.
- Alferink (1975) found that the duration of a postprime pause was a function of the size of the long FR component. This relation is also captured by the model, since V_I is determined only by the long FR component after the short FR requirement is met.

In summary, all of the most salient characteristics of behavior on mix FR FR schedules are consistent with the predictions of Eq. (3.8).

VR Schedules. When a mixed schedule contains a large number of FR components, it is usually called a VR schedule. Eqs. (3.6) and (3.8) predict that subjects will prefer VR schedules to FR schedules of equal average response requirements. Sherman and Thomas (1968: experiment 1) presented pigeons with a choice between FR and VR schedules. A subject could either respond on one key on a FR schedule, or peck at a second, switching key, which would change the schedule on the first key to VR 120. In separate conditions, the number of responses required on the switching key ranged from 1 to 90. By giving subjects a choice between VR 120 and FR schedules of different lengths, Sherman and Thomas could estimate the size of an FR schedule that was about equally preferred to the VR 120 plus the switching requirement. The results for each of their subjects are shown in Figure 3-9. The broken line shows the predictions of

Figure 3-9. For each subject in the experiment of Sherman and Thomas (1968: experiment 1), the circles show the sizes of FR schedules that were about equally preferred to the VR 120 schedule plus the switching requirement shown on the *x* axis. The broken line shows the predictions of an optimizing strategy, and the solid line shows the predictions of the model.

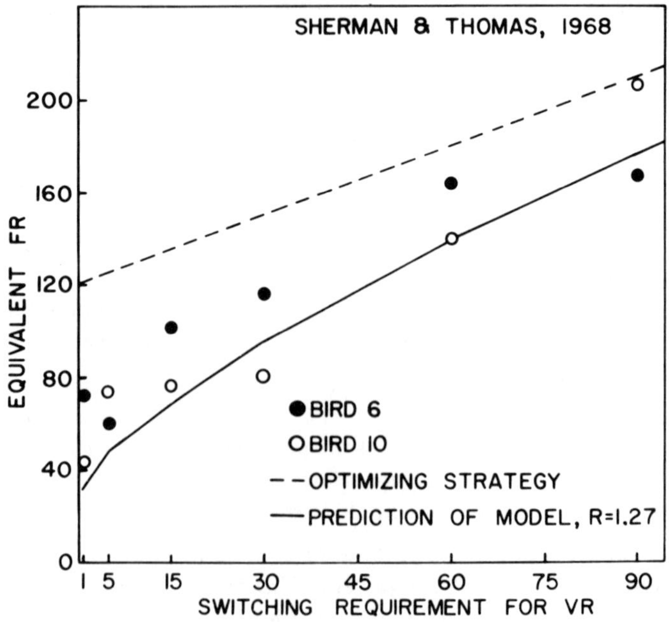

an optimizing strategy, in which a subject chooses whatever alternative requires, on the average, the smaller total number of pecks. The fact that the data points are below this line means that subjects showed a preference for the VR schedule. The solid line shows the predictions of the model, which were generated with no free parameters. R was set equal to 1.27 as in Powell's 1968 study; V_C was the same for each schedule so it cancelled out. The predictions of the molecular model were not perfect, but they were certainly more accurate than those of the optimizing strategy.

In all the schedules considered so far, $T_{I,t}$ changes as a function of the number of responses since the last reinforcer. In contrast, $T_{I,t}$ is constant from response to response on random-ratio (RR) schedules, in which each response has a constant probability of reinforcement regardless of the number of previous unreinforced responses. If the probability of reinforcement after any response is p, then the probability of a reinforcer occurring after exactly two responses is $(1-p)p$, and the probability of a reinforcer after exactly three responses is $(1-p)^2 p$. In general the probability of the next reinforcer occurring after exactly x responses is $(1-p)^{(x-1)}p$. By using this expression in Eq. (3.8), the following equation for RR schedules was obtained:

$$V_I = \sum_{x=1}^{\infty} (1-p)^{(x-1)} p V_C \left(\frac{T_C}{T_C + (T_{I,x} + 1)^R - 1} \right), \tag{3.9}$$

where p is the probability of a reinforcer for each response, and $T_{I,x}$ is the amount of time a subject uses in producing x instrumental responses. Because each response has an equal probability of reinforcement, the single value of V_I obtained from Eq. (3.9) applies to every response on a RR schedule, regardless of the number of responses since reinforcement.

Equation (3.9) was used to simulate the results of a study in which female Lashley rats, maintained at 80 percent of their free-feeding weights, pressed a 0.18-newton lever for milk reinforcement (Mazur, 1977). Each reinforcer was a 10-second presentation of the drinking tube. This experiment did not use actual RR schedules, but VR schedules in which the distribution of components was chosen to approximate an exponential distribution (Fleshler and Hoffman, 1962), the distribution produced by an RR schedule. Therefore, Eq. (3.9) was used to simulate the results of this study, which included five VR schedules ranging from VR 5 to VR 100 and an FR 1 schedule. The durations of drinking and lever pressing were measured as in the study of response effort described earlier.

Figure 3-10 shows the relative durations of lever pressing and of the two schedule behaviors during the last four sessions of each condition. The solid lines show the simulation results. Of potentially greater interest than the correspondence between simulation and data is the size of the difference predicted for VR versus FR schedules. The broken lines show the results of a simulation of FR

Figure 3-10. The relative durations of lever pressing and the combined schedule behaviors for Mazur's (1977) study of VR schedules. The solid line shows the results of the computer simulation, and the broken line shows the results of a simulation for FR schedules.

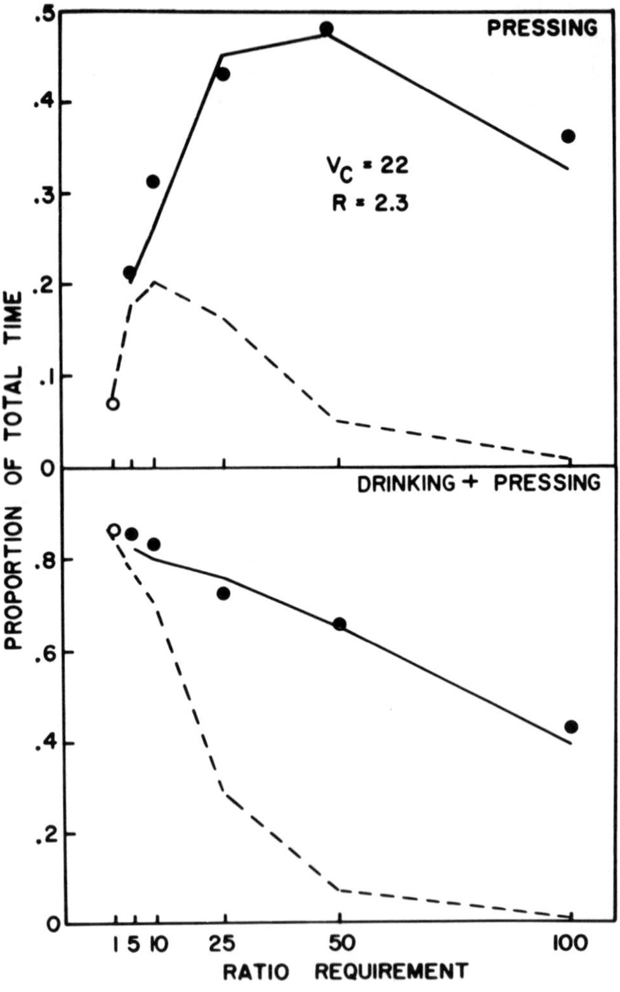

schedules using the same values for V_C and R. Unfortunately I know of no studies that systematically compared subjects' behavior on different FR schedules to their behavior on VR or RR schedules of equal average response requirements, so strong quantitative tests of the model's predictions are not now possible. Nevertheless, various studies provide qualitative support for some predictions of the model for these schedules. Subjects spend more time performing the instrumental response on VR schedules than on FR schedules (Ferster and Skinner, 1957; Webbe and Malagodi, 1978; Webbe, DeWeese, and Malagodi, 1974). The duration of instrumental responding is larger because postreinforcement pauses are shorter on VR schedules (Boren, 1973; Webbe and Malagodi, 1978). The model predicts that postreinforcement pauses with average durations of several seconds will occur with VR and RR schedules, although they will be smaller than on FR schedules. Several studies have found such postreinforcement pauses with VR schedules (Kintsch, 1965; Powell, 1972; Priddle-Higson, Lowe, and Harzem, 1976; Webbe, DeWeese, and Malagodi, 1974). As Figure 3-10 shows, the predicted differences in overall behavior durations between FR and VR schedules should become larger as ratio size increases. Similar differences in postreinforcement pauses are predicted. These predictions have not been tested. There is clearly a need for detailed, systematic comparisons of FR and VR or RR schedules, with respect to both molar and molecular features of behavior.

CHANGES IN REINFORCER VALUE

All simulations described so far have used estimates of V_C that were constant throughout a session and across conditions of a single experiment. This assumption was a convenient simplification, which seemed reasonable for studies where subjects were maintained at a constant level of deprivation from day to day and where the opportunity to engage in the contingent behavior during an experimental session was not great. However, when a reinforcement schedule allows extensive access to the contingent behavior, large decreases in instrumental responding may be observed across a session. Decreased instrumental responding can also be obtained by giving subjects free access to the reinforcer before a session. Accounting for the effects of satiation on the probabilities of the instrumental and contingent behaviors remains a major challenge for a molecular model. This section describes a few studies that examined the effects of changing reinforcer value.

After examining the behavior of rats, cats, and a monkey on FR schedules, Sidman and Stebbins (1954) reduced subjects' deprivation levels by presession feedings. These feedings resulted in much longer postreinforcement pauses, but once subjects started to respond their response rates were unchanged despite their reduced deprivation levels. This study therefore provides further support for the model's assumption of constancy of local rates on FR schedules. With their fairly short ratio sizes (FR 20 to FR 25), Sidman and Stebbins observed virtually no within-ratio pauses regardless of the subjects' deprivation levels. Malott (1966) obtained different results with pigeons on a FR 64 schedule. Like Sidman and Stebbins, Malott observed longer postreinforcement pauses when subjects were prefed, but he also found that running response rates were lower, especially near the beginning of a ratio run. To determine whether the present model could reconcile the findings of these two studies, a series of simulations was run in which R was set equal to 1.27 and V_C was varied between 200 and 8 to represent large changes in deprivation level. In the simulations of the longer schedule (FR 64), postreinforcement pauses grew longer as the value of V_C was lowered to 8, and numerous within-ratio pauses appeared (occupying over 25 percent of the total simulated session time). To parallel Malott's analysis, running response rates were calculated separately for successive 16-response blocks of each ratio run. As in Malott's results, running response rates were lowest during the first 16-response block, and successive blocks had increasing response rates. In a simulation of the shorter (FR 25) schedule of Sidman and Stebbins, very few within-ratio pauses occurred with $V_C = 8$, and they comprised less than 5 percent of the simulated session time. Thus the model can account for the major qualitative characteristics of FR performance with reduced levels of deprivation: (1) With short ratios, postreinforcement pauses become longer, but because of the inertia parameter and small values of $T_{I,t}$ there are few within-ratio pauses. (2) Because of the larger values of $T_{I,t}$ associated with longer ratios, postreinforcement pause durations increase greatly, and within-ratio pauses appear. (3) These within-ratio pauses are longest and most numerous near the beginning of a ratio run. As a result, running response rates are lowest at the beginning of a ratio run, and they increase steadily in successive portions of an average ratio run.

When satiation occurs within an experimental session, estimating the moment-to-moment value of V_C can be a complex task. A study I conducted illustrates this problem, for subjects' levels of motivation evidently changed drastically across a session in some conditions. Four male Lashley rats, maintained at 90 percent of their free-feeding body weights, pressed a 0.14-newton lever on six schedules ranging from FR 5 to FR 120. Each reinforcement period lasted until the subject drank milk for a cumulative total of 20 seconds. Each FR schedule was presented for nine 50-minute sessions. The combination of fairly long sessions and large reinforcer durations ensured that the subjects would drink considerable amounts of milk, especially with the short FR schedules. Besides the FR conditions, a baseline condition was included, where the milk was continuously available throughout each session. To examine within-session trends, the durations of drinking and lever pressing were recorded separately for each 5-minute block of a session. These block-by-block durations were averaged across the last four sessions of each condition.

Figure 3-11. For the study of satiation in FR schedules, the relative durations of the schedule behaviors (*upper*) and the conditional probabilities of drinking (*lower*), for successive tenths of a session in each condition.

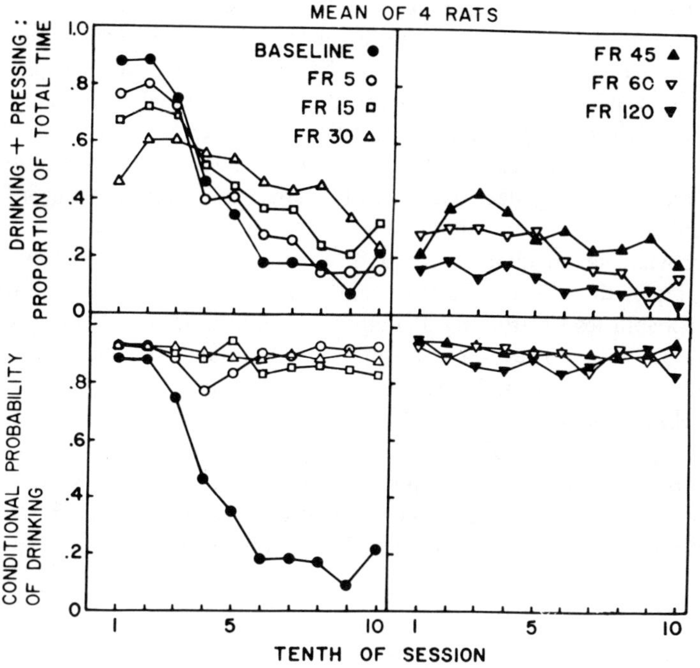

The results from each of the seven conditions are displayed in Figure 3-11. The top panels show the proportions of time occupied by the two schedule behaviors in successive tenths of a session. In the first three tenths of a session, the relation between ratio size and schedule behaviors was the same as in the studies of response effort and VR schedules reported above: With increasing ratio size, the probability of schedule behaviors decreased. However, different patterns emerged later in a session. With long ratios (FR 45 or greater) the probability of schedule behaviors exhibited a gradual decline across a session. With the smaller ratios a greater decrease in schedule behaviors occurred across a session such that less schedule behaviors occurred late in a session with the FR 5 schedule than with three larger ratio sizes. These within-session trends can be attributed to decreases in the value of drinking milk, because conditions with the greatest durations of drinking showed the largest within-session decreases in schedule behaviors. The within-session decreases were most extreme in the baseline condition.

The lower panels in Figure 3-11 show the conditional probabilities of drinking (the proportions of time subjects spent drinking of all the time the drinking tube was available). In the baseline condition the conditional probability is equal to the overall probability of drinking, since the tube was always available. The conditional probability of drinking was almost always above 0.80 in the FR conditions, and no declines in the conditional probability were evident across a session. This result is not surprising, since the value of a reinforcer might decrease to a level at which it cannot sustain much instrumental responding but still is high enough to make contingent behavior preferred over other behavior. Evidently this was the case in this experiment, because the probability of instrumental responding decreased greatly across a session in some of the FR conditions.

To provide an adequate account of this experiment, the molecular model should predict the effects of ratio size, the within-session trends, and the conditional probabilities of drinking. Perhaps the simplest approach is to assume that V_C decreases within a session as a function of cumulative drinking time and to attempt to discover a mathematical expression which describes this function. The baseline results in Figure 3-11 suggest that such an expression should describe fairly gradual initial decreases in V_C, followed by rapid decreases with further consumption, and a low but nonzero V_C with the highest levels of consumption. The following equation was chosen to approximate the relationship between cumulative drinking time and V_C, simply because it has the appropriate shape:

$$V_{C,t} = V_{C,0}\, e^{-A(d_t)^2}, \qquad (3.10)$$

where $V_{C,t}$ is the value of contingent behavior at time t, $V_{C,0}$ is its value at the start of a session, A is a parameter reflecting the rate of satiation, and d_t is the cumulative duration of drinking at time t in seconds.

Before this equation can be used in a simulation, the parameters $V_{C,0}$ and A must be specified. $V_{C,0}$ was set equal to 16, which was the value used for this parameter in the study of response effort, in which the rats were maintained at the same level of food deprivation. The results of the baseline session were used to estimate the value of A by adjusting this parameter so that a simulation of the baseline condition predicted a rapid decline in drinking between the third and fifth 3-minute blocks, as was the case in the actual baseline condition (see Figure 3-11). With $A = 0.0018$, Eqs. (3.1), (3.3), (3.4) and (3.10) generated the simulation of a baseline condition shown by the filled circles in Figure 3-12. This simulation of the baseline condition was not perfect, for the simulated probability of drinking in the last half of the session was lower than was observed in the experiment. Nevertheless, this rough approximation will suffice for the present illustration.

Figure 3-12. The results of a computer simulation for the study of FR schedules and satiation. The simulated results can be compared to the experimental results in Figure 3-11.

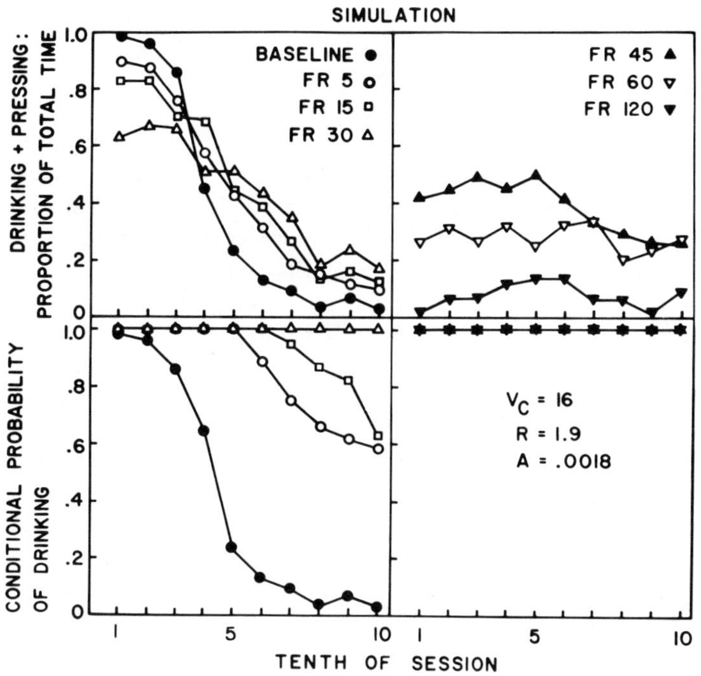

To simulate the FR conditions, R was treated as a free parameter, and with $R = 1.9$ the results shown in Figure 3-12 were generated. Equations (3.2)-(3.4), (3.6), and (3.10) were used to simulate behavior during nonreinforcement periods, and Eqs. (3.1), (3.3), (3.4), and (3.10) were used to simulate behavior during reinforcement periods. A comparison of the top panels of Figures 3-11 and 3-12 shows that the general patterns in the relative durations of the schedule behaviors were captured by the simulation. In the first three-tenths of a session, there was an inverse relation between ratio size and the probability of schedule behaviors. The probability of schedule behaviors decreased markedly through a session with short ratio requirements, but it was more or less constant through a session with the largest ratio requirements. Among the FR 5, FR 15, FR 30, and FR 45 conditions, the ordering of schedule behavior probabilities showed a complete reversal within a session—schedule behaviors were inversely related to ratio size at the start of a session but directly related to ratio size at the end of a session.

The simulation was less accurate in predicting the conditional probability of drinking. For most of the FR conditions, the predicted conditional probability of drinking was 1.0 throughout each session. Of course, the simulation did not take into account the time subjects needed to approach the milk before drinking or possible limitations on subjects' abilities to drink without stopping. The small proportion of available drinking time that subjects did not use for drinking might be due to these factors. However, in the simulation the conditional probability declined near the end of a session with the two shortest FR schedules. No such decline was evident in the subjects' behavior. This decline in the simulated conditional probability occurred because of the small value of V_C late in a session in the FR 5 and FR 15 conditions. It is possible that these predictions were inaccurate simply because the values of V_C predicted by Eq. (3.10) for large levels of consumption were somewhat too low. This possibility is supported by the observation that the predicted durations of the schedule behaviors for the end of a session in the baseline FR 5, FR 15, and FR 30 conditions were also lower than the obtained durations. In short, Eq. (3.10) is not the perfect satiation function. Much more work will have to be done before moment-to-moment changes in reinforcer value are well understood.

In addition to the monotonic decline in V_C across a session predicted by Eq. (3.10), the results of some studies indicate that temporary decreases in V_C may occur immediately after each reinforcement period. When the concentration of milk reinforcement varied unpredictably from one reinforcement period to the next, Lowe, Graham, and Harzem (1974) and Priddle-Higson, Lowe, and Harzem (1976) found longer postreinforcement pauses following higher concentrations of milk. Priddle-Higson et al. concluded that the occurrence of a reinforcer has an inhibitory aftereffect whose magnitude is proportional to the magnitude or concentration of the reinforcer. In the present molecular model, this aftereffect of reinforcement could be represented as a temporary decrease in V_C occurring after each reinforcement period. For reasons of parsimony these short-term changes in V_C were not considered in the simulations, but future research may show that reinforcement aftereffects cause, for example, a relative scarcity of very short postreinforcement pauses.

CONCLUSIONS

The major purpose of this chapter has been to emphasize the need for a quantitative molecular model of ratio schedule performance. My attempts to develop such a model and to test this model against the available empirical results have been described. Many different aspects of behavior on ratio schedules (the probability of within-ratio pauses at different points in a ratio run, the difference in postreinforcement pause durations on the long and short components of a mult FR FR schedule, the priming pattern on mix FR FR schedules, the difference between performance on FR versus VR schedules) strongly suggest that an adequate theory of behavior on these schedules must consider not only the overall schedule requirements but the size of the instrumental requirement that currently separates the subject from the next reinforcer. It is only because $T_{I,t}$ reflects the currently remaining instrumental requirement that the model can account for phenomena such as those just listed. The last experiment described shows that in some cases it is essential to consider the moment-to-moment changes in another variable—the value of the reinforcer—in order to account for changes in behavior across a session.

This model was designed to incorporate some of the major independent variables known to affect behavior on ratio schedules: the size of the instrumental requirement, the magnitude of the reinforcer, the value of the reinforcer, and the effort involved in the instrumental response. Evidence has been presented here that the model can account for changes in each of these independent variables by changing the appropriate parameter. Using the equations listed a computer program successfully simulated the results of a number of different studies. Although these initial simulations were successful, more stringent tests of this particular model will be possible when further studies are conducted. As just one example the careful measurement of postreinforcement pauses and behavior durations on various multiple FR, VR, and RR schedules may indicate whether Eqs. (3.8) and (3.9) provide adequate descriptions of behavior when the size of the remaining instrumental requirement cannot be known with certainty. The simple averaging rule described by the two equations may not be the best method for predicting behavior under schedules with variable instrumental requirements, and further research might suggest better alternatives.

It may be possible to expand this model to encompass a number of other factors that influence behavior on ratio schedules. Two of these factors are the baseline probability of the instrumental behavior and delay of reinforcement. If the baseline value of an instrumental behavior is greater than zero, can this baseline value be added to the value calculated from Eq. (3.6) to determine the behavior's total value in a contingency situation? Can a delay before reinforcement be treated as an additional "instrumental requirement" in Eq. (3.6) (but possibly one with a different value of R)? It will be useful to consider such questions in the future, because an important goal for research on ratio schedules should be the development of a molecular model that incorporates the widest possible range of independent variables in a unified framework.

REFERENCES

Ainslie, G. 1975. Specious reward: A behavioral theory of impulsiveness and impulse control. *Psychological Bulletin 82*: 463–96.

Alferink, L.A. 1975. Priming and the post-prime pause in mixed fixed-ratio schedules. Unpublished doctoral dissertation, Utah State University.

Allison, J. 1976. Contrast, induction, facilitation, suppression and conservation. *Journal of the Experimental Analysis of Behavior 25*: 185–98.

Atkinson, J.W., and D. Birch. 1970. *The Dynamics of Action.* New York: Wiley.

Boren, M.C.P. 1973. Fixed-ratio and variable-ratio schedules of brief stimuli in second-order schedules of matching to sample. *Journal of the Experimental Analysis of Behavior 20*: 219–33.

Chung, S.H., and R.J. Herrnstein. 1967. Choice and delay of reinforcement. *Journal of the Experimental Analysis of Behavior 10*: 67–74.

Crossman, E.K. 1968. Pause relationships in multiple and chained fixed-ratio schedules. *Journal of the Experimental Analysis of Behavior 11*: 117–26.

_____. 1971. The effects of fixed-ratio size in multiple and mixed fixed-ratio schedules. *Psychological Record 21*: 535–44.

Crossman, E.K., and L.T. Silverman. 1973. Alternating the proportion of components in mixed fixed-ratio schedule. *Journal of the Experimental Analysis of Behavior 20*: 273–79.

Davison, M.C., and R.F. Over. 1966. Interactions in a multiple fixed-ratio schedule. *Australian Psychologist 1*: 85.

Fantino, E. 1967. Preference for mixed- versus fixed-ratios. *Journal of the Experimental Analysis of Behavior 10*: 35–43.

Ferster, C.B., and B.F. Skinner. 1957. *Schedules of Reinforcement.* New York: Appleton–Century–Crofts.

Findley, J.D. 1962. An experimental outline for building and exploring multi-operant behavior repertoires. *Journal of the Experimental Analysis of Behavior 5*: 113–66.

Fleshler, M., and H.S. Hoffman. 1962. A progression for generating variable-interval schedules. *Journal of the Experimental Analysis of Behavior 5*: 529–30.

Herrnstein, R.J. 1961. Relative and absolute strength of response as a function of frequency of reinforcement. *Journal of the Experimental Analysis of Behavior 4*: 267–72.

Hull, C.L. 1943. *Principles of Behavior.* New York: Appleton–Century–Crofts.

Inman, D.P., and C.D. Cheney. 1974. Functional variables in fixed ratio pausing with rabbits. *Psychological Record 24*: 193–202.

Kaplan, M. 1956. The maintenance of escape behavior under fixed-ratio reinforcement. *Journal of Comparative and Physiological Psychology 49*: 153–57.

Keehn, J.D. 1965. Temporal alternation in the white rat? *Journal of the Experimental Analysis of Behavior 8*: 161–68.

Keehn, J.D., and V.A. Colotla. 1971. Stimulus and subject control of schedule induced drinking. *Journal of the Experimental Analysis of Behavior 16*: 257–262.

Kintsch, W. 1965. Frequency distribution of interresponse times during VI and VR reinforcement. *Journal of the Experimental Analysis of Behavior 8*: 347–52.

Logan, F.A. 1965. Decision-making by rats: Delay versus amount of reward. *Journal of Comparative and Physiological Psychology 59*: 1–12.

Lowe, C.F.; C.L.D. Graham; and P. Harzem. 1974. Effects of reinforcement magnitude on interval and ratio schedules. *Journal of the Experimental Analysis of Behavior 22*: 553–60.

Luce, R.D. 1959. *Individual Choice Behavior: A Theoretical Analysis.* New York: Wiley.

Malott, R.W. 1966. The effects of prefeeding in plain and chained fixed-ratio schedules of reinforcement. *Psychonomic Science 4*: 285–86.

Mazur, J.E. 1975. The matching law and quantifications related to Premack's principle. *Journal of Experimental Psychology: Animal Behavior Processes 1*: 374–86.

____. 1977. Quantitative studies of reinforcement relativity. Unpublished doctoral dissertation, Harvard University, part 3, experiment IV.

Mintz, D.E.; D.J. Mourer; and M. Gofseyeff. 1967. Sequential effects in fixed-ratio post-reinforcement pause duration. *Psychonomic Science 9*: 387–88.

Powell, R.W. 1968. The effect of small sequential changes in fixed-ratio size upon the post-reinforcement pause. *Journal of the Experimental Analysis of Behavior 11*: 589–93.

____. 1969. The effect of reinforcement magnitude upon responding under fixed-ratio schedules. *Journal of the Experimental Analysis of Behavior 12*: 605–608.

____. 1970. Probability of reinforcement and fixed ratio responding: A model. *Psychological Record 20*: 73–80.

____. 1972. Responding under basic schedules of reinforcement in the crow. *Journal of Comparative and Physiological Psychology 79*: 156–64.

Priddle-Higson, P.J.; C.F. Lowe; and P. Harzem. 1976. Aftereffects of reinforcement on variable-ratio schedules. *Journal of the Experimental Analysis of Behavior 25*: 347–54.

Rachlin, H. 1978, A molar theory of reinforcement schedules. *Journal of the Experimental Analysis of Behavior 30*: 345–60.

Schuster, C. 1959. Response latencies as a measure of the interaction of components on a multiple fixed-ratio schedule. *Journal of the Experimental Analysis of Behavior 2*: 259.

Schwartz, B. 1969. Effects of reinforcement magnitude on pigeons' preference for different fixed-ratio schedules of reinforcement. *Journal of the Experimental Analysis of Behavior 12*: 253–60.

Sherman, J.A., and J.R. Thomas. 1968. Some factors controlling preference between fixed-ratio and variable-ratio schedules of reinforcement. *Journal of the Experimental Analysis of Behavior 11*: 689–702.

Sidman, M., and W.C. Stebbins. 1954. Satiation effects under fixed-ratio schedules of reinforcement. *Journal of Comparative and Physiological Psychology 47*: 114–16.

Staddon, J.E.R. 1979. Operant behavior as adaptation to constraint. *Journal of Experimental Psychology: General 108*: 48–67.

Tversky, A. 1972. Elimination by aspects: A theory of choice. *Psychological Review 79*: 281–99.

Webbe, F.M.; J. DeWeese; and E.F. Malagodi. 1974. Induced attack during multiple fixed-ratio, variable-ratio schedules of reinforcement. *Journal of the Experimental Analysis of Behavior 22*: 197–206.

Webbe, F.M., and E.F. Malagodi. 1978. Second-order schedules of token reinforcement: Comparisons of performance under fixed-ratio and variable-ratio exchange schedules. *Journal of the Experimental Analysis of Behavior 20*: 219–24.

Winograd, E. 1965. Escape behavior under different fixed ratios and shock intensities. *Journal of the Experimental Analysis of Behavior 8*: 117–24.

4 REINFORCEMENT AND THE LOCAL ORGANIZATION OF BEHAVIOR

Charles P. Shimp

The topic of this volume, quantitative accounts of matching and maximizing, is a special case of more general questions in the experimental analysis of behavior. Some of these broader questions include the following: What precisely does a reinforcer do to behavior? On precisely what behavior does a reinforcer operate? In what ways do various reinforcement contingencies permit us to make progress in answering these questions? The present chapter attempts to address these broader questions and is organized into four parts: the relation between these broader questions and a molar analysis; a review of an alternative analysis that focuses greater attention of the local organization of behavior; a description of new data on how to control the local organization of behavior; and general remarks and tentative conclusions.

RELATIONS AMONG A "MOLAR" ANALYSIS OF OPERANT BEHAVIOR, THE LOCAL ORGANIZATION OF BEHAVIOR, AND MATCHING AND MAXIMIZING

Morse (1966) distinguished between strengthening and shaping properties of reinforcement. He reviewed a variety of different kinds of evidence and concluded that two different properties were needed to describe the effects of reinforcement and that both properties are at work in many schedule contingencies.

This research was supported in part by Grant 16928 from the National Institute of Mental Health. Stephen J. Hutton conducted the experiment with a thoughtfulness and conscientious attention to a myriad of details that contributed immensely to the success of the research.

In the next section, I shall suggest a specific interpretation of these two effects of reinforcement. For the moment all that is required is the idea that reinforcement both strengthens behavior and shapes it, that is, establishes a temporal patterning or structure to behavior.

Different investigators in the same scientific field may look at their field in different ways. Their favorite concepts and methods form a pair of cognitive spectacles through which they look at their field. This grid of concepts and methods helps scientists to make sense out of their field and at the same time tends to blind them to concepts and methods that are incompatible with theirs. As a result, different theorists may choose different data for their theories to assimilate (Hanson, 1958; Kuhn, 1970). In the present context, different theorists try to develop reasonably general theories while ignoring either the kinds of data upon which Morse based his strengthening notions or ignoring the data calling for a shaping notion. For example, in Shimp 1969a I proposed a pure shaping theory that did not extend to strengthening effects, and many theorists have articulated pure strengthening views that ignore data calling for the involvement of a shaping effect (Baum, 1973; Catania, 1973; Herrnstein, 1970, Rachlin, 1978). The latter theories apply to an empirical domain that does not include data on how reinforcement affects the local organization of behavior.

Often a theorist does not trouble to explain why a theory is designed to handle certain data and not others. Elsewhere, I have tried to explain in some detail why it is that molar accounts of operant behavior are designed to handle data of a certain type, mostly involving mean response rate (Shimp, 1975, 1976a). The explanation, too long to be repeated here, involves the nature of the historical context during which the cluster of molar laboratory methods, methods of data analysis, and concepts were developed, largely by Skinner in the 1930s. Suffice it to say here that the chief features of a molar analysis are those upon which one would look favorably if one subscribed to the kind of simple association theory that was dominant in the 1930s. I hasten to add that I am not saying a molar analysis was developed to be consistent in any explicit way with association theory. The connection between molar theory and association theory is much subtler. A simple association theory is built right into the experimental methods of a molar analysis. Therefore, although one might wish to take a more theoretically neutral and purely empirically oriented position (Skinner, 1950), the *use* of, say an ordinary variable-interval (VI) schedule is a commitment to a simple form of association theory (Shimp, 1975, 1976a, 1978).

So far I have only repeated the conclusion of an argument developed elsewhere, that a molar analysis is a pure strengthening account of reinforcement, the conceptual framework for which very closely resembles the views on learning subscribed to by nineteenth-century British empiricists and many American learning theorists in the 1930s. Like the association theory, with which it tends to be consistent, a molar analysis ignores structural concerns in general. Matters of patterning, organization, and so on are not viewed as matters that need to be considered in elementary, general laws. What this means for the present context is that molar analyses do not see anything particularly interesting in the local structure or organization of behavior. The picture a molar analysis gives of this organization is, as a matter of fact, the least interesting of all possible pictures: There is no order! It is random and chaotic. Reinforcement is assumed not to do anything interesting to the local organization of behavior. This view is usually implicit but is encapsulated explicitly in the assumption made by Rachlin (1978), to the effect that the steady-state distribution of interresponse times in a variable-interval schedule is a negative exponential distribution. That one might make this assumption is exceedingly interesting because it is, as Rachlin acknowledged, contrary to all relevant data. Every theorist probably at some point must sacrifice accuracy and generality on behalf of the elegance and coherence of a particular theoretical point of view. What is diagnostic about Rachlin's assumption is not that he sacrificed data on behalf of theory but, rather, *which* data he sacrificed. He made the sacrifice at the expense of the shaping property of reinforcement, at the expense that is, of the empirical literature on the local organization of behavior, reducing a sizable literature to, literally, chaos. This is another way of saying that a molar analysis finds little or nothing interesting or important in facts about the local organization of behavior. I have suggested elsewhere that this view of behavior derives from an unfortunate tendency to think about the local organization of behavior in terms of the picture of behavior given us by cumulative records (Shimp, 1979). These pictures of behavior are admirable when one wishes to compress large-scale patterning in behavior to a more manageable perceptual scale (Skinner, 1976). But the picture they give of the nature of the local organization of behavior could not be more misleading. The standard gearing of a cumulative recorder is designed to suppress local irregularities; it is designed to make a curve look smooth (Estes, 1959). Perhaps it is no surprise to find theoretical conceptions of operant behavior ignoring the local patterning of behavior when these patterns have been systematically obscured in cumulative records. The invisible nature of local patterning in a cumulative record also encourages an unfortunate tendency to think of local patterning as being on some other, more molecular "level" of analysis from the behavior one *can* see in a cumulative record. The conception of different, and independent, levels of analysis tends to be offered as a justification for why molar theories need not deal with the local organization of behavior.

This same molar view on the relevance of the local organization of behavior to a general understanding of reinforcement appears in many other forms besides the kind of explicit theoretical assumption adopted by Rachlin (1978). It appears, for example, in the kinds of data analyses offered by Nevin (1969), Herrnstein (reported in de Villiers, 1977), and Heyman (1979). These papers report data that are interpreted as showing that the local organization of behavior in concurrent VI VI schedules is for all practical purposes, random. According to these interpretations, certain local reinforcement contingencies automatically present in concurrent VI VI schedules have little or no effect on the local organization of behavior and the contingencies leave the random nature of this behavior unaffected. This interpretation is consistent with a search for general laws relating molar independent and dependent variables. Nevin's experiment has played an important role in determining scientific opinion on this matter (Nevin, 1979; Silberberg, Hamilton, Ziriax, and Casey, 1978) so it deserves our careful attention. I would like to call attention to two features of this experiment: I wish to argue that the experiment has logical problems and problems of mistaking statistical artifacts for empirical relations. Let us consider each of these problems in turn.

There is, it seems to me, a problem with the logic of Nevin's (1969, 1979) argument. His argument goes basically as follows: If a single case can be shown where the outcome compatible with a molar analysis occurs, local reinforcement contingencies fail to control behavior, and the local organization of behavior is random (in the sense corresponding to Rachlin's assumption), than it has been shown that the molar outcome is more general and important, and is independent of local reinforcement contingencies. The logic is straightforward, but the hand it deals can be overplayed. The problem is that the logic can be applied in other contexts to let one arrive at the opposite conclusion. A logical argument that leads to incompatible positions scarcely seems sufficient to settle the controversy. Consider the following: If a single case can be shown where the outcome compatible with a local analysis occurs and the molar contingencies fail to control behavior, then it has been shown that the local analysis is more general and important than the molar analysis. I am happy to be able to report that several experiments have produced results of the type required here (for example, see Shimp and Hawkes, 1974). According to Nevin's logic, then, molecular control is more general than molar control. But clearly this will not do: Each type of control appears to be more general than the other, and the whole argument can be based on a total of two experiments.

But suppose there were no problems with the logic. Then what would Nevin's results establish? The specific question the results were supposed to answer was whether switching from one alternative to the other in a two-alternative situation tends to occur more often at times when the local probability of reinforcement for switching is greater. A local analysis tends to suggest it will; a molar analysis tends to suggest it will not. Because of what might be called the ping-pong ball effect, results such as Nevin's are in principle incapable of answering the question (Shimp, 1979). Suppose that instead of putting a pigeon in the experimental chamber, a mischievous graduate student tosses in an incredibly elastic ping-pong ball, which manages to bounce around for a full hour, once in a while striking one or the other of the two keys in an experiment such as Nevin's. The student's advisor then analyzes the data. What will the advisor discover about the way changeovers from one key to the other at different times depend on the reinforcement for changeovers at those times? The percentage of time allocated to interchangeover times of a given duration (by the ping-pong ball) will be found to approximately equal, except for sampling fluctuation, the obtained relative reinforcements per hour for interchangeover times of that duration. There is a statistically forced time-allocation matching relation.[a] The argument has been presented in detail, elsewhere (Menlove, 1975; Shimp, 1979). Suffice it to say that the relation between the probability of a changeover at a given time since the last changeover and the obtained reinforcements per hour for changeovers at that point is not a very interesting statistic in a concurrent VI VI schedule.[b] Thus, even apart from the telling criticisms of Silberberg et al. (1978) and Staddon, Hinson, and Kram (1981), relations between emitted and reinforced interchangeover run lengths such as those reported by Nevin (1969) are not convincing evidence one way or the other for the issue, Does an organism switch when switching is more likely to be reinforced?

A similarly omnipresent time-allocation matching relation prevents all sorts of relations from being interesting in the case of emitted and reinforced interresponse times in a VI schedule (Reynolds and McLeod, 1970). This has been known and widely recognized for a long time (Anger, 1954). The situation is exactly analogous for emitted and reinforced interchangeover times in a concurrent VI VI schedule. One might sensibly ask what the distribution of interchangeover times looks like. Or one might sensibly ask what the distribution of reinforced interchangeover times looks like. But one might not sensibly ask what the *relation* is between these two distributions. The answer to this last question is fixed by the schedule. And it is this last question that is relevant to the problem we are confronting: Does the local reinforcement contingency affect the local organization of behavior or does it not? In short, a concurrent VI VI does not let us answer the question (Shimp, 1976b). All of this motivates the development of alternative methods free of the ping-pong ball syndrome.

SOME EFFECTS OF REINFORCEMENT OF THE LOCAL ORGANIZATION OF BEHAVIOR

Theorists who believe they can develop general theories of behavior without an appeal to a shaping effect of reinforcement on the local organization of behavior enjoy working with molar relations such as matching and are comfortable with

a. (Editors' note). The "matching relation" shown by the ping-pong ball would not be matching in the sense implied elsewhere in the book or by the matching law in general, since it counts only the reinforcements obtained for the first response following a changeover.

b. The time intervals entering into the time-allocation matching relation are not exactly equal to, but ordinarily may be assumed to be very highly correlated with, the time between successive changeovers, the interchangeover times. The relevant time intervals are those beginning with the last peck on a key before a changeover and ending with the next changeover back to the first key. This interval is an interchangeover time plus the time between the last peck on a key and the subsequent changeover.

the assumption that local structure is random. The specter of nonrandomness occasionally has been recognized in the development of molar theories (Herrnstein, 1970: 250; Herrnstein and Loveland, 1975: 115). The recognition has taken the form of acknowledging the possibility that local patterning may change the meaning of molar functions. However, the equations associated with not a single molar model have been modified to permit the assimilation of molecular data on interresponse times, interchangeover times, or runs of successive responses to a particular alternative. These molecular data remain outside the scope of molar models. If maximizing enters into a molar conceptual framework, it is in the form of a molar maximizing based on feedback functions defined in terms of molar variables (Baum, 1973; and several chapters in the present volume).

In sharp contrast to this view some theories attempt to assimilate data on the local structure of behavior on the grounds that there is a separate shaping effect of reinforcement. Why would anyone think this? How *can* one separate the strengthening and shaping effects? I have suggested the development of alternative contingencies free of the ping-pong ball problem and free of the unpleasant confusion between shaping and strengthening effects. Classical interval and ratio schedules clearly will not do (Anger, 1954; Shimp, 1967; Reynolds and McLeod, 1970). Anyone who takes seriously the possibility of a shaping effect will probably begin to tinker with these schedules to see how reinforcement affects local behavioral patterning. Some tinkering has involved synthetic VI schedules (Anger, 1954). The use of such contingencies led me to suggest tentatively a way of defining shaping and strengthening effects in terms of separate time-allocation functions (Shimp, 1974). It seems useful to define a strengthening effect as the percentage of time a subject allocates to all of the experimentally reinforced activities. A shaping effect may be defined as the way a subject allocates time across the various reinforced activities. These definitions require, of course, a reasonable way of measuring how much time a subject spends engaged in a particular reinforced activity. Classical schedules do not provide such measures, but synthetic schedules do.

It turns out to be possible to obtain orderly functions relating the local organization of behavior to parameters of reinforced temporal patterns and parameters of reinforcement associated with the patterns. At least it is possible when the temporal patterns are very simple, like interresponse times or short sequences of interresponse times (Shimp, 1967, 1968, 1969b, 1970, 1973; Shimp and Hawkes, 1974; Staddon, 1968). One can adopt as a working hypothesis that these results reveal something general about the shaping effects of reinforcement. If the results are tolerably general, they ought to be obtainable with temporal patterns other than interresponse times. And indeed they are. Recently I have shown that reinforcing interchangeover times in a two-key choice situation may produce local temporal patterning indistinguishable from the patterning produced by interresponse-time contingencies with a single key (Shimp, 1979).

Doubtless there are limits to this good fortune. But this success motivates one to continue seeking invariances in the ways local reinforcement contingencies control the local structure of behavior.

AN EXPERIMENT ON THE CONTROL OF BEHAVIORAL PATTERNS

The experiment to be described was motivated by a search for general empirical rules relating reinforced and emitted behavioral patterns. The purpose was to study the shaping property of reinforcement and not the strengthening property. The total number of reinforcements per hour was held approximately constant at a relatively high level, so that nearly all of each session was devoted to various activities resembling the reinforced ones. Like Nevin (1969) I used a discrete-trials procedure for simplicity. Also like Nevin I asked how the reinforcement of changeovers at particular points affects the likelihood of a changeover at that point. Unlike Nevin's experiment this one avoids the ping-pong ball syndrome because of the manner in which the reinforced patterns were experimentally controlled.

Method

Subjects. Eight experimentally naive White Carneaux pigeons were maintained at 80 percent, plus or minus 20 grams, of their free-feeding weights. Four additional subjects began the experiment but then either became ill or required more manual key-peck training than was convenient and hence were eliminated from the experiment.

Apparatus. Six standard two- or three-key Lehigh Valley Electronics pigeon chambers were interfaced to a Digital Equipment Corporation PDP-8e computer, which arranged all experimental contingencies and recorded data. White noise and a fan in each chamber helped to mask extraneous noises. The experiment used only the left and right keys in the experimental chambers.

Procedure

The procedure was a simplified, discrete-trials, synthetic concurrent VI VI schedule (Shimp, 1979) designed to permit the experimenter to control certain features of the behavioral patterns preceding reinforcement. It consisted of several parts, as follows:

Discrete-Trials Contingency. Each trial began with the illumination, with white light, of left and right keys. A response to either key turned off both key lights and initiated either reinforcement or an intertrial interval. Reinforcement consisted of mixed grain presented for 3 seconds during which only the light directly over the food hopper was on. The intertrial interval consisted of a 0.5-second blackout during which responses to any key had no programmed consequence.

Changeover Contingency. Only pecks on the right were reinforced. Furthermore, a peck on the right could be reinforced only if the preceding response was on the left. That is, the only responses reinforced in this experiment were changeover responses to the right key. But not all changeovers to the right were reinforced. Reinforcement for such a response depended on two separate contingencies: a VI contingency had to have arranged a reinforcer, and the number of preceding left-key responses had to satisfy a response-pattern contingency. Each of these contingencies is described in turn.

Variable-Interval Contingency. A constant-probability VI 20-second schedule (Catania and Reynolds, 1968) ran during the intertrial interval. This contingency arranged a reinforcer with probability 0.05 every 1 second. Once a reinforcer was arranged, the 1-second clock stopped until the reinforcer was delivered. And once arranged, a reinforcer had to be delivered before another could be arranged for some other behavior. This feature of the contingency permitted control over how reinforcers were allocated to different behavioral alternatives (Shimp, 1966; Stubbs and Pliskoff, 1969). The VI clock stopped also when the keylights were on. This feature was designed to reduce long pausing while the key lights were on.

Response-Pattern Contingency. Each reinforcer arranged by the VI schedule was assigned to a changeover following a particular number of consecutive pecks on the left key: Every reinforcer followed a changeover after a sequence of left-key pecks. For birds 1, 3, 4, and 6 all reinforcers followed changeovers after either two or four pecks on the left. For birds 7-10 all reinforcers followed changeovers after either two or eight pecks on the left. For each subject, therefore, a changeover was sometimes reinforced after a shorter sequence of left-key pecks and sometimes reinforced after a longer sequence of left-key pecks. The percentage of reinforcements assigned to the shorter sequence was varied over experimental conditions as shown in Table 4-1. In other words, the computer assigned each reinforcer arranged by the VI schedule either to the shorter or longer sequence, and its probability of assigning a reinforcer to the shorter sequence is shown in Table 4-1. Note that a reinforcer depended on the length of a sequence of left-key pecks, not on the time spent in emitting a sequence, although the number of responses in a sequence and the time spent emitting it were presumably rather closely correlated by virtue of the 0.5-second intertrial interval.

Table 4-1. Experimental conditions.

	Birds 1, 3, 4, 6 (Reinforced Sequences of 2 and 4 Left-Key Pecks)		*Birds 7-10 (Reinforced Sequences of 2 and 8 Left-Key Pecks)*	
Condition Number	*Number of Days*	*Programmed Relative Frequency of Reinforcement for Shorter Sequence*	*Number of Days*	*Programmed Relative Frequency of Reinforcement for Shorter Sequence*
1	59	.50	25 [b]	.50
2	21	.25	21	.20
3	30	.10	20	.05
4	30	.50	20	.00
5	27 [a]	.75	20	.05
6	—	—	20	.80

a. An incorrect parameter was used on day 7, so the condition was conducted for an extra 7 days.

b. This condition followed approximately 4 months of a variety of types of training.

Miscellaneous Details. Each session was 1 hour long. Sessions were conducted 6 days a week. It was not convenient to perform the analysis of sequential data on a daily basis and, therefore, decisions concerning when to change experimental conditions had to be based on some criterion oother than stability of the dependent variable. It was decided after the initial condition to conduct each condition for a standard 20 or 30 days.

Comments on the Procedure

A subject pecked on each trial on either a left or a right key. A switch from the left to the right key was in fact reinforced if a VI schedule had timed out and if the number of left-key pecks preceding the switch equaled the number the computer was looking for. This number was either 2 or 4, for some subjects, or 2 or 8, for the others. Notice that no exteroceptive stimulus at any time signaled either how many responses a subject had made on the left key or whether the VI schedule had arranged a reinforcer.

Although only a changeover was temporally contiguous with a reinforcer, it is helpful to think of the procedure as one where an entire sequence of several responses was reinforced as a single pattern. The experiment may then be thought of as involving the reinforcement of a shorter and of a longer behavioral pattern. Thus a peck on the left key seemed to mean something like, "I don't think I have yet made a run of left-key pecks that is as long as the run I selected, so I'll peck the left key again." The purpose of the experiment was to see how the imposition of such a contingency would affect the local organization of behavior. It was to see, that is, how choice, or switching, is a function of the local reinforcement contingencies such that sometimes the potential payoff for switching is high and sometimes it is low.

RESULTS

There are two preliminaries to the description of the principal results. The first is the need to check that the total reinforcement density remained throughout the experiment at a suitable level, so that the strengthening property of reinforcement may reasonably be assumed to have been approximately constant. The total reinforcements per hour varied from 31 to 53 or from 57 to 116, averaged over the four subjects with reinforced sequences of 2 and 8 or of 2 and 4, respectively. This variation is within the range according to which, on the basis of results obtained previously, it is reasonable to assume that in the present experiment variations in behavior are attributable to the shaping property of reinforcement, not to the strengthening property (Shimp, 1970, 1974). Second, since only changeovers to the right key were reinforced and sequences of more than one consecutive peck to the right key were never reinforced, there should have been very few occasions when a subject continued to respond on the right key after changing over to it. The relative frequency of sequences of pecks on the right key that consisted of more than a single key peck was often less than 0.01 and seldom more than 0.10, but occasionally, for birds 8 and 9, was as high as 0.30. It is generally correct to summarize these preliminaries in the following way: The experiment established the desired baseline situation. A subject spent nearly all the time responding some number of times on the left key, switching over to the right key for a single response and then returning to produce another sequence of responses on the left key.

We can now look at how the patterning of responding on the left key was affected by the experimental parameters. Figure 4-1 shows the relative frequency of occurrence of different sequences of left-key pecks, sequences with 1, 2, 3, . . . or 7 or more consecutive pecks on the left key. The samples selected for the figure are not representative of every single sample but show several properties found in nearly all of them. The left and right panels show results for subjects with reinforced sequences of length 2 and 4 and of 2 and 8, respectively. The data displayed in the figure were calculated as follows: For each of the last 5 days in a condition, the number of sequences of a given length was divided by the total number of left-key sequences. (A *sequence*, of course, refers simply to a number of left-key pecks preceded by and followed by right-key pecks.) This ratio was the relative frequency, for a single day, of a sequence of a particular length. If for a given subject there were 100 sequences of length 2 and there was a total of 1,000 sequences of all lengths, the relative frequency of length 2 would be 0.1. The 5 daily values of this statistic were averaged; the average is displayed in Figure 4-1.

Figure 4-1. The relative frequency of occurrence of sequences as a function of their lengths (the number of responses in them). The last category is for sequences of 7 or more (*left*) or 13 or more (*right*) responses. Four representative samples are displayed.

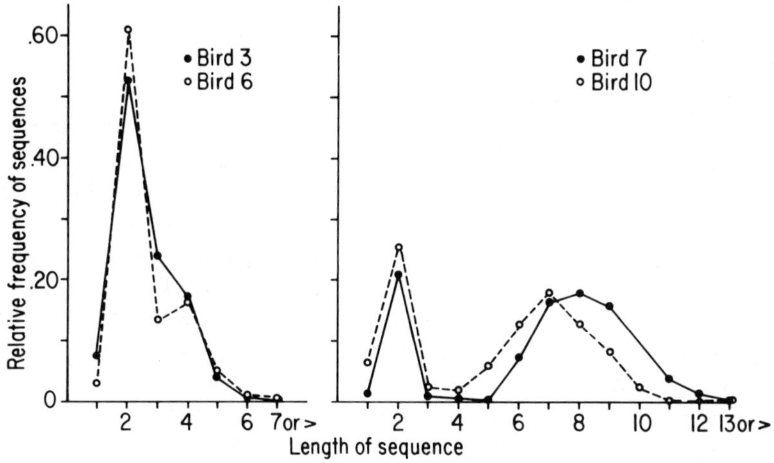

The left panel illustrates three different phenomena of interest. First, nearly all sequences resembled those, of length 2 and 4, that were reinforced. Some changeovers occurred after a single peck on the left key, but those of lengths 2, 3, and 4, and 5 virtually exhaust total behavioral output on the left key. The sequences the birds emitted closely resembled the reinforced ones. Second, not all the distributions were bimodal; enough sequences of length three were emitted so that some of the curves resembled that displayed for bird 3, for which the percentage for length 3 was greater than that for length 4. Presumably sequences of length 3 were sometimes confused with those of length 2 and sometimes with those of length 4. Third, even though the left panel represents a condition where sequences of lengths 2 and 4 were reinforced on the average equally often, there is a pronounced preference for the shorter of the two reinforced sequences.

The right panel shows the corresponding results for two subjects in a condition where sequences of lengths 2 and 8 were reinforced, with only 5 percent of the reinforcers going to the shorter sequence. Again, we can see three phenomena that occurred often. Here, as before, there is a clear tendency for switches to occur only when they would terminate sequences approximating the reinforced ones. Second, and presumably as a consequence of the first phenomenon, the curves are very clearly bimodal, with the shorter mode precisely at the shorter reinforced sequence and the longer mode either equal to or a little shorter than the longer reinforced sequence. Notice also that the second subdistribution near a length of 8 is a broader, flatter distribution than is the one near a length of 2. Third, we again see something like a bias for the shorter sequence in the sense that over 20 percent of all the sequences were of length 2 even though these received only 5 percent of the reinforcers.

In brief, Figure 4-1 shows that a changeover tended to occur where a changeover was more likely to be reinforced. Note here for future reference that the results described so far, for distributions of sequences of responses between successive changeovers, conform in all essentials to those obtained in corresponding experiments involving distributions of interresponse times when a shorter and a longer interresponse time is reinforced (Shimp, 1968, 1970). In fact, the curves in the figure could easily be mistaken for interresponse-time distributions.

Now we may turn to somewhat more quantitative questions. How does preference for one of two reinforced behavioral patterns depend on the reinforcement parameters associated with those patterns, and how does that preference depend on the structural parameters of the patterns themselves? In the present case the reinforcement parameter in question is the relative reinforcements per hour associated with a pattern, and the structural parameter is the relative length of the shorter of the two reinforced patterns.

To answer these questions requires dichotomizing a continuum, since some sequences must be called shorter and some longer. There are several ways to do this, all of which in corresponding situations appear to lead to the same or similar answers, so that the simplest solution seems appropriate here (Shimp, 1969b). The simplest solution is just to call sequences of length 2 the shorter sequence and to call those of length 4 (birds 1, 3, 4, and 6) or 8 (birds 7-10) the longer sequence. Thus for purposes of measuring preference between what in fact are two subdistributions of sequences, the shorter and longer sequences may simply be equated with the shorter and longer reinforced sequences. Then, of all sequences that were either the shorter or longer, what percentage were the shorter? And, how does this percentage depend on the corresponding relative reinforcements per hour and on the relative duration of the shorter sequence?

Figure 4-2 shows the relative frequency of the shorter pattern (the number of shorter patterns divided by the sum of shorter and longer patterns) for the four subjects for which shorter and longer patterns had lengths of 2 and 4. Each point is the mean of the daily relative frequencies over the last 5 days of a condition. The vertical line shows the range over those 5 days. Variability across days was clearly acceptable but with noticeable differences across birds. By and large the four curves closely resemble one another and the replication of the condition with equal reinforcements for the two patterns was satisfactory. Figure 4-2 shows that preferences for the shorter of two reinforced patterns was, approximately, a monotonically increasing, negatively accelerated function of the relative reinforcements per hour associated with that pattern. This result extends the conclusion derived from Figure 4-1 that changeovers tended to occur where changeovers were more likely to be reinforced: As the proportion

Figure 4-2. The relative frequency of the shorter of two reinforced sequences as a function of the relative reinforcements per hour for that sequence. Each panel shows results for a single subject. The shorter and longer sequences consist of 2 and 4 responses, respectively.

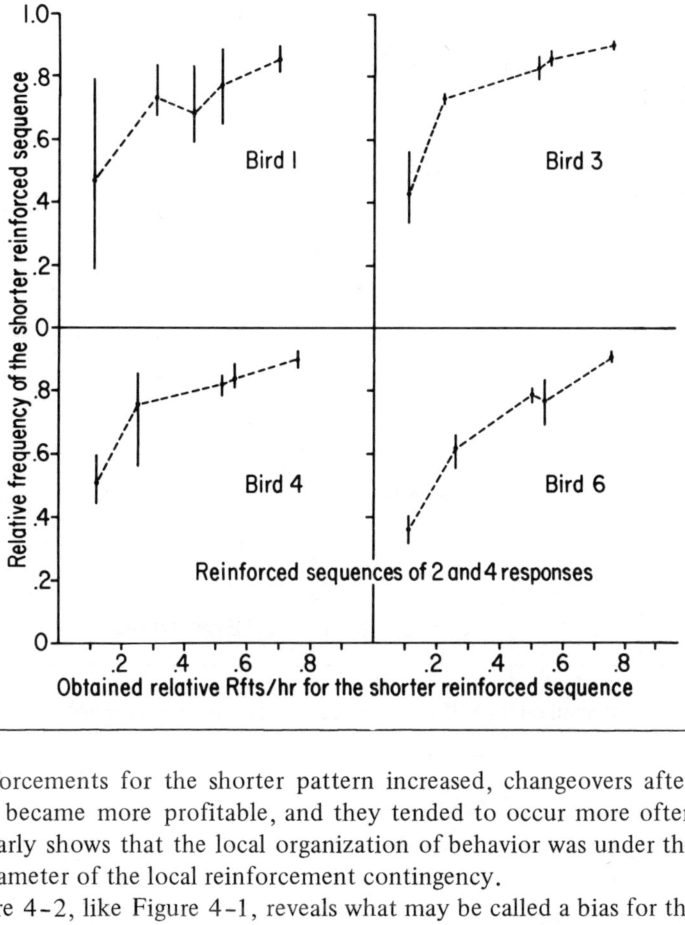

of reinforcements for the shorter pattern increased, changeovers after two responses became more profitable, and they tended to occur more often. Figure 4-2 clearly shows that the local organization of behavior was under the control of a parameter of the local reinforcement contingency.

Figure 4-2, like Figure 4-1, reveals what may be called a bias for the shorter pattern. When the two patterns were reinforced equally often, preference for the shorter ranged from about 0.7 to 0.8. Even when the shorter one received only about 10 percent of the reinforcers, it was emitted more than one time out of three.

Figure 4-3 shows the corresponding results for birds 7-10, for which the shorter and longer patterns involved two and eight left-key responses, respectively. Again, the curves are very orderly and may be described as approximately monotonically increasing and negatively accelerated. An even greater bias appears to exist in favor of the shorter pattern. This greater bias is more clearly

Figure 4-3. The relative frequency of the shorter of two reinforced sequences as a function of the relative reinforcements per hour for that sequence. Each panel shows results for a single subject. The shorter and longer sequences consisted of 2 and 8 responses, respectively.

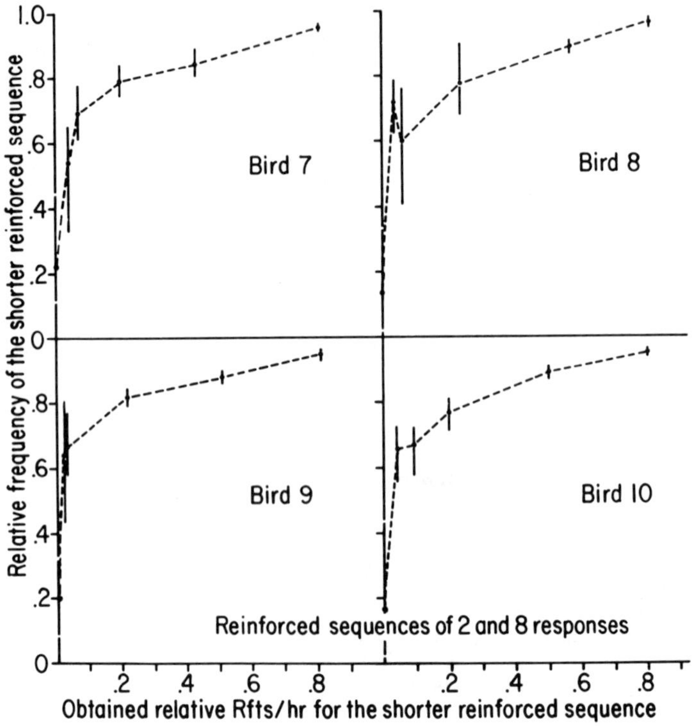

seen in Figure 4-4, which shows the means of each group of four subjects; the vertical lines show the ranges across the four subjects. The figure shows satisfactorily smooth curves with relatively little variability across subjects. The difference between the two curves may be attributed to a structural parameter of the reinforced patterns, namely, their relative lengths: The relative length of the shorter pattern was less in the context including reinforced sequences of 8 responses than in the context including reinforced sequences of 4 responses. A sequence of 2 was preferred more to a sequence of 8 than to a sequence of 4. Thus, Figure 4-4 shows the effects on preference between behavioral patterns both of a reinforcement parameter and of a structural parameter descriptive of the patterns themselves.

Figure 4-4. The relative frequency of the shorter of two reinforced sequences as a function of the relative reinforcements per hour for that sequence. Each curve is an average for four subjects.

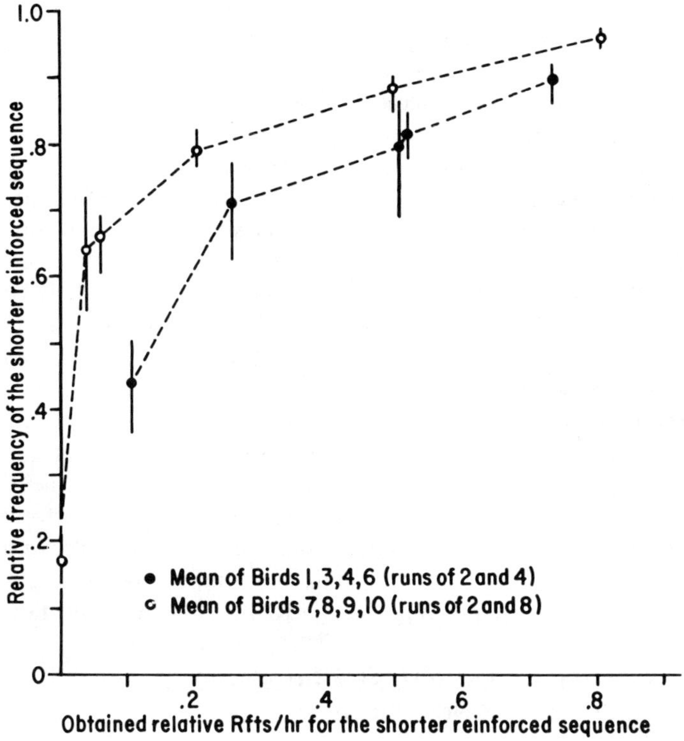

Discussion

The results just described are strikingly similar in several ways to results obtained previously in a different context, namely, the reinforcement of two behavioral patterns defined by a shorter and a longer class of reinforced interresponse times. The distributions in Figure 4-1 resemble interresponse-time distributions (Shimp, 1968, 1970; Staddon, 1968). The way preference for a shorter pattern depends on its relative reinforcements per hour seems similar whether the patterns are interresponse times or sequences of responses on one key preceding a changeover (Shimp, 1968; Staddon, 1968). And the way preference depends here on the relative number of responses in a sequence qualitatively resembles the way preference for an interresponse time depends on its relative temporal duration (Shimp, 1969b). It should be recalled that here reinforcement depended on the number of responses in a sequence rather than on temporal duration of a pattern as in earlier interresponse time studies, but that number of responses and sequence duration were presumably correlated in the present experiment by means of the intertrial interval that separated successive responses. Only further research will show the extent to which the similarity just described, between reinforcement of two different kinds of temporal patterns, sequences and interresponse times, depends on this experimentally induced correlation.

To take seriously the similarities among empirical results for reinforced interresponse times, reinforced interchangeover times, and reinforced sequences of responses is to make room in one's theorizing for a shaping effect of reinforcement. If such an effect is attributed to schedules such as VI and concurrent VI VI schedules, a more complex picture emerges of the behavior maintained by these schedules than is given by molar accounts such as the matching law (Shimp, 1975, 1976a, 1978, 1979). In general responding in a VI schedule would be expected to be importantly determined by which behavioral patterns, such as interresponse times, precede reinforcement. Similarly, responding in a component of a concurrent VI VI would depend on the reinforced patterns in that component. In addition, in a concurrent VI VI, the time spent responding in a component would depend on which interchangeover times preceded reinforcement. How the various multiple determinants would interact to determine performance is a largely unexplored question, but clearly one would not expect to find *general* relations between mean rate of responding and total reinforcements per hour or between relative rate of responding and relative reinforcements per hour.

SUMMARY, GENERAL REMARKS, AND TENTATIVE CONCLUSIONS

This chapter began with the assumption that a general account of reinforcement requires separate shaping and strengthening processes. Under this assumption most schedule effects would be interpreted as jointly determined by these two processes. In fact many common schedules turn out, again in terms of this assumption, to be incapable of doing anything other than confounding these separate effects. From this perspective the mean rate of responding maintained by a VI schedule and relative rate of responding in a concurrent VI VI schedule confound separate reinforcement effects and are therefore incapable of being involved in the expression of any general law about behavior or reinforcement (see also Gage, Evans, and Olton, 1979; Platt, 1979; Silberberg et al., 1978).

The assumption that there are separate strengthening and shaping effects therefore motivates the development of alternative reinforcement contingencies that permit the collecting of relatively unconfounded measures. Alternative contingencies have been developed and have led to the suggestion that the strengthening property of reinforcement be measured by the percentage of time when a subject engages in any of the reinforced activities and that the shaping property of reinforcement be measured in terms of the preference function over the various different reinforced activities (Shimp, 1974). The experiment reviewed was designed to investigate the latter effect, the shaping property of reinforcement. It was designed to do so without falling victim to the ping-pong ball effect (see the first section), according to which data descriptive of the schedule contingencies are mistaken for data descriptive of an empirical relation between behavior and reinforcement. The problem was to see how the local organization of behavior, its temporal patterning, could be controlled by local reinforcement contingencies. As noted elsewhere (Shimp, 1979) the same problem confronts investigators across a broad spectrum. The same problem arises in different guises in the study of miniature grammars, language comprehension, and speech production (Estes, 1978) and motor control in general (Stelmach, 1976, 1978). Notice that all of these analyses have in common a hierarchical view of behavior (Anderson and Bower, 1974; Shimp, 1978, 1979).

Some will want to pooh-pooh the relevance of the present data to the results obtained with plain, unvarnished VI schedules, concurrent VI VI schedules, and the like. These schedules have, after all, been tinkered with here and elsewhere in analyses of local effects. According to the present point of view, they *had* to be tinkered with to avoid the ping-pong ball effect and other problems. But since they have been tinkered with, the contingencies are no longer exactly like, say, the "real" concurrent VI VI contingency. Therefore, perhaps the resulting data do not have much, or anything, to do with the "real" thing.

Well, perhaps they do not. But I know of no data, as opposed to molar wishful thinking, that suggest we have two sets of laws, one set for VI schedules and the like and another for schedules that control local contingencies. Indeed, all the available evidence points in the other direction. Powerful similarities are found between the local organization of behavior in VI schedules and in synthetic VI schedules that control reinforced interresponse times (Shimp, 1979). Other results very convincingly argue for shaping effects on the local organization of behavior in VI schedules (Platt, 1979). A linear function is found between local reinforcement probability and time since some event, as exists in both VI and concurrent VI VI schedules, that powerfully controls local choice probability (Hale and Shimp, 1975). Data such as these do not leave much refuge for a molar theorist who wishes to assume the local organization of behavior is random and uncontrolled by reinforcement in classical schedules. Whatever it is that makes local reinforcement contingencies in a VI schedule fail to operate on behavior in the way they do elsewhere is unidentified and quite mysterious.

Schwartz (1980) has reported additional evidence for a pervasive shaping effect of reinforcement. He has shown that structure and patterning emerges in behavior even when it is not required by the reinforcement contingency. Such results are eminently compatible with the assumption that temporal patterning emerges in behavior maintained by VI and concurrent VI VI schedules, even though these schedules do not require it.

The main difference, it seems, between "ordinary" schedules like VI schedules on the one hand and schedules that control various features of the reinforced local organization of behavior on the other hand, is precisely that the former, the schedules with which we are all so familiar and comfortbale, leave much more up to chance, previous training, biological predispositons, or whatever, in the determination of the behavior that precedes the delivery of a reinforcer. I continue to hope that someone will explain to me why a science ostensibly dedicated to the development of methods to control the behavior of a single organism remains enamored of methods that control behavior in only a sloppy sense, the sense, that is, in which we pay no attention to the local organization of behavior, average over it all, and come up with a single number, mean response rate. If the day ever comes when a general and quantitative behavioral law involving the effects of reinforcement can be expressed, I wish to have gone on record as having predicted it will not be expressed in terms of mean response rate.

REFERENCES

Anderson, J.R., and G.H. Bower. 1974. *Human Associative Memory.* New York: John Wiley.

Anger, D. 1954. The effect upon simple animal behavior of different frequencies of reinforcement. Document 7779, ADI Auxiliary Publications Project. Library of Congress.

Baum, W.M. 1973. The correlation-based-law of effect. *Journal of the Experimental Analysis of Behavior 20*: 137–53.

Catania, A.C. 1973. Self-inhibiting effects of reinforcement. *Journal of the Experimental Analysis of Behavior 19*: 517–26.

Catania, A.C., and G.S. Reynolds. 1968. A quantitative analysis of the behavior maintained by interval schedules of reinforcement. *Journal of the Experimental Analysis of Behavior 11*: 327–83.

de Villiers, P.A. 1977. Choice in concurrent schedules and a quantitative formulation of the law of effect. In W.K. Honig and J.E.R. Staddon, eds., *Handbook of Operant Behavior*, Englewood Cliffs, N.J.: Prentice-Hall.

Estes, W.K. 1959. The statistical approach to learning theory. In S. Koch, ed., *Psychology: A Study of a Science*, vol. 2, New York: McGraw-Hill.

_____. 1978. *Handbook of Learning and Cognitive Processes,* vol. 6, *Linguistic Functions in Cognitive Theory*, Hillsdale, N.J.: Lawrence Erlbaum.

Gage, F.H.; S.H. Evans; and D.S. Olton. 1979. Multivariate analyses of performances in a DRL paradigm. *Animal Learning and Behavior 7*: 323–27.

Hale, J.M., and C.P. Shimp. 1975. Molecular contingencies: Reinforcement probability. *Journal of the Experimental Analysis of Behavior 24*: 315–21.

Hanson, N.R. 1958. *Patterns of Discovery.* Cambridge, England: Cambridge University Press.

Herrnstein, R.J. 1961. Relative and absolute strength of response as a function of frequency of reinforcement. *Journal of the Experimental Analysis of Behavior 4*: 267–72.

____. 1970. On the law of effect. *Journal of the Experimental Analysis of Behavior 13*: 243–66.

Herrnstein, R.J., and D.H. Loveland. 1975. Maximizing and matching on concurrent ratio schedules. *Journal of the Experimental Analysis of Behavior 24*: 107–16.

Heyman, G.M. 1979. Markov model description of changeover probabilities on concurrent variable-interval schedules. *Journal of the Experimental Analysis of Behavior 31*: 41–51.

Kuhn, T.S. 1970. *The Structure of Scientific Revolutions*, 2nd ed. Chicago: University of Chicago Press.

Menlove, R.L. 1975. Local patterns of response maintained by concurrent and multiple schedules. *Journal of the Experimental Analysis of Behavior 23*: 309–37.

Morse, W.H. 1966. Intermittent reinforcement. In W.K. Honig, ed., *Operant Behavior: Areas of Research and Application*, New York: Appleton–Century–Crofts.

Nevin, J.A. 1969. Interval reinforcement of choice in discrete trials. *Journal of the Experimental Analysis of Behavior 12*: 875–85.

____. 1979. Overall matching versus momentary maximizing: Nevin (1969) revisited. *Journal of Experimental Psychology: Animal Behavior Processes 5*: 300–306.

Platt, J.R. 1979. Interresponse-time shaping by variable-interval-like interresponse-time reinforcement contingencies. *Journal of the Experimental Analysis of Behavior 31*: 3–14.

Rachlin, H. 1978. A molar theory of reinforcement schedules. *Journal of the Experimental Analysis of Behavior 30*: 345–60.

Reynolds, G.S., and A. McLeod. 1970. On the theory of interresponse time reinforcement. In G.H. Bower, ed., *The Psychology of Learning and Motivation*, New York: Academic Press.

Schwartz, B. 1980. Development of complex, stereotyped behavior in pigeons. *Journal of the Experimental Analysis of Behavior 33*: 153–66.

Shimp, C.P. 1966. Probabilistically reinforced choice behavior in pigeons. *Journal of the Experimental Analysis of Behavior 9*: 443–55.

____. 1967. Reinforcement of short interresponse times. *Journal of the Experimental Analysis of Behavior 10*: 425–34.

____. 1968. Magnitude and frequency of reinforcement and frequencies of interresponse times. *Journal of the Experimental Analysis of Behavior 11*: 525–35.

____. 1969a. Optimal behavior in free-operant experiments. *Psychological Review 76*: 97–112.

____. 1969b. Concurrent reinforcement of two IRTs: the relative frequency of an IRT equals its relative harmonic length. *Journal of the Experimental Analysis of Behavior 12*: 403–11.

____. 1970. Concurrent reinforcement of two IRTs: Absolute rate of reinforcement. *Journal of the Experimental Analysis of Behavior 13*: 1–8.

____. 1973. Sequential dependencies in free responding. *Journal of the Experimental Analysis of Behavior 19*: 491–97.

____. 1974. Time allocation and response rate. *Journal of the Experimental Analysis of Behavior 21*: 491–99.

____. 1975. Perspectives on the behavioral unit: Choice behavior in animals. In W.K. Estes, ed., *Handbook of Learning and Cognitive Processes*, vol. 2, Hillsdale, N.J.: Lawrence Erlbaum, 1975.

____. 1976a. Organization in memory and behavior. *Journal of the Experimental Analysis of Behavior 26*: 113–30.

____. 1976b. Short-term memory in the pigeon: The previously reinforced response. *Journal of the Experimental Analysis of Behavior 26*: 487–93.

____. 1978. Memory, temporal discrimination, and learned structure in behavior. In G.H. Bower, ed., *The Psychology of Learning and Motivation*, vol. 12, New York: Academic Press.

____. 1979. The local organization of behavior: method and theory. In P. Harzem and M.D. Zeiler, eds., *Advances in Behavior*, vol. 1, New York: Wiley.

Shimp, C.P., and L. Hawkes. 1974. Time-allocation, matching and contrast. *Journal of the Experimental Analysis of Behavior 22*: 1–10.

Silberberg, A.; B. Hamilton; J.M. Ziriax; and J. Casey. 1978. The structure of choice. *Journal of Experimental Psychology: Animal Behavior Processes 4*: 368–398.

Skinner, B.F. 1950. Are theories of learning necessary? *Psychological Review 57*: 193–216.

____. 1976. Farewell my lovely. *Journal of the Experimental Analysis of Behavior 25*: 218.

Staddon, J.E.R. 1968. Spaced responding and choice: A preliminary analysis. *Journal of the Experimental Analysis of Behavior 11*: 669–82.

Staddon, J.E.R.; J.M. Hinson; and R. Kram. 1981. Optimal choice. *Journal of Experimental Analysis of Behavior 35*: 397–412.

Stelmach, G.E. 1976. *Motor Control: Issues and Trends.* New York: Academic Press.

____. 1978. *Information Processing in Motor Control and Learning.* New York: Academic Press.

Stubbs, D.A., and S.S. Pliskoff. 1969. Concurrent responding with fixed relative rate of reinforcement. *Journal of the Experimental Analysis of Behavior 12*: 887–95.

5 THE INTERCHANGEOVER TIME AS A MOLECULAR DEPENDENT VARIABLE IN CONCURRENT SCHEDULES

Alan Silberberg and
John M. Ziriax

Measures of choice behavior can be split into two categories: molar dependent variables that index preference among choice alternatives and molecular dependent variables that decompose preference into its putative constituents. Measures such as relative response rate (responses to an alternative divided by all choices; see Herrnstein, 1961) exemplify the molar category; sequential statistics (probability of a choice as function of the composition of prior choices; see Shimp, 1966) exemplifies the molecular category. Although both types of measure have been used to assess choice, the large majority of work has used molar measures exclusively. We view this as an imbalance and have argued for a heavier emphasis on molecular measures (Silberberg, Hamilton, Ziriax, and Casey, 1978). In the present chapter we elaborate this view by reviewing some of the history of molar and molecular analysis in choice; discussing the inadequacies of the two major response-based dependent variables that have been used in molecular analysis; and advancing a new molecular measure—one based on the distribution of time allocations to an alternative—that may be more successful than the response-based molecular measures in accounting for the dynamics of choice allocation.

A SELECT REVIEW OF MOLAR AND MOLECULAR ANALYSIS

For the last 20 years, choice behavior has been a major topic of interest among operant-oriented researchers. Given the variety of issues to which these students

This research was supported in part by Grant 1-R03-MH35476-01 from the National Institute of Mental Health to The American University.

of behavior might devote their efforts, it is reasonable to ask why this heavy an emphasis on choice behavior. Part of the answer is no doubt empirical: Choice behavior varies in orderly, reliable, and reproducible ways with manipulations in schedules of reinforcement.

Consider for example the prototypic choice study of Herrnstein (1961). In that study food-deprived pigeons were exposed to two variable-interval (VI) schedules of food reinforcement, each of which was associated with a separate response key. A changeover delay (COD) that specified a minimum interval during which reinforcement was unavailable after a changeover from responding on either key was used to minimize schedule interactions. Herrnstein found across a range of concurrent VI VI schedules that the proportion of responses to a key (responses to a key divided by response total) equaled or "matched" the proportion of reinforcements delivered for responding to that key (that is, reinforcements to a key divided by reinforcement total).

Although this finding of matching—that animals distribute their responses in direct proportion to the reinforcements those responses produce—explains much of the interest in concurrent performances, there is also an important theoretical reason for studying choice behavior. As Herrnstein (1970) has noted, a corollary of obtaining matching is that response rate on one of two keys is proportional to the reinforcement frequency on that key (assuming the total reinforcement for both keys is unchanged). This one-to-one correspondence between response rate and reinforcement frequency makes the relative-response-rate measure a viable alternative to absolute rate as the dependent variable for operant research. In fact, Baum and Rachlin (1969) have forcefully argued that relative response rate in a choice procedure equals the relative *value* of the reinforcers maintaining behavior.

Equating relative response rate with relative response strength (Herrnstein, 1970) or relative reinforcing value (Baum and Rachlin, 1969) boldly extends the conceptual domain of the matching law. It suggests that choice theorists have answered a historically significant question in the psychology of learning: What is the appropriate measure of the strength of a reflex (Hull, 1943; Skinner, 1953)? The answer is provided by the dependent variable of the matching law: relative response rate. Moreover, the equation of relative rate with value or strength need not be limited to choice procedures. As Herrnstein (1970) has shown, viewing single-schedule performances as choice situations where the animal selects between the experimenter-arranged reinforcement schedule (say, a VI 1-minute schedule of food reinforcement) and some implicit "environmental" schedule (including preening, defecating, and other activities), permits the extension of the relative-rate metric to describing a wide range of molar response-rate data outside the context of concurrent schedules (see de Villiers and Herrnstein, 1976, for a review).

The matching law's interpretive breadth and power represent a major integrative step toward developing an effective measure of the strength of behavior. For this very reason considerable research effort has been devoted to defining the limits of the procedures and data to which it can be usefully applied. In most of these tests the central question has been whether some version of the matching law (such as that of Baum, 1974, or Herrnstein, 1970) is violated by the relative or absolute response-rate data generated in a particular procedure. The remarkable finding from these tests is that descriptive inadequacies are seldom uncovered (see de Villiers, 1977, for a review).

The consistent confirmation of the matching law offered by these tests establishes the predictive adequacy of the matching law, at least in the contexts in which it has been applied. Yet the reader should keep in mind that the matching law is often viewed as more than a descriptive statistic: It is often claimed that embedding single-key and two-key performances within the matching equation leads to the definition of a reinforcer's relative strength or value (Baum and Rachlin, 1969; Herrnstein, 1970). This definition may, of course, be correct; however, one note of caution is in order. Most of the tests of the matching law discussed previously assess this model's adequacy in predicting response rates, relative response rates or some other *molar* dimension of behavior in single- and concurrent-schedule procedures. The main feature of these molar analyses is that behavior is integrated over time, ignoring local perturbations in, say, the interresponse time (IRT) distributions composing response rate or the choice sequences composing relative response rate. And therein resides a possible problem: The relative-rate statistic used by the matching law is based on an average of an organism's individual choices. For this measure to be more than a descriptive statistic—that is, for it to be isomorphic with the psychological processes of choice allocation—requires a demonstration that matching itself is not due to some more molecular process such as an explicit patterning of response or time allocations to the choice alternatives. For this reason it will be useful to review extant data on molecular assays of choice to see whether matching can be attributable to more molecular processes.

Two Molecular Measures of Choice: Shimp (1966) and Nevin (1969)

Two studies are regularly cited by others in any discussion of molecular tests of the matching law: Nevin (1969), and Shimp (1966). In the Shimp study all response sequences between choice alternatives were recorded and reported as the probability of a left- or right-key choice given a particular prior sequence of choices, such as the probability of a left(L) given the prior response sequence had been LR (right), RL, RR, or LL. Shimp found matching in his procedure at the molar level of relative response rate; however, he claimed his sequential-statistics measure showed this matching result to be due to birds ordering their successive choices so as to maximize the local or momentary likelihood of reinforcement (momentary maximizing). Based on this finding, Shimp argued that matching was a consequence of maximizing.

The Nevin choice study differed not only procedurally from Shimp's but also in terms of the molecular measure used: Where Shimp recorded the probability of a choice given all combinations of prior sequences, Nevin recorded the probability of a changeover between keys as a function of run length to a given key. He found that these changeover-probability curves had a slight negative slope even though the probability of reinforcement for a changeover increased the longer a bird responded to one key in his study. In other words, birds showed a small tendency toward key perseveration with run length; and even though this small perseverative tendency failed to track the local reinforcement probabilities as dictated by momentary maximizing, matching obtained.

Quite obviously the results of these two experiments are contradictory: The Shimp study found evidence that matching was a consequence of molecular control of choice, whereas Nevin found matching with only weak evidence of statistical dependencies in his choice data. Since matching at the level of relative response rate was the only statistic that held invariant across both choice studies, Herrnstein and Loveland (1975) argued that matching is the primary principle of choice allocation even though it might occasionally appear to be a by-product of some molecular response rule (as in Shimp, 1966).

Although the logic of Herrnstein and Loveland's argument is sound, its empirical justification is now suspect. In a 1978 study by Silberberg, Hamilton, Ziriax, and Casey, both the Nevin and Shimp procedures were replicated using two alternative schemes for recording response sequences: probability of a changeover as a function of run length to a key—the molecular statistic used by Nevin; and probability of a left-key choice given all possible combinations of prior response sequences between successive reinforcements—the measure used by Shimp. In Silberberg et al. (1978), it found that when Nevin's changeover-probability measure was used, matching in both the Nevin and Shimp studies seemed to occur without evidence of maximizing (Nevin's result); when, however, Shimp's detailed sequential statistics were used, matching in both studies seemed to be the outcome of a molecular response rule based on maximizing principles (Shimp's result).

This unexpected finding was attributed to the fact that Nevin's molecular measure excluded sequences that did not terminate in a changeover—that is, the very sequences necessary to demonstrate an underlying dependency of choice were the sequences Nevin's measure did not record. Only when all sequences are recorded, as was done by Shimp's sequential statistics, can the true statistical relationship among successive choices be determined. And when this is done both the Nevin and Shimp procedures evidence response patterning consistent with a reinforcement-maximizing molecular response rule.

This interpretation has been questioned by Nevin (1979). In response to the Silberberg et al. report, Nevin reanalyzed the data from his 1969 study in terms of Shimp's sequential-statistics measure. He found strong response patterning but in a fashion consistent neither with the characterization of his data in Silberberg et al. (1978), nor with molecular maximizing. He drew two conclusions from his reanalysis: First, despite the many similarities between his study and the Silberberg et al. replication, some difference between these studies accounts for the different response patterning obtained. And, second, maximizing is not a requisite for matching, since his animals produced matching results without maximizing being evidenced in their choice sequences.

In light of Nevin (1979) we replicated his study a second time, this time with a procedure that was duplicative of the one he used in his original study (Ziriax and Silberberg, unpublished data). We found that early in training birds showed sequential dependencies akin to those reported by Nevin (1979), but that as the number of sessions increased, these probabilities changed to the form predicted by molecular maximizing and reported by Silberberg et al. in the Nevin replication. These cross-session changes in response patterning are consistent with both the Silberberg et al. data and the data from Nevin (1979). Nevin's reanalysis was based on sessions 7–14 of a reconditioning procedure following extinction; the data from Silberberg et al., on the other hand, were from sessions 61–70. Thus, the variable accounting for the empirical discrepancies between Nevin and Silberberg et al. seems to be length of training. If choice is measured preasymptotically during a reconditioning phase following extinction, it is possible to get matching without evidence of maximizing in the microstructure of choice. However, if conditioning is continued to steady-state levels, maximizing emerges as the proper characterization of the microstructure of choice. (See Houston and McNamara, 1981: 386–87, for another explanation of the discrepancy between the Nevin and Silberberg et al. studies.)

The importance of this discussion is that it suggests that molecular maximizing does play a role in governing choice in the Nevin procedure, at least in the steady state. Nevertheless, whether or not choice is explained by molecular maximizing is not the central concern. Rather, it is in demonstrating the superiority of sequential statistics over changeover-probability plots as molecular assays of choice. Regardless of the interpretation of the processes accounting for choice allocation in the Nevin procedure, our claim is that changeover-probability plots obscure the depiction of choice's microstructure by only looking at sequences that end in changeovers. Were one to look only at Nevin's (1969) changeover plots, one might incorrectly conclude that choice allocation was largely stochastic, with a small tendency to response perseveration to an alternative as run length increased. Yet Nevin's reanalysis, like both of our replications of it, showed strong sequential dependencies, a finding demonstrating local structure to choice; and since only sequential statistics demonstrate this molecular control, we argue that they are a better molecular measure of choice than changeover-probability plots.

The point we make here is not new, for our 1978 paper (Silberberg et al.) offered both data (Shimp and Nevin replications) and a simulation (maximizing-with-errors model) showing that molecular maximizing can occur despite changeover-probability curves that show only weak dependencies among choices. Despite these prior demonstrations, enthusiasm for the use and citation of the changeover statistic as a molecular assay appears unabated (Herrnstein, 1979; Herrnstein and Heyman, 1979; Heyman, 1979; Heyman and Luce, 1979; Nevin and Baum, 1980). For this reason we wish to offer another demonstration of its inadequacy. In the next section a new molecular-maximizing model is presented that can produce changeover-probability curves comparable in slope to Nevin's (1969), which demonstrates once again that there is no incompatibility between maximizing at a molecular level and changeover probability curves that fail to be successful at indicating this fact.

Molecular-Maximizing Model that Produces Changeover Curves Similar to Nevin's (1969)

In Silberberg et al. (1978) it was argued that matching in that study's replications of the Nevin and Shimp studies was due to a molecular maximizing process, even though the changeover-probability curves this process produced appeared similar to Nevin's. If this assertion is correct, it should be possible to produce models where choice follows local reinforcement probabilities in a fashion consistent with maximizing and where changeover-probability curves fail to demonstrate this molecular control.

One model with these properties was described in detail in Experiment 1 of Silberberg et al. In this chapter we present a second maximizing model producing matching without strong trends evidenced in its changeover-probability curves. In the second model a stat bird's response rate on concurrent VI 1-minute VI 4-minute schedules without a COD is controlled by a probability generator set at 0.3 that is interrogated by a 0.33-second clock. The output of this probability generator defines the stat bird's free-operant response rate (60 choices/minute). In this model the stat bird has perfect recall of the distributions of interreinforcement intervals for both choice schedules and of the time since the last response to each schedule. With this information the stat bird calculates the relative likelihood of a reinforcer being available for each alternative at each moment it makes a choice. Its choice rule dictates that it always select that alternative more likely to provide a reinforcement. In the case where a choice occurs when neither schedule can provide reinforcement, it selects the same schedule that it had responded to in its prior choice.

This model is called OMNI 1, a title that acknowledges the stat bird's omniscience regarding all past reinforcers and choices. By having perfect recall of these events OMNI bird can optimize its time rate of reinforcement far more effectively than can a bird that has simply learned, say, a momentary-maximizing sequence. Of course, it is not our intent to argue that OMNI bird mimics a real bird's performance. Obviously no animal can share this stat bird's omniscience. Rather, we offer this example as a tour de force, a demonstration that the ultimate molecular maximizer (1) matches, (2) demonstrates the presence of its molecular response strategy through strong sequential dependencies, and (3) cannot demonstrate the presence of a response strategy in terms of a changeover analysis. The consequences of OMNI bird's maximizing-based choice behavior for matching relations, changeover-probability curves, and sequential statistics can be seen in Figure 5-1.

The top panel of Figure 5-1 presents the matching data from this simulation. The solid, striped, and empty bars present respectively the relative reinforcement rate, relative response rate, and relative time to the VI 1-minute schedule. As is clear from this panel, the quality of response and time matching is excellent.

The second panel of the figure presents the changeover-probability curves from this simulation. Two things are noteworthy in these data: First, they are quite representative of the kinds of curves often generated in procedures that produce matching; that is, they show a small tendency toward greater perseveration with increasing run length (as in Silberberg et al., 1978: figure 1, and Nevin, 1969: figure 4). Second, except for perseveration, these curves offer no suggestion that choice is statistically dependent on prior choices.

A different characterization of choice follows from analysis of the sequential statistics from OMNI bird (bottom of Figure 5-1). Unlike the changeover-probability curves, these data show powerful sequential dependencies, demonstrating what we know to be true—that OMNI bird has a response strategy strongly determined by the composition of its prior response sequence.[a] The results from OMNI bird show once again that a simple molecular model can produce matching that by definition is due to molecular control even though little evidence of its operation can be discerned in the changeover-probability data this model generates.

a. The results from this simulation vary from those typically seen in maximizing simulations. For example, the probability of an A is greater after an A than after a B, a result that seems unexpected considering how local reinforcement probability changes with run length to a key. The reason this difference occurs is that OMNI bird responds much more rapidly than the average time to reinforcement availability on either key. Since this stat bird often "knows" that reinforcement is unavailable for a response (remember, it has learned and can recall the duration of all interreinforcement intervals), it allocates these unreinforceable choices to the same key it had responded to last. This strategy leads to high frequencies of response perseveration. This perseveration causes the probability of an A given A to remain high and the probability of an A given B to be lower.

A second result that may perplex is that the probability of an A given AB is higher than given BB. This outcome is due to both perseveration and how sequence probabilities are calculated (the "all trials" technique from Silberberg et al., 1978). For example, in emitting an ABBBBB sequence, the stat bird emits four BB sequences. This increment is not matched by the AB sequence counter, for while there are four ways to get BB from the foregoing sequence, there is only one way to get an AB sequence. Consequently, the probability of an A is lower following a BB sequence than an AB sequence.

Figure 5-1. Matching Data (*top*), changeover-probability curves (*middle*) and sequential statistics (*bottom*) from OMNI 1 simulation.

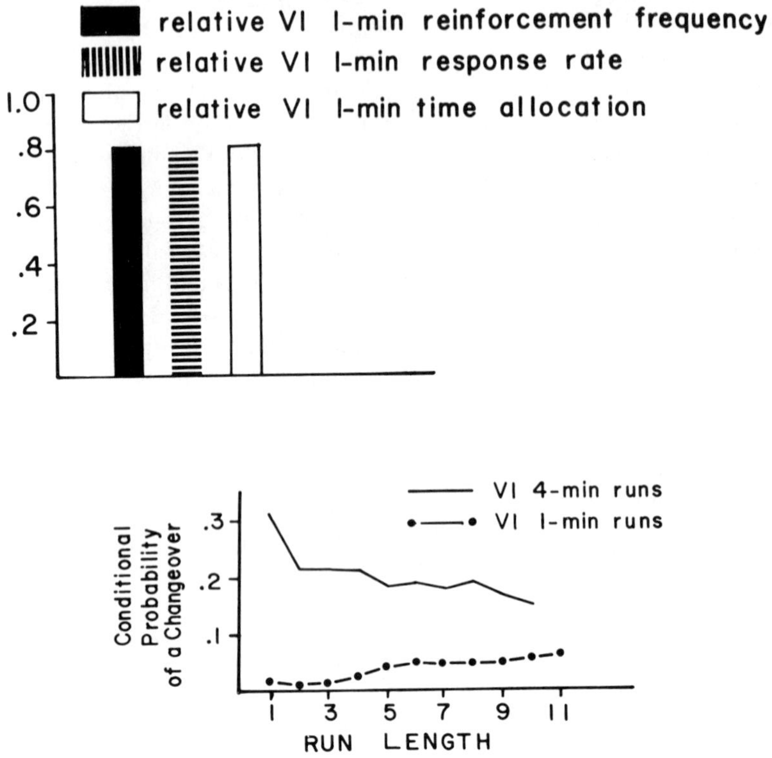

SEQUENTIAL STATISTICS

Probability of an A given:	A	.95
	B	.17
A = VI l-min choice	A A	.95
B = VI 4-min choice	A B	.32
	B A	.98
	B B	.14
	A A A	.95
	B B B	.13

On the Sufficiency of Sequential Statistics. The superiority of sequential statistics as a dependent variable in describing the data from OMNI bird is evident in Figure 5-1. Unlike the changeover-probability measure, these statistics showed in convincing fashion that choice allocation was not probabilistic in this simulation, a result that compels the search for the microstructural bases of the matching relations obtained. Nevertheless, this evidence for the superiority of sequential statistics should not be interpreted as evidence for their sufficiency in defining the molecular processes of choice. The challenge for those interested in the molecular bases of concurrent performances is that on the basis of sequential statistics alone it does not seem possible to synthesize the choice algorithm by which these choice data were generated. In other words, although sequential statistics show that the data from OMNI bird were generated by a response strategy, it is not apparent how these data enable one to arrive at the response rule by which these data were produced.

This problem was also apparent in the five choice procedures presented in Silberberg et al. (1978). In all of these studies statistical dependencies of choice were evident in the sequential-statistics data. This result suggests that the relative response rates produced were due to some molecular response rule, but it was not possible to claim with confidence that their model for synthesizing their molecular and molar data (maximizing with errors and perseveration) was isomorphic with the pigeon's choice rule, even though it reproduced important features of the sequential-statistics data. This conclusion is somewhat embarrassing, for although it shows that sequential statistics can be used to determine whether explicit response patterning occurs between alternatives, it cannot be used to define conclusively the rule by which it occurs.

What is clearly needed is a new dependent variable, one not only sensitive to whether choice is stochastic but also suggesting the response rule by which choice is governed. If properly defined this alternate molecular assay holds the promise of resolving not only whether matching is molecularly controlled but also of defining what that molecular rule might be.

A molecular measure likely to meet these goals is called the interchangeover time (ICT). This measure defines the interval between the first peck to one key and the first peck to the alternate key. Consequently the ICT equals the time spent responding to a key plus the time spent in switching the locus of choice. When presented in terms of a distribution this measure offers a molecular portrayal of the time-allocation functions for each alternative in a choice situation. In the research to be described, we hope to show that ICT distributions have two crucial characteristics required to define the microstructure of choice: (1) ICT distributions are largely invariant in form across different choice procedures used to study matching relations, a result suggesting the operation of a common molecular process; and (2) the form of these functions is similar to the probability-of-reinforcement functions ICTs produce, a result indicating how local reinforcement probability might control the allocation of ICTs between choice alternatives.

As a first step toward illustrating the invariance in form of ICT distributions, Figure 5-2 presents four birds' ICT distributions on concurrent VI 60-second VI 180-second schedules without a COD from Experiment 3 of Silberberg et al. (1978). These data, plotted so as to present the conditional probability of a changeover as a function of time since the last changeover, show that there is a modal time for staying on a key before switching, that the mode is either the same for both keys (B62, B64) or longer for the preferred key (B61, B63), and that the variance in the ICT distributions is generally substantially less for the unpreferred key.

What causes these between-key differences in the form of the ICT distributions? An explanation consistent with molecular maximizing principles would be that responding on the VI 180-second key is of brief duration because only the first peck to the lean key is likely to be reinforced; and conversely, ICTs to the preferred VI 60-second key are of more variable duration because there are no sharp changes in the local reinforcement probability for response runs to that key. In order to assess the adequacy of this maximizing account, concurrent VI 60-second VI 180-second schedules were programmed on a computer. The interreinforcement intervals of each schedule were identical to those from Experiment 3 of Silberberg et al. A stat bird was programmed to peck at the rate of 1 per second, allocating 75 percent of its responses to the VI 60-second key and 25 percent of its responses to the VI 180-second key. Successive choices were statistically independent. After 300 minutes of simulated session time, the average probability of reinforcement for runs to a key of lengths up to 6 was calculated.

The probability of reinforcement for an ICT was calculated by dividing the number of reinforcements provided for a particular ICT by the number of ICTs that were of that duration or longer. For example, if a bird responding on key A were to switch to key B for a period of time and then switched back to key A, the ICT for key B would equal the time between the first peck to key B and the first peck to key A. This B-key ICT distribution would be recorded on a "per opportunity" basis. For example, if the number of ICTs of 1-, 2-, and 3-seconds duration were, respectively, 15, 10, and 5, then the per opportunity distribution would be 30 (15 + 10 + 5), 15 (10 + 5) and 5. Each of these frequencies within the distribution is then divided into the number of reinforcers that occurred during a particular ICT. For example, if 3, 4, and 2 reinforcers, respectively, occurred during the 1-, 2, and 3-second ICTs, then the probability of reinforcement for each of these intervals would be 0.1 (3/30), 0.27 (4/15), and 0.4 (2/5). When plotted in terms of ICT classes, these proportions define the probability-of-reinforcement function.

The results of this first simulation can be found in Figure 5-3 which presents the probability of reinforcement on the VI 60-second and VI 180-second keys as a function of time in the presence of a key (time between changeovers). A comparison of the slopes of this figure's curves with those from Figure 5-2

Figure 5-2. Conditional probability of a changeover as a function of time between changeovers for concurrent VI 60-second VI 180-second schedules from Silberberg et al., 1978: experiment 3.

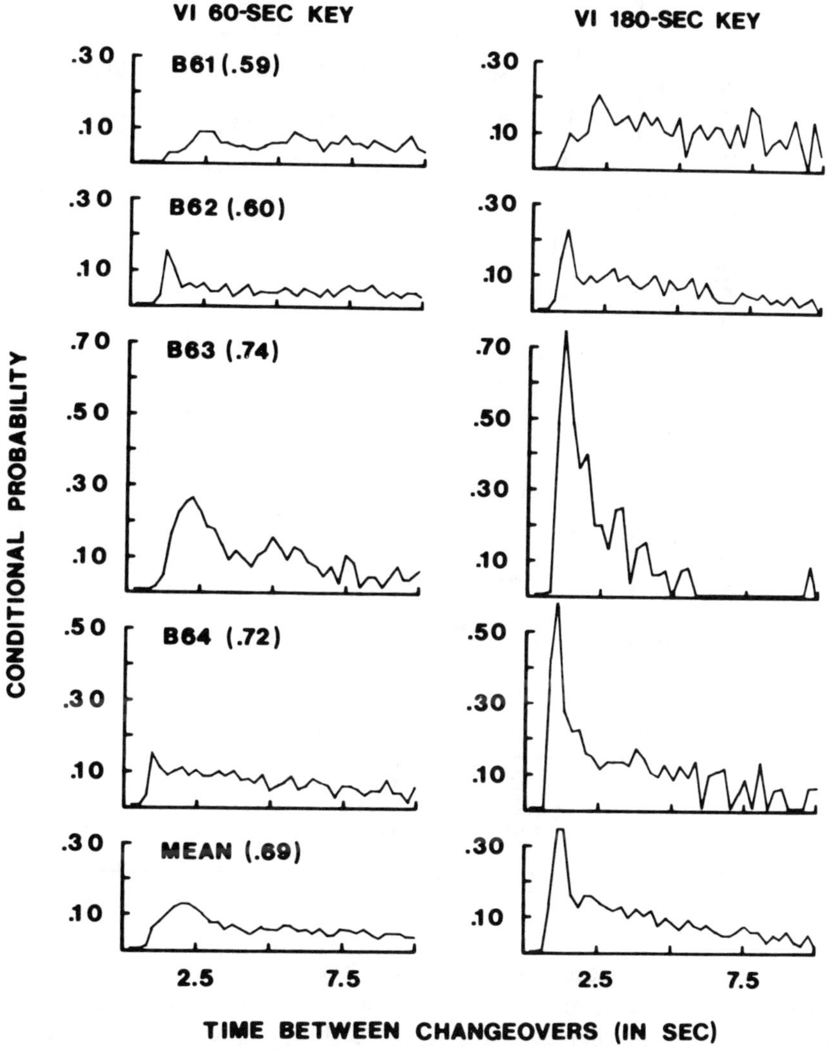

Figure 5-3. Probability of reinforcement on concurrent VI 60-second VI 180-second schedules as a function of time between changeovers from a computer simulation.

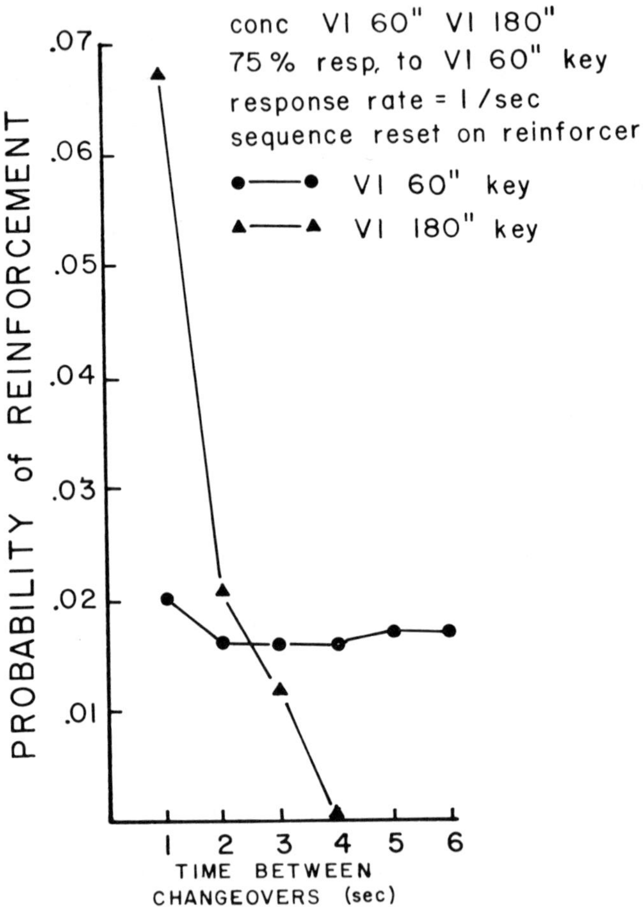

underscores an interesting correspondence: The probability of an ICT (Figure 5-2) on a key appears to follow the local probability of reinforcement on a key (Figure 5-3). If appearance conforms with reality, this is a new demonstration of the molecular control of choice: Birds terminate lean-key runs immediately because the local probability of reinforcement for longer runs is nil; conversely, the conditional probability of an ICT on the rich key drops slowly with time in the presence of that key because the probability of reinforcement for longer times also drops slowly.[b]

In Experiment 4 of Silberberg et al. ICT distributions were plotted during the initial links of concurrent chain VI 120-second VI 40-second versus chain VI 120-second VI 80-second schedules. These data are presented in Figure 5-4. These results are of interest for two reasons: They duplicate the between-key asymmetry evident in Figure 5-2's ICT distribution on another choice procedure, a result suggesting that the finding of peaked ICTs only on the lean key is of some generality, and they suggest that these ICT asymmetries are not a direct consequence of between-key differences in the distribution of interreinforcement intervals composing the choice alternatives. This conclusion follows from the fact that in concurrent chains ICTs are measured only during the initial links where the choice schedules are identical.

The relation we are noting between the distribution of ICTs and the reinforcements they produce has also been noted by Menlove (1975). However, Menlove viewed the correspondence as artifactual, because it seemed to him the obtained distribution of reinforcements must follow the distribution of ICTs. For example, if 90 percent of a bird's ICTs are of 2-seconds duration, it may seem to follow that about 90 percent of the reinforcers will occur for 2-second ICTs. Nevertheless, we believe this conclusion was incorrect, for although there is a dependence between ICTs and reinforcement, the data from Figure 5-3 show it is by no means as forced as the preceding example suggests. In Figure 5-3 the probability-of-reinforcement curves were generated by stochastic choice allocation. As a consequence of stochastic choice the conditional probability of an ICT was constant as a function of time spent responding to a key, a circumstance that would be reflected by straight lines with a slope of zero were the simulated ICT distributions presented. Yet the probability-of-reinforcement curves in Figure 5-3 are decidedly different from these flat ICT curves. And although these ICT curves differ from the probability-of-reinforcement curves when choice is independent, this difference is not maintained in well-trained animals. As a comparison of Figures 5-2 and 5-3 makes clear, steady-state allocation of ICTs closely conforms with the local probability of reinforcement for times spent responding to a key. This result strongly suggests that the ICT functions characterizing choice in free-operant concurrent chains schedules (see Figures 5-2 and 5-4) are due to control by the local probability of reinforcement.

b. Figure 5-3 probability-of-reinforcement curves were produced by stochastic choice allocation—a response rule consistent not with molecular control of choice but with molar matching. Because these curves are based on *mean* probabilities, their form is far more sensitive to the degree of preference in the choice data than to the order of choices that produced this preference. Consequently, a molar matching choice algorithm was used not because it characterizes the microstructure of choice, but simply because it is easy to program and produces curves not substantially different from those produced by an acceptable alternate molecular response rule.

Figure 5–4. Conditional probability of a changeover as a function of time between changeovers for concurrent chain VI 120-second VI 40-second versus chain VI 120-second VI 80-second schedules from Silberberg et al., 1978: experiment 4.

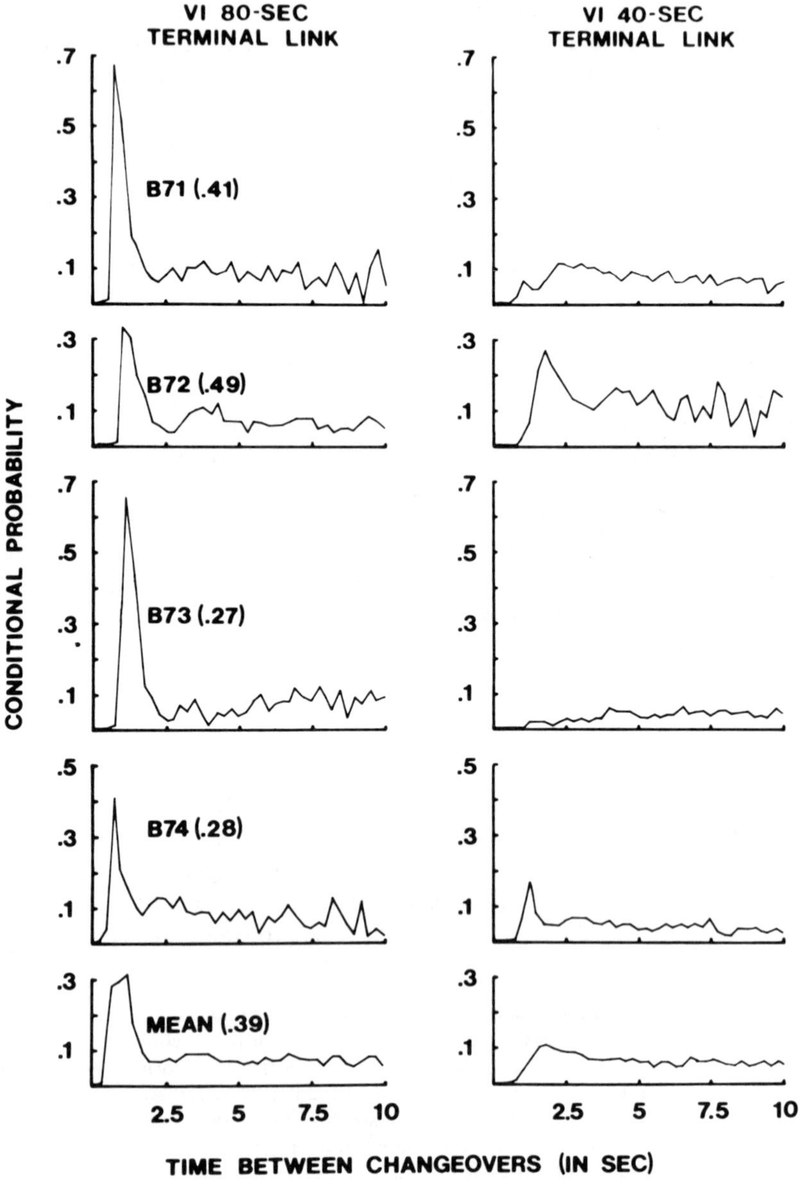

Still unanswered is why the probability-of-reinforcement curves show the between-key asymmetries evident in Figure 5-3. That the preferred-key curve is relatively flat seems consistent with a priori expectations because it is the property of VI schedules that local reinforcement probabilities do not vary substantially as a function of time spent in their presence. Needing more evaluation, however, is the sharply peaked curve for the lean key in Figure 5-3. This peaked curve is a direct consequence of preference for the rich key (this statement says nothing about the response rule by which this preference is produced; see footnote b). This outcome seems reasonable in terms of how concurrent VI VI schedules are typically arranged, with both tapes running regardless of which choice alternative is being sampled. The greater the preference for the rich key, the longer the rich key's mean ICT, and, consequently, the higher the mean probability of reinforcement for a changeover to the lean key. Yet this elevated lean-key reinforcement probability obtains only for the first changeover peck. Subsequent responding on the key is only infrequently reinforced because the associated VI schedule is itself lean. Therefore, the greater the preference for the rich key, the higher the average probability of reinforcement for the first lean-key peck and the more precipitous the drop in reinforcement probability for subsequent responding on that key.

In order to test this notion more thoroughly, the probability of reinforcement for ICTs to each of two choice alternatives was assessed by computer simulation. Four simulations were run that were identical to the simulation of concurrent VI 60-second VI 180-second schedules described except for the changes noted in Table 5-1. For the first three simulations listed in the table, reinforcement for both choice alternatives was provided by two independent VI 60-second schedules while the relative duration of hopper presentation was varied. The stat birds' relative response rate was programmed to match in terms of relative hopper duration (see Catania, 1963). The fourth simulation in Table 5-1 arranged a right-key preference even though the left key provided twice as many reinforcers. A right-key preference of 0.67 was scheduled because the hopper duration was four times greater on the VI 120-second key than on the VI 60-second key.

Table 5-1. Simulated schedule contingencies and simulated relative rate.

	VI Schedule (seconds)		*Hopper Duration (seconds)*		
Simulation Number	*Left Key*	*Right Key*	*Left Key*	*Right Key*	*Relative Left-Key Response Rate*
1	60	60	4	4	.5
2	60	60	6	3	.67
3	60	60	9	3	.75
4	60	120	3	12	.33

Figure 5-5. Probability-of-reinforcement curves from a computer simulation where preference is varied.

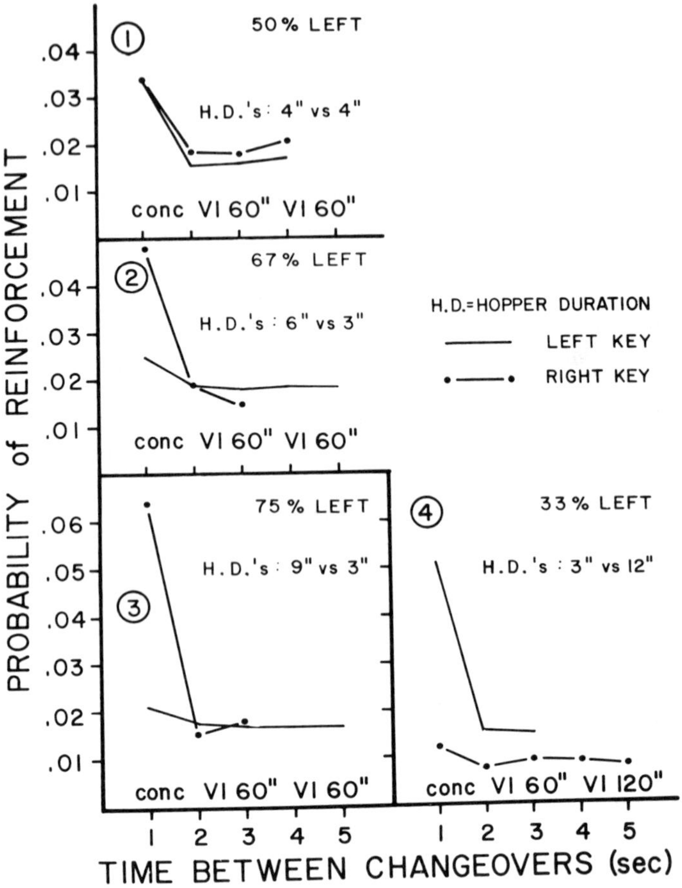

Plots of the probability of reinforcement for ICTs through lengths of 5 seconds for each key are presented in Figure 5-5. In the instances where data are not plotted for ICTs of less than 5 seconds, there were insufficient data in the simulated session length (300 minutes) to produce stable estimates of the probability of reinforcement. The obtained probability of reinforcement on the left and right keys is signified, respectively, by solid lines and dots joined by lines in the curves of Figure 5-5. The programmed relative left-key response rate, the duration of hopper presentations (H.D. in Figure 5-5) for a left- and right-key response, and the values of the concurrent VI VI schedules in effect are presented within each panel. The circled numbers in each panel conform with the simulation number listed in Table 5-1.

As the curves in Figure 5-5 make clear, there is an interaction between preference and the local probability of reinforcement: The greater the preference for a VI schedule, the smaller the local changes in reinforcement probability on that key and the greater the local changes on the alternate key. If our thesis concerning these curves is correct—that changes in their form produce equivalent changes in their associated ICT distributions, the following predictions can be made: (1) Preferred-key ICT distributions will be flat or have a small mode at short ICT values, whereas the unpreferred-key distributions will have a negative slope with a sharp peak at short ICT values. (2) The greater the preference the more exaggerated this asymmetry will be.

These predictions are verified in the ICT data from Figures 5-2 and 5-4. In Figure 5-2 B61 and B62 show the smallest preference for the VI 60-second key (the relative VI 60-second response rate is in parentheses beside the bird number), and these birds appear to have the smallest between-key asymmetries with each distribution's curves being relatively flat. B63 and B65, on the other hand, show stronger preferences for the VI 60-second key, and their curves are more asymmetric with minor modes on the preferred key while the lean-key curves are sharply peaked. The same argument seems to hold for the data in Figure 5-4. Only one bird, B72, was essentially indifferent between choice alternatives, and this bird was the only subject in the concurrent-chains study not to show clear-cut asymmetries in between-key ICT-distribution comparisons.

That the ICT distributions in Figures 5-2 and 5-4 are likely to conform in shape with the probability-of-reinforcement curves each subject's preference generates suggests that the local characteristics of time allocations to the two choice alternatives are governed by their local likelihoods of reinforcement. A problem with this analysis, however, is that it compares pigeon data (ICT curves) with simulated data (probability-of-reinforcement curves). Before concluding that the putative dependent and independent variables correspond, we must show these correspondences in an actual choice experiment. Toward this end we programmed ICT and reinforcement-distribution measures into an ongoing experiment in our laboratory. The experiment is a literal duplication of the Nevin (1969) study mentioned earlier. In brief, concurrent arithmetic VI 1-minute VI 3-minute schedules are presented on a 6-second trials basis. Every 3 minutes, key colors and their associated schedules alternate between the two side keys. Trials terminate with a peck or after 2 seconds, whichever occurs first. After completing 81 30-minute sessions, the ICT and reinforcement-distribution measures were introduced for the next 10 sessions.

The four-bird group data using these measures can be found in Figure 5-6. This figure presents the probability of reinforcement (top two panels) and the conditional probability of a changeover (bottom two panels) as a function of time in the presence of a key. The left-side panels present these data from a

Figure 5-6. Probability of reinforcement (*top*) and the conditional probability of a changeover (*bottom*) as a function of time between changeovers, from Nevin replication and a computer simulation.

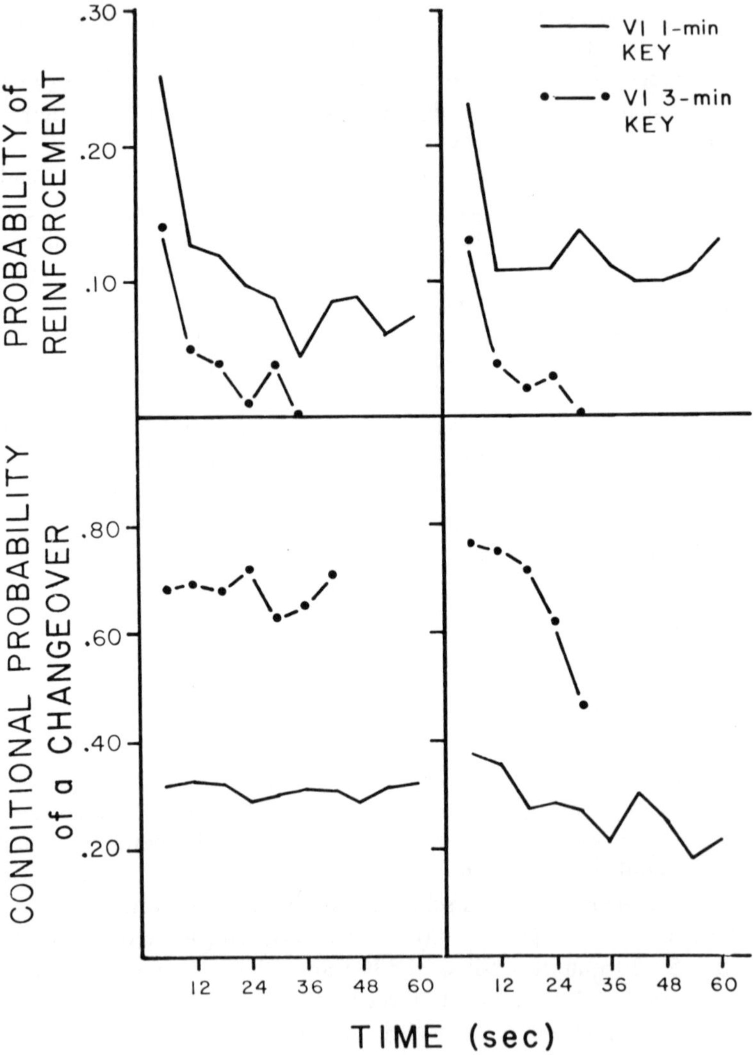

100-session computer simulation where choices in the Nevin procedure were allocated on a response-independent basis so that the simulated relative response rate equaled the four-bird group with a mean preference of 0.71. The right-side panels present the group data. ICT and probability-of-reinforcement curves representing VI 1-minute and VI 3-minute performances are signified, respectively, by solid lines and dots joined by lines.

The curves that map the probability of reinforcement for the stat bird and the group data are similar in form. Both sets of data show that the probability of reinforcement drops with increasing ICTs and that the drop is somewhat sharper for the lean key. This finding is of interest because it shows that the simulated probability-of-reinforcement curves do correspond closely with those produced by real subjects. As regards the ICT data, the slopes of the ICT curves from the group data bear a rough correspondence with the slopes of their probability-of-reinforcement curves; however, this correspondence is not mirrored in the simulation data, where the ICT curves are flat while the reinforcement curves trend downward. These findings establish there is no necessary correspondence between the slopes of the probability-of-reinforcement and ICT functions—a point that contradicts Menlove's 1975 interpretation of the forced relation between these variables and lends credence to our claim that the form of the ICT function depends on the form of the probability-of-reinforcement function.

Summary

We have noted that the two major measures used to characterize the microstructures of choice—changeover-probability curves and detailed sequential statistics—are inadequate. As regards the changeover-probability statistic, it is often impossible to answer whether explicit response patterning is occurring with this measure. We showed this in a computer simulation where choice was governed by a maximizing molecular response rule (see also maximizing-with-errors model from Silberberg et al., 1978). In this simulation matching obtained with changeover-probability curves that were relatively flat—a result that has been interpreted by others (like Heyman, 1979) as evidence that successive choices are statistically independent. Yet this interpretation must be in error because this simulation's response rule did not allocate choice stochastically. A clearer picture of the microstructure of choice was provided by detailed sequential statistics, for when the same data were analyzed in terms of this measure, the true, nonstochastic nature of the response-allocation rule was apparent (see bottom of Figure 5-1). Nevertheless, this measure was also inadequate, for while it correctly demonstrated the presence of explicit response patterning, it shed no light on the molecular choice algorithm by which these data were generated. In other words, sequential statistics tell half the story—whether or not a molecular strategy is likely to be present. Unfortunately, they give few clues as to what that strategy might be.

Because of the inadequacies of these prior measures, a new assay based on the conditional probability of an ICT was advanced. This measure proved capable not only of uncovering molecular order in choice data but also of providing a context for its interpretation: The consistent congruence between ICT and probability-of-reinforcement functions suggested that the times allocated to a schedule are controlled by their local likelihoods of reinforcement, and that this correspondence might define the molecular basis of choice.

REFERENCES

Baum, W.M. 1974. On two types of deviation from the matching law: Bias and undermatching. *Journal of the Experimental Analysis of Behavior 22*: 231–42.

Baum, W.M., and H.C. Rachlin. 1969. Choice as time allocation. *Journal of the Experimental Analysis of Behavior 12*: 861–74.

Catania, A.C. 1963. Concurrent performances: A baseline for the study of reinforcement magnitude. *Journal of the Experimental Analysis of Behavior 6*: 299–300.

de Villiers, P.A. 1977. Choice in concurrent schedules and a quantitative formulation of the law of effect. In W.K. Honig and J.E.R. Staddon, eds., *Handbook of Operant Behavior*, Englewood Cliffs, N.J.: Prentice-Hall.

de Villiers, P.A., and R.J. Herrnstein. 1976. Toward a law of response strength. *Psychological Bulletin 83*: 1130–53.

Herrnstein, R.J. 1961. Relative and absolute strength of response as a function of frequency of reinforcement. *Journal of Experimental Analysis of Behavior 4*: 267–72.

_____. 1970. On the law of effect. *Journal of the Experimental Analysis of Behavior 13*: 243–66.

_____. 1979. Derivatives of matching. *Psychological Review 86*, 486–95.

Herrnstein, R.J., and G.M. Heyman. 1979. Is matching compatible with reinforcement rate maximization on concurrent VI VR? *Journal of the Experimental Analysis of Behavior 31*: 209–23.

Herrnstein, R.J., and D.H. Loveland. 1975. Maximizing and matching on concurrent ratio schedules. *Journal of the Experimental Analysis of Behavior 24*: 107–16.

Heyman, G.M. 1979. A Markov model description of changeover probabilities on concurrent variable-interval schedules. *Journal of the Experimental Analysis of Behavior 31*: 41–51.

Heyman, G.M., and R.D. Luce. 1979. Operant matching is not a logical consequence of maximizing reinforcement rate. *Animal Learning and Behavior 7*: 133–40.

Houston, A.I., and J. McNamara. 1981. How to maximize reward rate on two variable-interval paradigms. *Journal of the Experimental Analysis of Behavior 35*: 367–96.

Hull, C.L. 1943. *Principles of Behavior.* New York: Appleton-Century-Crofts.

Menlove, R.L. 1975. Local patterns of responding maintained by concurrent and multiple schedules. *Journal of the Experimental Analysis of Behavior 23*: 309–37.

Nevin, J.A. 1969. Interval reinforcement of choice behavior in discrete trials. *Journal of the Experimental Analysis of Behavior 12*: 875–85.

____. 1979. Overall matching versus momentary maximizing: Nevin (1969) revisited. *Journal of Experimental Psychology: Animal Behavior Processes 5*: 300–306.

Nevin, J.A., and W.M. Baum. 1980. Feedback functions for variable-interval reinforcement. *Journal of the Experimental Analysis of Behavior 34*: 207–17.

Shimp, C.P. 1966. Probabilistically reinforced choice behavior in pigeons. *Journal of the Experimental Analysis of Behavior 9*: 433–55.

Silberberg, A.; B. Hamilton; J.M. Ziriax; and J. Casey. 1978. The structure of choice. *Journal of Experimental Psychology: Animal Behavior Processes 4*: 368–98.

Skinner, B.F. 1953. *Science and Human Behavior.* New York: Macmillan.

6 SOME PERSISTENT ISSUES IN THE STUDY OF MATCHING AND MAXIMIZING

John A. Nevin

Herrnstein (1961) arranged that hungry pigeons could obtain food by pecking either of two continuously available keys. Reinforcers were scheduled independently on each key by concurrent variable-interval (VI) schedules, in which the first response after an interval of varied length is reinforced. A brief changeover delay prevented reinforcement for a peck at one key immediately after switching away from the other. On each of several pairs of VI schedules, the relative frequency of pecks at one key matched the relative frequency of reinforcement obtained on that key. This simple and elegant result has been replicated in several species, including human beings, with a variety of responses and reinforcers (see de Villiers, 1977, for review). Accordingly, it has taken on the status of a general law of behavior.

A seemingly different result emerges in situations where two responses are reinforced according to variable-ratio (VR) schedules, in which the first response after a variable number of responses is reinforced. For example, Herrnstein and Loveland (1975) arranged different VR schedules concurrently for pecks at two response keys. When the ratios differed by a factor of 2 or more, they observed virtually exclusive preference for the key with the shorter average ratio and hence the higher probability of reinforcement for each response. The latter result, which maximizes both momentary and overall payoff, has been obtained with several species, including human beings (as by Siegel and Goldstein, 1959), in a variety of experimental settings (see Shimp, 1966, for ex-

Supported in part by National Science Foundation Grant 76-11028 to the University of New Hampshire. I am indebted to my colleague W. M. Baum for discussion of the ideas presented here.

ample). Such findings reinforce the well-nigh universal assumption that organisms act to maximize reinforcement (see Herrnstein, Chapter 19 of this volume, for interpretation of such findings in relation to matching).

The question addressed by authors in this volume, in one form or another, is the relation between matching and maximizing. My approach will be to ask whether matching is derivative from maximizing processes that most experimenters assume, at least informally, to be fundamental. To begin, two levels of analysis, widely termed molar and molecular, will be distinguished. In a molar analysis, responses and reinforcers are aggregated over large blocks of time (e.g., entire sessions), without regard for particular sequences of responses, times since reinforcement, or other local properties of choices between concurrently available reinforcers. This is the level of aggregation at which the matching law is stated. A molecular analysis, by contrast, aggregates data according to particular sequences of events, such as prior responses or reinforcers, to ascertain the nature and determiners of local patterns of responding. Both levels use the same data—responses and reinforcers—so they are better termed levels of aggregation than levels of analysis. A molar aggregation can be constructed from molecular data, but it is rarely possible to infer the properties of molecular aggregations from molar data.

At the molar level, maximizing implies an overall allocation of responding that yields the maximum possible overall rate of reinforcement (See Chapters 16 and 18). In the standard concurrent VI VI experiment, the subject must distribute its behavior across both keys in order to achieve the maximum rate of reinforcement, because exclusive responding produces only those reinforcers assigned to one alternative. A number of researchers have shown that the maximum total rate of reinforcement may coincide with matching to relative reinforcement in concurrent VI VI schedules (Baum, 1981; Rachlin, Green, Kagel, and Battalio, 1976; Staddon and Motheral, 1978). This coincidence raises the question of whether maximizing or matching is the primary behavioral process at the molar level. Several researchers in the present volume have addressed the issue experimentally (Herrnstein, Heyman, Vaughan) and have found that matching occurs when overall reinforcement rate is constant or when a different allocation of behavior would achieve a still higher rate of reinforcement. This line of work suggests that the coincidence of matching with the maximizing of overall rate of reinforcement in concurrent VI VI schedules is fortuitous, and that molar maximizing is not necessary for matching.

The molar results do not rule out the possibility that overall matching arises from more molecular maximizing processes. Shimp (1966) proposed that subjects may allocate each successive response to that alternative with the higher momentary probability of reinforcement; he demonstrated by computer simulation that a momentary maximizing strategy of this sort would lead to overall matching in concurrent VI VI schedules. Moreover, he analyzed sequences of responses in a discrete-trial choice experiment that mimicked interdependent concurrent VI VI schedules (cf. Stubbs and Pliskoff, 1969), and found that pigeons may roughly approximate the ideal momentary maximizing strategy. Silberberg, Hamilton, Ziriax, and Casey (1978) replicated Shimp's (1966) results, and suggested that processes such as perseveration and forgetting could explain departures from perfect conformity to momentary maximizing. Both Shimp and Silberberg et al. have argued that the usual molar matching result is an artifact of pooling response sequences structured at a local level according to a maximizing principle.

To see how local maximizing might operate in concurrent VI VI schedules, consider what happens while a subject is responding to one alternative (key A). The probability of reinforcement for a switch to the other alternative (key B) grows steadily because its schedule runs continuously, and the likelihood that a key B reinforcer is available increases with time. At some point the probability of reinforcement for switching will exceed that for continued responding to key A, and a switch to key B should occur. Then, the probability of reinforcement for switching back to key A will grow while the subject responds on key B, eventually making reinforcement for switching back more probable than for continuing, and so forth. This aspect of concurrent VI VI schedules suggests that if responding is controlled by local reinforcement probabilities, switching should become more probable as time spent responding to one alternative increases.

Relevant data do not always show such a pattern, however. Herrnstein (described in de Villiers, 1977) and Heyman (1979) reported that the probability of switching was essentially stationary as a function of number of pecks on one key, and I found that it tended to decrease (Nevin, 1969). All three studies obtained excellent overall matching, suggesting that this sort of local maximizing strategy is not necessary for matching to occur. My 1969 study will be described briefly, because it has become the focus of disagreement in this problem area. Three pigeons were trained on a two-key, discrete-trial choice procedure in which reinforcers were assigned to a green key by a VI 1-minute schedule and to a red key by an independent VI 3-minute schedule. The keys were lighted after a 6-second intertrial interval and went off after a peck at either key or after 2 seconds if no peck occurred. Dark-key pecks reset the intertrial interval. Red and green alternated from left to right every 3 minutes. Sessions lasted 30 minutes. The pigeons received 60 sessions of training, 10 sessions of extinction, and 15 sessions of retraining. Detailed data were taken for the final 8 sessions. The major result was that the probability of switching from green to red decreased with successive pecks at green, even though the probability of reinforcement actually obtained for switching increased markedly relative to that for continuing on green (see Nevin, 1969: fig. 4). At that time I concluded, perhaps prematurely, that momentary maximizing was not necessary for matching.

Two kinds of arguments have been leveled against my analysis and conclusions. First, Silberberg et al. (1978) pointed out, quite correctly, that the 1969 analysis was based on only some response sequences—namely, those that ended with a switch. In Chapter 5 Silberberg and Ziriax argue convincingly that a horizontal or decreasing function relating the probability of switching to the number of successive pecks at a key does not imply the absence of a maximizing strategy. Shimp (1966) had examined a larger set of sequential statistics, and Silberberg et al. (1978) had examined the full set of possible dependencies in sequences up to 5 in length in arriving at the conclusion that their birds tended to maximize. Moreover, Silberberg et al. (1978) confirmed my observation that the probability of switching decreased with consecutive pecks at one alternative and found that a momentary maximizing pattern coexisted with this trend. Thus, both the logic of computer simulation and the behavior of Silberberg's pigeons demonstrate that more complete analyses than the one originally reported are required to identify maximizing patterns. Accordingly the original data were reanalyzed and all uninterrupted response sequences up to 4 in length were extracted (Nevin, 1979). The patterns were quite different from those reported by Silberberg et al. (1978), and gave no evidence of maximizing trends. For example, the average probability of pecking green after a green-key peck was 0.79, and the average probability of pecking green after a red-key peck was 0.72, while the overall average probability of pecking green was 0.75. These data suggest that the birds tended to perseverate on whichever key they had been pecking. Perseveration of this sort is explicitly contrary to a momentary maximizing expectation. To appreciate this it is necessary to know the probabilities of reinforcement for perseverating and for switching. Pecking green after green was reinforced with probability 0.15; pecking red after green was reinforced with probability 0.20. Conversely, pecking red after red was reinforced with probability 0.03, and pecking green after red was reinforced with probability 0.25. (These data are summarized in Table 6-1.) Clearly, the local probabilities of reinforcement for switching were higher than for perseverating; nevertheless the birds perseverated. My 1979 reanalysis considered all possible sequences up

Table 6-1. First-order sequential statistics for responding and reinforcement in a simulation of Nevin's (1969) study, assuming that responding always follows the pattern GGGRGGGR. . . . Results of the study itself, as reanalyzed by Nevin (1979), are included for comparison.

Prior Response	*Probability of Pecking G*	*Probability of Reinforcement*		*Relative Probability of Reinforcement*	*Relative Frequency of Reinforcement*
		G	*R*		
Simulation					
G	.65	.18	.18	.50	.65
R	1.00	.26	.00	1.00	1.00
Nevin data					
G	.79	.15	.20	.42	.74
R	.72	.25	.03	.88	.96

to length 4 and showed that the probability of emitting a sequence was negatively correlated with the relative probability of reinforcement for that sequence, as in the foregoing example. This confirmed my earlier conclusion that conformity to a momentary maximizing strategy is not necessary for overall matching.

The issue is complicated by the fact that there is a forced relation between perseveration and momentary maximizing, given that overall matching occurs. To see this, consider the matrix in Table 6-2, for first-order dependencies in all two-response sequences. By definition, the overall probability of pecking green is

$$\frac{a+c}{a+b+c+d},$$

where a is the number of green pecks following a green peck, c is the number of green pecks following a red peck, and so on. The conditional probability of pecking green on trial $N+1$, given a peck at green on trial N, is $a/(a+b)$, and the conditional probability of pecking green on trial $N+1$, given a peck at red on trial N, is $c/(c+d)$. The following constraints must operate:

1. $a+b+c+d$ = number of trials with a peck;
2. $b = c \pm 1$, because the total number of switches must be the same in both directions (± 1); and
3. $\dfrac{a+c}{a+b+c+d}$ = relative number of reinforcers obtained on green, because overall matching is assumed.

To exemplify the consequences of these constraints, assume 1,000 trials with a peck, and 0.75 for the relative number of reinforcers obtained on green, as expected with concurrent VI 1-minute, VI 3-minute schedules. (The value of 0.75 is based on the assumption that 60 reinforcers per hour are obtained on green, and 20 per hour on red. Actual number would depend somewhat on

Table 6-2. Matrix of first-order dependencies.

Response on Trial N		Response on Trial $N+1$: G	R
	G	a	b
	R	c	d

switching rates and patterns as will be discussed.) If $(a + c)/(a + b + c + d) = 0.75$, and $a + b + c + d = 1{,}000$, then $a + c = 750$ and $b + d = 250$. Thus b cannot exceed 250, and the maximum possible number of switches per 1,000 trials, $b + c$, is 500. This leads to the first-order transition matrix of Table 6-3. If the 500 switches are distributed uniformly throughout the 1,000-trial session, behavior must follow the sequence GGGRGGGRGGGR . . . , which is the momentary maximizing sequence when interruptions by reinforcement are ignored and reinforcers are scheduled by two independent VI tapes. The resulting conditional probabilities of pecking green given green and pecking green given red are 0.67 and 1.0, respectively.

Variation in the overall rate of switching leads to systematic changes in the conditional probabilities of responding. The conditional probability of pecking green given green is $a/(a + b)$. Because $a + b = 750$, this may be rewritten as $(750 - b)/750$, which decreases linearly from 1.0 to 0.67 as b goes from 0 to 250. The conditional probability of pecking green given red is $c/(c + d)$. Because $c + d = 250$, this may be rewritten simply as $c/250$, which increases linearly from 0.0 to 1.0 as c goes from 0 to 250. The relations between the two conditional probabilities of pecking green and the overall switching rate are portrayed in Figure 6-1. Clearly, the overall rate of switching will determine whether the data will exhibit momentary maximizing, with $p(G|G) < p(G|R)$, or perseveration, with $p(G|G) > p(G|R)$.

High switching rates may occur for reasons other than momentary maximizing. Heyman and Luce (1979) have demonstrated that the sum of rates of reinforcement obtained on independent concurrent VI VI schedules must increase as switching rate increases. This may be seen intuitively by noting that if the subject switches only once, three-quarters of the way through a session, it will obtain about three-quarters of all the reinforcers available per hour on the VI 1-minute schedule (say, 45), and one-quarter of the reinforcers available per hour on the VI 3-minute schedule (say, 5). The total, 50 reinforcers per hour, is

Table 6-3. Matrix of first-order dependencies assuming 500 switches in 1,000 trials and matching to relative reinforcement at 0.75.

Response on Trial N		Response on Trial $N + 1$: G	R
	G	500	250
	R	250	0

Figure 6-1. The conditional probability of pecking green given a prior peck to green and of pecking green given a prior peck to red, as functions of the total number of switches. The shaded area represents momentary maximizing. The constraints determining these functions are described in the text.

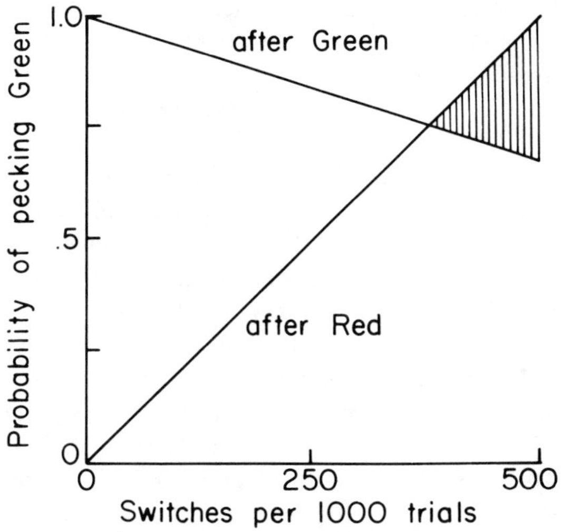

substantially less than the 80 per hour that could be obtained by switching frequently. Thus a molar maximizing tendency of the sort described would operate to establish and maintain high switching rates.

Earlier it was argued that the overall allocation of choice responding is not controlled by molar maximizing. It is a separate, empirical question as to whether switching rates are controlled by the maximizing of overall reinforcement rate. The possibility also exists that overall switching rates are determined in part by tendencies toward overall matching. In the foregoing example there is a substantial departure from matching when only one switch occurs per session: the relative allocation of responding is 0.75, but relative obtained reinforcement is 0.90. High switching rates may be established and maintained because very low rates force departures from matching.

It is tempting to suggest that momentary maximizing is nothing more than an artifact of whatever factors determine switching rate. It is more correct, though, simply to observe that first-order sequential dependencies and overall switching rates are inextricably linked. The empirical question is whether overall matching is necessarily correlated with high switching rates and hence with local maximizing patterns. The answer, suggested in Nevin (1969 and 1979), is that overall matching may appear despite relatively low switching rates and hence perseveration.

A second line of argument against the interpretation of my 1969 data has been advanced by Shimp (1979 and Chapter 4). He suggests that the overall proportion of session time allocated to a particular number of responses between switches must match the frequency of reinforcement for a switch after that number of responses, relative to all reinforcers for switching.[a] In this respect the behavior of a pigeon cannot be distinguished from the behavior of a ping-pong ball that bounces about and occasionally strikes the keys. In view of this "ping-pong-ball syndrome," Shimp argues that no empirically meaningful statements can be made about the relation between switching and the relative rate of reinforcement for switching.

Shimp's argument is correct if only those reinforcers obtained for switching are counted but may not be if reinforcers obtained for continued responding at a key are also counted. One way to test Shimp's argument is to ask how different patterns of key-pecking affect probabilities of reinforcement for switching relative to continued responding at a key. To this end, a simulation of my 1969 study was conducted. It assumed that key pecking was always patterned according to the momentary maximizing sequence GGGRGGGR . . . . Specifically, independent concurrent arithmetic VI 1-minute and VI 3-minute schedules of reinforcement with 10-second minimum intervals were simulated. It was assumed that trials occurred every 10 seconds (an approximation to my 1969 procedure in which dark-key pecks reset the 6-second intertrial interval). The simulation ran until 240 reinforcers were obtained. The simulation data were analyzed for all uninterrupted sequences of two key pecks and the frequencies of reinforcement that they produced. The outcome of the simulation is presented in Table 6-1, which also shows the mean data of the three pigeons that served in my 1969 study. Clearly, the response data differ radically (as they must given the differences between assumed and actual patterns of behavior), while the obtained reinforcement data differ only trivially.

The comparison of the real and simulated reinforcement data suggests that the obtained probabilities and frequencies of reinforcement for switching, relative to those for continuing on a key, are largely determined by the contingencies of concurrent VI VI schedules, and that changes in response patterns, given overall matching, have relatively minor effects. To the extent that local relative reinforcement probability or frequency is independent of performance, it constitutes an independent variable in the study of choice. The failure of this variable to control local patterns of behavior is an empirically meaningful result. (Also see Pear, Rector, and Legris, Chapter 1.) Momentary maximizing specifi-

a. See also Menlove (1975). Actually, this assertion is correct only if the average time between switches is appreciably shorter than the average time between reinforcers for switching, as it usually is. I am indebted to J.E.R. Staddon for pointing this out in a personal communication. Silberberg and Ziriax (Chapter 5) also describe cases in which the relation between time between switches and the probability of reinforcement for switches is not forced.

cally predicts that switching probabilities will conform to relative probability of reinforcement for switching relative to continuing, not for switching after one run of responses relative to switching after a different run length. Thus the ping-pong ball syndrome does not preclude the possibility of falsifying the predictions of momentary maximizing.

Finally, comments on some more general aspects of Shimp's discussion are in order. In Nevin (1969 and 1979) I observed overall matching in the absence of momentary maximizing trends, as did Heyman (1979) and Herrnstein (described in de Villiers, 1977), whereas Shimp (1966) and Silberberg et al. (1978) observed overall matching accompanied by momentary maximizing. On the basis of all five studies I concluded that overall matching is more general than momentary maximizing and suggested that "to the extent that overall matching is found in separate studies with different sequential response patterns, the molar matching law is reaffirmed as a general law of behavior, not an artifact or by-product of more molecular processes" (Nevin, 1979: 325). There is no reason to change these conclusions.

In Chapter 4 Shimp counters my argument that overall matching is more general than momentary maximizing by pointing out an apparent counterinstance. He refers to the study by Shimp and Hawkes (1974), in which key pecking was reinforced if the interresponse time (IRT) was between 1.5 and 2.5 seconds or between 3.5 and 4.5 seconds. The relative frequency of reinforcement for the shorter IRT was controlled by interdependent concurrent VI VI schedules. When relative reinforcement was varied the proportion of short IRTs changed according to a function resembling those in Shimp's Figure 4-4. Indeed if interchange-over responses are viewed as analogous to IRTs, the two experiments are equivalent. Clearly this is a molar result: Responses and reinforcers are summed for each reinforced class over large blocks of time, without regard for particular sequences of events. Moreover the result resembles those reported in studies concerned exclusively with molar aggregations of pecks at two keys. In Figure 6-2 the Shimp and Hawkes (1974) data have been replotted from their Table 2 in the form first used by Staddon (1968) in a similar study of concurrent reinforcement of different IRTs and later used by Baum (1974) to identify deviations from matching in a variety of two-response concurrent-schedule experiments. The data do not conform exactly to matching but may be described by the generalized matching law (Baum, 1974):

$$\frac{R_1}{R_2} = c\left(\frac{r_1}{r_2}\right)^a ,$$

where c represents bias toward one or the other response class and a represents sensitivity to reinforcement. The parameters of the fitted line shown in Figure 4-2 indicate slight undermatching and substantial bias toward the shorter IRT class.

Figure 6-2. The data of Shimp and Hawkes (1974) are replotted to show the ratio of short to long interresponse times as a function of the ratio of reinforcers received for these interresponse times. The light line indicates exact matching. The parameters of the heavy line, fitted by the method of least squares, are indicated.

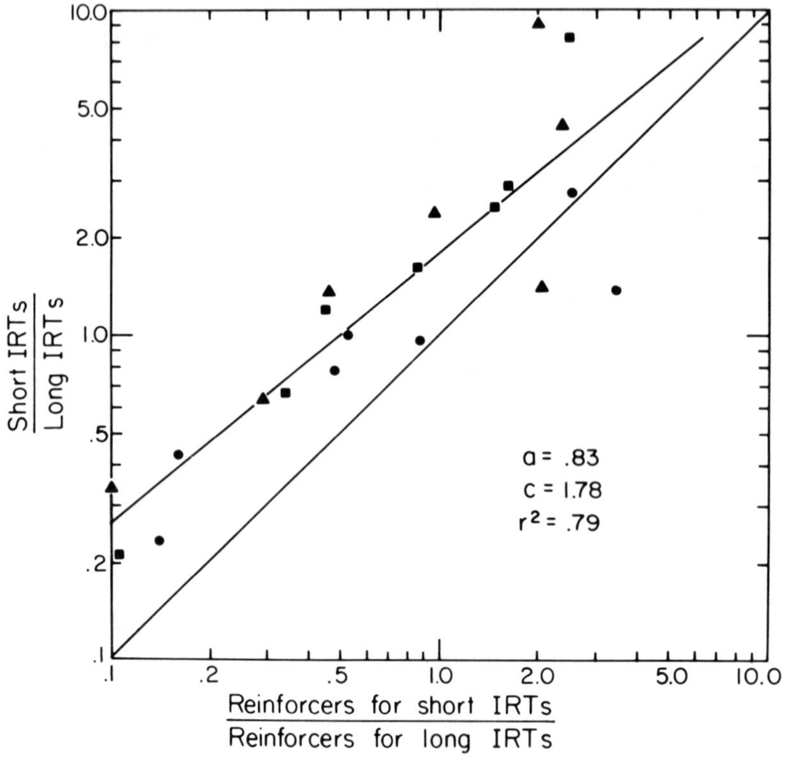

Can these deviations be explained by a local analysis? No local or sequential data are offered to indicate whether a momentary maximizing process controlled alternation between reinforced classes in such a way as to generate undermatching in the overall function. However, a local analysis may help to explain bias. The reinforcement contingencies specified two response classes requiring different lengths of time to execute. Bias toward the less time-consuming class is plausible. Elsewhere Shimp (e.g. 1969) has suggested that when reinforcement rates are equated, the frequency of one IRT relative to a different IRT matches the relative reciprocal of their lengths, and the Shimp and Hawkes (1974) data conform approximately to this finding. In this respect the local analysis urged by Shimp pertains to the definition and properties of two response classes in relation to the contingencies of reinforcement and extends the analysis of concurrent operants to cases in which the operants are defined by temporal contingencies. Although interesting and important, this issue is irrelevant to the argument that overall matching is more general than momentary maximizing in concurrent VI VI schedule experiments.

Apparent failure of control by molar contingencies in the Shimp and Hawkes study is evident at a still more molar level of aggregation: The average rate of key pecking, where all pecks are pooled regardless of IRT, is not an orderly function of the overall obtained rate of reinforcement, where all reinforcers are pooled. This is hardly surprising, because the two defined response classes require different amounts of time to emit, as noted earlier. No theorist of any persuasion would expect the average rate of various blends of two distinct classes of action differing in duration to be orderly. The real question is, When are we in fact pooling different response classes in constructing molar aggregations, and when are we counting instances of a single class? Molar aggregations are suspect whenever different classes are pooled; in this I agree fully with Shimp. One important task facing the experimental analysis of behavior is to determine when local reinforcement contingencies actually establish different response classes.

A response class may be defined operationally by reference to a contingency of reinforcement, but it is by no means obvious that this operational definition is behaviorally significant. To decide whether an aggregation of responses constitutes a behaviorally meaningful class, it is necessary to change the frequency of one member of the class and determine whether other presumed members of the class are similarly affected. For example, one might withhold food or present shock for a given duration of key pecking and ask whether the frequency of all durations decreases in proportion. If frequency does decrease, pecking is a single class, regardless of duration. If it does not, duration defines separate response classes, perhaps because differential reinforcement for duration is implied by the programming apparatus or because there is more than one kind of key peck (e.g., Schwartz, 1977). The question of whether differential reinforcement establishes distinct response classes is empirical. Evidently it did in the Shimp and Hawkes study; otherwise it is unlikely that the relation between the allocation of responding to different IRT classes and relative reinforcement would have been orderly.

The fact that explicit differential reinforcement of IRTs or interchangeover responses may sometimes establish different response classes does not imply that it always does so. With respect to my 1969 experiment, I would argue that differential reinforcement of various response sequences, which is inevitable in concurrent VI VI schedules, did not suffice to establish these sequences as different response classes. I say this because relative frequency or probability of reinforcement did not control the relative frequency of occurrence of different sequential patterns. More specifically, the frequency of sequences terminating with pecks to green, relative to identical sequences terminating with pecks to red, was negatively correlated with the relative frequency or probability of reinforcement (Nevin, 1979). It is true that I did not intervene experimentally to differentially reinforce or extinguish one particular sequence, so my observations are not entirely conclusive in this regard. Nevertheless, given the failure of local control that I observed, it seems legitimate to aggregate responses to the green and red keys without regard for the sequences in which they occurred and ask whether their overall allocation matched the overall allocation of reinforcers.

A final comment: Although a rebuttal of molecular arguments against a molar interpretation of my 1969 study has been set forth, there is no reason to see molar and molecular approaches to the study of behavior as fundamentally opposed. A molecular approach is necessary to understand the determiners and structure of response classes. Once such classes are defined, their overall allocation is determined by the molar matching law. Herrnstein and Heyman (1979) arrived at the same conclusion in their study of concurrent interval and ratio schedules. In this respect there is evidence of convergence of different approaches to the study of maximizing and matching in operant behavior.

REFERENCES

Baum, W.M. 1974. On two types of deviation from the matching law: Bias and undermatching. *Journal of the Experimental Analysis of Behavior 22*: 231–42.

_____. 1981. Optimization and the matching law as accounts of instrumental behavior. *Journal of the Experimental Analysis of Behavior 36*: 387–403.

de Villiers, P.A. 1977. Choice in concurrent schedules and a quantitative formulation of the law of effect. In W.K. Honig and J.E.R. Staddon, eds., *Handbook of Operant Behavior*, Englewood Cliffs, N.J.: Prentice-Hall.

Herrnstein, R.J. 1961. Relative and absolute strength of response as a function of frequency of reinforcement. *Journal of the Experimental Analysis of Behavior 4*: 267–72.

Herrnstein, R.J., and G.M. Heyman. 1979. Is matching compatible with reinforcement maximization on concurrent variable interval variable ratio? *Journal of the Experimental Analysis of Behavior 31*: 209–23.

Herrnstein, R.J., and D.H. Loveland. 1975. Maximizing and matching on concurrent ratio schedules. *Journal of the Experimental Analysis of Behavior 24*: 107–16.

Heyman, G.M. 1979. Markov model description of changeover probabilities on concurrent variable-interval schedules. *Journal of the Experimental Analysis of Behavior 31*: 41–51.

Heyman, G.M., and R.D. Luce. 1979. Operant matching is not a logical consequence of maximizing reinforcement rate. *Animal Learning and Behavior 7*: 133–40.

Menlove, R.L. 1975. Local patterns of response maintained by concurrent and multiple schedules. *Journal of the Experimental Analysis of Behavior 23*: 309–37.

Nevin, J.A. 1969. Interval reinforcement of choice behavior in discrete trials. *Journal of the Experimental Analysis of Behavior 12*: 875–85.

_____. 1979. Overall matching versus momentary maximizing: Nevin (1969) revisited. *Journal of Experimental Psychology: Animal Behavior Processes 5*: 300–306.

Rachlin, H.; L. Green; J.H. Kagel; and R.C. Battalio. 1976. Economic demand theory and psychological studies of choice. In G.H. Bower, ed., *The Psychology of Learning and Motivation*, vol. 10, New York: Academic Press.

Schwartz, B. 1977. Two types of pigeon key-pecking: Suppression of long- but not short-duration key pecks by duration-dependent shock. *Journal of the Experimental Analysis of Behavior 27*: 393–98.

Shimp, C.P. 1966. Probabilistically reinforced choice behavior in pigeons. *Journal of the Experimental Analysis of Behavior 9*: 443–55.

_____. 1969. The concurrent reinforcement of two interresponse times: The relative frequency of an interresponse time equals its relative harmonic length. *Journal of the Experimental Analysis of Behavior 12*: 403–11.

_____. 1979. The local organization of behaviour: Method and theory. In M.D. Zeiler and P. Harzem, eds., *Reinforcement and the Organization of Behaviour*, Chichester, England: Wiley, 1979.

Shimp, C.P., and L. Hawkes. 1974. Time-allocation, matching, and contrast. *Journal of the Experimental Analysis of Behavior 22*: 1–10.

Siegel, S., and D.A. Goldstein. 1959. Decision-making behavior in a two-choice uncertain outcome situation. *Journal of Experimental Psychology 57*: 37–42.

Silberberg, A.; B. Hamilton; J.M. Ziriax; and J. Casey. 1978. The structure of choice. *Journal of Experimental Psychology: Animal Behavior Processes 4*: 368–98.

Staddon, J.E.R. 1968. Spaced responding and choice: A preliminary analysis. *Journal of the Experimental Analysis of Behavior 11*: 669–82.

Staddon, J.E.R., and S. Motheral. 1978. On matching and maximizing in operant choice experiments. *Psychological Review 85*: 436–44.

Stubbs, D.A., and S.S. Pliskoff. 1969. Concurrent responding with fixed relative rates of reinforcement. *Journal of the Experimental Analysis of Behavior 12*: 887–95.

II ECOLOGY OF INSTRUMENTAL RESPONDING

7 THE MECHANISM OF OPTIMALITY IN FORAGING

S. E. G. Lea

Many scientists working in operant conditioning are aware of a fast-growing literature on a subject usually called "optimal foraging," and aware of a familiar ring to some of the problems discussed under that heading. This chapter explores some aspects of the relation between foraging and conditioning.

Foraging is one of the major concerns of behavioral ecology. It is a very general concept. A cow wandering through a field taking the occasional mouthful of grass, a redshank wading along the mud of an estuary occasionally probing with its beak and taking out a worm, a thrush running across a lawn, a flycatcher sitting on a perch waiting for a fly to pass, a cat sitting by a mousehole, even the country children who scour the English hedgerows for blackberries each autumn—all can be described as foragers; ecologically speaking, prey may be either animal or vegetable. Two points in common between foraging and the behavior shown in operant conditioning experiments are striking. First, at least formally, both involve instrumental behavior; second, what is produced by that instrumental behavior is nearly always food, though in principle it need not be in either case.

The current explosion of literature on foraging stems very largely from an idea first stated clearly by MacArthur and Pianka (1966): that, within constraints imposed by the kind of animal it is (or, what is very nearly the same thing, by the environmental niche it occupies), natural selection should tend to turn every species of animals into *optimal foragers*, and useful predictions can therefore be made about foraging behavior by considering what the best possible behavior in the circumstances would be. It is worth pausing for a moment to think why this should be so. More precisely, why is it reasonable to accept the caveat "within the constraints imposed the kind of animal it is"? Too easy acceptance of this kind of clause can lead to lax argument from natural selection, as Sahlins (1977: 81 ff) suggests in his critique of sociobiology. What is being said is that, while it might indeed be optimal for the lion to evolve so as to eat grass like the ox, it is actually more likely to evolve so as to hunt oxen more efficiently. There are three reasons why this is plausible. First, optimizing ox hunting involves a smaller, and therefore presumably more probable, evolutionary change than switching to grass eating. Second, the lion may well be in or approaching a local maximum of adaptiveness: While grass eating would be better than ox eating, certain intermediate stages would be worse than either. Third, the ox niche is already occupied by another species; the lion niche is not; Gause's (1934) principle of competitive exclusion therefore says that it is better to go on being a lion.

Foraging is thus one more area in which to ask whether instrumental behavior is generally optimal, a question that appears in several other contexts elsewhere in this book. The links between the optimal foraging literature and operant conditioning are, however, much stronger than that. The argument has three major points:

1. Some of the behavioral phenomena that have been reported in the ecological literature on foraging seem analogous to well-known conditioning effects.
2. It turns out that foraging-like phenomena can be produced in the laboratory, when the responses involved are not the species' characteristic "instinctive" foraging patterns but conditioned operants.
3. It is therefore possible to argue that the behavior patterns producing optimal, or near optimal, foraging whether in the laboratory or in the wild may be produced by operant conditioning; in other words, operant conditioning is the mechanism of optimal foraging. But if that is so, it is also quite likely that the selective advantage of foraging optimally explains the form of the principles of operant conditioning.

The present chapter is not a general exposition of foraging theories or of the data supporting them. A number of expositions are available in the literature, such as Krebs (1978), a general introduction; Pyke, Pulliam, and Charnov (1977), which is ecologically authoritative; Collier and Rovee-Collier (1981), with special reference to the general theory of feeding; and Lea (1982), which expands on the relation between foraging and operant conditioning. This chapter aims to illustrate the three points just listed by reference to a single theme that appears in one form or another in other chapters in this book, the idea of contrast.

ANALOGIES BETWEEN FORAGING AND CONDITIONING DATA ON CONTRAST

In an operant context contrast is defined as any interaction between schedules of reinforcement in which an increase in reinforcement rate on one schedule produces a decrease in response rate on another, or vice versa. Alternation between the two schedules may be under the subject's control (concurrent schedules) or the experimenter's (multiple schedules), but in either case some difference of stimulus conditions or manipulandum location is necessarily involved. It is reasonable to talk about contrast in both concurrent and multiple schedules (cf. Rachlin, 1973).

A number of papers in the literature of optimal foraging are concerned with the following situation, which has something of the nature of both concurrent and multiple schedules. A forager is working through an environment containing many prey items, which are of two or more different types (they could form a graded series). Every so often it comes across a prey item. On a particular occasion it meets one of the *i*th type. Now, will it, or will it not, launch an attack on that prey item? Behavioral ecologists have established what the optimal forager should do. The following terms and symbols are needed to describe the theory:

1. The *net energy worth* of each perceptually distinguishable prey type is E; E_i is the worth of the *i*th prey type. In the style of classical decision theory E_i is taken net of the energy costs of chasing, catching, eating, and digesting that type of prey and is discounted by the probability of capture where that is less than 1. Thus if a serval spotted a bird of type i, from which it could extract 50,000 joules of energy, with the expenditure of 1,000 joules of energy in leaping to catch it, and a probability of 0.1 of successful capture if it does leap, then $E_i = (0.1 \times 50{,}000) - 1{,}000 = 4{,}000$ joules.

2. The *handling time h* for each prey type includes all time during which the forager could not be looking out for a different kind of prey, and like net energy worth it is averaged if handling times for a prey type are variable. So, to continue with the mythical serval, leaping at the bird and missing might perhaps cost it 60 seconds (57 seconds stalking into position and 3 seconds to leap up, miss, and recover). Leaping successfully might cost it 600 seconds – 1 minute as before and another 9 to eat the bird. Then h_i for that kind of bird would be $(0.9 \times 60) + (0.1 \times 600) = 114$ seconds.

3. The *encounter frequency* for each prey type is ν. More precisely, ν_i is the frequency with which the *i*th prey type is encountered, per time spent foraging.

Equipped with all this information the optimal forager can calculate two further parameters for each prey type. Both have the dimensions of rate of energy intake.

The first is the ratio E_i/h_i, net energy worth per handling time for the *i*th prey type. Clearly prey types can be rank ordered on the basis of E/h; the prey of highest E/h is the best in terms of energy flow.

The second derived parameter the optimal forager must calculate can be symbolized as $(E/T)_i$. It is the rate at which the forager would extract energy from the environment per time spent foraging if he never pursued the *i*th prey type or any other prey type (*j*, say) for which E_j/h_j is less than E_i/h_i. $(E/T)_i$ is a function of the E/h values *and* the encounter rates ν for all the prey types of *higher* E/h than the *i*th, but it is not affected by the encounter rate ν_i for the *i*th type or for any type of *lower* E/h.

The last point may seem counterintuitive but is in fact obvious: Since $(E/T)_i$ is by definition the energy intake when the *i*th prey type and all those of lower E/h are excluded from the diet, their parameters must be irrelevant to it.

Having calculated E_i and $(E/T)_i$ for every prey type, the optimal forager now follows the following very simple decision rule. The *i*th prey type is always pursued if $E_i/h_i > (E/T)_i$ and never pursued if $E_i/h_i < (E/T)_i$; if $E_i/h_i = (E/T)_i$ it does not matter what the forager does. To see why this is optimal simply think in terms of opportunity costs; $(E/T)_i$ is the rate of gain if the forager does not pursue the *i*th type, hence what it is giving up if it does pursue; pursuit is worthwhile only if the rate of gain it produces, E_i/h_i, is greater.

This decision rule obviously assumes that the forager is trying to maximize energy intake. Equally obviously it makes no difference if any other *single* factor is to be maximized, but quite different principles are introduced if the forager is trying to maximize intake of two distinct factors at once (cf. Rapport 1971). The upshot is that for an optimal forager pursuit of any prey type except the very best will depend on that type's own net energy worth and handling time and on the worth and handling time of all better prey types in the environment. It will also depend on the densities of all better prey types but *not* on its own density. Worth, handling time and densities of any *worse* prey type will all be irrelevant. An important corollary is that if the densities of all prey types change together, so that the absolute prey densities all increase without any change in relative density, less favorable prey types will progressively drop out of the diet; conversely, a simultaneous decrease in all prey densities will cause less favorable prey types to become progressively more acceptable. This is a particularly easy prediction to test, especially in natural or quasi-natural conditions; it is also consistent with common sense and lay psychology, forming the basis for the remark with which my grandfather, a First World War veteran, so often irritated me when I was a child: "If you were out in France, young feller-me-lad, you'd eat up your cabbage and be glad of it!"

Behavior broadly consistent with this optimal strategy has been observed in carp (Ivlev, 1961: chap. 4), bluegill sunfish (Werner and Hall, 1974), rainbow trout (Ware, 1972), great tits (Krebs, Erichsen, Webber, and Charnov, 1977), wagtails (Davies, 1977a), flycatchers (Davies, 1977b), and redshank (Goss-Custard, 1977), to mention only the clearest demonstrations. Of these studies, only those of Davies and Goss-Custard were in truly natural conditions, but in all of them natural prey or a reasonable approximation were used and the subjects pursued them in a natural way. In no case was real foraging behavior exactly like that of an optimal forager, and in particular the "all-or-none" property of the optimal behavior has never been observed. The absence of mammals from this list of observations is curious, but it does not reflect any failure of optimal foraging predictions in mammalian studies, it is simply that there are no reports bearing on this kind of choice in mammals (given the dominance of the optimizing paradigm, that may reflect some unpublished failures of prediction).

The basic analogy on which this chapter turns is as follows: Each prey type, with its attendant parameters (E, h, and ν), can be thought of as a naturally occurring schedule of reinforcement, and pursuit of that prey as instrumental behavior governed by that schedule. The ecological literature shows that increasing the density of better prey types decreases the probability of pursuing worse types. This effect of the density of one kind of prey on the probability of pursuing another is therefore analogous to a contrast effect of one schedule of reinforcement on behavior maintained by another; an increase in reinforcement rate from one schedule (prey type) causes a reduction in response rate to another schedule (prey type). Thus contrast is a feature of both the predicted behavior of optimal foragers and the observed behavior of real foragers. It is also a very well known feature of behavior in operant conditioning experiments; to consider only the present volume, it is at the heart of well over half of the chapters.

"FORAGING" USING LEARNED INSTRUMENTAL BEHAVIOR

The fact that phenomena meeting the definition of contrast are found in both foraging and operant conditioning is interesting. But it need be no more than coincidence; on the face of it there are substantial differences between the situations involved and between the preoccupations of those investigating them. To oversimplify a little, conditioning experiments concern arbitrary responses occurring in man-made environments, and have as object the investigation of the principles determining the behavior of individual organisms as a function of events within the life of those individuals. Foraging experiments involve species-typical responses performed in the natural environment, and their object is investigation of the principles determining the behavior of a species as a function of the ecological pressures affecting the entire gene pool and indeed the gene pools of other species in the predator-prey system. Furthermore, the prey-choice situation described above has, as already noted, no exact parallel in the operant literature: It is neither a concurrent nor a multiple schedule, nor is it even an exact match to a "changeover key" concurrent (Findley, 1958) or an "advance key" multiple schedule (Honig, Beale, Seraganian, Lander, and Muir, 1972).

The contrast between the two situations is not, in truth, quite as stark as the last paragraph suggests. In practice operants are far from arbitrary (Jenkins and Moore, 1973; Moore, 1973). And some studies of foraging have in fact used responses that had to some degree to be learned; for example Krebs, Ryan, and Charnov (1974) taught chickadees to peck through paper windows in artificial pine cones (compare Ferster and Skinner's "punchboard" for key-training pigeons, 1957: 32). However, if the analogy between foraging and conditioning is to be plausible, it is important to demonstrate that they are not two isolated paradigms but, rather, points on a continuum, not separated by any large changes in the nature of behavior. One profitable research strategy, then, is to fill in some of the intermediate points on the continuum. A relatively easy starting point is the structure of the choice situation. Suppose that a schedule of reinforcement were constructed that used conventional operant apparatus, responses, subjects, and so forth but mimicked the contingencies found in foraging. What kind of behavior would be observed? Would the predictions of optimal foraging theory be borne out? And would behavior be consistent with what is seen using the same apparatus and so on but more conventional schedules of reinforcement?

These questions seem to have occurred to a number of people more or less independently. Collier and Rovee-Collier (1981) describe an experiment by Collier and Kaufman, using rats in a free behavior paradigm (that is, the subjects lived in the operant chamber and obtained all their food from schedules of reinforcement, which were permanently available; cf. Logan, 1964). Two kinds of behavior were required. The first ("procurement") made "prey" of varying types available; the second ("consumption") actually led to the delivery of food pellets. That sounds quite like real foraging, but both procurement and consumption were done by pressing levers under fixed-ratio schedules. Collier and Kaufman observed behavior that was qualitatively consistent with optimal foraging theory and also with their general theory of feeding behavior. In an experiment (Lea, 1979) done along similar lines but perhaps somewhat more conventionally, using sessions of fixed lengths, pigeons were the subjects, and to facilitate comparison with predictions both from optimal foraging theory and from psychological theories such as Herrnstein's (1970) "matching law," interval rather than ratio schedules were used throughout. Figure 7-1 is a flowchart of the contingencies used. At the beginning of each daily session the pigeon entered the *search* state, with just the central key of a conventional three-key chamber illuminated, with white light. It could then work on a fixed-interval schedule of t seconds. The reinforcement for the first peck after t seconds, however, was not access to food, but the lighting up of a side key, which was green with probability p and red with probability $(1 - p)$. The white center key light remained on, and the schedule was now in the *choice* state, in which the pigeon could peck

Figure 7-1. Flowchart of the schedule of reinforcement used to simulate a foraging situation in Lea, 1979. In the present experiment h_L was always 20 seconds, h_S was 5 seconds, E_L and E_S were both 2.5 seconds, and d was zero. The color assignments shown were reversed for half the birds used in the experiment.

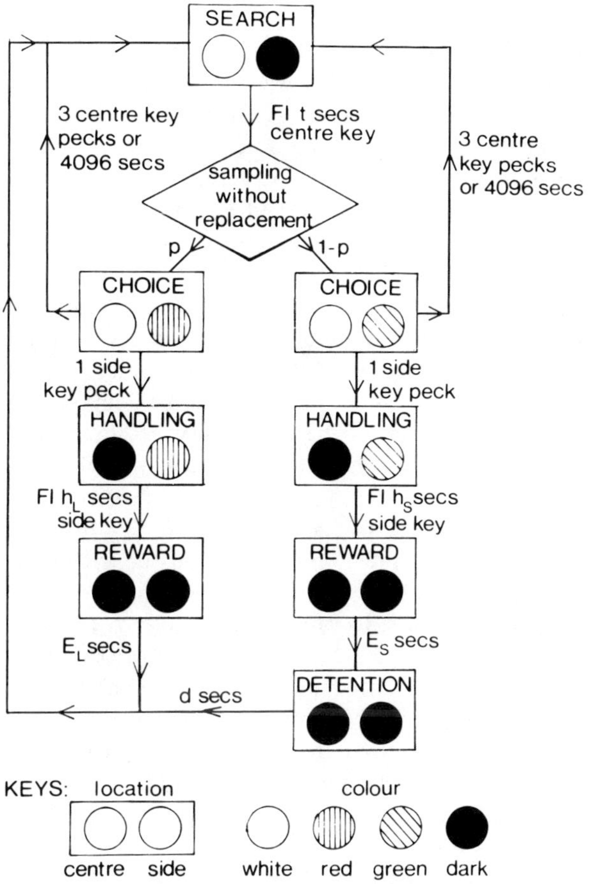

either key. A single peck at the side key turned off the center key and put the pigeon into the *handling* state, in which it was committed to getting food reinforcement through an FI schedule, which has h_S seconds (in practice always 5 seconds) if the key was green and h_L seconds (in practice always 20 seconds) if it was red. Alternatively three pecks to the center key (or just doing nothing for a long time, but this only happened once in the entire experiment) turned off the side key and put the pigeon back into the search state. If the pigeon chose to enter the handling state, reinforcement of E_S or E_L seconds' access to grain eventually followed. Reinforcement on the green key was followed by d seconds in the *detention* state, during which both keys were dark and inoperative.

From a foraging point of view, then, the pigeon's pecks on the center key moved it through an environment containing two sorts of prey. One of them, the green key, took 5 seconds to "handle"; the other, the red key, took 20 seconds. I shall call these the "short prey" and "long prey." (For half the birds, the color allocations were reversed, so the red key was the long prey and the green key the short prey). In principle the hopper access times E_S and E_L, should be added to these handling times and so should the detention time d, imposed after consumption of the short prey. In practice, however, in the conditions discussed in the present chapter, d was always zero and E_S and E_L were both fixed at 2.5 seconds, and so these parameters can be ignored (for the effects of varying them see Lea, 1979). Whenever the pigeon encountered a prey item, it could either accept it, by pecking the side key, or reject it, by pecking the center key. The prediction from optimal foraging theory is that the short prey should be accepted whenever it is encountered but that the probability of accepting the long prey should switch from zero to one, at a predictable point, as the frequency of encountering the short prey falls. The probability of accepting the long prey should not be affected at all by the rate at which it is itself encountered. As a corollary, if the encounter frequencies of both prey types were to fall together, the pigeons should switch (again at a predictable point) from always rejecting to always accepting the long prey. In the light of the ecological literature we might expect the switchover to be something other than all or none.

The simplest manipulation is to vary the parameter t, the search-state fixed interval (see Figure 7-1). This reduces the encounter rates with both prey types together. Figure 7-2 shows what happened to the probability of accepting the long prey when this was done (these data are described in more detail in Lea, 1979). Qualitatively the prediction from optimal foraging theory was upheld: The pigeons switched from rejection to acceptance of the long prey type as t was increased. Quantitatively, however, optimality was not upheld. First the switch did not come at the point at which the rate of energy intake, E_i/h_i, for the long prey was just less than the best rate attainable without it, $(E/T)_i$. The pigeons went on rejecting the long prey at t values that were much too high from that point of view (the optimal switch point is marked with a dashed line in Figure 7-2). Moreover, the switchover was gradual rather than all or none.

The probability of accepting the short prey type never varied from 1.0 in any condition mentioned in this chapter, so these data are not shown. In that respect at least the predictions of optimal foraging theory were magnificently accurate.

However, these results do not speak to the only really counterintuitive prediction from optimal foraging theory, the suggestion that the probability of accept-

Figure 7-2. Simulated foraging (Lea, 1979), showing the effect of varying the search-state fixed-interval schedule parameter (t in Figure 7-1) on the probability of accepting the "long-prey" schedule when it was offered. Note the logarithmic scale on the abscissa.

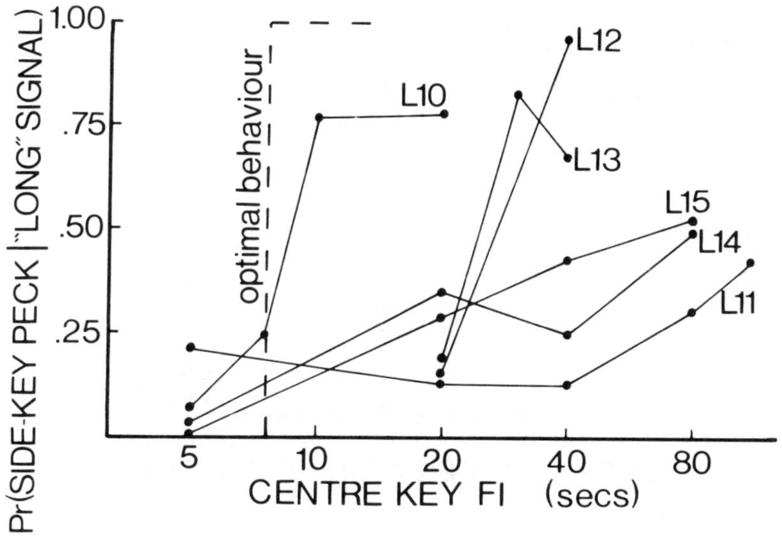

ing the long prey should be affected by variations in the density of the short prey but not by variations in its own density. To vary the two densities (that is, encounter rates) independently, it was necessary to manipulate both t, the search-state fixed-interval parameter, and p the probability that a prey once encountered would be long. Figure 7-3 shows the effects of short-prey encounter rate on long-prey acceptance, with the long-prey encounter rate held constant: there was indeed a drastic switch from acceptance to rejection of the long prey as the short prey became more common. Figure 7-4 shows the effects of long-prey encounter rate. They were clearly much less marked, although bird L12 was more affected by long than short prey density, so that I had to resort to regression techniques (described in Lea, 1979) to confirm that, overall, short-prey density had the greater effect. Once again the data are in qualitative but not quantitative agreement with optimal foraging theory.

But what is the relation between these results and the usual principles of operant conditioning? At first glance something unusual seems to be going on here. First an asymmetric contrast effect exists—one schedule (short-prey density) affects performance on another (long-prey acceptance)—but no converse effect exists, whereas in operant experiments, contrast is most usually symmetri-

Figure 7-3. Effect of varying the rate of occurrence (per unit time spent foraging) of the short prey on probability of accepting the long prey; long-prey density was held constant at 0.025 hertz. Note the logarithmic scale on the abscissa.

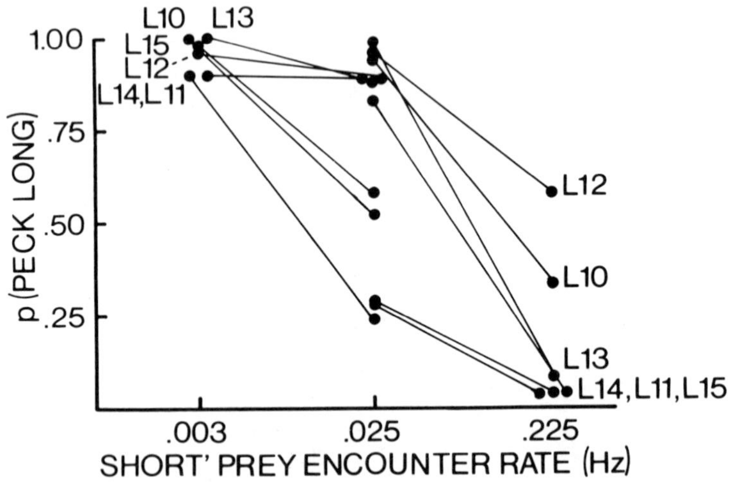

cal. Second the extent of the contrast observed is surprising. Contrast effects tend to be considered as secondary relative to the direct effects of schedule parameters, and even in "matching" under concurrent interval shcedules, all we observe is that the parameter of a concurrently available schedule affects performance as much as the parameter of that schedule itself. It seems odd for a contrast effect to be larger than a direct effect.

The oddness of these results is only superficial, for a large number of reasons. In runway experiments asymmetric contrast, with the richer schedule affecting the leaner but not vice versa, is now claimed to be the rule rather than the exception (e.g., Black, 1968; Mackintosh, 1974: 384-90); besides, performance on the short-prey schedule was subject to a ceiling effect, and maybe the constant acceptance probability of 1.0 conceals a range of preferences. However, these possibilities need not be pursued. For the purpose of comparisons with conventional concurrent schedules the appropriate thing to do is to concentrate on behavior in the choice state, during which two schedules were concurrently available. Both were in effect chained schedules, in which an interval requirement followed a ratio requirement. On the side key, a single peck led to a stimulus change (the darkening of the center key) and made available a fixed interval of 20 seconds (assuming that a long prey had been presented). On the center key, a fixed ratio of 3 gave access to a compound, variable-interval schedule. It

Figure 7-4. Effect of varying the rate of occurrence of the long prey (per unit time spent foraging) on probability of accepting it when encountered. The density of short prey was held constant at 0.025 hertz. Note the logarithmic scale on the abscissa.

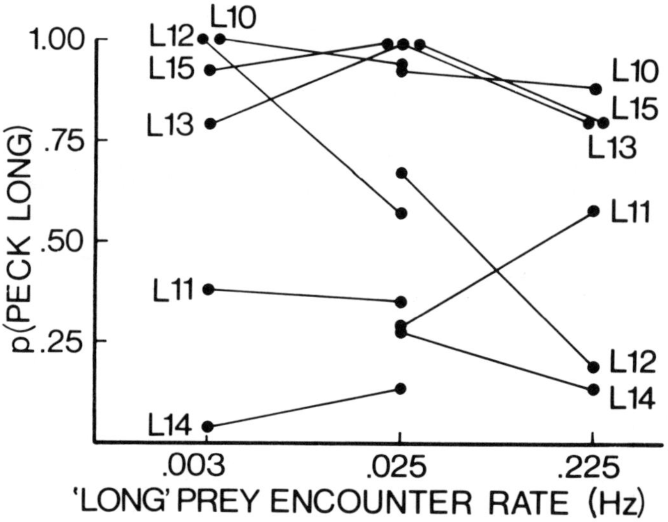

is possible to work out what the mean time to the next reinforcement on this schedule ought to be, by assuming that the pigeons always completed all fixed intervals as fast as possible; in practice, however, they did sometimes show quite long postreinforcement pauses in the search schedule, and they also had quite long and variable latencies in the choice state (associated with a good deal of bobbing from one key to the other and pecking just in front of the keys, that is, "vicarious trial and error," Muenzinger, 1938). Accordingly it seemed better to calculate the obtained mean delays to the next reinforcement, from leaving the choice state by either route. This was done from a new series of sessions (Lea, Note 1) in which the probability of the long prey was kept at 0.5 and *t*, the search-state fixed interval, was varied rather more systematically than in the series whose results are shown in Figure 7-2. All birds were exposed to four blocks of sessions during which *t* took values of 20, 5, and 7.5 seconds (twice).

In Figure 7-5 the results of these sessions are plotted in the manner recommended by Baum (1974), as a log response ratio plotted against the log of a quantity with the dimensions of a reinforcement rate ratio. To be precise, the dependent variable is the ratio of the number of responses made to the side key during choice states when the long prey was available, to the number of responses made to the center key during such states, plotted on a logarithmic

Figure 7–5. Effect of observed ratio of reinforcement immediacy on observed response rate ratio in the choice state. Both ratio values are expressed as (side key)/(center key). Note that the scales on both axis are logarithmic but that the ordinate scale is compressed by a factor of 10. Data collected by Lea (Note 1) using the procedures of Lea (1979). The solid line connects geometric means across birds of data points for each of the three search-state fixed-interval values used.

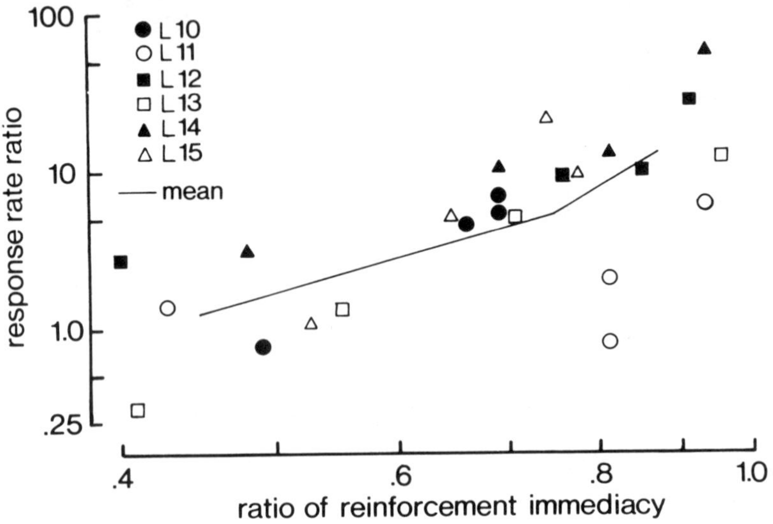

scale. The independent variable is the ratio of reinforcement immediacies (reciprocals of delays) by the two routes from such states, also on a logarithmic scale. Reinforcement immediacy from the side key peck was, of course, very nearly 1/20 seconds^{-1}. Reinforcement immediacy from a center-key peck is not so easy to. grasp. Suppose that the search-state fixed interval t was set at 7.5 seconds and the following sequence of events occurred: The pigeon rejected a long prey on a given trial; worked through the search schedule taking 7.8 seconds to do so; found another long prey; rejected that after spending 6.3 seconds in the choice state; worked through the search schedule again, taking 8.1 seconds this time; found a short prey; accepted it with a latency of 0.8 seconds in the choice state; and completed the short prey handling schedule in 5.2 seconds. This sequence of events contributes two different values of the delay of reinforcement from center-key pecks in the choice state. The first of these is simply the total of all the times mentioned; that is, 7.8 + 6.3 + 8.1 + 0.8 + 5.2 = 28.2 seconds. The second is the sum of the second search state time, the second choice state time, and the short-prey handling time; that is, 8.1 + 0.8 + 5.2 = 14.1 seconds. Both of these contribute to the mean delay of reinforcement from the choice state from a center peck.

The vertical axis of Figure 7-5 is compressed by a factor of 10. Thus the slope of the line drawn through the mean is approximately 3.8. This slope is an estimate of the parameter a in Baum's (1974) generalized matching law,

$$\log(P_1/P_2) = a \log(R_1/R_2) + b \quad , \tag{7.1}$$

where P_i and R_i are response and reinforcement rates on the ith schedule. The parameter a assesses the sensitivity of responding to reinforcement (Lobb and Davison, 1975); a value of 3.8 indicates very high sensitivity, or, to put it another way, extreme overmatching. This is very much higher than is typically observed on simple concurrent schedules, where a values in the region of 0.8 are the rule (Lobb and Davison, 1975; Myers and Myers, 1977). But the situation in the choice state of the present schedule is more strictly analogous to the initial link of concurrent chain schedules (Autor, 1969), in which the responses whose rate ratio is being studied obtain access to other schedules of reinforcement rather than to reinforcement directly (Fantino and Logan, 1979: 493-94, make a similar point about Collier and Kaufmann's foraging simulation). In a number of studies of concurrent chain schedules in which the terminal links were interval schedules or delays of noncontingent food presentation, it turns out that overmatching in the initial links is the rule rather than the exception and tends to become more marked as the initial link schedules become shorter (Lea, 1982: table 2). This tendency is in fact expressed directly in the mathematical models for concurrent chain performance that have been proposed by Davison and Temple (1973) and Squires and Fantino (1971). Although the extent of overmatching shown in Figure 7-5 is exceptional, the initial link schedules were exceptionally short (FR1 on the side key, FR3 on the center key), and the data are in fact consistent with the general trend. Another possible source of extreme overmatching might be the fact that the initial link schedules were ratios, whereas most published studies have used intervals; however, at the low parameter values used here, schedule types converge; "continuous reinforcement" is both fixed ratio 1 and fixed interval 0 seconds.

From these data it may be concluded that behavior in the simulated foraging situation is both qualitatively and quantitatively consistent with previous operant results on contrast between reinforcement schedules. There is nothing surprising about this conclusion in itself, of course; like most previous data on contrast, these were obtained from pigeons in highly conventionalized apparatus. What is important is that the same results are recognizably like those found in field studies of foraging; and so they contribute the continuum of data that are needed if the analogy between foraging and conditioning is to be more than a mere analogy.

CONDITIONING AS THE MECHANISM OF OPTIMAL FORAGING

Given two kinds of behavior that seem to obey the same principles, with no proven discontinuity between them, it would be uneconomical to suppose that different principles governed them. Yet that is what we seem to do at present: Behavior in operant chambers is referred to principles of conditioning, while foraging behavior is referred to principles of natural selection. Of course there is no real conflict between these two, because one is concerned with proximate causation, the mechanisms of behavior within the life of individual organisms, and the other is concerned with ultimate causation, the selective advantages that have led to the evolution of behavior. Both principles need each other: Foraging needs a mechanism within the life of the individual, and conditioning requires an evolutionary explanation.

While foraging calls for some kind of mechanism, however, it remains an open question whether conditioning as we know it in the laboratory can provide that mechanism. Probably most of those who have worked on instrumental conditioning have assumed like Small (1900-01) that they were examining the processes by which animals learn to obtain their food under natural conditions, but the grounds for that tacit assumption have rarely been examined. Indeed, the drift in recent years has been to presume that conditioning is largely a laboratory artifact of little general interest. The discovery of "constraints on learning" (see Hinde and Stevenson-Hinde, 1973) and the consequential attacks on "general process learning theory" (e.g., Seligman, 1970) have led to a *Zeitgeist* supposing that each species is prepared by evolution to learn certain rather specific associations or kinds of association and that the laws of each of these special kinds of learning will in turn be special. In contrast, I should like to argue here (admittedly somewhat speculatively) that foraging probably does involve general processes of conditioning rather than any special kinds of learning. The major steps in the argument are as follows:

1. Optimal foraging necessarily involves some learning.
2. The kinds of things that have to be learned about in foraging are the kinds of things studied in conditioning experiments; parameter values are also comparable.
3. Some aspects of the mechanisms of foraging ought to be evolutionarily conservative, that is, similar in a wide range of species.
4. Known conditioning principles predict both that behavior will be nearly optimal in the foraging situations that have been studied and also that optimality will fail in some situations.
5. There is no discontinuity of observed behavior between naturalistic studies of foraging and laboratory studies of conditioning.

Optimal Foraging Must Involve Learning. The optimal forager must react to the E, h, and v values of each kind of prey. It is conceivable that in an unchanging environment the E and h values of different prey types might be coded genetically; and in the limit that must be true to some extent of all but the most extreme omnivores—most animals "know" without having to find out that certain kinds of prey are too small to bother with (E is too low to justify the likely h) or too large to attempt (so that even if E were high h would be disproportionately higher). But it is very unlikely that v values could be acquired other than by individual experience. In practice the same is probably true of at least some E and h values too: cases of wild animals learning to take new prey are of course on record (e.g., Fisher and Hinde, 1949; Itani, 1958; Poirier, 1967).

Foraging and Conditioning Involve the Same Kinds of Learning. The parameters E, h, and v are precisely the kinds that crop up in schedules of reinforcement, whereas the other major kinds of learning involved in foraging, learning about prey densities in different places, are precisely what the early learning theorists harnessed when they used mazes to study rats' learning, and some of the most interesting recent work on animal learning has returned to this paradigm (e.g., Menzel, 1978; Olton, Chapter 9 of this volume).

The Mechanism of Foraging Ought to be Evolutionarily Conservative. Some aspects of animals' structure and behavior vary greatly between species, either to fit the species to its particular niche or in response to competition from otherwise similar species. Others are evolutionarily conservative. To take a nonbehavioral example, the number and nature of teeth varies greatly between mammals; their location (in the mouth) does not—elephants, of course, excepted. Similarly, the sources of energy exploited by different predators vary greatly; their fate when digested is invariably to be converted to glucose for intercellular purposes and adenosine triphosphate for intracellular purposes. It is plausible to suppose that a similar situation holds with respect to learning about food sources. The kinds of information collected by different predators are variable, but there is no reason why the mechanism for handling it should be. The situation is quite different when a species has evolved a special skill like migration or song, when it is quite plausible for a special learning mechanism to have evolved with it. In the case of foraging, each species' foraging skills have presumably evolved from an ancestor's foraging skills and the simplest course to postulate for that evolution involves retaining the learning mechanism but changing the information input to it.

Conditioning Principles Predict Optimality. This proposition is argued at length elsewhere in this book. Some of the unresolved questions raised in other chapters are highly relevant to the present point. For example, optimal foraging requires animals to be sensitive to overall rates of reinforcement rather than individual reinforcing events, which is relevant to the dispute between Silberberg and Ziriax (Chapter 5) and Heyman (Chapter 20); See also Commons, Woodford and Ducheny (Chapter 2). On the other hand, Herrnstein's "melioration" principle (Chapter 19), which requires an intermediate degree of temporal integration and so cannot always produce optimal behavior, tends to do well in situations governed by opportunity cost, such as the prey selection problem on which my foraging simulation experiments were based. The very fact that optimization is a controversial topic within conditioning theory, however, shows that conditioning principles can predict *nearly* optimal behavior, and that is quite likely to be good enough for evolutionary purposes. Consider a case where many conditioning theories predict nonoptimal behavior, the question of stochastic rather than all-or-none choice between probablistic sources of reinforcement. The selective advantage to an animal of choosing a source that delivers reinforcement on 75 percent of trials over a 25 percent source, all the time rather than most of the time is not very large. In this case indeed the situation is quite complicated, because stochastic preference is typically observed in foraging data but is not predicted by some conditoning theories (e.g., the matching law; cf. Herrnstein, Chapter 19); at least it can be said, however, that this is not a case where natural behavior is optimal and conditioning predicts nonoptimality.

Continuity Between Foraging Studies and Conditioning Experiments. The major issue of earlier sections of the present chapter needs no further analysis here except to point out that the example of contrast is only one of a number of cases that could have been taken. In my laboratory, for example, Susan Dow has been extending the study of unpredictable prey densities started by Krebs, Kacelnik, and Taylor (1978); but whereas Krebs et al. used great tits, which flew between patches of prey, Dow uses pigeons, which perform learned key peck responses to switch between schedules of reinforcement. Yet our results are quite consistent with those of the somewhat more naturalistic situation.

The Evolution of Conditioning Mechanisms

If conditioning principles do provide the mechanism by which animals forage more or less optimally, it is likely that a reverse causality applies on the evolutionary time scale: Conditioning principles are as they are *because* they produce optimal foraging or a good approximation to it. It is possible, of course, that more general selective pressures could have produced a more general learning mechanism, which is applied to foraging along with other problems; but there are a number of arguments against this possibility:

1. As already mentioned, certain special skills such as migration or song seem to involve their own special laws of learning; this suggests that there is no

general process for learning that is readily applied to any new need (new on the evolutionary time scale, that is).

2. At least at the level of parameter values, the principles of conditioning with food reinforcers do not hold uniformly for other kinds of reinforcement (Hogan and Roper, 1978).

3. It is hard to think of any kind of learning that would be needed as often or as regularly as learning about the values and densities of prey types, unless it should be learning about the social roles of other members of a group—and it has been suggested that this poses problems so special and so severe as to be capable of accounting for the evolution of human intelligence (Humphrey, 1976). At least in less social species, therefore, the requirements of foraging seem likely to be the dominant selective pressure on any putative general learning process.

If conditioning principles have evolved under the pressure of the need to forage optimally, that should give the comfort of some ecological validity to those who investigate them; it refutes the facile presumption that conditioning is nothing but a laboratory phenomenon with laws of no general interest. The comfort is limited, however. Although foraging is the most plausible source of a general learning mechanism, and although that is likely to be reflected in laboratory conditioning, in making that argument I was forced to concede that a species' special skills might well have their own special modes and mechanisms of learning. This would be the likely course of evolution whenever the skill involved was not derived from a foraging behavior. Precisely because foraging (hence, conditioning) mechanisms seem likely to be evolutionarily conservative, they are unlikely to be reflected in types of learning that are restricted to a single species; unfortunately that includes many of the most interesting kinds of human learning.

These conclusions are by no means certain; indeed, the whole of this final section has been speculative, and loosely speculative at that. What does seem certain is that in the past too little systematic thought has been given to the evolutionary origins of the principles of conditioning, and that the rise of the study of foraging provides a new opportunity to put this right.

REFERENCE NOTE

1. Lea, S.E.G. *Foraging, matching and delay of reinforcement in pigeons.* Manuscript submitted for publication.

REFERENCES

Autor, S.M. 1969. The strength of conditioned reinforcers as a function of frequency and probability of reinforcement. In D.P. Hendry, ed., *Conditioned Reinforcement*, Homewood, Ill.: Dorsey Press, pp. 127–62.

Baum, W.M. 1974. On two types of deviation from the matching law: Bias and undermatching. *Journal of the Experimental Analysis of Behavior 22*: 231–42.

Black, R.W. 1968. Shifts in magnitude of reward and contrast effects in instrumental and selective learning: A reinterpretation. *Psychological Review 75*: 114–26.

Collier, G.H., and C.K. Rovee-Collier. 1981. A comparative analysis of optimal foraging behavior: Laboratory simulations. In A.C. Kamil and T.D. Sargent, eds., *Mechanisms of Foraging Behavior*, New York: Garland.

Davies, N.B. 1977a. Prey selection and social behavior in wagtails (*Aves: Motacillidae*). *Journal of Animal Ecology 46*: 37–57.

_____. 1977b. Prey selection and the search strategy of the spotted flycatcher (*Muscicapa striata*): A field study of optimal foraging. *Animal Behaviour 25*: 1016–33.

Davison, M.C., and W. Temple. 1973. Preference for fixed-interval schedules: An alternative model. *Journal of the Experimental Analysis of Behavior 20*: 393–403.

Fantino, E., and C.A. Logan. 1979. *The Experimental Analysis of Behavior: A Biological Perspective.* San Francisco: Freeman.

Ferster, C.B., and B.F. Skinner. 1957. *Schedules of Reinforcement.* New York: Appleton-Century-Crofts.

Findley, J.D. 1958. Preference and switching under concurrent scheduling. *Journal of the Experimental Analysis of Behavior 1*: 123–44.

Fisher, J., and R.A. Hinde. 1949. The opening of milk bottles by tits. *British Birds 42*: 347–57.

Gause, G.F. 1934. *The Struggle for Existence.* Baltimore: Williams and Wilkins.

Goss-Custard, J.D. 1977. Optimal foraging and the size selection of worms by redshank, *Tringa totanus*, in the field. *Animal Behaviour 25*: 10–29.

Herrnstein, R.J. 1970. On the law of effect. *Journal of the Experimental Analysis of Behavior 13*: 243–66.

Hinde, R.A., and J. Stevenson-Hinde, eds. 1973. *Constraints on Learning.* London: Academic Press.

Hogan, J.A., and T.J. Roper. 1978. A comparison of the properties of different reinforcers. In J.S. Rosenblatt, R.A. Hinde, E. Shaw, and C. Beer, eds., *Advances in the Study of Behavior*, vol. 8, New York: Academic Press, pp. 155–255.

Honig, W.K.; I. Beale; P. Seraganian; D. Lander; and D. Muir. 1972. Stimulus and response reduction: Two aspects of inhibitory control in learning. In R.A. Boakes and M.S. Halliday, eds., *Inhibition and Learning.* London: Academic Press, pp. 41–71.

Humphrey, N.K. 1976. The social function of intellect. In P.P.G. Bateson and R.A. Hinde, eds., *Growing Points in Ethology*, Cambridge, England: Cambridge University Press, pp. 303–317.

Itani, J. 1958. On the acquisition and propagation of a new food habit in the natural group of the Japanese monkey at Takasaki-Yama. *Primates 1*: 84–98.

Ivlev, V.A. 1961. *Experimental Ecology of the Feeding of Fishes.* New Haven, Conn.: Yale University Press.

Jenkins, H.M., and B.R. Moore. 1973. The form of the auto-shaped response with food or water reinforcers. *Journal of the Experimental Analysis of Behavior 20*: 163–81.

Krebs, J.R. 1978. Optimal foraging: decision rules for predators. In J.R. Krebs and N.B. Davies, eds., *Behavioural Ecology*, Oxford: Blackwell, pp. 23–63.

Krebs, J.R.; J.R. Erichson; M.I. Webber; and E.L. Charnov. 1977. Optimal prey selection in the great tit (*Parus major*). *Animal Behaviour 25*: 30–38.

Krebs, J.R.; A. Kacelnik; and P. Taylor. 1978. Optimal sampling by foraging birds: An experiment with great tits (*Parus major*). *Nature 276*: 27–31.

Krebs, J.R.; J.C. Ryan; and E.L. Charnov. 1974. Hunting by expectation or optimal foraging? A study of patch use by chickadees. *Animal Behaviour 22*: 953–64.

Lea, S.E.G. 1979. Foraging and reinforcement schedules in the pigeon: Optimal and non-optimal aspects of choice. *Animal Behaviour 27*: 875–86.

_____. 1982. Correlation and contiguity in foraging behaviour. In P. Harzem and M.D. Zeiler, eds., *Advances in the Analysis of Behavior*, vol. 2, *Predictability, Correlation, and Contiguity*, Chichester, England: Wiley.

Lobb, B., and M.C. Davison. 1975. Performance on concurrent variable-interval schedules: a systematic replication. *Journal of the Experimental Analysis of Behavior 24*: 191–97.

Logan, F.A. 1964. The free behavior situation. In D. Levine, ed., *Nebraska Symposium on Motivation*, vol. 12, Lincoln: University of Nebraska Press, pp. 99–131.

MacArthur, R.H., and E.R. Pianka. 1966. On optimal use of a patchy environment. *American Naturalist 100*: 603–609.

Mackintosh, N.J. 1974. *The Psychology of Animal Learning.* London: Academic Press.

Menzel, E.W. 1978. Cognitive mapping in chimpanzees. In S.H. Hulse, H. Fowler, and W.K. Honig, eds., *Cognitive Processes in Animal Behavior*, Hillsdale, N.J.: Lawrence Erlbaum.

Moore, B.R. 1973. The role of directed Pavlovian reactions in simple instrumental learning in the pigeon. In R.A. Hinde and J. Stevenson-Hinde, eds., *Constraints on Learning*, London: Academic Press, pp. 159–88.

Muenzinger, K.F. 1938. Vicarious trial and error at a point of choice. I. A general survey of its relation to learning efficiency. *Journal of Genetic Psychology 53*; 75–86.

Myers, D.L., and L.E. Myers. 1977. Undermatching: A reappraisal of performance on concurrent variable-interval schedules of reinforcement. *Journal of the Experimental Analysis of Behavior 27*: 203–214.

Poirier, F.E. 1969. Behavioral flexibility and intertroop variation among Nilgiri Langurs (*Presbytis johnii*) of South India. *Folia Primatologica 11*; 119–33.

Pyke, G.H.; H.R. Pulliam; and E.L. Charnov. 1977. Optimal foraging: A selective review of theory and tests. *Quarterly Review of Biology 52*: 137–54.

Rachlin, H. 1973. Contrast and matching. *Psychological Review 80*: 217–34.

Rapport, D.J. 1971. An optimization model of food selection. *American Naturalist 105*: 575–87.

Sahlins, M. 1977. *The Use and Abuse of Biology*. London: Tavistock.

Seligman, M.E.P. 1970. On the generality of the laws of learning. *Psychological Review 77*: 406–418.

Small, W.S. 1900–01. Experimental study of the mental processes of the rat, II. *American Journal of Psychology 12*: 206–239.

Squires, N., and E. Fantino. 1971. A model for choice in simple concurrent and concurrent-chain schedules. *Journal of the Experimental Analysis of Behavior 15*: 27–38.

Ware, D.M. 1972. Predation by rainbow trout (*Salmo gairdneri*): The influence of hunger, prey density, and prey size. *Journal of the Fisheries Research Board of Canada 29*: 1193–1201.

Werner, E.E., and D.J. Hall. 1974. Optimal foraging and the size selection of prey by the bluegill sunfish (*Lepomis macrochirus*). *Ecology 55*: 1042–52.

8 AN ECOLOGICAL PERSPECTIVE ON THE STUDY OF THE ALLOCATION OF BEHAVIOR

Alan C. Kamil, Jerrilynn Peters, and Frederick J. Lindstrom

Different approaches to the study of the allocation of behavior among alternatives have been developed by psychologists and ecologists. These approaches are integrated in the research described in this chapter. For example, ecological theory predicts that when a predator is hunting for food that is distributed in discrete patches, the greater the distance between patches, the more persistent the predator should be within patches. We confirmed this prediction in laboratory simulation of the prey detection problem, in which blue jays were trained to search projected imates for cryptic moths. When this experiment was replicated as a red/green discrimination problem, however, no effect was obtained. These results imply that many phenomena may be apparent only under ecologically valid conditions.

The basic theme of this volume can be expressed quite simply: When an animal is faced with response alternatives that pay off differently (with differing probabilities, amounts of reinforcement, delays, effort, and so on), how will the animal distribute its behavior among the alternatives? This question is of obvious interest to psychologists, as can be seen in any textbook on learning. In recent years this question has also received considerable attention from ecologists and evolutionary biologists interested in foraging behavior and predator-prey interactions. This interest is the result of the realization that the allocation of behavior among alternatives determines foraging efficiency and has important implica-

We thank Ted Sargent for providing the *Catocala* specimens for the preparation of stimulus materials and for his advice. We also thank R.J. Herrnstein and H.L. Roitblat for their comments on a preliminary draft of this chapter. Supported by Grant BNS 76-80275 from the National Science Foundation.

tions for a variety of ecological issues. For example, the decisions made while foraging play a role in the control of prey populations by predators (e.g., Holling, 1959; Tinbergen, 1960), in the coevolution of plants and their animal pollinators (e.g., Heinrich, 1979; Pyke, 1980), and in changes in the breadth of diet of species in response to competition (MacArthur and Pianka, 1966).

As might be expected, the hypotheses and methods that biologists have brought to bear on the problem of the allocation of behavior are different from those that psychologists have applied to the problem. Ecologists particularly emphasize the relation between different foraging strategies and Darwinian fitness (Pyke, Pulliam, and Charnov, 1977). Other things being equal, animals with greater foraging efficiencies should leave more offspring on the average than animals with lower efficiencies. To the extent that differences in foraging behavior are heritable (Arnold, 1980; Gray, 1979), this differential reproduction should lead to the evolution of highly efficient foraging. In its most extreme form this view leads to the prediction that when an animal is tested appropriately, in the environment in which the species evolved or an environment similar to it, the foraging strategy of the animal should approach the optimal in terms of efficiency. This logic has produced a theoretical approach called optimal foraging theory (see reviews by Krebs, 1978; Pyke, Pulliam, and Charnov, 1977).

Therefore, although ecologists have carried out many laboratory experiments on foraging behavior, their emphasis is consistently upon understanding behavior occurring under natural conditions in the field. The basic message of this chapter is that despite these differences ecological work on foraging behavior, both empirical and theoretical, is very relevant to psychological research on the allocation of behavior. Psychologists need to broaden their perspectives so as to include more of the information and challenges that ecology offers. Too often psychological research on the allocation of behavior appears to be very narrowly conceived and defined and largely irrelevant to many of the problems animals face in nature.

For example, consider the traditional approach to the study of probability learning in discrete trials situations. In the typical probability learning experiment, the subject is faced with two response alternatives, each of which is reinforced with particular probability. Reinforcement of either alternative is unpredictable on any given trial, and the sum of the two probabilities is usually 1.0. This kind of research should be directly relevant to natural foraging problems, since foragers must distribute their foraging between different places (often called patches in the ecological literature), which differ in the probability that they will contain prey (e.g., Royama, 1970). But in fact the probability learning literature is largely irrelevant to this ecological problem. Part of this irrelevance stems from some basic assumptions psychologists tend to make about the nature of the probability learning problem, assumptions that limit the behavioral questions asked through experimentation.

For instance, what is the best strategy available to a subject in a probability learning experiment? Psychologists usually assume that the optimal strategy is to maximize the frequency of reinforcement by choosing the alternative with the higher overall probability of reinforcement on each and every trial (e.g., Bitterman, 1975). This strategy is the best available by one peculiar criterion: It maximizes the number of reinforcements obtained within a particular session.

An ecologist would probably disagree with this answer, because in an ecological context the animal's problem is more broadly defined. The maximizing strategy leaves the subject completely insensitive to any changes in the distribution of reinforcement at the unchosen alternative. Although such changes may be rare in probability experiments, they are probably common in nature. Krebs, Kacelnik, and Taylor (1978) indicate how some behavioral ecologists have attempted to deal with frequent variations in food availability.

On another level it is assumed in most probability learning experiments that the sum of the probabilities of reinforcement across the alternatives should be unity. This places a subtle constraint upon the events that can take place within an experiment. As long as the sum of the probabilities is a constant there can be no change in the probability of reinforcement at one of the alternatives without changes at the other. This constraint does not exist in nature. A bird feeding in a Scotch pine woods, for instance, may have several different prey types simultaneously available. Any one of these may increase or decrease in abundance while the others remain constant (Tinbergen, 1960). In addition this assumption results in reinforcement densities that may be much higher than those experienced under natural conditions.

Some of the differences between the psychological and ecological approaches to the problem of allocating behavior among alternatives are illustrated by the experiments of Smith and Sweatman (1974). These behavioral ecologists were interested in how birds distribute their foraging behavior among areas that differ in prey density. To investigate this, they set up six separate foraging areas, or patches, in a large aviary. Each of these six patches contained 256 small holes capped with aluminum foil, under which a mealworm might have been concealed. Great tits (*Parus major*), which had learned to uncap the holes to find mealworms, were allowed to forage within the aviary.

In the first experiment each patch was designated as a 0, 4, 8, 16, or 20-mealworm density area. Under these constant density conditions the six tits rapidly learned to concentrate their foraging behavior in the high-density patches. Although most foraging occurred in the most profitable patches, three tits continued to visit the patch that never contained food throughout the 30 days of the experiment. This behavior appears inefficient, since it never resulted in reinforcement and it certainly did not maximize the rate of reinforcement. However, poor foraging areas may improve, and this behavior might increase the sensitivity of the birds to any such changes.

A second experiment was then conducted to investigate the effects of changing the spatial distribution of the food. The number of mealworms in the zero-density patch was gradually increased to 20 while the number of mealworms in the 20-patch was gradually decreased to 0. Two of the three "inefficient" tits learned to utilize the new high-density patch, whereas none of the three "efficient" tits learned to utilize it even after 22 sessions. An inefficient behavior in the short run was quite efficient in the long run. This result suggests that the psychological view of the problem is relatively limited in scope. The theoretical models of ecologists (e.g., Hughes, 1979; Oaten, 1977) and the laboratory and field data upon which they are based (e.g., Gill and Wolf, 1977; Goss-Custard, 1980; Tinbergen, 1960) indicate that the range of problems faced by foraging animals is much broader than the range of problems psychologists have investigated. This in turn suggests that many important factors affecting the allocation of behavior between alternatives remain undetected in the psychological literature because of the restricted range of psychological research. This restricted range need not exist. Many of the issues raised by the ecological study of foraging behavior can be profitably attacked with techniques psychologists have developed.

Some of our research demonstrates both the utility of operant techniques for the investigation of foraging strategies and the probability that the narrow focus of psychological research has produced an incomplete understanding of the factors affecting the allocation of behavior. The basic approach has been to use operant techniques to simulate a foraging problem known to occur under natural conditions. The problem of particular interest is how a predator hunts for cryptic prey that vary in appearance and density. The specific predator-prey interaction to be simulated is between blue jays (*Cyanocitta cristata*) and *Catocala* moths, highly palatable and cryptic insects upon which blue jays prey in nature (Sargent, 1976). These moths are nocturnal and spend the day resting on substrates, usually tree trunks, which closely resemble their disruptively colored forewings (see Figure 8-1). They are also usually highly dispersed, so that any specific length of tree trunk is unlikely to contain more than one moth (Sargent, 1976). The foraging behavior of a blue jay hunting for *Catocala*, can be divided into three segments: (1) traveling to an area that may contain a moth (a patch); (2) searching that patch for a moth and then either deciding there is no moth present (in which case, the bird would leave and travel to the next patch to be searched) or deciding there is a moth present; and (3) attacking, which would result in ingestion only if the stimulus attacked really was a moth and if the attack was successful.

Our simulation of this foraging situation is based upon the concept formation technique first developed by Herrnstein and Loveland (1964). Blue jays were trained to search images projected on a pecking key for *Catocala* moths (Pietrewicz and Kamil, 1977, 1979). The apparatus used was an operant chamber with two keys, the first a large rectangular key upon which slides could be

Figure 8–1. A *Catocala relicta* on a white birch tree trunk, photographed with a subject-to-camera distance of 0.6 meters.

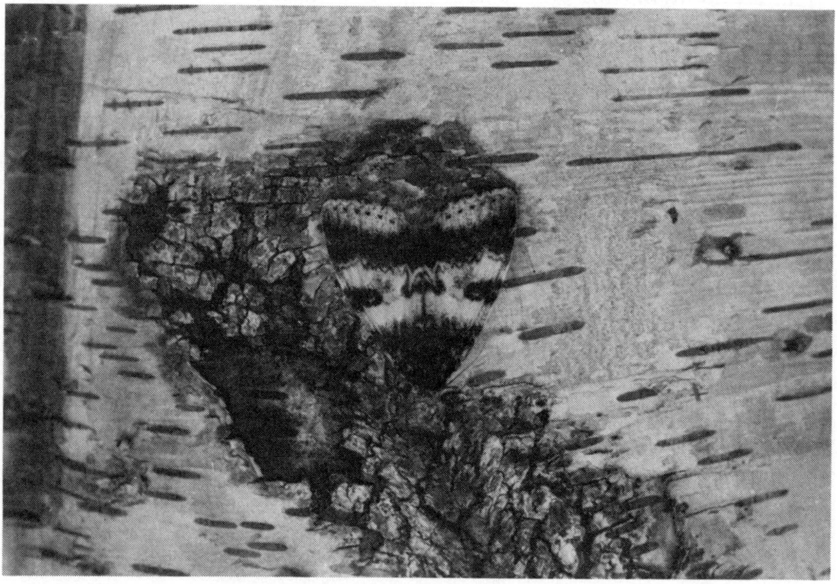

projected and the second a small round key upon which geometric stimuli could be projected was used. The large key was called the stimulus key; the small key served as both the start key and the advance key during different parts of every trial.

Each trial began with the illumination of the start key with a white plus sign. After a few seconds, called travel time, the display on the start key changed to a red circle. The next peck at the start key after the red circle came on caused a slide to be projected on the stimulus key. The display on the round key changed to a yellow circle, which indicated it was now an advance key. During this search stage of the trial, there were two response alternatives: peck the advance key or peck the stimulus key. If the jay pecked at the advance key, both keys were turned off and the trial ended. There was a 2-second intertrial interval during which the slide projector behind the stimulus key advanced, and then the next trial began. If the jay pecked at the stimulus key, then the display on the advance key was turned off and that key became inoperable while the display on the stimulus key was unchanged. The first peck at the stimulus key after 30 seconds (attack time) had passed since the first peck at the stimulus key produced the end of the trial. If there had been a moth in the slide projected on the stimulus key reinforcement, a piece of mealworm (*Tenebrio* larva) was delivered. If there had been no moth, no mealworm was delivered. In either case the next trial began after a 2-second intertrial interval during which the projector advanced.

If each slide projected on the stimulus key is considered a patch that might contain a prey item, this procedure modeled the natural situation closely. There were three stages to each trial. The travel time, during which the start key was illuminated with the white plus sign, was the minimum time between patches, analogous to traveling from one patch to the next. Once the image was projected on the stimulus key, the jay could search for a moth and terminate the search by either progressing to the next patch by pecking the advance key, which started the next travel time interval, or commence an attack by pecking at the stimulus key. This attack would be reinforced only if a moth was present.

In psychological terms this was a successive discrimination procedure with the addition of both a fixed-interval schedule on the start key preceding each trial, and an advance key. In essence the blue jay was asked if a moth was present in the image on the stimulus key. The jay could answer no by pecking the advance key, or answer yes by pecking at the stimulus key. Correct positives were reinforced whereas incorrect positives were discouraged by the "attack time" requirement. As in nature, there was no feedback provided on the accuracy of decisions to proceed to the next patch by pecking the advance key.

Many different kinds of hypotheses can be tested with this paradigm. In one series of experiments we have tested some of the predictions of optimal foraging theory (often referred to as OFT in the ecological literature). These are theories that assume that animals behave so as to maximize their foraging efficiency, usually by maximizing their average rate of energy gain per unit foraging time (Charnov, 1976; Krebs, 1978; Pyke, Pulliam, and Charnov, 1977; Schoener, 1971; see also papers in Kamil and Sargent, 1980). The OFT approach proceeds by constructing a mathematical or graphical model of a particular foraging problem and solving for the solution that maximizes intake rate. This optimal solution usually varies as a function of such parameters as prey density, ease of capture of the prey, spatial distribution of the prey, or caloric value of the prey. Given the assumption that animals forage optimally, these variations in the optimal solution can be treated as predictions about foraging behavior.

One problem that has been analyzed extensively in the OFT literature is hunting for prey that are found in discrete patches. The classic natural case is that of a bird hunting for insect larvae that burrow beneath the surface of pine cones (Gibb, 1962; Krebs, Ryan, and Charnov, 1974). Each pine cone is a discrete patch containing an unknown number of larvae. When should the bird leave the pine cone it is currently searching and travel to another? According to the type of OFT model most frequently applied to this situation, the marginal value theorem (Charnov, 1976), the decision to give up and leave a patch should be a function of two expected rates of intake: the overall rate for the environment as a whole and the rate to be expected if search within the current patch is continued. This within-patch rate would be relatively high when the bird first arrives in a new patch but would decline as time in the patch increases, because of prey depletion. The marginal value theorem predicts that the bird should leave when the within-patch expected rate has dropped to the level of the overall rate, since this will maximize the intake rate across the entire foraging bout (Charnov, 1976).

This simple model makes several predictions about how an animal foraging for patchily distributed prey should respond to various parameters. One prediction, which has been confirmed in several experiments (e.g., Cowie, 1977), is that the farther apart the patches within which prey may be found are located, the more persistent and thorough the predator should be in searching any particular patch. This is because increases in travel time decrease the overall rate of intake in the environment as a whole without affecting the expected within-patch rate. Therefore, more search time within an individual patch is required before the expected within-patch rate will drop to the level of the overall expected rate.

Since this travel time effect was well established, it seemed a good place to begin testing the applicability of OFT to our paradigm. However, the marginal value theorem cannot be applied directly since in our procedure each patch contains only exactly one or zero prey whereas the marginal value theorem assumes a continuous distribution of prey per patch. Fitzpatrick (1978) has shown that a similar prediction holds for the 0 or 1 prey/patch situation. Fitzpatrick predicts that a forager should leave a patch when

$$p(0) - p(t) = B/C ,$$

where

$p(t)$ is the probability of detecting a prey in a patch in which t seconds have been spent in unsuccessful search;

$p(0)$ is the probability of detecting a prey upon arrival in a new patch, when $t = 0$;

B is the benefit of consuming a prey, in nutritional units; and

C is the cost of moving between patches, in the same units as B (e.g., calories).

The value of B must be relatively large compared to C, or else the predator would soon starve to death, using more energy moving between patches than it can gain from the prey it may find. The quantity $p(0) - p(t)$ starts at 0 for $t = 0$ and gradually increases as search proceeds. Since increasing the distance between patches increases C without changing B, increasing interpatch distances are predicted to increase the persistence of the predator before giving up on any particular empty patch.

This OFT prediction is basically a prediction about how animals will allocate their behavior between alternatives. Consider a blue jay searching a tree trunk for a cryptic *Catocala* moth. At any given moment, if the jay has failed to find a moth on the tree trunk, the jay has two alternatives; it can continue to search or it can go on to the next tree trunk (or patch). According to Fitzpatrick's optimization model, the farther apart the patches are, the longer the jays should stay on a tree trunk containing no moth before giving up. In the first experiment this prediction was tested by varying the travel time, the duration of the signaled fixed-interval schedule on the start key. The greater this travel time, the longer the interval between images on the stimulus key and the greater the average number of pecks directed at the start key. It seems reasonable to suppose that increases in travel time so defined increase the cost of "moving" from one trial to the next. Therefore, if Fitzpatrick's model can be applied to our situation, increases in travel time should produce increased persistence in searching mothless images.

The first experiment tested this prediction. There were five blue jays (*Cyanocitta cristata*) in this experiment, all highly experienced at finding moths in projected images. Every slide showed two birch logs standing vertically, most without moths, photographed at a subject-to-camera distance of 1.3 meters. Slides that contained a moth had a *Catocala relicta* on one of the logs (the position of the moth varied randomly, with the restriction that it appear equally often on each log and equally often in each vertical third of the slides). *C. relicta* is a white moth with black disruptive coloration. In the northeastern United States, *C. relicta* usually rests on birch trees upon which it is quite cryptic (Figure 8-1; Pietrewicz and Kamil, 1977; Sargent, 1976). Eighty-one slides were used. Of these 27 (1/3) contained *C. relicta.* Session duration was fixed at 45 minutes, so that the speed with which the jays responded affected the numbers of trials they received. Three travel times were used: 5, 15, and 30 seconds. Each jay received each of the three travel times in a different sequence, with five consecutive daily sessions at each of the travel times.

The analysis has concentrated upon those aspects of the results most relevant to the prediction of Fitzpatrick. On each trial the latency of the first response after the stimulus key was illuminated was recorded. The basic measure of persistence was giving-up time (GUT), defined as the latency of advance key pecks on trials on which no moth was present. Also recorded was attack time (ATT), defined as the latency of the first peck at the stimulus key on trials on which a moth was present. (Trials on which the jays made errors were omitted from this analysis.)

Since it usually took the jays a few sessions to adjust to a new travel time, GUT and ATT were analyzed over the last two sessions at each travel time. Increasing travel times resulted in significant increases in GUT, $F(2, 28) = 6.12$, $p < 0.05$, as shown in Figure 8-2. These increases in GUT were not accompanied by increases in ATT, which tended to decrease as travel time increased

Figure 8-2. The mean giving-up and attack times across the last two sessions at each travel time for jays search for cryptic moths. The lower right panel shows the means for the group; the other panels show data for the individual jays.

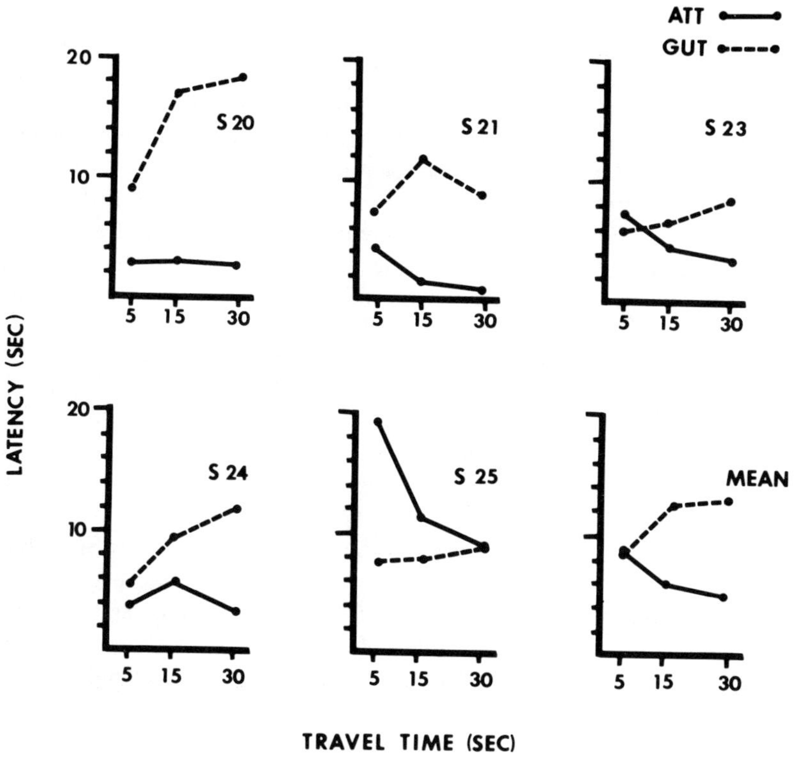

($p < 0.10$). Thus, the increases in GUT were not the result of a general tendency to respond slowly at high travel time values.

From the point of view of the ecologist these results offer good qualitative support for Fitzpatrick's model. They also demonstrate the contribution psychologists can make to testing predictions of ecological models of foraging behavior, as does the other research by psychologists such as that of Lea (1979 and Chapter 7 of this volume) and Collier (Collier and Rovee-Collier, 1980). This contribution could and should be substantial since most ecological models of foraging behavior conceptualize foraging as consisting of sequences of decisions, and psychologists have developed a wide array of techniques for the study of decisionmaking by animals. In fact some ecologists have suggested that the efforts of psychologists will be of critical importance in further development of foraging theory (Pulliam, 1980; Schoener, 1977).

From the point of view of a psychologist these results have somewhat different implication. The effect of travel time on the allocation of behavior is not predicted directly by any psychological theory we are aware of, although it is not inconsistent with traditional psychological views. For example, the travel time requirement might be considered analogous to the changeover delay (COD) requirement used in many studies of behavior on concurrent schedules (e.g., Shull and Pliskoff, 1967). Since increasing the COD requirement increases the time spent in a single component before switching, the results of the two types of experiments appear to be similar. However, this analogy is very tentative. The probability distributions of reinforcement as a function of time in the patch or in the schedule component are very different in the two types of experiments. But the travel time effect does follow very directly and compellingly from an ecological perspective. This raises the possibility that there are many parameters like travel time, which affect behavior allocation in significant ways, but are most evident from an ecological, functional perspective. The scope of psychological research could broaden beneficially by heightened awareness of ecology.

The ecological perspective suggests many changes in psychological research: new parameters to study, the development of new procedures, or important modifications of old procedures. It can even prompt psychologists to do field research (Kamil, 1978). In the case of the travel-time experiment the procedures used were explicitly designed to simulate the real world problems of a blue jay hunting for cryptic prey. These procedures were developed because of our feeling that some important phenomena are most likely to be detected in experiments run under relatively natural conditions.

Although no operant chamber can be a New England woodland, the use of real prey items photographed on their natural substrates, which probably accurately simulate the detection problem *Catocala* present to predators (Pietrewicz and Kamil, 1977), is a critical part of this simulation. The natural concept literature (e.g., Cerella, 1979; Herrnstein, 1979) shows that discrimination and generalization experiments with complex natural stimuli may produce very different results than experiments using simple stimuli. The strong travel-time effect found with natural stimuli provided an opportunity to test the value of these stimuli by testing the effects of travel time with artificial stimuli more typical of psychological research.

Five blue jays were trained on a red/green discrimination problem, with red as the positive stimulus, just as the jays in the previous study had been trained on the cryptic moth problem. They next received the same sequence of travel-time tests as the jays of the previous experiment, except that red slides were substituted for images with moths and green slides for images without moths. As shown in Figure 8-3, travel time had very different effects in this experiment than in the previous experiment. There were no significant effects of travel time

Figure 8-3. The mean giving-up and attack times across the last two sessions at each travel time for jays on a red-green discrimination. The lower right panel shows the means for the group; the other panels show the data for the individual jays.

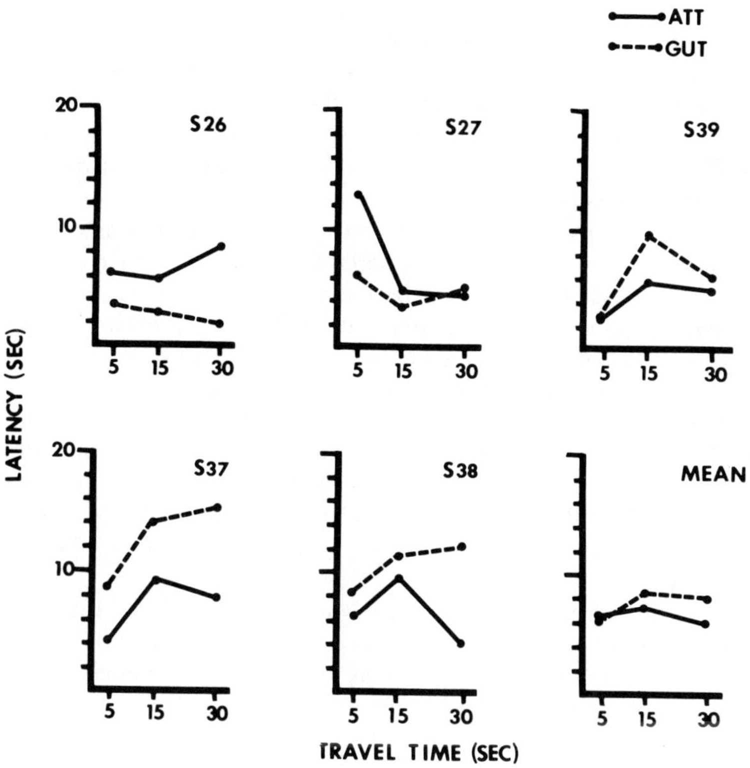

on GUT, $F(2, 8) < 1$. Furthermore, individual jays that did show increases in GUT at high travel times also showed increases in ATT, indicating a general decrement in the speed of response at high travel times. In order to compare the experiments statistically, GUT minus ATT was calculated and a significant result was found, the experiment by travel-time interaction, $F(2, 16) = 4.06$, $p < 0.05$. It is clear that the change in the stimuli used altered the effects of travel time.

The difference in the results of the two travel-time experiments suggests that the effects of important variables may be most apparent under ecologically valid conditions. It can be argued that this research only provides a suggestive example of this contention, not a proof. But no single set of experiments could prove a point of this type. The advantage of laboratory research over naturalistic field work is that the laboratory permits precise experimental control. But the laboratory also requires simplification, with the concomitant elimination of many factors. In the course of this simplification important dimensions of the natural environment of the animal may be lost. Many important effects may remain inaccessible to psychologists because inappropriate and arbitrary experimental settings are employed too frequently.

The solution to this problem is not to end laboratory research. What is needed is an increase in researchers' awareness of the natural world of the animals studied, particularly of the problems they face that may be solved through learning. This information could contribute, where appropriate, toward laboratory analyses of learned behavior. This strategy is a large part of the ecological approach recommended here.

Historically this may be the right time for an increased emphasis on the functional approach in psychology, particularly with regard to foraging behavior. A great deal of useful information is becoming available to guide the laboratory experimenter interested in simulating natural foraging problems. In the first place many field studies are now available in the literature in which the foraging behavior of individually marked or recognizable animals has been recorded. Many of these studies have been summarized or referenced in two recent books (Curio, 1976; Kamil and Sargent, 1980). In addition, the following examples may be of interest to the reader: Davies (1977), Glander (1978), Gill and Wolf (1977), Goss-Custard (1977), Heinrich (1976), Kamil (1978), Murton (1971), Partridge (1976), Reichman (1977), Royama (1970), Smith (1974 a, b), Tinbergen (1960).

In the second place, many field studies in which very little or no behavioral data were collected offer indirect evidence about the behavior of foragers. For example, our own work on the *Catocala* is facilitated by the large amount of available data about these animals reported by lepidopterists (Sargent, 1976). In general, information about the characteristics of the food for which predators are searching often carries important implication about foraging behavior. This seems particularly true in predator-prey interactions (e.g., prey polymorphisms and the search image hypothesis, Croze, 1970), and in systems where there has been coevolution, such as nectar feeding animals pollinating the plants whose flowers they feed on (e.g., Kamil, 1978; Pyke, 1980).

The data reported here indicated that travel time interacts with some aspect of the perceptual problem with which the animal is presented. Two specific aspects of the difference between the moth problem and the red/green problem seem most likely to be important. The moth detection task was a moderately difficult detection problem whereas the red/green task was an easy discrimination problem. Either task difficulty or the difference between detection and discrimination problems could be responsible. Further research is planned to identify the critical variable to producing the travel-time effect.

Travel-time effects probably will appear only in difficult detection situations. In discrimination problems there is a distinct negative stimulus. Once this negative stimulus has been perceived on any particular trial, further search time is unlikely to be productive. But in a detection problem, particularly a difficult one such as that presented by highly cryptic prey, there is no distinct, unambiguous negative stimulus. An apparently empty birch tree always might contain a moth. With this uncertainty about negative evidence, continuing to search is a viable option in the allocation of behavior.

Anyone who studies behavior knows that most independent variables interact with others. It seems reasonable to suppose that many such interactions are structured so that natural, ecologically valid conditions will be most revealing. Our research supports this view. We are not arguing that ecological relevance explains behavior but, rather, that such relevance can be an important contribution to doing informative research. Using operant techniques to study effects derived from the ecological literature will be a very productive strategy, one that is beginning to be adopted by some behavioral ecologists (e.g., Krebs, Kacelnik, and Taylor, 1978). Such cross-disciplinary effort will be essential to developing a complete understanding of the ways in which animals allocate their behavior among alternatives, a problem of great interest to both psychologists and ecologists. Thus we urge psychologists interested in the allocation of behavior and in animal learning in general to consider the importance of studies of behavior in the field to their own laboratory research.

REFERENCES

Arnold. S. 1980. Microevolution of feeding behavior. In A.C. Kamil and T.D. Sargent, eds., *Foraging Behavior: Ecological, Ethological, and Psychological Approaches*, New York: Garland.

Bitterman, M.E. 1975. The comparative analysis of learning. *Science 188*: 699–709.

Cerella, J. 1979. Visual classes and natural categories in the pigeon. *Journal of Experimental Psychology: Human Perception and Performance 5*: 68–77.

Charnov, E.L. 1976. Optimal foraging: The marginal value theorem. *Theoretical Population Biology 9*: 129–36.

Collier, G.H., and C.K. Rovee-Collier. 1980. A comparative analysis of optimal foraging behavior: Laboratory simulations. In A.C. Kamil and T.D. Sargent, eds., *Foraging Behavior: Ecological, Ethological, and Psychological Approaches*, New York: Garland.

Cowie, R.J. 1977. Optimal foraging in great tits (*Parus major*). *Nature 268*: 137–39.

Croze, H.J. 1970. Searching images in carrion crows. *Zeitschrift fur Tierpsychologie 5*;1–85.

Curio, E. 1976. *The Ethology of Predation.* Berlin: Springer-Verlag.

Davies, N.B. 1977. Prey selection and the search strategy of the spotted flycatcher (*Muscicapa straita*): A field study on optimal foraging. *Animal Behaviour 25*:1016–33.

Fitzpatrick, J.W. 1978. Foraging behavior and adaptive radiation in the avian family Tyrannidae. Unpublished doctoral dissertation, Princeton University.

Gibb, J.A. 1962. Tinbergen's hypothesis of the role of specific searching images, *Ibis 104*:106–11.

Gill, F.B., and L.L. Wolf. 1977. Nonrandom foraging by sunbirds in a patchy environment. *Ecology 58*:1284–96.

Glander, K.E. 1978. Howling monkey feeding behavior and plant secondary compounds: A study of strategies. In G.G. Montgomery, ed., *The Ecology of Arboreal Folivores*, Washington, D.C.: Smithsonian Press.

Goss-Custard, J.D. 1977. The energetics of prey selection, *Tringa tatanus* (L.), in relation to prey density. *Journal of Animal Ecology 46*:1–19.

____. 1980. Feeding behavior of Redshank, Tringa totanus, and optimal foraging behavior. In A.C. Kamil and T.D. Sargent, eds., *Foraging Behavior: Ecological, Ethological, and Psychological Approaches*, New York: Garland.

Gray, L. 1979. Feeding diversity in deer mice. *Journal of Comparative and Psysiological Psychology 93*:1118–26.

Heinrich, B. 1976. The foraging specializations of individual bumblebees. *Ecological Monographs 46*;105–28.

____. 1979. *Bumblebee Economics.* Cambridge, Mass.: Harvard University Press.

Herrnstein, R.J. 1979. Acquisition, generalization, and discrimination reversal of a natural concept. *Journal of Experimental Psychology: Animal Behavior Processes 5*:116–29.

Herrnstein, R.J., and D.H. Loveland. 1964. Complex visual concept in the pigeon. *Science 146*: 549–51.

Holling, C.S. 1959. The components of predation as revealed by study of small mammal predation of the European pine sawfly. *The Canadian Entomologist 91*;293–320.

Hughes, R.N. 1979. Optimal diets under the energy optimization premise: The effects of recognition time and learning. *The American Naturalist 113*: 209–21.

Kamil, A.C. 1978. Systematic foraging by a nectar-feeding bird, the Amakihi (*Loxops virens*). *Journal of Comparative and Physiological Psychology 92*: 388–96.

Kamil, A.C., and T.D. Sargent, eds. 1980. *Foraging Behavior: Ecological, Ethological, and Psychological Approaches.* New York: Garland.

Krebs, J.R. 1978. Optimal foraging: Decision rules for predators. In J.R. Krebs and N.B. Davies, eds., *Behavioral Ecology: An Evolutionary Approach*, London: Blackwell.

Krebs, J.R.; A. Kacelnik; and P.J. Taylor. 1978. Test of optimal sampling by foraging great tits. *Nature 275*: 27–31.

Krebs, J.R.; J.C. Ryan; and E.L. Charnov. 1974. Hunting by expectation or optimal foraging? A study of patch use by chickadees. *Animal Behaviour 22*: 953–64.

Lea, S.E.G. 1979. Foraging and reinforcement schedules in the pigeon: Optimal and non-optimal aspects of choice. *Animal Behaviour 27*: 875–86.

MacArthur, R.H., and E.R. Pianka. 1966. On the optimal use of a patchy environment. *American Naturalist 100*: 603–609.

Murton, R.K. 1971. The significance of a specific search image in the feeding behaviour of the wood-pigeon. *Behaviour 40*: 10–42.

Oaten, A. 1977. Optimal foraging in patches: A case for stochasticity. *Theoretical Population Biology 12*: 263–85.

Partridge, L. 1976. Field and laboratory observations on the foraging and feeding techniques of blue tits (*Parus ceruleus*) and coal tits (*Parus ater*) in relation to their habitats. *Animal Behaviour 24*: 534–44.

Pietrewicz, A.T., and A.C. Kamil. 1977. Visual detection of cryptic prey by blue jays (*Cyanocitta cristata*). *Science 195*: 580–82.

_____. 1979. Search image formation in the blue jay (*Cyanocitta cristata*). *Science 204*: 1332–33.

Pulliam, H.R. 1980. Learning to forage optimally. In A.C. Kamil and T.D. Sargent, eds., *Foraging Behavior: Ecological, Ethological, and Psychological Approaches*, New York: Garland.

Pyke, G.H. 1980. Optimal foraging in nectar-feeding animals and coevolution with their plants. In A.C. Kamil and T.D. Sargent, eds., *Foraging Behavior: Ecological, Ethological, and Psychological Approaches*, New York: Garland.

Pyke, G.H.; H.R. Pulliam; and E.L. Charnov. 1977. Optimal foraging: A selective review of theory and tests. *Quarterly Review of Biology 52*: 137–54.

Reichman, O.J. 1977. Optimization of diets through food preferences by heteromyid rodents. *Ecology 58*: 454–57.

Royama, T. 1970. Factors governing the hunting behaviour and selection of food by the Great Tit (*Parus major L.*). *Journal of Animal Ecology 39*: 619–68.

Sargent, T.D. 1976. *Legion of the Night: The Underwing Moths.* Amherst: University of Massachusetts Press.

Schoener, T.W. 1971. Theory of feeding strategies. *Annual Review of Systematics and Evolution 2*: 123–41.

_____. 1977. Feeding behavior. (Review of *The Ethology of Predation* by E. Curio). *Science 196*: 157–58.

Shull, R.L., and S.S. Pliskoff. 1967. Changeover delay and concurrent schedules: Some effects on relative performance measures. *Journal of the Experimental Analysis of Behavior 10*: 517–27.

Smith, J.N.M. 1974a. The food searching behaviour of two European thrushes. I. Description and analysis of search paths. *Behaviour 48*: 276–302.

_____. 1974b. The food searching behaviour of two European thrushes. II. The adaptiveness of search pattern. *Behaviour 49*: 1–61.

Smith, J.N.M., and H.P. Sweatman. 1974. Food searching behavior of titmice in patchy environments. *Ecology 55*: 1216–32.

Tinbergen, L. 1960. The natural control of insects in pine woods. I. Factors influencing the intensity of predation in song birds. *Archives Neerlandaise Zoologie 13*: 235–43.

9 STAYING AND SHIFTING
Their Effects on Discrimination Learning

David S. Olton

A fundamental problem for any animal is finding food. Many animals solve this problem with a good memory and an appropriate food-searching strategy. The memory provides information about places that have been visited previously and the results of those visits. The strategy takes this information and directs subsequent searches to the most profitable areas, maximizing the benefits and minimizing the costs.

This chapter discusses both memory and food-searching strategies in the context of discrimination learning in laboratory experiments and optimal foraging in natural habitats. The first series of experiments demonstrates that rats have a very good memory for the places they have been and that they use this information as if it were a list of items stored in working memory. The second series of experiments focuses on food-searching strategies, with special emphasis on *win-stay* and *win-shift* strategies. These are introduced in the context of predators searching for prey in natural environments and then used to describe the behavior of rats solving discrimination problems in the laboratory. This analysis assumes that optimal foraging and sucessful discrimination performance have much in common. Indeed, one of the goals of this research is to determine the extent to which an appreciation of the behavior of an animal in the role of pre-

This research was supported in part by Research Grant NSF-BNS-7824160 from the National Science Foundation and by a Biomedical Sciences Training Grant from The Johns Hopkins University. The author thanks S. Glickman, G. Handelmann, A. Kamil, A. Riley, J. Reichman, J. Walker, and T. Whitham for helping to refine these ideas, B. Taylor for testing the rats in the Maier three-table experiment, and M. Weigel and J. Krach for typing the manuscript.

dator searching for prey can help in understanding the performance of that same animal when placed in the role of a subject solving a discrimination problem in the laboratory (cf. Kamil, Peters, and Lindstrom in the preceding chapter).

SPATIAL MEMORY

Animals consistently exhibit an excellent memory for spatial information (see reviews in Olton, 1978, 1979). Consequently, spatial tasks can be used to address questions about animal memory that have been difficult to answer with nonspatial procedures. For example, one set of experiments was designed to determine the characteristics of working memory in rats. The test procedure required rats to visit each of a series of places once and not return there for the remainder of each test session. The apparatus was a radial-arm maze, with each of the arms extending from a central platform like a spoke on a wheel. At the beginning of each test, one pellet of food was placed at the end of each arm. A food-deprived rat was placed on the central platform and allowed to choose freely among the arms. In order to perform correctly and avoid returning to an arm that had already been chosen (and from which the food had been removed), the rat had to determine which arms had already been chosen on that test.

The rats learned rapidly and performed well. On a maze with 4 arms, after about 10 test sessions they performed perfectly, going to each arm once in the first 4 choices. On a maze with 17 arms, they reached asympotic performance after about 50 test sessions, going to 15 different arms in the first 17 choices. The rats did not rely on response chains or intramaze cues to choose correctly. Instead they identified and remembered each arm on the basis of extramaze cues associated with the position of the arm in the testing room (Beatty, 1979; Beatty and Shavalia, 1980; Black, Augerinos, and Suzuki, Note 1; Maki, Brokofsky, and Berg, 1978; Olton, 1978, 1979; Olton and Collison, 1979; Olton, Collison, and Werz, 1978; Olton, Handelmann, and Walker, 1980; Olton and Samuelson, 1976; Olton and Schlosberg, 1978; Roberts, 1979; Suzuki, Augerinos, and Black, 1980; Walker and Olton, 1979; Zoladek and Roberts, 1978).

A model of this behavior (Olton, 1978) has emphasized the development of a list of items in working memory (Honig, 1978), with each item representing one of the arms chosen by the rat. This list has the following characteristics.

- The maximum amount of information that can be recalled is approximately 12 items.
- Interference occurs among items in a given list; as the number of items to be remembered increases, the accuracy of recall decreases.
- In contrast, interference among items in different lists is minimal (Beatty and Shavalia, in press; Maki, Brokofsky, and Berg, 1979; Olton and Samuelson, 1976; Olton, Becker, and Handelman, 1980b).

- A recency effect is present; items just entered into working memory are recalled more accurately than those entered previously.
- There is no evidence of a primacy effect; the first items entered into memory are recalled no more accurately than items entered subsequently.
- Decay of the memory does not take place for extended periods of time; recall of items from working memory is accurate even when several hours separate the time at which the items were entered in memory and the time when the rat recalls these items.
- Working memory can be reset; interference among items in one list is not carried over to items in a different list.

The test procedure using the radial-arm maze requires "working memory" (see Olton, Becker, and Handelmann, 1980 a, b, for a discussion of the characteristics of working memory). An important point is that this terminology is based on the definition offered by Honig (1978), which is an operational one emphasizing that this type of memory is useful for only a single test session. No stipulations are made about the types of cognitive processes that might be taking place while working memory is active, so that this definition is considerably different from that offered by Baddley and Hitch (1974).

A common example of a task requiring working memory is delayed matching to sample. At the beginning of each trial a stimulus is presented as the sample and removed. After a delay interval it is presented again along with another stimulus. The animal is reinforced for responding only to the sample stimulus. The stimulus presented as the sample varies from trial to trial. Consequently the only way the animal can determine which stimulus should be chosen at the end of a trial is to remember which stimulus was presented as the sample at the beginning of that trial. After the end of a given trial the animal should forget the sample stimulus because the memory of it will not help (and will actually interfere with) performance on the next trial (D'Amato, 1973).

The radial-arm maze procedure has a working-memory component because the list of choices developed during each test session is useful for that session but not for subsequent sessions. As will be seen from the experiments to be discussed, many types of discrimination problems, both in the laboratory and in natural habitats, have working-memory components.

FORAGING IN NATURAL HABITATS

Consider a predator who has just found prey and has returned to some central place. This situation commonly occurs with *central-place foragers* (Hamilton and Watt, 1970) who eat their food in a protected location, store it for future use, or feed it to their young. On the next foraging trip the predator is faced with a choice between returning to the place where the prey has just been obtained or searching elsewhere. Returning to the place of a previously successful foraging trip is an example of a *win-stay* strategy, while going to a different place is an example of a *win-shift* strategy. In both cases, the predator has *won* on the initial trip and obtained food. The question is whether he should *stay* with the previously successful area and search there or *shift* to an alternative area and search there.

This type of foraging situation obviously has a working-memory component and is very similar to a delayed matching to sample procedure. The initially visited place may be considered as the sample stimulus. The return to the central place imposes a delay, after which the predator has a choice between responding to the sample stimulus or to some other stimulus (another place in which prey might be obtained). Obvious differences exist between the parameters used for delayed matching to sample tasks in the laboratory and those found in examples of predators foraging for prey in natural habitats, but the two situations are formally analogous and may be analyzed with the same conceptual framework.

The foregoing example assumes that the predator removes the prey from the area where it was caught. Some predators consume their prey where they find them. In such cases the stay-shift question has been addressed in terms of "giving-up-time," the length of time that the predator should continue to search in the same area before leaving it and going to a different area (Charnov, 1976; Gibb, 1962; Krebs, 1973; Krebs, Ryan, and Charnov, 1974; Zach and Falls, 1976b). Obviously the relative costs of staying and shifting differ in these two examples. A predator remaining in a given area will have to spend less energy to resume searching in that area than a predator who has left the area and must return to it. As a result, the optimal point for shifting will occur sooner for a predator who has left the original area than for a predator who remains there. Nonetheless the basic question for the predator remains the same, and both of these cases will be analyzed in terms of win-stay and win-shift strategies.

The stay-shift decision would have few consequences if prey were distributed randomly in the predator's habitat. But such is not the case. Prey usually occur in "patches," discrete areas within the habitat having a greater density of prey than the habitat as a whole. Thus the optimal strategy for a predator clearly depends on the way in which the prey are distributed in the patches.

An example of a system supporting a win-shift strategy has been described by Kamil (1978). The Hawaiian honeycreeper, *Loxops virens*, obtains nectar from flowers of the mamane tree, *Sophora chrysophylla*. On each visit to a flower the honeycreeper depletes that flower of nectar. Because the flower takes several hours to replenish the nectar, the bird will not be able to obtain the maximum amount of nectar until sufficient time has passed for nectar production. Meanwhile the most appropriate strategy for the bird is to shift and visit other flowers before returning to this one. Honeycreepers are very good at following this shift strategy. Furthermore, they do not mark the flowers when they visit them, nor do they follow a response sequence (such as beginning at the north end of their territory and moving south) when feeding. The available evidence suggests the birds avoid repeating flowers by remembering the ones they have previously chosen, storing the information in working memory.

A more complicated example of an environment supporting a win-shift strategy has been described by Whitham (1977). Bumblebees, *Bombus sonorus*, obtain nectar from *Chilopsis linearis* flowers. These flowers have two sources of nectar. Some of the nectar is in a pool at the bottom of the flower. This source is large and the nectar can be obtained easily and quickly. The rest of the nectar is in grooves on the side of the flower. This source is relatively small and the nectar difficult to obtain. The bees forage in a systematic fashion, visiting the flowers twice each day. On the first trip to a group of flowers a bee obtains only the pool nectar from each flower. On the second trip the bee obtains the groove nectar. This pattern is repeated each morning, giving two sequences of win-shift behaviors; the first occurs while the bees obtain the pool nectar, the second occurs while they obtain the groove nectar.

Environments can also support win-stay strategies. J.N.M. Smith has described the search patterns of two species of English thrush (Smith, 1974 a, b; Smith and Sweatman, 1974). These birds hunt for earthworms and other small invertebrates in meadows. When they find prey their rate of forward movement decreases while their rate of turning increases. These two changes combine to keep the birds in the vicinity of the successful catch. This strategy is adaptive because the worms and grubs they eat tend to be grouped together so that finding some prey increases the probability of finding additional prey in the same area.

Ovenbirds, *Seirus aurocapillus*, exhibited a win-stay strategy when searching for mealworms (Zach and Falls, 1976 a, b). The test environment was an enclosure 6 meters square placed on a forest floor. The arena was subdivided into several patches and mealworms distributed unequally among them. The ovenbirds were released in the enclosure and allowed to search for the worms. Finding a mealworm had two effects on subsequent search behavior, increasing both the probability of the bird remaining in the area in which the mealworm was found and the probability of returning to that area if the bird left it. Consequently, the birds spent most of their time searching in the patches where the density of mealworms was the greatest.

The purpose of this brief discussion of optimal foraging is to point out that different environments support different search strategies. (A summary of other relevant examples may be found in Kamil and Sargent, 1980.) Individuals adopting a search strategy appropriate to the distribution of food in their environment should have an adaptive advantage over individuals who adopt an inappropriate strategy. If the characteristics of the prey remain constant for many generations and the strategy adopted by an individual animal is influenced by genetic variables, then natural selection should gradually lead to the evolution of a species-specific tendency to follow the most adaptive strategy. Thus any given animal tested in a laboratory task might be expected to have a predisposition to search for food in a particular way.

SEARCHING FOR FOOD IN LABORATORY EXPERIMENTS

The behavior of rats in laboratory tests suggests that they have a strong predisposition to go to new places rather than return to previously investigated ones. A well-known example of this preference is spontaneous alternation, usually tested in a T maze (see reviews by Dember and Fowler, 1958; Douglas, 1966). The rat is placed in the start box of the T maze and allowed to choose one arm. It is then picked up, returned to the start box, and allowed to make another choice. The question of interest is whether it goes back to the arm it initially entered or goes to the other arm. Rats consistently prefer to alternate their choices, going to the arm not entered on the previous choice.

This tendency to avoid already visited places appears in more complicated mazes, such as the Dashiell maze (Dashiell, 1930) and the Bättig maze (Bättig, Driscoll, Schlatter, and Uster, 1976; Bättig and Schlatter, in press; Uster, 1977; Uster, Bättig, and Nägeli, 1976). The results from the Bättig maze are of particular interest because they demonstrate the efficiency with which rats can explore an environment, going to places where they have not been recently and avoiding places where they have been. The maze was constructed of three concentric hexagonal alleys interconnected by other alleys perpendicular to the hexagonal ones. Inserts were placed in the alleys to block them, creating many different configurations and allowing rats to get from one area to another only through the remaining unblocked alleys. Rats were placed in the maze once a day and allowed to explore it for several minutes. During the first few tests the rats moved around the maze unsystematically and often returned to sections previously visited during that test. Very quickly, however, they developed an efficient patrol of the entire maze, beginning at one point and rarely returning to a previously visited place until they had explored all the other available places.

Three Tower Enclosure

Rats searching for food in a relatively unstructured environment shifted their choices among all possible food sources before responding only to the source that actually had food (Olton, Walker, Gage, and Johnson, 1977). The test apparatus was an arena, 2 meters by 3 meters, filled with a variety of objects. Three towers served as sources of food. The top of a tower was used as a food source for two reasons. First, the response of running up the tower to get food was easily seen and clearly distinguished from other responses that the rats made. Second, the top of the tower was relatively unpleasant for the rats, so that they remained there only long enough to pick up a few food pellets, which were subsequently taken to a protected place in the arena to be eaten. Consequently the rats made many trips to the towers to obtain food, giving many responses for analysis.

On any given day the top of only one tower had food on it. The tower that actually had food changed from day to day so that the rat was unable to predict which of the three towers might actually have food. Thus the optimal strategy for the rat was to choose among the towers until finding the one that contained food and then to continue to go only to that tower, removing the food from it and eating it elsewhere.

The rats did not follow this strategy, however. Instead they visited all three towers every day, even when they had found the single food source on the very first choice. Indeed, the probability of returning to the tower with food remained at approximately chance levels until the other two towers had each been visited once. In contrast, the probability of returning to a tower without food was reduced to zero for the remainder of a day's test by a single visit to that tower. Thus a necessary and sufficient condition for correct responding (choosing only the tower that had food on it) was a visit to each of the towers that might have had food. In short, the rats did not stay with the original food source until they had shifted and visited all possible food sources.

These and other data (see Olton, 1978, 1979, for reviews) suggest that rats have a consistent predisposition to go to unvisited places in which they might find food rather than return to visited places from which they have already gotten food. Thus they might be expected to prefer to adopt a win-shift strategy when learning a discrimination instead of a win-stay strategy. Such a preference should increase their ability to learn and/or perform a win-shift discrimination and decrease their ability to learn and/or perform a win-stay discrimination. The two experiments to be described were designed to test this hypothesis.

Radial-Arm Maze

The original experiments with the radial-arm maze reinforced a win-shift strategy; after choosing an arm the rat obtained additional food only by going to a different arm. The rapidity of learning and the high level of accuracy in this procedure suggest that rats are predisposed to follow this win-shift strategy. If such is the case then changing the procedure so that it provides reinforcement for returning to previously chosen arms should lead to slower learning or less accurate performance.

The experiments described here evaluated the validity of this hypothesis (Olton and Schlosberg, 1978). Rats were tested on the radial-arm maze and given either a win-stay or win-shift discrimination. We predicted that rats given the win-shift discrimination would learn faster and/or perform better than those given the win-stay discrimination.

The apparatus was a radial-arm maze with eight arms. A wall surrounded the central platform. At the point where each arm was connected to the central platform a hole was cut into the wall. A guillotine door was placed over each hole. By lowering these doors the experimenter could confine the rat to the central platform.

One group of rats was trained to follow a win-shift strategy using the same procedure as described in the original radial-arm maze experiments. At the beginning of each test, one pellet of food was placed at the end of each arm. The guillotine doors were closed and the rat placed on the central platform. The doors were then raised and the rat allowed to choose freely until all eight pieces of food had been obtained. This group was called "Choose 8" (C8), indicating that the optimal strategy was to choose each of the eight arms once, with no delay period imposed during the test procedure.

For two other groups of rats each test session began as described. One pellet of food was placed at the end of each arm and the rat was allowed to choose freely until it had gone to four different arms. When the rat returned to the central platform after the fourth choice, however, the doors were lowered, confining the rat there for approximately 1 minute while the experimenter walked around the maze and moved the remaining food pellets if necessary. The doors were then raised and the rat allowed to choose freely until the other four food pellets on the maze had been obtained.

One group of rats tested with the confinement procedure was reinforced for following a win-shift strategy. While the rat was confined to the central platform following a response to the fourth different arm, the food was left untouched. Consequently, when the doors were raised for the continuation of the test session, the rat was reinforced for the first response to each arm not chosen before confinement. This group was called "Choose any four arms; then choose the different four arms" (CA4-CD4). The only difference between this procedure and that of the C8 group was the confinement after the fourth choice, reflected in the hyphen between CA4 and CD4.

The second group of rats tested with the confinement procedure was reinforced for following a win-stay procedure. While the rat was confined to the central platform following a response to the fourth different arm, the food pellets were removed from the arms that had not been chosen and placed on the arms that had been chosen. Consequently, when the doors were raised for the continuation of the test session, the rat was reinforced for the first response to each arm that had already been chosen during the first part of that session. This group was called "Choose any four arms; then choose the same four arms" (CA4-CS4). The only difference between this procedure and that of the CA4-CD4 group was the arms on which food was placed following the confinement.

All the rats were removed from their mothers at 21 days of age, placed in individual cages, and deprived of food. They were then given one test each day. Several hours following testing they were given free access to Purina Laboratory

Figure 9-1. The results of the radial-arm maze test. The number of correct responses in the first four choices following a response to the fourth different arm is indicated on the vertical axis, the blocks of test days is indicated on the horizontal axis. Abbreviations: C8, chose all eight arms; CA4-CD4, chose any four arms prior to confinement and the four different arms after confinement; CA4-CS4, chose any four arms prior to confinement and the same four arms after confinement.

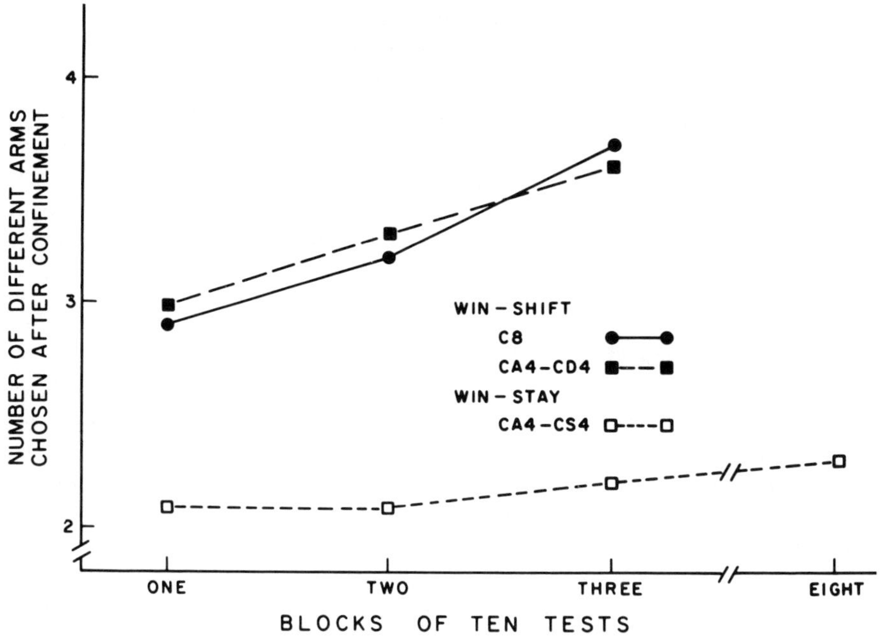

Chow for approximately 5 hours. The food was removed from the home cages and the rats deprived until after the next day's testing.

Rats in the C8 Group learned rapidly and performed well. During the first 10 days of testing they had a mean of 2.9 correct responses in the first 4 choices following a response to the fourth different arm (at which point rats in the other two groups were confined to the central platform). As illustrated in Figure 9-1, this level of performance improved so that during the last block of 10 days the rats had a mean of 3.7 correct responses in the first 4 choices following a response to the fourth different arm. These results confirm those of the previous radial-arm maze experiments, demonstrating rapid learning and accurate performance.

Rats in the other win-shift group, CA4-CD4, performed similarly. The mean number of correct responses in the first 4 choices following confinement for these rats was 2.8 during the first 10 days of testing and 3.6 during the last 10 days. Thus the confinement procedure after the fourth choice did not influence the accuracy of subsequent choice behavior.

In contrast, rats in the win-stay group, CA4–CS4, performed poorly. During the first 10 tests, they had a mean of only 2.0 correct responses in the first 4 choices following confinement, a value that did not change substantially throughout testing. Even after 80 tests these rats gave little indication that they were able to learn the procedure.

This pattern of results is consistent with the hypothesis that rats have a predisposition to follow a win-shift strategy, but there is some difficulty with this interpretation. Rats in the win-shift groups followed a win-shift strategy throughout testing. Rats in the win-stay group had to change strategies because prior to confinement they were reinforced for following a win-shift strategy. Thus their poor performance may have been due to the change in strategies, rather than just to the presence of a win-stay requirement. The next experiment avoided this difficulty by having the win-stay strategy apply to all components of testing.

Three Table Maze

The experimental procedure was an adaptation of the three-table "reasoning" problem used by Maier (1929; see also Stahl and Ellen, 1974). The maze had three arms extending from a central choice point. Each arm was 10 centimeters wide and 100 cm long with an edge 3 cm high along each side. At the end of each arm was a table 30 cm square. A block of wood, 10 cm long and 3 cm high, was used to prevent the rat from seeing the food until he had gone to a table. The entire apparatus was elevated 1 meter above the floor and placed in a well-lit room with many stimuli.

At the beginning of each trial, 190 milligram Noyes pellets were put on one of the tables, the "placement table." The rat was carried to this table by hand, put on it, and allowed to eat some food there. It was then picked up and taken to a holding cage and left for 30 seconds while the experimenter made adjustments to the maze. The rat was then carried to a different table, one that did not have food on it. This table functioned as the "start table" from which the rat moved to the central choice point from which to choose between the two "goal tables," either returning to the placement table or going to the other goal table, the one on which it had not been placed on that trial. Choice of the placement table would indicate that the rat was following a win-stay strategy, because it returned to the place where food had originally been obtained. Choice of the other goal table would indicate that the rat was following a win-shift strategy, because it went to a place different from the one where food had originally been obtained.

The rats were derived from a Sprague-Dawley strain, born and raised in the Hopkins colony. When 3 months old they were deprived to 85 percent of their body weight and maintained at that level, plus 5 grams per week to allow for normal weight gain, throughout the experiment. They were given 3 or 4 trials each day, 5 days a week, with at least 1 hour between trials. The table used as the start table was always the table that the rat had chosen on the previous trial. The table used as the placement table was chosen randomly, with the constraint that each table be used as the placement table equally often. Thus in order to choose correctly when put on the start table, the rat had to remember which table was used as the placement table for that particular trial, a process that requires working memory.

Two groups of rats were reinforced for following a win-stay strategy. For these rats the goal table containing food was the same as the placement table, so that they received food only if they returned to the table on which they had been placed at the start of that trial. Two other groups of rats were reinforced for following a win-shift strategy. For these rats the goal table that contained food was the table that had not been used as the placement table, so that they received food only if they avoided the placement table and went to the table on which they had not been placed at the start of that trial.

Rats were given one of two different amounts of food on the placement table and either allowed to eat all of it or removed from the table while some food still remained. Two groups of rats were given 3 food pellets on the placement table and allowed to eat all of them before they were removed. The other groups of rats were given 13 food pellets on the placement table and removed after they had eaten 3 of them. Thus all rats were allowed to eat 3 pellets. However, rats in the first two groups left no food on the placement table, whereas rats in the second two groups had 10 pellets remaining.

As illustrated in Figure 9-2, this design produced four groups of rats: stay-depletion, stay-nondepletion, shift-depletion, and shift-nondepletion. Note that the memory requirements for the rats in all four groups were the same. Each rat had to remember the table on which it was placed at the start of that trial and then either return to it or go to the other one. The only differences in procedure were the strategy that was reinforced and the state of the food supply on the placement table when the rats were removed from it.

The following predictions were made. The first two were based on the results of the laboratory experiments demonstrating that rats prefer to go to unvisited places rather than return to visited places. This preference should lead them to approach any spatial task with a preference for shifting. Consequently rats in all groups should follow a win-shift procedure during the initial tests and rats given the win-shift discrimination should learn more rapidly and/or perform better than rats given the win-stay discrimination.

The third prediction was based on two pieces of information. First, descriptions of optimal foraging assume that predators use information about the state

Figure 9-2. A summary of the procedure for the discrimination test. The four groups of rats are indicated on the left, the stages of the testing procedure across the top. The maze itself is indicated by the lines with the circles. The top two circles represent the placement table (on the right) and the other goal table (on the left). The bottom circle represents the starting table. The table containing food is indicated by the black circle on the maze and the number of pellets on that table is indicated by the number under the circle. At the beginning of each trial, a rat was carried to the placement table that had food on it (the first column), allowed to eat 3 pellets (second column), and given 10 pellets if it made a correct choice (third column). For rats in the two win-stay groups, the correct choice was returning to the placement table, for rats in the two win-shift groups, the correct choice was going to the other goal table. Rats in the two depletion groups were given 3 pellets on the placement table and allowed to eat them all, rats in the two nondepletion groups were given 13 pellets and removed after they had eaten 3. In the actual test procedure, of course, the tables used for the placement and start tables changed from test to test.

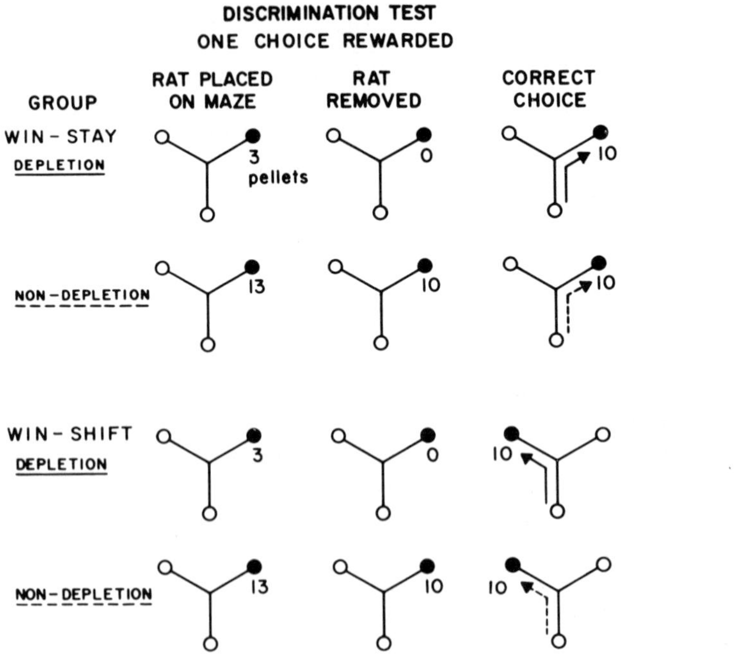

Figure 9-3. The results of the three-table maze test. The percentage of shift responses is indicated on the vertical axis. Because only one of the two goal tables had food on it, chance performance was 0.50. For rats in the two win-shift groups, performance more accurate than that expected by chance is indicated by scores greater than 0.50 with perfect choice accuracy indicated by a score of 1. For rats in the two win-stay groups, performance more accurate than that expected by chance is indicated by scores lower than 0.50 with perfect choice accuracy indicated by a score of 0.

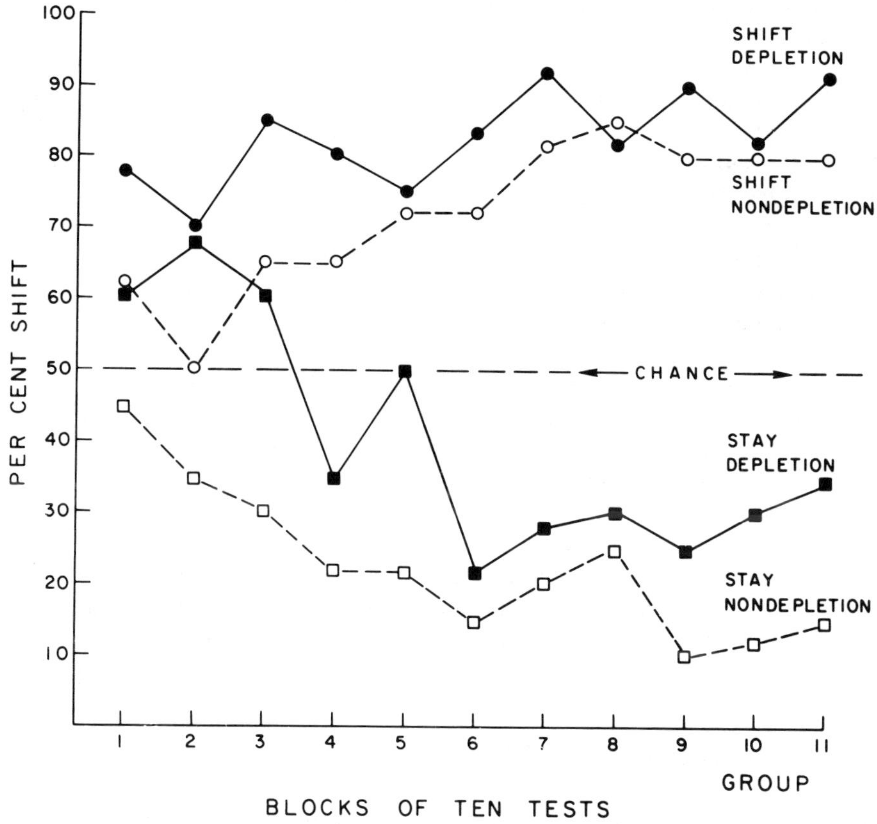

of the initially visited food source to determine whether they should return to it or search elsewhere (Charnov, 1976; Gibb, 1962; Krebs, 1973; Krebs, Ryan, and Charnov, 1974). Second, common sense suggests that a rat should notice whether or not food remains on the placement table when removed from it and should use that information to guide subsequent choice behavior. Thus rats who depleted the food on the placement table were expected to be more likely to shift than rats who were removed from the placement table while food still remained on it. This effect should interact with the strategy being reinforced by the discrimination procedure; depletion of the food on the placement table should enhance the performance of rats learning the win-shift discrimination, but it should interfere with the performance or rats learning the win-stay discrimination.

The results are summarized in Figure 9-3. The vertical axis is the percentage of trials in which the rat shifted and went to the table on which it had not been placed. Because there were two goal tables, chance performance was 0.50. For rats learning the win-shift discrimination, accurate performance is indicated by scores greater than 0.50 with perfect choice accuracy indicated by a score of 1. For rats learning the win-stay discrimination, accurate performance is indicated by scores less than 0.50, with perfect performance indicated by a score of 0.

All three predictions were confirmed. First, the rats began the experiment with an initial preference to shift. Of the 16 rats, 13 had scores greater than 0.50 during the first 10 trials. Second, rats given the win-shift discrimination learned more quickly and performed more accurately than rats given the win-stay discrimination. Indeed, rats in the stay-depletion group did not perform consistently better than expected by chance until the fourth block of 10 trials. Third, the amount of food remaining on the placement table when rats were removed from it influenced their subsequent choice behavior. Rats who ate all of the food were more likely to follow a win-shift strategy than rats who left some of the food behind them. This effect interacted with the discrimination requirements. For the rats learning the win-shift discrimination, those who depleted the food on the placement table performed better than those who left some food behind, while for rats learning the win-stay discrimination, the opposite was the case; those who depleted the food on the placement table performed worse than those who left some food behind.

DISCUSSION

Rats in laboratory mazes consistently exhibit a predisposition to shift their responses among all relevant goal locations rather than restrict them to some subset of those locations (see reviews in Olton, 1978, 1979). The results of the experiments reported here are consistent with this conclusion. Of particular importance are the data from the Maier three-table maze. At the start of testing, rats were more likely to go to the unvisited table than to return to the one on which they had been placed at the start of that trial. Furthermore, rats trained in the win-shift discriminations learned them faster than rats trained in the win-stay discriminations. Thus both the initial choice behavior and the rate of learning reflected the rats' preference to shift responses between the two goal locations available during any given trial.

This predisposition to shift was not immutable, of course, and in the three-table experiments it was influenced by two factors. The first was the amount of food remaining on the placement table when the rat was removed from that table; rats removed while food remained on the table (nondepletion condition) were more likely to return to the placement table than rats removed after they had eaten all of the food (depletion condition). The second was the reinforcement history of the rat in the experiment; at the end of testing, rats that had been given food only if they returned to the placement table (win-stay discriminations) had a greater tendency to return to that table than rats that were given food only if they went to the unvisited goal table (win-shift discriminations).

The stay-shift analysis described here is most appropriate for discrete trial testing procedures. Consider, for example, the Maier three-table maze experiments. Each trial was composed of two parts. In the first part the rat was given information about the location of the food on the maze and the amount of food in that location. In the second part the rat was given a choice between returning to the table where food had been received (staying) or going to the other table (shifting). At the end of each trial the rat was removed from the maze and returned to the home cage before being given the next trial.

This discrete trial procedure assumes that a meaningful unit of analysis is the single trial, both at a behavioral level and at a cognitive level (Honig, 1978; Olton, 1978). The accurate performance of rats in the three-table maze procedure supports these assumptions and demonstrates that a single stimulus event (information about the amount of food in a particular location) can have a substantial influence on the probability of the rat returning to that location at the next opportunity. The effectiveness of a single experience is not unusual in spatial testing procedures. Indeed behavior in mazes is generally characterized by its flexibility and adaptiveness (Olton, 1978, 1979). Coupled with a conditional probability analysis, which can provide a quantitative estimate of the influence of a single response on subsequent choice behavior (see Olton, Walker, Gage, and Johnson, 1977), spatial testing procedures can provide a powerful tool for examining the cognitive processes underlying memory and food searching strategies.

Because of this level of analysis the concepts of shifting and staying are most easily applied to situations that can be analyzed in terms of discrete trials. In the laboratory relevant examples include delayed matching to sample, spontaneous alternation, rewarded alternation, and other procedures in which working memory (Honig, 1978; Olton, 1978) is required. In natural habitats the most relevant examples come from central-place foragers.

If the tendency to shift found in laboratory mazes also appears during the performance of the operant schedules used to study matching and maximizing, then rats should tend to match rather than maximize. Maximizing is a type of stay strategy because in order to maximize an animal must restrict himself to a single response, both when this response has been followed by reinforcement (win-stay) and when it has been followed by nonreinforcement (lose-stay). In contrast, matching is a type of shift strategy, because in order to match the animal must shift among the response alternatives, distributing responses among them in proportion to their reinforcement rates. The maze data (Olton, Walker, Gage, and Johnson, 1977) suggest that this shifting should be more likely following nonreinforcement (lose-shift) than following reinforcement (win-shift), but no particular pattern of shifting is necessary to generate an overall pattern of matching.

This extension of the stay-shift analysis to matching and maximizing is complicated by several factors. First, most analyses of performance in operant schedules have considered long chains of responses (but see Herrnstein, 1979; Shimp, 1976) rather than single trials as in the discrete trial procedure. Thus information about the effects of a single event is generally unavailable. Second, the costs of staying and shifting are usually unequal. Because the animal in an operant procedure often remains at the place of the previous response (rather than returning to a central choice location), the time and energy required to shift and make an alternative response are greater than those required to continue making the previous response. This experimental bias should obviously decrease the tendency to shift and increase the tendency to stay. Finally the response-reinforcement contingencies in the operant schedules are probabilistic. As Kamil, Peters, and Lindstrom (Chapter 8) point out, this characteristic may lead animals to shift responses in order to gather information rather than to gather food. Consequently a shift might occur for a variety of reasons, and determining whether or not two shift responses are homologous in terms of their underlying mechanism poses a formidable challenge.

Placing the analysis of discrimination learning in laboratory tasks within the context of optimal foraging in natural habitats assumes that these two situations have common elements (cf. Chapter 8). One way of testing that assumption is to examine the search strategies of rats in their natural environments and determine the extent to which these strategies match those observed in laboratory mazes. Although there are some relevant descriptions of the ways in which rats gather food (Boice, 1954; Calhoun, 1963), none of these is sufficiently detailed to provide information about the behavior of individual rats. Consequently the validity of this assumption cannot be evaluated at the present time. Nonetheless the predictions are clear: If rats search for food in their natural habitats using the same strategies that they exhibit in laboratory mazes, they ought to shift among all possible food sources rather than concentrate on a single one (cf. Olton, Walker, Gage, and Johnson, 1977).

The fact that rats have a predisposition to follow a particular search strategy even when they have been raised in a laboratory environment suggests that there may be a genetic component to this behavior. Optimal foraging theory predicts species-specific search strategies, especially ones reflecting the reinforcement contingencies of the animals' natural habitat. The descriptions of the foraging patterns of adult animals cannot address this issue because the adult patterns may have developed from experience in the particular environment rather than from any genetic influence. Thus animals must be raised away from their natural habitats. Again the predictions are clear. These animals should exhibit a preference for the search strategy that is most adaptive in their natural environment, even though they have never had an opportunity to be in that environment. This statement is not meant to deny the effects of experience. Such a denial would by myopic in the extreme and would not even be consistent with the data already obtained. Still, the important point is that food searching strategies should be influenced by genetic factors as well as experience.

This type of approach has already been used successfully with two subspecies of *Peromyscus maniculatis* (Gray, 1979; Gray and Tardiff, 1979). *P. m. blandus* was trapped in the Arizona desert, *P. m. borealis* in the prairies of Alberta. They were brought to the laboratory and given a choice among the places to get food, the type of food to eat, and the type of response required to get the food. *P. m. borealis* was a generalist, choosing to shift among all the available alternatives. *P. m. blandus* was a specialist, choosing to concentrate upon only a few of the alternatives. An important point is that the most preferred alternatives were not the same for all *P. m. blandus.* Thus the tendency of individuals to specialize did not arise from any general preference or aversion of the group as a whole. A subsequent experiment (Gray and Tardiff, 1979) raised mice from these two strains in the laboratory. The results of that experiment demonstrated that the tendency to be a generalist or a specialist was influenced by both genetic and experiential factors.

The analysis presented here is rudimentary and needs to be extended in many ways. Some extensions are relatively obvious. For example, how does the stay/shift preference change after an animal "loses" and does not find food in a place where it might have been present? A previous analysis showed that at least for rats, not finding food in an expected food location increased the probability of shifting to its maximum levels (Olton, Walker, Gage, and Johnson, 1977). The same may or may not be true for other animals, and one can conceive of environments (both in natural habitats and in the laboratory) in which the optimal strategy is to return to the place in which the previous response was not reinforced. Other extensions are more difficult, and the overall issue is not just staying or shifting, but the allocation of time and energy among all possible behaviors (cf. Kamil, Peters, and Lindstrom, Chapter 8). The analysis begun here will not be complete until it can deal with these much broader problems.

The description of staying and shifting offered here is still a qualitative one. The results of these experiments could be quantified, of course, in which case the probability of shifting would be a weighted product of a species-specific tendency to follow a particular search strategy, the reinforcement contingencies in the environment, the state of the initially visited food source, and the number of trials that the animal has received. Given the limited set of data available, such a quantified analysis seems premature. An important point, however, is that the variables mentioned did have a significant effect on the rats' behavior and can be incorporated into a mathematical analysis as more data become available.

In the experiment with the three-table maze, the stimulus directing the choice behavior was a single event (eating food on the placement table) and the response to this stimulus was also a single event (returning to that table or going to the other goal table). This emphasis on discrete stimuli and responses is unusual in operant tasks (but see Herrnstein, 1979; Shimp, 1976). It is common in spatial tasks, however, because a single event often has a profound influence on subsequent choice behavior (Olton, 1978, 1979; Olton, Walker, Gage, and Johnson, 1977). The effectiveness of a single visit is important for two reasons. First, it suggests that this event is a meaningful unit of information to the rat and can be treated as such in a behavioral analysis. Second, it provides a means of accurately specifying the time at which information was given to the rat so that parametric analyses of the temporal characteristics of working memory can be carried out (cf. Beatty and Shavalia, 1980; D'Amato, 1973).

The way in which rats search for food is markedly different from the way in which they treat food once they find it. The win-shift strategy is a gambling one; the rat has learned that a previous response produced reinforcement but elects to try an alternative response instead. In contrast, the ingestive behavior of rats is extremely conservative; rats prefer eating familiar foods, react cautiously to new foods, and very quickly learn to avoid foods that make them sick. This is a striking difference in behavior, and the reasons for it are not readily apparent.

In summary, then, the tendency to stay and shift influences the performance of rats in laboratory mazes. The experiments described here demonstrate that the concepts developed within the framework of optimal foraging can be useful in designing and interpreting the results of experiments in the laboratory. They are consistent with the basic theme of this chapter, that appreciating the behavior of an animal in the role of predator searching for prey in the predator's natural habitat will help one to understand the behavior of the same type of animal in the role of a subject solving a discrimination task in the laboratory.

REFERENCE NOTES

1. Black, A.H.; G. Augerinos; and S. Suzuki. Stimulus control of behavior on the 8 arm maze. Paper delivered at the Canadian Psychological Association, Ottawa, June 6, 1978.

REFERENCES

Baddley, A.D., and G. Hitch. 1974. Working memory. *The Psychology of Learning and Memory 8*: 47-89.

Bättig, K.; P. Driscoll; J. Schlatter; and H.J. Uster. 1976. Effects of nicotine on the exploratory locomotion patterns of female Roman high- and low-avoidance rats. *Pharmacology, Biochemistry and Behavior 4*: 435-39.

Bättig, K., and J. Schlatter. 1979. Effects of sex and strain on exploratory locomotion and development of nonreinforced maze patrolling. *Animal Learning and Behavior 7*: 99-105.

Beatty, W.W. 1979. Very long short term memory in rats: Effects of hippocampal lesions. *Society for Neuroscience Abstracts 5*:313.

Beatty, W.W., and R.T. Shavalia. 1980. Spatial memory in rats: Time course of working memory. *Behavioral and Neural Biology 28*: 454-462.

Boice, R. 1977. Burrows of wild and albino rats: Effects of domestication, outdoor raising, age, experience, and maternal state. *Journal of Comparative and Physiological Psychology 91*: 647-61.

Calhoun, J.B. 1963. *The Ecology and Sociology of the Norway Rat.* U.S. Public Health Service Pub. No. 1008. Washington, D.C.: U.S. Government Printing Office.

Charnov, E.L. 1974. Hunting by expectation or optimal foraging? A study of patch use by chickadees. *Animal Behaviour 22*: 953-64.

_____. 1976. Optimal foraging: The marginal value theorem. *Theoretical Population Biology 9*: 129-36.

D'Amato, M.R. 1973. Delayed matching short-term memory in monkeys. In G.H. Bower, ed., *The Psychology of Learning and Motivation*, vol. 7, *Advances in Research Theory*, New York: Academic Press.

Dashiell, J.F. 1930. Direction orientation in maze running. *Comparative Psychology Monographs 7*, no. 32: 1-36.

Dember, W.N., and H. Fowler. 1958. Spontaneous alternation behavior. *Psychological Bulletin 55*: 412-28.

Douglas, R.J. 1966. Cues of spontaneous alternation. *Journal of Comparative and Physiological Psychology 62*: 171-83.

Gibb, J.A. 1962. Tits and their food supply in English pine woods: A problem in applied ornithology. In *Festchreift Vegelschutzwarte Hessen*, Rheinland: Pfalz und Saarland, pp. 58-66. Discussed in J.R. Krebs, J.C. Ryan and E.L. Charnov, 1974, *Animal Behaviour 22*: 953-64.

Gray, L. 1979. Feeding diversity in deer mice. *Journal of Comparative and Physiological Psychology 93*: 1118-26.

Gray, L., and R.R. Tradiff. 1979. Development of feeding diversity in deer mice. *Journal of Comparative and Physiological Psychology 93*: 1127-35.

Hamilton, W.J., III, and K.E.F. Watt. 1970. Refuging. *Annual Review of Ecology and Systematics 1*: 263-86.

Herrnstein, R.J. 1979. Derivatives of matching. *Psychological Review 86*: 486-95.

Honig, W.K. 1978. Working memory in pigeons. In S.H. Hulse, W.K. Honig, and H. Fowler, eds., *Cognitive Aspects of Animal Behavior*, Hillsdale, N.J.: Lawrence Erlbaum.

Kamil, A.C. 1978. Systematic foraging by a nectar-feeding bird, the Amakihi (*Loxops virens*). *Journal of Comparative and Physiological Psychology 92*: 388–96.

Kamil, A. and T. Sargent, eds. 1980. *Optimal Foraging.* New York: Garland.

Krebs, J.R. 1973. Behavioral aspects of predation. In P.G. Bateson and P. Klopfer, eds., *Perspectives in Ethology*, vol. 1, New York: Plenum Press.

Krebs, J.R.; J.C. Ryan; and E.L. Charnov. 1974. Hunting by expectation or optimal foraging: A study of patch use by chickadees. *Animal Behaviour 22*: 953–64.

Maier, N.R.F. 1929. Reasoning in which rats. *Comparative Psychology Monograph 6*: 1–93.

Maier, W. 1932. The effect of cerebral destruction on reasoning and learning in rats. *Journal of Comparative Psychology 54*: 45–75.

Maki, W.S.; S. Brokofsky; and B. Berg. 1979. Spatial memory in rats: Resistance to retroactive interference. *Animal Learning and Behavior 7*: 25–30.

Olton, D.S. 1972. Discrimination behavior in the rat: Differential effects of reinforcement and non-reinforcement. *Journal of Comparative and Physiological Psychology 79*: 450–456.

_____. 1978. Characteristics of spatial memory. In S.H. Hulse, H.F. Fowler, and W.K. Honig, eds., *Cognitive Aspects of Animal Behavior*, Hillsdale, N.J.: Lawrence Erlbaum; pp. 341–73.

_____. 1979. Mazes, maps and memory. *American Psychologist 34*: 588–96.

Olton, D.S.; J.T. Becker; and G.E. Handelmann. 1980a. Hippocampus, space, and memory. *The Behavioral and Brain Sciences 2*: 313–22.

_____. 1980b. A re-examination of the role of hippocampus in working memory. *The Behavioral and Brain Sciences 2*: 352–59.

Olton, D.S., and C. Collison. 1979. Intramaze cues and "odor trails" fail to direct choice behavior on an elevated maze. *Animal Learning and Behavior 7*: 221–323.

Olton, D.S.; C. Collison; and W.A. Werz. 1977. Spatial memory and radial arm maze performance by rats. *Learning and Motivation 8*: 289–314.

Olton, D.S.; G.E. Handelmann; and J.A. Walker. 1980. Characteristics of spatial memory and optimal foraging in rodents. In A. Kamil and T. Sargent, eds., *Optimal Foraging*, New York: Garland. pp. 333–54.

Olton, D.S., and R.J. Samuelson. 1976. Remembrance of places passed: Spatial memory in rats. *Journal of Experimental Psychology: Animal Behavior Processes 2*: 97–116.

Olton, D.S., and P. Schlosberg. 1978. Food-searching strategies in young rats: Win-shift predominates over win-stay. *Journal of Comparative and Physiological Psychology 92*: 609–18.

Olton, D.S.; J.A. Walker; F.H. Gage; and C.T. Johnson. 1977. Choice behavior of rats searching for food. *Learning and Motivation 8*: 315–31.

Roberts, W.A. 1979. Spatial memory in the rat on a hierarchial maze. *Learning and Motivation 10*: 117–40.

Schoener, T.W. 1971. Theory of feeding strategies. *Annual Review of Ecology and Systematics 2*: 369–404.

Shimp, C.P. 1976. Short-term memory in the pigeon: The previously reinforced response. *Journal of the Experimental Analysis of Behavior 26*: 487–93.

Smith, J.N.M. 1974a. The food searching behavior of two European thrushes. I. Description and analysis of search paths. *Behaviour 48*: 276–302.

____. 1974b. The food searching behavior of two European thrushes. II. The adaptiveness of the search patterns. *Behaviour 49*: 1–61.

Smith, J.N.M., and H.P.A. Sweatman. 1974. Food searching behavior of tit mice in patchy environments. *Ecology 55*: 1216–32.

Stahl, J.M., and P. Ellen. 1974. Factors in the reasoning performance of the rat. *Journal of Comparative and Physiological Psychology 87*: 598–614.

Suzuki, S., G. Augerinos; and A.H. Black. 1980. Stimulus control of spatial behavior on the eight arm maze in rats. *Learning and Motivation 11*: 1–18.

Uster, H.J. 1977. Analyse des esplorationverhaltens von ratten in labyrinthen unterschiedlicher kimplexitaet. Doctoral Dissertation, Eidgenoessichen Technischen Hochscule, Zurich, Switzerland.

Uster, H.J., K. Bättig; and H.H. Nägeli. 1976. Effects of maze geometry and experience on exploratory behavior in the rat. *Animal Learning and Behavior 4*: 84–88.

Walker, J.A., and D.S. Olton. 1979. The role of response and reward in spatial memory. *Learning and Motivation 10*: 73–84.

Whitham, T.G. 1977. Coevolution of foraging in Bombus and nectar dispensing in Chilopsis: A last dreg theory. *Science 197*: 593–95.

Wells, M.J., and S.K.L. Buckley. 1972. Snails and trails. *Animal Behaviour 20*: 345–55.

Zach, R., and J.B. Falls. 1976a. Foraging behavior, learning, and exploration by captive ovenbirds (*Aves: Parulidae*). *Canadian Journal of Zoology 54*: 1880–93.

____. 1976b. Do ovenbirds (*Aves: Parulidae*) hunt by expectation? *Canadian Journal of Zoology 54*: 1894–1903.

Zoladek, L., and W.A. Roberts. 1978. The sensory basis of spatial memory in the rat. *Animal Learning and Behavior 6*: 77–81.

10 INSTRUMENTAL BEHAVIOR AND FORAGING IN THE WILD

William M. Baum

Recent years have seen the development in both biology and psychology of theories of behavior based on the notion that organisms follow optimizing strategies (e.g., Baum, 1981; Houston and McNamara, 1981; Krebs and Davies, 1978; Pyke, Pulliam, and Charnov, 1977; Rachlin, Battalio, Kagel, and Green, 1981; Staddon, Hinson, and Kram, 1981). The psychological theories concern instrumental behavior in relation to its consequences, reinforcement and punishment; the biological theories concern foraging in relation to predation. Although the two efforts proceeded independently at first, recently researchers on both sides have begun to recognize a commonality of interest (e.g., Collier and Rovee-Collier, 1979; Killeen, Smith, and Hanson, 1981; Pulliam, 1979).

Although conventional schedules of reinforcement may fail to incorporate some basic features of the wild, nevertheless instrumental behavior in the laboratory can be viewed as an analog to foraging in the wild. Both are motor activities, and both typically vary according to their consequences. Reinforcement constitutes the analog to predation, be it finding of food or a parasite's host (Krebs, 1978; Hassell, 1978).

Optimizing theories of instrumental behavior compete with theories based on the matching law, which states that the proportion of behavior allocated to an activity matches the proportion of reinforcement, out of the total obtained from the environment, that it provides (Herrnstein, 1974). This simple principle has substantial support from experiments in the laboratory (e.g., Baum, 1974b).

Preparation of this chapter was partially supported by Grant BNS-7906852 from the National Science Foundation to the University of New Hampshire.

Although optimization and matching make similar predictions for some situations, they diverge often enough to make experimental tests possible. On balance, however, no decision between the two can be made at present, because both approaches have strengths and weaknesses: The matching law accounts for experiments on choice better, whereas optimization accounts for experiments on single schedules (one reinforced alternative) better (Baum, 1981).

A similar theoretical dichotomy appears in theories of foraging. Optimizing models compete with functional response theory, which focuses on describing the form of the function relating predation to parameters of the environment such as size and density of prey (Hassell, 1978; Holling, 1965). Research on functional response seeks simple rules like the matching relation that a predator might follow in foraging—rules that, if not perfectly optimal, would work reasonably well and without the presumably complex mechanisms organisms would have to possess to optimize. As with instrumental behavior, it is too soon to decide which approach is more fruitful or whether the two can somehow be combined.

The attractiveness of optimization lies in its coincidence with modern economic theory, on the one hand, and modern evolutionary theory, on the other. Both provide broad organizing frameworks built around an assumed goal of optimization. For economics the consumption of goods by the community must move toward an optimum, as long as a variety of goods is available, and hence selection is possible. For evolution the consumption of resources by a population must move toward an optimum, as long as natural selection is possible.

Both the matching law and functional response theory contrast with optimization in that they assume that individual organisms can follow a relatively simple fixed rule that works in most situations, even if it is sometimes nonoptimal. They suggest that natural selection may have produced individuals that follow less complicated principles than optimization, even though natural selection tends to optimize. Simple approximations to optimality require less effort and less machinery.

The equilibrium situation for a population of predators and a population of prey is *management*, in which predation simultaneously satisfies the need of the predators and maintains the prey at a density that allows replenishment just to meet this need. Predation beyond this results in overexploitation; the prey population falls, and some predators must die, switch prey, or emigrate. Predation below this—underexploitation—results in growth of the prey population, and allows growth of the population of predators. In practice, such equilibrium, though common, may sometimes fail to occur. Oscillation in predator populations is well documented (Hutchinson, 1978). In the terms of control theory, however, oscillation results from phase lag between prey and predator growth, and the set point of the system can still correspond to management.

A MODEL FOR RELATING INSTRUMENTAL BEHAVIOR AND FORAGING

The organism and its environment together constitute a feedback system. Figure 10-1 illustrates the basic components. The boxes marked Beneficial Relations and Costly Relations represent the feedback functions that the environment arranges to produce positive and negative consequences. The input, or set point, represents a rule, such as matching or optimizing, that allows balancing of cost and benefit; it constitutes the main contribution of the organism to the situation. The box marked Organism represents the transfer function by which the behavior (output) is adjusted on the basis of the comparison between input, cost, and benefit. Viewed in this way the researcher's problem consists in identifying the input, transfer function, and feedback functions. The focus in this chapter is on the input and beneficial relations; costly relations are left for future consideration.

Laboratory research enjoys the advantage that the beneficial relations can be specified and manipulated. Figure 10-2 shows, for example, feedback functions imposed by several different variable-interval schedules. On such a schedule reinforcement becomes available for responses only at irregular intervals, with the result that reinforcement is unpredictable, occurs at a rate that depends on rate of responding, but has an upper limit. The matching relation was originally

Figure 10-1. The organism and environment represented as a feedback system.

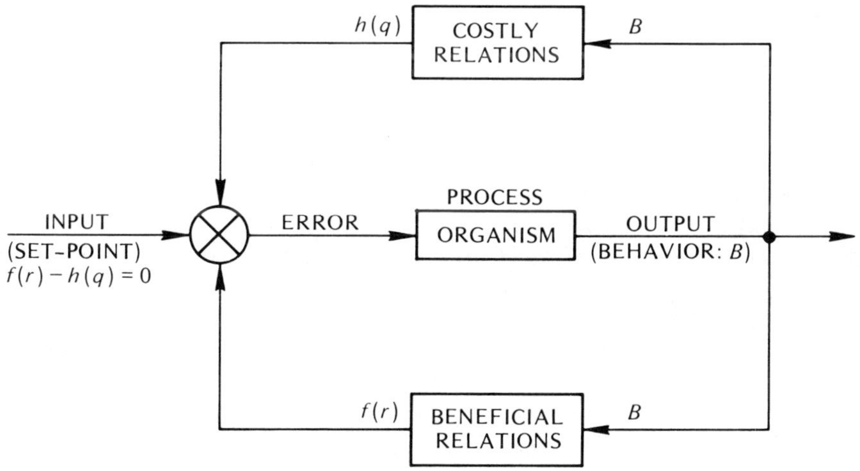

Figure 10-2. The dependence of rate of reinforcement on response rate imposed by several conventional variable-interval schedules (feedback functions). The equation of each curve is $r = 60/(t + 1/B)$, where r is reinforcers per hour, B is responses per minute, and t is the VI parameter in minutes.

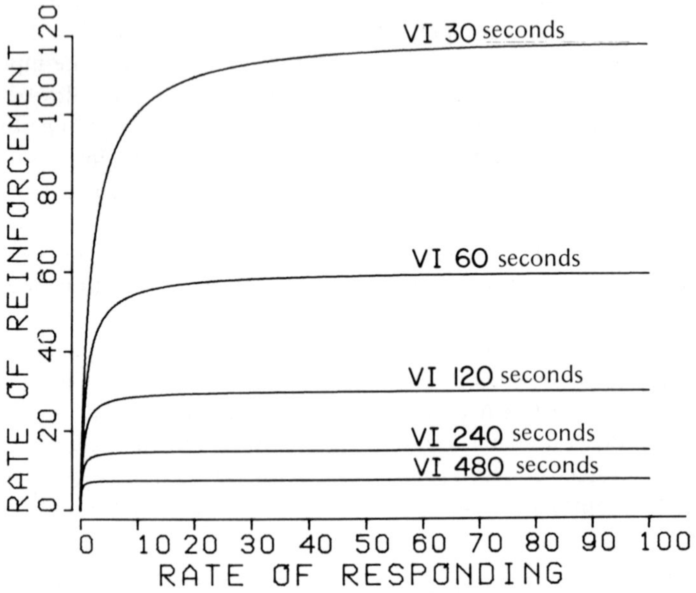

found to apply to responding distributed between pairs of such schedules (Herrnstein, 1961). It can be represented as

$$\frac{B_1}{B_2} = \frac{r_1}{r_2} , \tag{10.1}$$

where B_1 and B_2 represent responding on the two schedules, and r_1 and r_2 represent rates of reinforcement obtained from them. Sometimes the two alternatives differ in the size or value of the reinforcer or in the nature of the response required. Such asymmetries introduce a bias toward one or the other alternative, which, as long as conditions serve only to vary the rates of reinforcement, can be represented by a multiplicative parameter w,

$$\frac{B_1}{B_2} = w \frac{r_1}{r_2} . \tag{10.2}$$

Equation (10.2) can be called biased matching (Baum, 1974a). It has been found to apply not only to pairs of variable-interval schedules but also to pairs in which one schedule is an interval schedule and the other a ratio schedule, which requires a certain amount of responding for each reinforcement and hence imposes a feedback function that consists in direct proportionality between rate of reinforcement and response rate (Baum, 1974a; Herrnstein and Heyman, 1979).

To liken foraging to instrumental behavior requires specifying the feedback function by which predation (rate of reinforcement for foraging) depends on rate of foraging. In a patchy environment, prey populations occur in several separate locations. Research and theory have been directed toward this type of environment, probably because a nonpatchy environment can be thought of as one patch. The problem, then, is to understand what sort of a schedule of reinforcement characterizes a patch.

Let us assume a patch with prey population n and growth rate given by

$$\frac{dn}{dt} = gn - r \quad , \tag{10.3}$$

where g is the standard growth factor of population biology and r is rate of predation (reinforcement). If the prey are uniformly distributed throughout the patch and foraging is uniform across the patch and at random with respect to locations of prey—that is, prey are found only by searching—then rate of predation should depend on rate of foraging and prey density as follows:

$$r = n \cdot D \cdot B \quad , \tag{10.4}$$

where D is the reciprocal of the area of the patch and B is the rate of foraging (in units of area per time). Just as when average response rate is calculated the time may include several episodes of responding alternating with pauses, so B, the average rate of foraging, spans time that may include several visits to and absences from a patch. Combining Eqs. (10.3) and (10.4) and solving for n yields

$$n = n_0 \, e^{(g - DB)t} \tag{10.5}$$

$$r = BDn_0 \, e^{(g - DB)t} \quad , \tag{10.6}$$

where n_0 is the number of prey on first entry into the patch, and t is time since that first entry. Equation (10.6) specifies a feedback function relating reinforcement (predation) to foraging. Unlike feedback functions for conventional schedules, it contains time as a variable. This means the relation between predation (r) and foraging (B) is not stationary through time for most rates of foraging. The schedule can be thought of as a sort of adjusting ratio, because it implies a proportionality between rate of reinforcement and response rate (Equation (10.4)) that shifts through time, depending on previous behavior. When B is less than g/D, the expression $g - DB$ is positive, the prey population n increases, and the proportionality between predation and foraging (Equation (10.4)) improves.

When B exceeds g/D, the expression $g - DB$ is negative, n decreases, and the patch becomes poorer.

When rate of foraging B exactly equals g/D, the expression $g - DB$ equals zero, time disappears from the equation, and the prey population (n) remains stable. This rate of foraging therefore constitutes management.

When B equals g/D, Eq. (10.6) reduces to

$$r = n_0 DB \quad . \tag{10.7}$$

If two such patches are both managed, two equations result:

$$r_1 = n_1 D_1 B_1 \quad ; \tag{10.8}$$

$$r_2 = n_2 D_2 B_2 \quad . \tag{10.9}$$

The ratio of Eqs. (10.8) and (10.9) can be written as

$$\frac{B_1}{B_2} = \frac{n_2 D_2}{n_1 D_1} \cdot \frac{r_1}{r_2} \quad . \tag{10.10}$$

The terms $n_1 D_1$ and $n_2 D_2$ represent the densities of prey in the two patches. If the densities in the patches are unequal, Eq. (10.10) represents biased matching (Eq. (10.2)). If the densities are equal, bias disappears and Eq. (10.10) represents simple matching (Eq. (10.1)). Choice between conventional ratio schedules produces exclusive preference for the smaller ratio, a trivial matching, in which the two sides of Eq. (10.1) equal zero or infinity, and an optimal solution. Since the adjusting ratio schedule represented by Eq. (10.6) mimics the essential feature of a variable-interval schedule—that likelihood of reinforcement rises with the absence of responding or foraging—nontrivial matching can result.

Although management produces matching, the converse is not necessarily true. Both underexploitation and overexploitation in time lead to trivial matching, because the ratio of feeding to foraging (r/B), must approach 1.0 for underexploitation and 0.0 for overexploitation. Other patterns might produce nontrivial matching, but only management would be stable indefinitely. Conceivably selection favors organisms that achieve stable matching, because in nature this generally coincides with management. Management would have to be optimal in some sense. To explore the relation between stability (management) and optimality requires taking account of the main constraint on the predator's behavior: meeting its need.

The Predator's Need

Thus far the major factor that limits foraging and predation has been ignored: A predator will not forage beyond its need. Over some time interval T, which could be an hour, day, month, or year, depending on the situation, a predator must meet its need or perish. More generally, the task (survival) consists in maintaining a rate of reinforcement, averaged over some time interval T, that equals the average rate of reinforcement that meets its need through time.

The duration of T determines the sensitivity of foraging to short-term variations in rate of reinforcement. If T is too short the predator will shift its foraging needlessly on the basis of short-term changes, wasting time and energy. If T is long foraging will be more stable. If T is too long, however, the predator might starve before adapting to a long-term change in abundance of prey. Although it may be clear that natural selection might produce predators for whom T is neither too short nor too long, to get a precise idea of the influence of T on foraging, we must incorporate it into our model. Integrating Eq. (10.6) over the interval T and dividing by T to obtain the average rate of reinforcement yields

$$r = \frac{n_0 \, DB}{(DB - g)\,T} \cdot \left\{1 - e^{(g - DB)T}\right\} . \tag{10.11}$$

This equation specifies a feedback function for equal intervals of time. Figure 10-3 shows curves relating r and B generated by Eq. (10.11) with various values of g, n_0, and D. The interval T was held constant throughout. Two sets of assumptions were made: The top half of the figure (panel A) shows curves for patches in which the density of prey, $n_0 D$, was assumed constant from patch to patch, making D and n_0 vary inversely; panel B shows curves for patches assumed all equal in area (D constant), in which n_0 varied from patch to patch.

Creating these graphs entailed the additional assumption that there is an upper limit on the prey population that can live in a patch. Although in time underexploitation ($B < g/D$) would bring the population to this level, n_0 was assumed less than it—as if the predator arrived initially to find a growing or managed prey population. Each curve approaches an asymptote n_0/T because, with overexploitation, as rate of foraging increases, predation approaches complete depletion of the patch. Figure 10-3 omits the information that, were overexploitation to persist day after day, all these asymptotes would have to drop to zero. Indeed the curves in Figure 10-3 cannot be considered stable, because foraging below the management level causes n_0 to increase in successive intervals T, and foraging above the management level causes n_0 to decrease. It is as if through time the feedback function would shift upward or downward in Figure 10-3B—upward for underexploitation and downward for overexploitation.

Each curve in Figure 10-3 has an optimum. Low rates of foraging mean low rates of reinforcement, but, as rate of foraging increases, depletion while the predator is in the patch eventually turns the rate of reinforcement downward for high rates of foraging. The optimal rate of foraging provides a high rate of reinforcement while the predator is in the patch but enough time out of the patch

Figure 10-3. Feedback functions for foraging showing predation (reinforcement) as a function of rate of foraging. Prey density constant, patch size varying (*top*, A); patch area constant, prey density varying (*bottom*, B). Symbols indicate maximal rates of reinforcement.

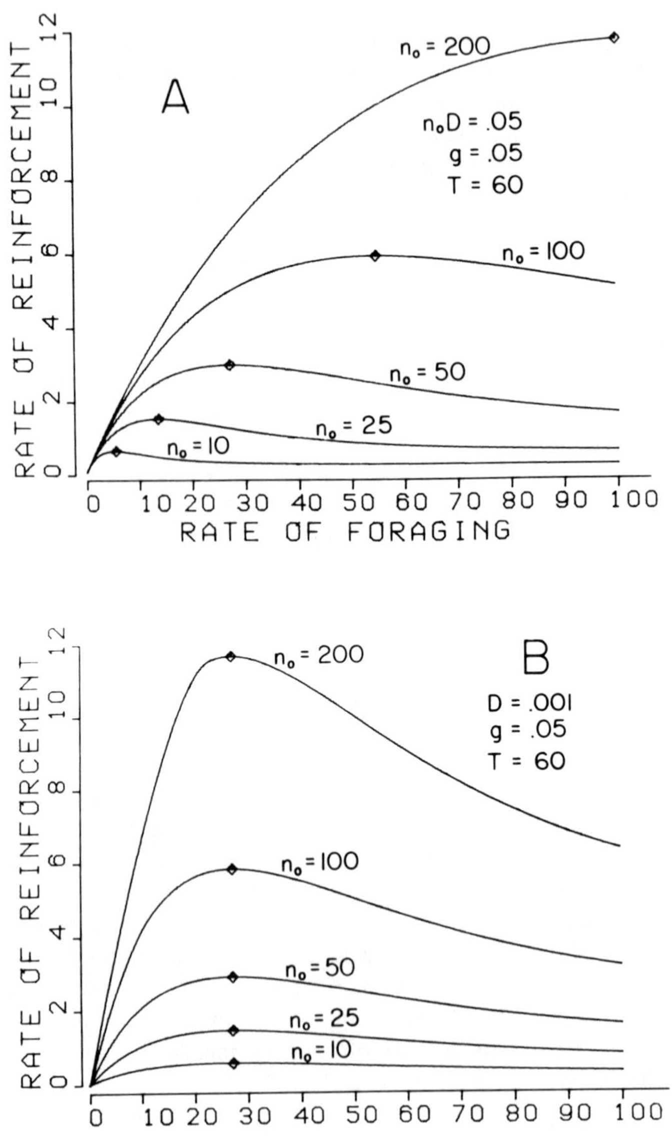

(or in the patch but not foraging) to allow some replenishment. Since the rate of foraging is an average that includes time spent in other activities (perhaps foraging in other patches or sleeping), the optima shown in Figure 10–3 should be distinguished from previous optimal solutions, such as Charnov's (1976) optimal giving-up times. Although there may exist an optimal time when a predator ought to leave a patch after having entered and begun foraging, the average rate of foraging, B in Eqs. (10.6) and (10.11) and Figure 10–3, can arise from many different combinations of times spent in and out of the patch. If B' represents the rate of foraging while in the patch, then

$$B = B' \frac{t_i}{t_i + t_o} \tag{10.12}$$

where t_i is the time in the patch (giving-up time) and t_o is the time out of the patch (time between visits). Since the time over which the rate B is averaged can include numerous visits to the patch, this model may provide a means to reconcile Charnov's theory with Oaten's (1977) criticism that giving-up time must vary stochastically. When travel time between patches and other factors relating to time away from the patch are considered, there may well be optimal combinations of t_i and t_o. The model is silent on the matter at present.

For a single predator with fixed need, a single patch may or may not suffice to fill the need. Figure 10–4 illustrates the situation in which the patch does suffice. The horizontal line, intersecting the curve at points a and c, represents the predator's needed average rate of reinforcement. There is no reason, except chance, by which it would intersect at point b, the optimum, or point d, the rate for management (given by Eq. (10.7)). Point a represents the most economical performance; need is met for the least foraging. At point b foraging produces more than the need, and at point c the need is met but at a higher rate of foraging. Regardless of whether the predator's performance lies at point a, b, or c, however, its rate of foraging falls below that for management (point d).

Foraging below management means a superabundance of prey in time. For the situation in Figure 10–4, if the individual predator optimizes, it cannot manage the patch. Moreover, if the individual predator manages the patch, it cannot meet its need. The two requirements are incompatible for the individual predator.

These incompatibilities depend on the size of the interval of averaging, T, however. Figure 10–4 shows the relations that hold when T is relatively long. When T is short they reverse. Figure 10–5 shows feedback curves for several durations of T, with g, D, and n_0 held fixed. The longer is T, the lower the optimal rate of foraging; the shorter is T, the higher the optimal rate of foraging. When T is relatively long replenishment between visits can be substantial if rate of foraging is low, with the result that rate of reinforcement is highest for a relatively low rate of foraging—not too low, however, because predation must de-

Figure 10–4. Feedback function for foraging when averaging interval T is relatively long. Horizontal line represents predator's need. Vertical line indicates rate of foraging for management.

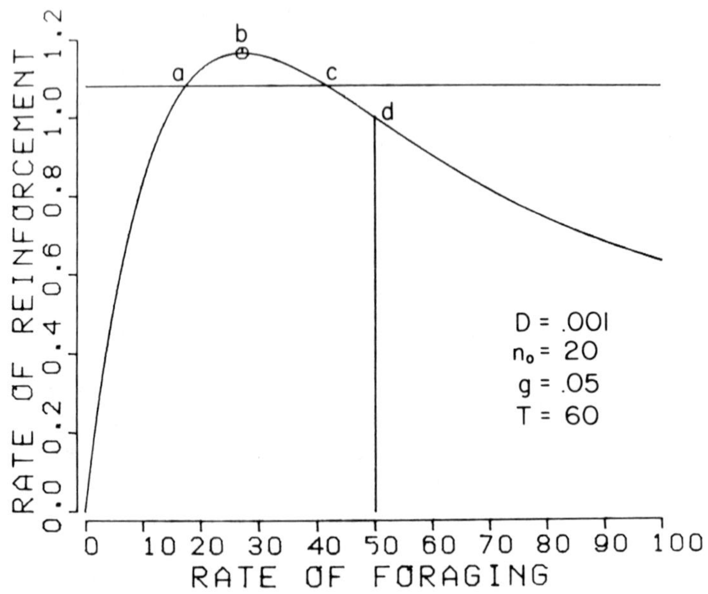

crease for very low rates of foraging. When T is relatively short less replenishment can occur, with the result that overexploitation becomes optimal. What little replenishment does occur limits the optimal rate of foraging, but the less replenishment, the more overexploitation is optimal.

When the optimal rate of foraging exceeds the rate for management, if the predator optimizes, the patch will be depleted in time. That is, n_0 will decrease across successive intervals T. Figure 10–6 illustrates the situation. If the predator cannot meet its need from the patch, optimal performance tends to point b. If its need corresponds to the horizontal line shown, its performance tends to point a, because point c is wasteful. Whether it tends to a or b, however, n_0 will fall across successive intervals T, with the result that the curve falls as in Figure 10–3B. With continued overexploitation the patch eventually would fail to meet the predator's need and the optimal rate of foraging would correspond to point b.

Since such short-term optimizing would eventually exterminate the prey population, there must exist a longer term optimizing strategy that would be superior in that it would maintain the prey population and allow predation (reinforcement) to continue. "Longer term" or "global" means a longer value of T.

Figure 10-5. Feedback functions for foraging in the same patch when various averaging intervals *T* are assumed. Symbols indicate maximal rates of reinforcement.

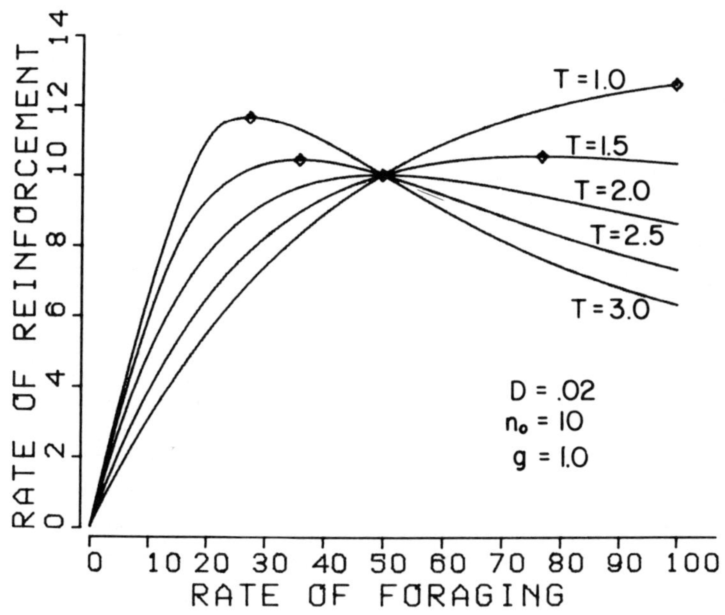

If the interval *T* instead of remaining fixed tended to increase when the prey population was overexploited, that change would tend to decrease the optimal rate of foraging and hence to correct the overexploitation. In other words, if *T* tended to increase in response to overexploitation, optimal foraging would tend toward management.

Figure 10-5 makes it clear that there exists one value of *T* for which optimization and management coincide. When *T* equals $2/g$, the optimal rate of foraging is the same as the rate for management. For *T* less than $2/g$ the optimal rate exceeds management (Figure 10-6), and for *T* greater than $2/g$ the optimal rate falls below management (Figure 10-4). The size of the predator's averaging interval *T* relative to the replenishing capability of the prey population *g* determines which situation occurs. If the averaging interval is long and replenishment rapid, then the optimal rate of foraging falls below management (Figure 10-4). If the averaging interval is short and replenishment slow, then the optimal rate exceeds management (Figure 10-6).

Although the question of when *T* exceeds, equals, or falls below $2/g$ requires an empirical answer, the sorts of factors that should apply can be imagined. Whales feeding on plankton present a different situation from wolves feeding on

Figure 10-6. Feedback function for foraging when averaging interval T is relatively short. Horizontal line represents predator's need. Vertical line indicates rate of foraging for management.

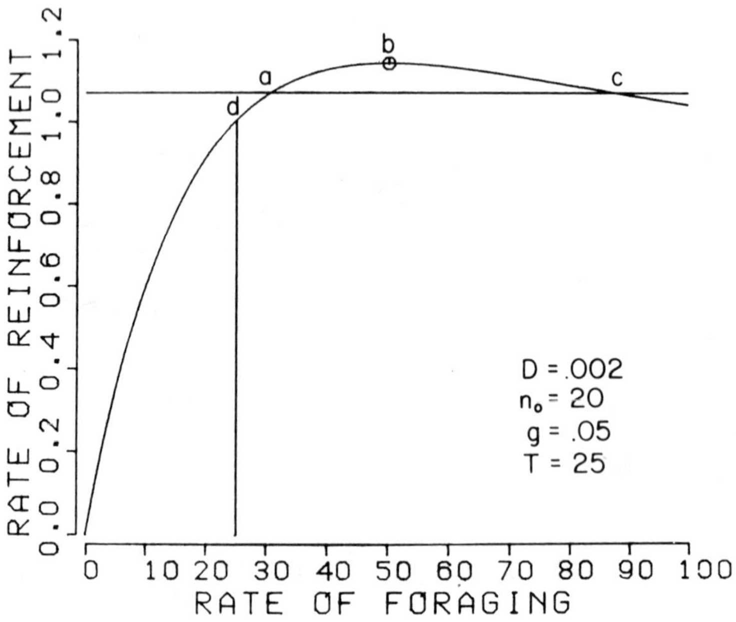

moose. If a predator feeds on more than one type of prey, the relation between g and T might even differ from one type to another. It is conceivable that T might decrease in a rich environment and increase in a sparse one. If there are other predators for the same prey, the possibility of replenishment is reduced. Competition therefore effectively decreases g. For this reason one might guess that the situation shown in Figure 10-6 ($T < 2/g$; optimization means over-exploitation) would be common. These are questions for future research.

Two or More Patches

Does the situation change if the predator forages in more than one patch? If the horizontal line in Figures 10-4 or 10-6 fell above the maximum possible rate of reinforcement, the predator could meet its need only by foraging in other patches besides the one shown. Indeed, it might forage in a large number of patches. In almost all, it could either optimize or manage (that is, match; Eqs. (10.1) and (10.2)), obtaining part of its need from each. It is improbable, however, that its need would be met exactly by all the patches together following either strategy. For one patch, at least—presumably the poorest, if the predator prefers all better ones—the situation shown in Figures 10-4 or 10-6 would hold; the only difference from the single-patch situation would be that the horizontal line would represent *remaining* need instead of total need.

Optimization, Management, and Matching

Figures 10-4 – 10-6 suggest that the world may present organisms with situations in which management and optimization are incompatible. For some niches, selection might favor individuals that optimize; for others, individuals that manage by following the matching relation. Selection might also produce individuals that shift strategy according to prey species or prey abundance. If some mechanism could vary the duration of the interval of averaging, T, all three strategies could be made to coincide.

To what sort of environment the experimental chamber of the psychological laboratory ought to correspond remains a challenging question. It might be construed as sometimes a rich environment, one in which a lot of food is available for a short time. If T depended on abundance this would make for short-term averaging—as if it were an instance of the relations in Figure 10-6, favoring a high response rate if the organism optimizes. This may be the reason that performance on single schedules seems best described as optimization (Baum, 1981).

Faced with choice, on the other hand—that is, an environment with more than one patch—organisms may tend to shift from short-term optimizing to a strategy appropriate to management, possibly matching. This may be the reason that performance in choice situations seems best described by the matching relation (Baum, 1981).

That understanding of the strategies that guide instrumental behavior should be sought follows from the correspondence between reinforcement and the consumption of resources (predation). The effects of natural selection on the dependence between organisms' foraging and its varying consequences (predation) must carry over into the laboratory. Only by understanding the relation between foraging and predation in an organism's ecological niche will a full understanding be gained of how food and other reinforcers change and maintain the responding that produces them. The type of model described in this chapter may provide a beginning for future research considering features of the natural environment, such as depletion, replenishment, and separation between patches, as well as features of organisms, such as averaging through time.

REFERENCES

Baum, W.M. 1974a. On two types of deviation from the matching law: Bias and undermatching. *Journal of the Experimental Analysis of Behavior 22*: 231-42.

____. 1974b. Choice in free-ranging wild pigeons. *Science 185*: 78-79.

____. 1981. Optimization and the matching law as accounts of instrumental behavior. *Journal of the Experimental Analysis of Behavior 36*: 387-403.

Charnov, E.L. 1976. Optimal foraging; the marginal value theorem. *Theoretical Population Biology 9*: 129-36.

Collier, G.H., and C.K. Rovee-Collier. 1979. A comparative analysis of optimal foraging behavior: Laboratory simulation. In A.C. Kamil and T.D. Sargent, eds., *Foraging Behavior*, New York: Garland.

Hassell, M.P. 1978. *The Dynamics of Arthropod Predator-Prey Systems.* Princeton, N.J.: Princeton University Press.

Herrnstein, R.J. 1961. Relative and absolute strength of response as a function of frequency of reinforcement. *Journal of Experimental Analysis of Behavior 4*: 267-72.

____. 1974. Formal properties of the matching law. *Journal of the Experimental Analysis of Behavior 21*: 159-64.

Herrnstein, R.J., and G.M. Heyman. 1979. Is matching compatible with reinforcement maximization on concurrent variable interval, variable ratio? *Journal of the Experimental Analysis of Behavior 31*: 209-23.

Holling, C.S. 1965. The functional response of predators to prey density and its role in mimicry and population regulation. *Memoires of the Entomological Society of Canada 45*: 1-60.

Houston, A.I., and J. McNamara. 1981. How to maximize reward rate on two variable-interval paradigms. *Journal of the Experimental Analysis of Behavior 35*: 367-96.

Hutchinson, G.E. 1978. *An Introduction to Population Ecology.* New Haven, Conn.: Yale University Press.

Killeen, P.R., J.P. Smith; and S.J. Hanson. 1981. Central place foraging in *Rattus norwegicus. Animal Behavior 29*: 64-70.

Krebs, J.R. 1978. Optimal foraging: Decision rules for predators. In J.R. Krebs and N.B. Davies, eds., *Behavioural Ecology*, Oxford: Blackwell.

Krebs, J.R., and N.B. Davies. 1978. *Behavioural Ecology.* Oxford: Blackwell.

Oaten, A. 1977. Optimal foraging in patches: A case for stochasticity. *Theoretical Population Biology 12*: 263-85.

Pulliam, H.R. 1979. Learning to forage optimally. In A.C. Kamil and T.D. Sargent, eds., *Foraging Behavior.* New York: Garland.

Pyke, G.H.; H.R. Pulliam; and E.L. Charnov. 1977. Optimal foraging: A Selective review of theory and tests. *Quarterly Review of Biology 52*: 137-54.

Rachlin, H.; R.C. Battalio; J.H. Kagel; and L. Green. 1981. Maximization theory in behavioral psychology. *The Behavioral and Brain Sciences 4*: 371-419.

Staddon, J.E.R.; J.M. Hinson; and R. Kram. 1981. Optimal choice. *Journal of the Experimental Analysis of Behavior 35*: 397-412.

III MULTIPLE AND CONCURRENT SCHEDULES

11 BEHAVIORAL COMPETITION, CONTRAST AND MATCHING

J. E. R. Staddon

The aim of this chapter is to show three things: (1) The empirical fact of *matching* on concurrent variable-interval, variable-interval schedules derives from behavioral competition, where *competitiveness* is inversely proportional to the rate of an activity and directly proportional to its frequency of reinforcement. (2) Qualitative predictions of undermatching and behavioral contrast on multiple-schedules derived from this analysis are more accurate than predictions of a modified form of Herrnstein's equation. (3) All static analyses of multiple and related schedules are subject to an intrinsic paradox, so that full understanding of quantitative relations must await advances in knowledge about dynamic processes.

MULTIPLE SCHEDULES—HERRNSTEIN'S ANALYSIS

On concurrent variable-interval, variable-interval schedules pigeons and rats approximately match the ratio of responses to each alternative to the ratio of reinforcements obtained (Herrnstein, 1961). On multiple VI VI schedules, however, animals generally fail to show matching of this simple sort. Instead, the ratio of responses is usually closer to indifference than the ratio of reinforcements. This is termed *undermatching* (Baum, 1974).

This research was supported by grants from the National Science Foundation to Duke University. I thank Richard Ettinger, Richard Herrnstein, John Hinson, and Howard Rachlin for comments on an earlier version of this chapter.

Herrnstein (1970), in a well-known analysis of simple concurrent and multiple variable-interval schedules, derived matching from the fundamental theoretical relation

$$x = kR(x)/[R(x) + R(\bar{x})] \quad , \qquad (11.1)$$

where x is the rate of response X, $R(x)$ obtained reinforcement rate for that response, k a scale constant equal to the maximum value for x, and $R(\bar{x})$ a constant hypothesized to represent reinforcement rate for responses other than X. The generalized form of Eq. (11.1) is

$$x_i = kR(x_i)/\sum^{N} R(x_i) \quad , \qquad (11.2)$$

for a set of N activities. Matching, $x/y = R(x)/R(y)$, can immediately be deduced for two similar, concurrently reinforced responses, since k and the denominator of Eq. (11.2) are the same for both.

To accommodate undermatching on multiple schedules, Herrnstein (1970) proposed the following modification of Eq. (11.1):

$$x = kR(x)/[R(x) + R(z) + mR(y)] \quad , \quad 0 < m < 1 \quad , \qquad (11.3)$$

where x and y are response rates in the two multiple-schedule components, $R(z)$ is the (constant) reinforcement rate for responses other than X and Y, and m is a constant representing the degree of interaction between components. The matching relation is then

$$x/y = \{R(x)/R(y)\}[R(y) + R(z) + mR(x)]/[R(x) + R(z) + mR(y)] \quad . \qquad (11.4)$$

Equation (11.4) predicts undermatching because when the reinforcement-ratio term on the right-hand side (the term in { }) is greater than 1, the other term (in [ ]) is less than 1, and vice versa, so that the response ratio x/y is always closer to 1 (indifference) than the reinforcement ratio $R(x)/R(y)$.

Although successful in dealing with undermatching, Eq. (11.3) runs into difficulties in accounting for absolute-rate effects. For example, a paradoxical prediction of Eq. (11.3) is that response rate in a multiple VI VI schedule with two identical components should be less than response rate in the comparable simple schedule, since the denominator for Eq. (11.3) contains the term $mR(y)$ only in the multiple case. Experimental results on this point are generally negative. Some studies have found lower rates in the multiple case, but all were conducted for other purposes and two are still unpublished (cf. McSweeney, 1980). However, McSweeney (1980) explicitly designed an experiment to look for this effect and failed to find it.

Interaction between multiple-schedule components, measured by the parameter m, should increase as component duration decreases. Hence, from Eq. (11.3), absolute response rate in a component should be inversely related to component duration. But, as Edmon (1978) pointed out, the opposite prediction is closer to the truth.

Equation (11.3) also makes counterfactual predictions about behavioral contrast. For present purposes contrast can be defined as the change in response rate in one component of a multiple schedule associated with a change in the conditions of reinforcement in the other component. An increase in response rate in the unchanged component is termed *positive contrast*; a decrease, *negative contrast.* The ratio of post- and prechange response rates in the unchanged component provides a measure of contrast. From Eq. (11.3) this ratio is

$$x_{\text{post}}/x_{\text{pre}} = C_o = [R(x) + R(z) + mR(y)]/[R(x) + R(z)] \quad , \qquad (11.5)$$

since the term involving $R(y)$ vanishes when reinforcement is abolished in the changed component in the "post" condition. An obvious prediction of Eq. (11.5) is that the magnitude of contrast (C_o) is inversely related to the magnitude of $R(z)$, reinforcement rate for "other" activities: As $R(z)$ increases, C_o converges on unity (no contrast). $R(z)$ can be manipulated experimentally by adding a competing activity, such as wheel running for a rat. Since rats will spend much time running in a wheel if one is available (implying a substantial value for $R(z)$), Eq. (11.5) must predict that rats should show smaller behavioral contrast effects with a wheel available than without. Since rats usually show rather small contrast effects even without a wheel (cf. Schwartz and Gamzu, 1977), such a prediction is already quite plausible. Yet Hinson and Staddon (1978) showed it to be false: Under unexceptional conditions, contrast effects can be substantially *increased* by the addition of a competing activity.

It seems clear that although Eq. (11.1) remains useful, its generalization in the form of Eq. (11.3) is not. Perhaps Eq. (11.1) can itself be derived from some still more general rule—but one that will allow a natural extension to multiple schedules that is free of the flaws of Eq. (11.3).

COMPETITION AND THE EQUATION OF MARGINALS

In most operant conditioning experiments the animals keep very busy; they are almost never inactive. In consequence, activities must compete for the available time. There is much evidence for this. For example, a plot of response rate to one alternative in a concurrent experiment against response rate to the other is always a curve of negative slope and very often simply a straight line (cf. Staddon and Motheral, 1978). To a reasonable approximation, therefore, the weighted sum of response rates to each alternative can be considered as constant (cf. Rachlin and Burkhard, 1978; Staddon, 1979). This *time-allocation constraint* will turn out to be important to understanding the difference between concurrent and multiple schedules that shows up (for example) as undermatching, rather than matching, in multiple schedules.

Equation (11.1), Herrnstein's equation, can be derived from a simple dynamic model (Staddon, 1977) that assumes just two things: (1) Reinforcement for one response suppresses other responses in proportion to their rate of occurrence, so the higher the rate, the greater the suppression. (2) In simple VI schedules, the reinforced response and other responses take up the total time, which is the time-allocation constraint. The first assumption can be expressed formally as follows[a]:

$$dx/dt = -a_x R(z)x \quad , \tag{11.6}$$

and

$$dz/dt = -a_z R(x)z \quad , \tag{11.7}$$

where a_i is a constant representing the sensitivity to reinforcement of each activity and z is the rate of responses not explicitly reinforced; call these *interim* activities (Staddon and Simmelhag, 1971). The second assumption is just

$$x + z = 1 \tag{11.8}$$

for the unit session, where x (the reinforced response) and z (interim activities) are scaled in time units.[b]

Equations (11.6) and (11.7) represent a reversible process whereby x is transformed into z at a rate $R(z)$ and z into x at a rate $R(x)$. In the steady state, the rates of these two opposed reactions must be equal. Hence, the following relation can be deduced at once:

$$a_z R(x)z = a_x R(z)x \quad ,$$

which is equivalent to matching between x and z and the reinforcement rates for x and z – *biased* matching if $a_x \neq a_z$, simple matching otherwise –

$$x/z = [R(x)/R(z)]\,[a_z/a_x] \quad ,$$

or the relation

$$R(x)/x = [a_x/a_z]\,R(z)/z = qR(z)/z \,, \quad q > 0 \;. \tag{11.9}$$

a. The term z in Eq. (11.8) need not represent a single activity; under certain not too restrictive conditions it can be an aggregate of many different activities, each with its own reinforcement rate.

b. This model and Eq. (11.1) treat other behavior as if it were reinforced according to a VI schedule, whereas a ratio schedule might seem more appropriate for activities entirely under the animal's control. However, most activities seem to gain in attractiveness as time elapses since they were last performed and to lose attractiveness as they continue to occur. Satiation-deprivation processes of this sort suggest that an interval feedback function may not be inappropriate even for intrinsic activities. I return to this topic in the section on dynamics.

Equation (11.1) can be immediately deduced from Eqs. (11.8) and (11.9). But the latter equations represent the more fundamental law, since they can be applied in a natural way to multiple schedules and do not lead to the counterfactual predictions of Eq. (11.3) already discussed.

Equation (11.9) can also be deduced from optimality theory, either in the form of simple reinforcement maximization, given a VI feedback function whose derivative can be expressed solely in terms of the ratio $R(x)/x$ (cf. Staddon, 1980: appendix; Staddon and Motheral, 1978) or, given a costless instrumental response, from the minimum-distance model (see Staddon, 1979, Appendix, for the derivation of the simple case where $q = 1$). In this sense, Eq. (11.9) can be regarded as the equation of *marginal utilities* of competing responses. Equation (11.9) is a mathematization of the diminishing-returns principle proposed by Hinson and Staddon (1978: note 10) to underlie behavioral contrast, namely that the competitiveness of an activity is inversely related to its rate, so the higher the rate, the less the competitiveness.

Multiple and Concurrent Schedules. In concurrent schedules, from Eq. (11.9) (assuming identical responses so that $q = 1$, since this does not affect the conclusions), $R(x)/x = R(z)/z = R(y)/y$, which implies matching between x and y and their reinforcement rates.

The distinctive difference between multiple and concurrent schedules is that on concurrent schedules, as the reinforcement rate for one response increases, the time devoted to that response (hence its overall rate) is limited only by the total time available. On multiple schedules, however, the time that can be devoted to one response is limited by the fraction of the total time associated with that schedule *component*—50 percent of the total time when components are of equal duration. In one sense this asymmetry does not matter, since the response rates of interest, x and y, are usually computed relative to the time available within a component, not relative to total session time. But the restriction of response rates to their respective components does *not* apply to interim activities, which are free to compete with X and Y in both components. Thus, an appropriate time denominator for response Z is the total time, whereas the denominators for X and Y are restricted to (usually) half the total. It is this feature that is responsible for undermatching and the other properties of multiple schedules. The problem of the time denominator implies that no static model can provide a fully adequate account of multiple schedules. Nevertheless, some basic, qualitative conclusions can be derived easily.

For example, suppose in a particular multiple schedule the reinforcement rate in each component is 50/hour and the constraint on total responding within the session is 1,000 (for a 30-minute session). Let $R(z)$ be 10/hour. Then from Eq. (11.9), $x/z = y/z = 5$. Given the time constraint $x + y + z = 1{,}000$, with x, y, and z scaled in equal-time units, x and y must each equal 455 and z 91. Suppose now that the reinforcement rate in component Y is reduced from 50 to 10; the predicted value for x is now 857. But x cannot take up more than half the total time, so that its maximum value is 500. In the matching relation, therefore, the ratio x/y will be closer to indifference than the ratio of reinforcement rates for X and Y, that is, undermatching. Thus, by taking Eq. (11.9) rather than Eq. (11.1) as fundamental—and taking into account the constraints imposed by time allocation—undermatching on multiple schedules follows immediately.

A simple way to visualize this derivation of the matching law is depicted in Figure 11-1. The figure shows an airtight cylinder of unit volume containing two frictionless free pistons that divide it into three compartments of volume x, y, and z, respectively. Clearly, $x + y + z = 1$, as required by the time-allocation constraint. In equilibrium the pressure of air in each of the three compartments must be the same, or else the pistons would move. Hence we can assume that $P(x) = P(y) = P(z)$. From Boyle's law we know that for each compartment, $P(i) \cdot i = NT$, where i is the volume, N is the number of molecules per unit volume, and T is the absolute temperature. Suppose we identify N with the reinforcement rate for activity i, so that we can write $P(i) \cdot i = R(i) \cdot T$, or $P(i) = TR(i)/i$ for all three activities. Equating the values of P leads at once to Eq. (11.9): $R(x)/x = R(y)/y = R(z)/z$.

If the pistons in Figure 11-1 are completely free to move anywhere in the cylinder, the model is appropriate for concurrent VI VI schedules, since the component duration is then entirely under the control of the animal. For multiple schedules the model must be modified by the addition of a *detent* in the middle (for equal-duration components) of the cylinder (A in Figure 11-1) that prevents x or y from taking up more than half the total volume. It is obvious from this analogue that perfect matching can hold in the multiple case only if $R(z)$ is sufficiently large that neither x nor y expands to more than half the total volume. Formally this amounts to the requirement that the absolute value of the difference between the reinforcement rates for X and Y must not exceed $R(z)$, the reinforcement rate for interim behavior.

Figure 11-1. Pressure model for equilibria in concurrent and multiple VI VI schedules; x, y, and z represent rates of (times taken up by) the two instrumental responses and the interim activity.

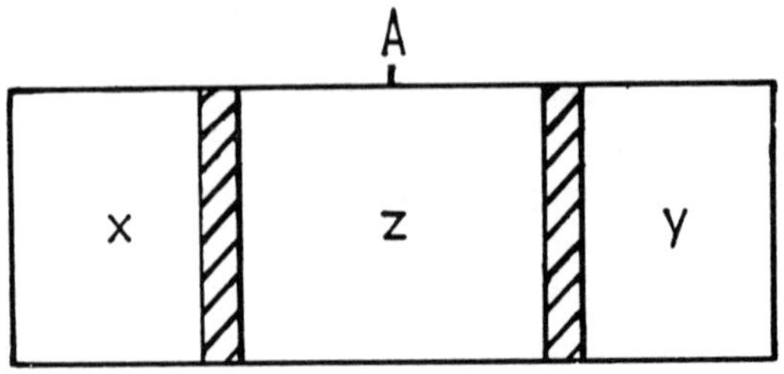

For obvious reasons I shall term $R(x)/x$, the quantity corresponding to pressure, the *competitiveness* of X; the term *response strength*, as it has been used by Nevin (e.g., 1974), is very similar in meaning.[c]

Formal Derivation. Behavioral contrast is defined by the ratio of response rates in component X of a multiple schedule under two conditions: a *prediscrimination* phase, in which responses X and Y receive equal VI reinforcement, and a *postdiscrimination* phase, in which reinforcement for X continues as before but reinforcement for Y is abolished. The limitation on the maximum value of x and y relative to z, the level of the interim activity, means that the analysis must consider separately the two cases where $R(z)$ is less than or equal to or else greater than $R(x)$. For simplicity assume identical sensitivities to reinforcement for all responses ($q = 1$ in Eq. (11.9)), but the same conclusions follow when this condition is relaxed.

1. $R(z) \leq R(x)$: In the prediscrimination phase, $R(x) = R(y)$; hence it is obvious that z takes up equal volumes in compartments x and y. If $R(z) > 0$, neither x nor y can "saturate," so that the simple matching law applies. Thus,

$$x_{\text{pre}} = R(x)/[2R(x) + R(z)] \quad , \tag{11.10}$$

in the prediscrimination phase. When reinforcement for Y is abolished in the postdiscrimination phase, however,

$$x_{\text{post}} = R(x)/[R(x) + R(z)] \quad , \tag{11.11}$$

but since $R(z) \leq R(x)$, this expression is ≥ 0.5, the maximum possible value for x. Consequently, $x_{\text{post}} = 0.5$ (x has saturated) and the magnitude of contrast is therefore given by

$$x_{\text{post}}/x_{\text{pre}} = C_o = [R(x) + 0.5R(z)]/R(x) =$$

$$= 1 + R(z)/2R(x) \quad , \qquad R(z) \leq R(x) \quad , \tag{11.12}$$

which is a positive function in $R(z)$ and negative in $R(x)$. Contrast is at a maximum, at $C_o = 1.5$, when $R(x) = R(z)$.

2. $R(z) > R(x)$: As before, $x_{\text{pre}} = R(x)/[2R(x) + R(z)]$. However, the matching relation yields a value for x_{post} that is less than the maximum possible, because $R(x) < R(z)$; consequently x_{post} is given by Eq. (11.11) and contrast is therefore given by

$$C_o = [2R(x) + R(z)]/[R(x) + R(z)] \quad , \qquad R(z) > R(x) \quad . \tag{11.13}$$

c. The basis for this assertion is a paper (Staddon, 1978) showing that Nevin's quantitative results can be derived from the competitive assumption that animals equate the quantity $a_i\, dx_i/x_i$, for all activities, where x_i is the rate of the ith response and a_i approximates its reinforcement rate. The comparable expression in the present analysis is just a_i/x_i.

Equation (11.13) is a negative, hyperbolic function of $R(z)$.

Figure 11-2 shows these relations. The top graph shows the dependence of contrast on $R(z)$, with $R(x)$ held fixed. The bottom panel shows contrast as a function of $R(x)$, with $R(z)$ fixed. Both functions are discontinuous and bitonic, with maxima at $R(x) = R(z)$.

The predictions for matching in multiple schedules must be derived in a similar piecemeal way, except that there are now three cases to consider:

a. If $R(x) \geq R(y) + R(z)$, then $x = 0.5$ (X saturates) and

$$y = 0.5R(y)/[R(y) + (z)] \ , \tag{11.14}$$

that is, matching between Y and Z in a compartment of duration 0.5.

b. If $R(y) \geq R(x) + R(z)$, then $y = 0.5$ (Y saturates) and

$$x = 0.5R(x)/[R(x) + R(z)] \ . \tag{11.15}$$

c. If neither (a) nor (b) is true, then $R(z) > | R(x) - R(y) | R$; neither X nor Y saturates, so that overall matching applies: $x/y = R(x)/R(y)$.

Figure 11-2. Contrast (x_{post}/x_{pre}) as a function of the reinforcement for the interim activity, $R(z)$ (*top*) or for the instrumental response in the unchanged component, $R(x)$ (*bottom*) deduced from the pressure model.

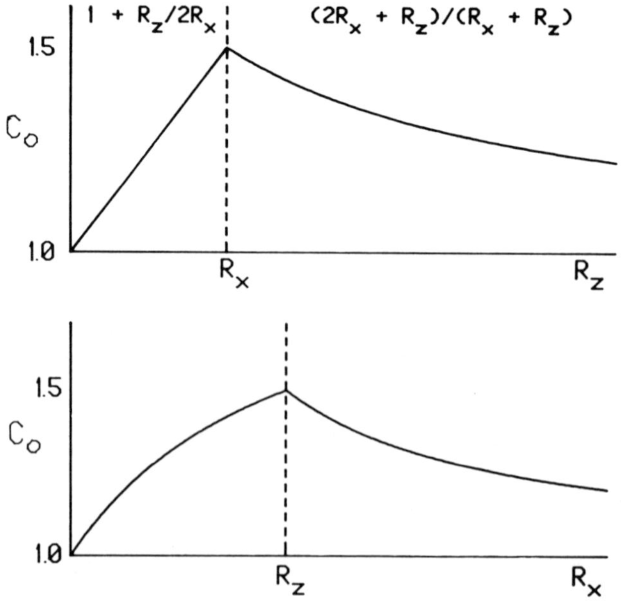

If a constraint such as $R(x) + R(y) = K$ is placed on $R(x)$ and $R(y)$, then the function relating $x/(x + y)$ to $R(x)/[R(x) + R(y)]$ can be derived. Examples are shown in Figure 11-3. Each function has three regions corresponding to cases a–c. Simple matching holds in the central region defined by $|R(x) - R(y)| < R(z)$ (that is, $R(z)$ greater than the absolute value of the difference between $R(x)$ and $R(y)$). As the value of $R(z)$ increases relative to $|R(x) - R(y)|$, region c increases relative to regions a and b, the curvature of the function in regions a and b decreases, and the function as a whole more and more closely approximates matching.

EMPIRICAL RESULTS

The present model (Eqs. (11.8) and (11.9)) easily accounts for undermatching on multiple schedules, as already described. Indeed, the model predicts undermatching whenever there is any constraint on the proportion of time devoted to any response, as in concurrent VI FI schedules, for example, where the time devoted to the VI choice is often restricted to the postreinforcement pause on the

Figure 11-3. Matching relations on multiple VI VI schedules, deduced from the pressure model.

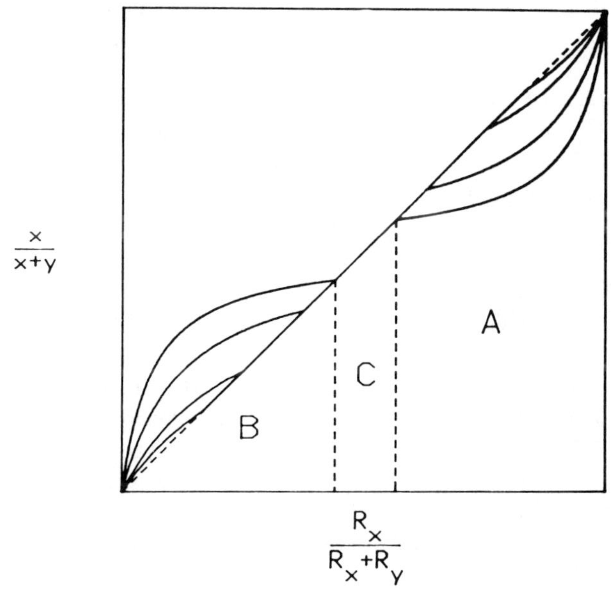

FI choice (Nevin, 1971). Baum (1979) has reviewed numerous other instances of undermatching that can be interpreted in this way.

The model predicts that response rate in a multiple schedule with equal VIs should be the same as rate in the comparable simple schedule, not lower as implied by Eq. (11.3). Consideration of dynamic effects allows this approach to account not only for the generally higher rates in multiple schedules with short components but also for local contrast effects.

Equations (11.11)-(11.13) predict that contrast, defined as the ratio of x_{post}/x_{pre}, is at a maximum for an intermediate level of $R(z)$, the "strength" of competing activity. Equation (11.3), on the other hand, predicts that contrast is at a maximum when $R(z)$ is zero. The data of Hinson and Staddon (1978), showing that contrast can be increased by adding a competing activity (increasing $R(z)$), clearly support the former prediction.

Both the present model and Eq. (11.4) predict that matching on multiple schedules should be better the larger the value of $R(z)$ relative to $R(x)$ and $R(y)$. In the case of Eq. (11.3), increasing $R(z)$ reduces the curvature of the predicted ogive until it increasingly approximates the matching line. In the case of the present model, the linear region of perfect matching in Figure 11-3 is progressively increased and the curvature of the regions of undermatching is decreased. One way to increase $R(z)$ relative to $R(x)$ and $R(y)$ is by allowing the animal's body weight to increase: If the animal is less hungry, presumably activities other than eating and food-reinforced behavior will gain in relative strength. Herrnstein and Loveland (1974) have shown that less hungry animals do indeed yield better matching on multiple schedules, as predicted by both models.

Several studies have shown that contrast effects are small with a treadle-pressing response (for pigeons) and, often, with lever pressing by rats (cf. Schwartz and Gamzu, 1977). The rat data have already been interpreted as owing to lack of a substantial competing activity. The pigeon treadle-pressing data can be interpreted in the opposite way, as owing to a competing activity that is relatively too strong. Pigeons press treadles much more slowly than they peck a key at comparable rates of food reinforcement, suggesting that a_i (in Eq. (11.9)) for treadle pressing is smaller than a_i for key pecking: Treadle pressing is less sensitive than pecking to food reinforcement. Hence, in a comparison with a constant competing activity, $R(z)$ will be relatively larger for treadle pressing than for key pecking. Thus, at the same reinforcement rate, treadle pressing may be further from the maximum in Figure 11-2 than key pecking.

Although couched in terms of response rate (on the assumption that rate and time spent are roughly proportional), the present model depends critically on time allocation. Several contrast studies have studied time allocation directly, however, and all find that time allocated to the food-related activity in the unchanged component increases when reinforcement rate in the other component is decreased, whether reinforcement in the unchanged component is delivered independently of responding, for standing on one side or other of a chamber, or for a response such as key pecking (Bouzas and Baum, 1976; White, 1978).

The detailed predictions of undermatching shown in Figure 11-3 are obviously implausible in one respect: the discontinuity of the predicted functions. Nature may not abhor a vacuum, but biology seems rarely to give rise to empirical functions with sharp corners. Nevertheless, the ogival form of the functions is consistent with published data (e.g., Reynolds, 1963, replotted in Herrnstein, 1970), and there are both reasons why real discontinuities might escape detection and reasons why the predicted sharp corners should be smoothed out.

Many matching studies present only three data points, symmetrically arranged around equality. For example, ratios of $R(x)/R(y)$ of 3, 1, and 1/3 are commonly studied. It is easy to see from Figure 11-3 that for all such experiments the predicted locus of points is a straight line, and linear functions have not infrequently been reported (e.g., Menlove, 1975).

Simple variability would tend to make any kind of discontinuity difficult to detect in practice. In addition a consistent application of the present approach implies dynamic processes that should smooth the sharp corners predicted from the static approximation.

An experiment by Pliskoff, Shull, and Gollub (1968) affords a partial comparison of the present model and Eq. (11.3). Pliskoff et al. trained pigeons in a two-key apparatus in which a multiple VI VI schedule was programmed on one key and a fixed VI on the other.[d] Thus responding on the two multiple-schedule components corresponds to responses X and Y in the equations, and responding on the alternate VI corresponds to response Z. For the conditions of interest here, the scheduled reinforcement rate for response X was 40/minute, for response Y 10/minute and for response Z, 20/minute. Since $R(z) < |R(x) - R(y)|$ and $R(x) > R(z) + R(y)$ the relation between x/y and the reinforcement rate for the concurrent VI, $R(z)$, can be obtained from Eq. (11.14): X is saturated, hence, in time units, $x = 0.5$; from Eq. (11.14), $y = 0.5R(y)/[R(y) + R(z)]$, so that

$$x/y = 1 + R(z)/R(y) \quad , \tag{11.15}$$

that is, a linear function with intercept at $x/y = 1$ and slope $1/R(y)$ (see Figure 11-4).

d. This description is something of an idealization because the "common" VI was signaled by different stimuli within a single session and there was a changeover delay affecting switches between the multiple and common keys. Since it is not clear what effect these procedural features might have on the predictions of the present model, neither confirmation nor disconfirmation of the model by these data can carry great weight. I discuss the experiment here because it provides the only available parametric test of the effect of a common source of reinforcement on the allocation of behavior between components of a multiple VI VI schedule.

Figure 11–4. Ratio of response rates in a multiple schedule (x/y) as a function of reinforcement rate on a concurrent VI schedule, $R(z)$, deduced from the pressure model.

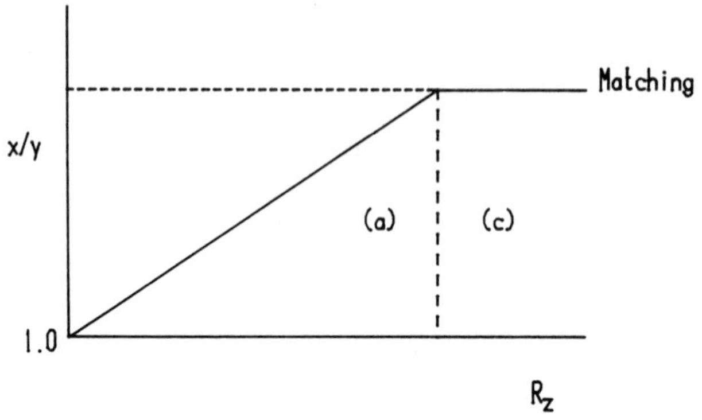

The data are in partial agreement with Eq. (11.15): The function (only three points from Pliskoff et al. 1968: figure 5) is not consistently different from a linear one and has an intercept close to 1.0 for three pigeons. But its slope is on the order of 0.025 rather than 0.1 as predicted. The data are clearly inconsistent both with Eq. (11.3) and with a model due to Catania (1963), discussed by Pliskoff et al., both of which predict a curvilinear relation between x/y and $R(z)$. The present model also predicts that x should be approximately constant, and again the data are in qualified agreement. The changes in x as $R(z)$ increases are much smaller than the changes in y (Pliskoff et al. 1968: table 2): For the three birds x goes from 66 to 52, 31 to 26, and 86 to 36 per minute as $R(z)$ goes from 0 to 60 scheduled reinforcements per hour. At the same time the changes in y are 33 to 15, 32 to 10, and 53 to 11 per minute. These discrepancies and others mentioned earlier may reflect dynamic processes.

An objection sometimes raised to a competition analysis of contrast is the results of signaled-VI and DRO experiments, in which contrast is found even though no instrumental response is required in the changed component. There are two answers to this. The first is that this objection applies equally to any account derived from Eq. (11.1), which is most naturally interpreted in competition terms. The second is to refer to numerous observational studies (e.g., Staddon and Simmelhag, 1971) showing that animals spend about as much time engaging in terminal activities on procedures such as fixed time and DRO that require no instrumental response as on procedures such as fixed interval that do require a response. The competition account depends only upon time-allocation constraints and diminishing returns; the existence or nonexistence of a response requirement is irrelevant providing animals spend time in terminal responding.

DYNAMIC EFFECTS

A static model can never be fully adequate to account for equilibria on multiple schedules because static models necessarily involve relations among averages, average response and reinforcement rates. An average rate implies a time denominator over which the averaged events are counted. So long as the denominator is the same for all competing activities, our ignorance about the method the animal actually uses to strike averages does not matter: Whatever it is, it is the same for all. On multiple schedules it is *not* the same, because activities other than the instrumental responses are free to occur at any time, whereas the responses are confined to their components. Thus there is no nonarbitrary way to decide between contradictory applications of matching. For example, in a two-component multiple schedule, different conclusions are reached depending upon whether components are considered separately or together. Within a component matching implies that $z = R(z)/[R(x) + R(z)]$; whereas for the session as a whole matching implies that $z = R(z)/[R(x) + R(y) + R(z)]$. There is no static resolution of this contradiction.

The present explanation for contrast is that Z, the competing, interim activity, is reallocated to the extinction component (or the component with lowered reinforcement rate if reinforcement rate is merely reduced); with more total time devoted to it, Z becomes less competitive in the unchanged component allowing response rate x to rise (positive contrast). The fine print here is the word *total*: Over *what period* is this total to be measured? Animals appear to regulate the rate of activities not just over days but even over periods of a few minutes. Yet the static model assumes that the competitiveness of Z (which is proportional to $R(z)$ in the model) is constant and depends upon its average across a whole experimental session or at least across one component cycle. This is obviously wrong when component duration becomes longer than a few seconds. For example, if component duration is 30 minutes, say, the competitiveness of interim activities at the end of a "rich" component, when there has been little time to engage in anything but the instrumental response, is likely to be considerably higher than at the end of the alternating "lean" component, when there has been much free time. Thus the assumption of constant $R(z)$ will increasingly be violated at long component durations. What effects might this have?

There are at least two ways to answer this question. One is to make some assumption about a dynamic averaging process, a possibility most extensively explored by Killeen (e.g., Killeen, 1981). I will not pursue that line here, although I believe it can be interpreted so as to converge on the second possibility, which is to consider the second-by-second processes that determine the strength of an activity. What is known here is unsatisfactory because it is practically impossible to study activities in isolation. Nevertheless, some reasonable guesses can be made based on the idea that the longer an animal engages in an activity, the less attractive (competitive, in the present analysis) it becomes. Some such *satiation-deprivation* process must underlie all nonexclusive choice and has often been proposed (as in Atkinson and Birch, 1970, and Staddon, 1980). It is implicit in Eqs. (11.6)-(11.8). One way such an analysis might be applied is as follows.

In multiple schedules, there are three activities to consider: the instrumental responses in each component, X and Y, and the interim activities Z. Since animals on these schedules are generally very hungry, within a session relatively little weakening of the tendency to make the food-related response (X or Y) might be expected. The interim activities are presumably less urgent than eating, however, so should be more subject to satiation. If the rates of deprivation and satiation are relatively rapid, changes might then be expected in the competitiveness (strength) of Z (reflecting changes in $R(z)$) from component to component of the sort depicted in Figure 11-5. The center panel shows the changes in the strength of Z, on the assumption that it occurs only at a low rate in the *rich* component of a multiple VI VI schedule, so that deprivation causes an increase in strength; Z can presumably occur at a higher rate in the *lean* component, so that satiation causes a decrease in strength. In the rich component the strength of the instrumental response is high; in the lean component it is low, as shown in the top panel. Since $z + x$ or y is constant, because of time allocation, the actual values of x and y will be the complements of z, as shown in the bottom panel.

The changes shown in Figure 11-5 are the familiar *local contrast* effects reported at an intermediate stage of multiple VI VI training by Nevin and Shettleworth (1966) and others. The present analysis gives few clues as to the transience of such effects but does suggest that they should be larger (in the sense that the difference between initial and terminal rates within a component should be greater) in schedules with longer components. No systematic analysis seems to have been done, but large and apparently stable effects in a schedule with 8-minute components have been reported (Rachlin, 1973). Conversely, short components (e.g., 30 seconds) are often associated with weak or opposite effects (e.g., Hamilton and Silberberg, 1978). John Hinson and I, in an experiment with a multiple VI EXT schedule in which three different EXT durations alternated with a fixed-duration VI component (Hinson and Staddon, 1981), found the magnitude of positive local contrast in the VI to be directly related to the duration of the preceding EXT component—as expected from a satiation-deprivation analysis.

Relative response and reinforcement rates in multiple schedules conform better to matching when component duration is short than when it is long (Shimp and Wheatley, 1971; Todorov, 1972). Dynamics suggest why: The longer the components, the less valid the assumption (necessary to the derivation of matching) that $R(z)$ is constant throughout. $R(z)$ tends to increase during a long, rich component and to decrease during a long lean one. Hence, Z is more com-

Figure 11-5. A competition model for local contrast effects. "Strength" of terminal response (*top*), high in rich component, low in lean component. Changes in strength of interim (competing) activities owing to satiation-deprivation (*middle*), increasing in rich component, decreasing in lean. Resultant level of terminal response (*bottom, top* minus *middle*), decreasing in rich component, increasing in lean component.

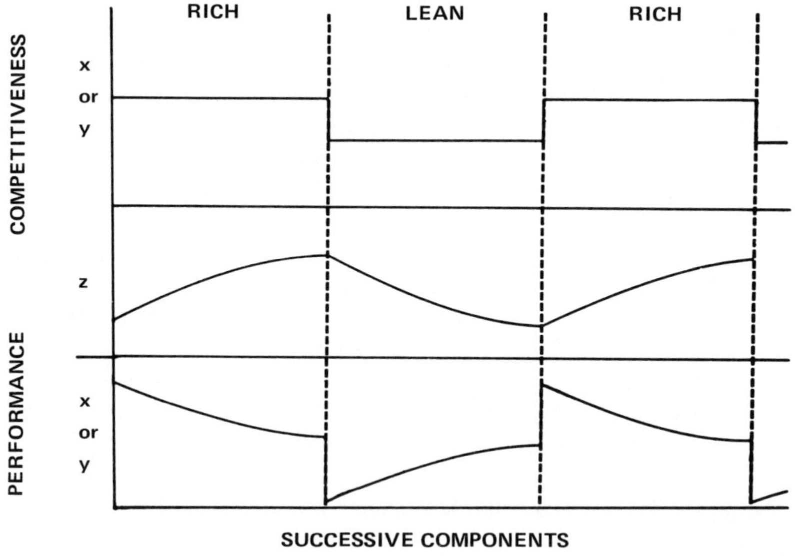

petitive during the rich component than during the lean, so that rate is lower at the end of the rich component and higher at the end of the lean than required by matching. This process leads to undermatching, the usual result on multiple schedules.

Satiation-deprivation changes in the competitiveness of interim activities provide an obvious explanation for pausing on fixed-ratio schedules. If, as might be expected, the instrumental response is always dominant, the tendency to engage in interim activities will always be highest after food, that is, after a period when they have not been occurring. The time taken up by eating may account for the (smaller) tendency to pause after food on variable-ratio schedules. The process may be less effective on variable-interval schedules, because the tendency to make the instrumental response is weak just after a response (reflecting the lowered probability of food at that time), allowing time there for the occurrence of interim activities and correspondingly weakening their tendency to occur after food.

One prediction of this analysis is that the postfood pause on fixed-ratio schedules should be reduced by postfood timeouts longer than the typical pause; this seems to be the case, unless the timeouts are very long (Richards and Blackman, 1981).

Under some conditions local contrast effects may fail to occur at all, or occur in a different way—being correlated with the succeeding rather than the preceding component, for example (e.g., Buck, Rothstein, and Williams, 1975). The static analysis suggests some boundary conditions. For example, if $R(x) > R(y) + R(z)$ and component duration is short, response X may saturate throughout the component so that positive local contrast is impossible. Both local and behavioral contrast effects may be weakened or abolished if the schedule contains many extinction or timeout periods when Z can occur unopposed, so that it competes only weakly with the instrumental response in other components (cf. Buck et al., 1975). Where these properties are not attributable to the complexity of the procedures, they may reflect metastability, a dependence on the particular training sequence.

Given the kind of complex individual dynamics described, it is far from obvious what should be expected when several components are used and their durations are close to the periodicity of the hypothesized satiation-deprivation processes. It is not too difficult to provide a formal model for the simple case in which only the single competing activity Z is subject to satiation-deprivation effects. If there is more than one such activity and the time constants for each are different, or if the instrumental responses are subject to satiation-deprivation,[e] it is not trivial to provide an exact account, yet without such an account it is impossible to be sure what form local contrast effects should take. Our lack of understanding of the transience of these effects further emphasizes our ignorance of the molecular processes involved.

SUMMARY

Herrnstein's equation for simple variable-interval schedules and matching on concurrent variable-interval schedules can be deduced from two assumptions: that activities compete for time and that in equilibrium the ratio reinforcement rate/response rate is equal for all activities. This same static model accounts for undermatching on multiple schedules as well as other qualitative effects, such as enhanced contrast associated with the addition of a competing activity (such as

e. As suggested, for example, by Catania's "freedom" experiments, where pigeons prefer a situation in which several identically programmed response keys are available to a situation with a single key (e.g., Catania, 1980): If we assume that the competitiveness (or value) of responding on a given key decreases even slightly with time spent responding on it, then the competitiveness of responding on any of the keys in the multiple-key situation will generally be higher than competitiveness of responding on the single key. Hence, the multiple-key situation should be chosen over the single-key situation.

wheel running) and improved matching in satiated animals. This model is also free of the paradoxical features of Herrnstein's multiple-schedule equation. These static relations appear to reflect a fundamental dynamic property of competing activities: that they become relatively more attractive if their rate is suppressed and relatively less attractive as they occur at a high rate (satiation-deprivation). Contradictions involving choice of time denominator for response and reinforcement rates imply that no static analysis is fully adequate to account for performance on multiple schedules.

APPENDIX 11A

The formal argument deriving matching from the minimum-distance model is as follows: The objective function for three activity classes (Staddon, 1979: Eq. (5)) is

$$C[x, R(x), z] = a(x_0 - x)^2 + b(R_0 - R(x))^2 + c(z_0 - z)^2 \quad , \qquad (11A.1)$$

where x is the rate of instrumental responding, z is the level of interim activity, and $R(x)$ is the obtained rate of reinforcement. The constraints are due to time allocation:

$$x + R(x) + z = 1 \quad , \qquad (11A.2)$$

and the VI feedback function:

$$R(x) = Ax/(A + x) \quad , \qquad (11A.3)$$

where A is the programmed (maximum) VI reinforcement rate. These three equations are then combined to form the Lagrangian:

$$L[x, R(x), z, \lambda_1, \lambda_2] = a(x_0 - x)^2 + b(R_0 - R(x))^2 + c(z_0 - z)^2 -$$

$$-\lambda_1(1 - x - R(x) - z) - \lambda_2(R(x) - Ax/(A + x)) \quad . \qquad (11A.4)$$

If we assume that the cost of x is negligible, then $a = 0$ and the first term in Eq. (11A.4) vanishes. Taking partial derivatives to find the minimum then yields

$$\partial L/\partial x = \lambda_1 + \lambda_2 A^2/(A + x)^2 \quad ,$$

which can be rewritten as

$$\partial L/\partial x = \lambda_1 + \lambda_2 [R(x)/x]^2 \quad , \qquad (11A.5)$$

plus four other expressions. A similar exercise for the other instrumental response yields an expression exactly parallel to this equation for response y. By the hypothesis these two marginals must be equal to the fixed marginal for interim response z; hence they can be equated, which yields $R(x)/x = R(y)/y$, the matching relation. This argument holds for any feedback function whose derivative with respect to x can be expressed as a function of $R(x)/x$ (Staddon, 1980).

REFERENCES

Atkinson, J.W., and D. Birch. 1970. *The Dynamics of Action.* New York: Wiley.

Baum, W.M. 1974. On two types of deviation from the matching law: Bias and undermatching. *Journal of the Experimental Analysis of Behavior 22*: 231–42.

_____. 1979. Matching, undermatching, and overmatching in studies of choice. *Journal of the Experimental Analysis of Behavior 32*: 269–81.

Bouzas, A., and W.M. Baum. 1976. Behavioral contrast of time allocation. *Journal of the Experimental Analysis of Behavior 25*: 179–84.

Buck, S.L.; B. Rothstein; and B.A. Williams. 1975. A re-examination of local contrast in multiple schedules. *Journal of the Experimental Analysis of Behavior 24*: 291–301.

Catania, A.C. 1963. Concurrent performances: Reinforcement interaction and response independence. *Journal of the Experimental Analysis of Behavior 6*: 253–64.

_____. 1980. Preference for free choice over forced choice in pigeons. *Journal of the Experimental Analysis of Behavior 34*: 77–86.

Edmon, E.L. 1978. Multiple schedule component duration: A reanalysis of Shimp and Wheatley (1971) and Todorov (1972). *Journal of the Experimental Analysis of Behavior 30*: 239–41.

Hamilton, B.E., and A. Silberberg. 1978. Contrast and autoshaping in multiple schedules varying reinforcer rate and duration. *Journal of the Experimental Analysis of Behavior 30*: 107–22.

Herrnstein, R.J. 1961. Relative and absolute strength of response as a function of frequency of reinforcement. *Journal of the Experimental Analysis of Behavior 4*: 267–72.

_____. 1970. On the law of effect. *Journal of the Experimental Analysis of Behavior 13*: 243–56.

Herrnstein, R.J., and D.H. Loveland. 1974. Hunger and contrast in a multiple schedule. *Journal of the Experimental Analysis of Behavior 21*: 511–17.

Hinson, J.M., and J.E.R. Staddon. 1978. Behavioral competition: A mechanism for schedule interactions. *Science 202*: 432–34.

_____. 1981. Some temporal properties of local contrast. *Behaviour Analysis Letters 5*: 275–81.

Killeen, P.R. 1981. Averaging theory. In C.M. Bradshaw, ed., *International Symposium on Recent Developments in the Quantification of Steady-State Operant Behaviour.* Amsterdam: Elsevier North–Holland.

McSweeney, F.K. 1980. Differences between rates of responding emitted during simple and multiple schedules. *Animal Learning and Behavior 8*: 392–400.

Menlove, R.L. 1975. Local patterns of responding maintained by concurrent and multiple schedules. *Journal of the Experimental Analysis of Behavior 23*: 309–37.

Nevin, J.A. 1971. Rates and patterns of responding with concurrent fixed-interval and variable-interval reinforcement. *Journal of the Experimental Analysis of Behavior 16*: 241–47.

_____ . 1974. Response strength in multiple schedules. *Journal of the Experimental Analysis of Behavior 21*: 389–408.

Nevin, J.A., and S.J. Shettleworth. 1966. An analysis of contrast effects in multiple schedules. *Journal of the Experimental Analysis of Behavior 9*: 305–15.

Pliskoff, S.S.; R.L. Shull; and L.R. Gollub. 1968. The relation between response rates and reinforcement rates in a multiple schedule. *Journal of the Experimental Analysis of Behavior 11*: 271–84.

Rachlin, H. 1973. Contrast and matching. *Psychological Review 80*: 217–34.

Rachlin, H., and B. Burkhard. 1978. The temporal triangle: Response substitution in instrumental conditioning. *Psychological Review 85*: 22–47.

Reynolds, G.S. 1963. Some limitations on behavioral contrast and induction during successive discrimination. *Journal of the Experimental Analysis of Behavior 6*: 131–39.

Richards, J.J., and D.E. Blackman. 1981. Effect of time-out duration after food reinforcement in a fixed-ratio schedule with pigeons. *Behaviour Analysis Letters 1*: 81–87.

Schwartz, B., and E. Gamzu. 1977. Pavlovian control of operant behavior. In W.K. Honig and J.E.R. Staddon, eds., *Handbook of Operant Behavior*, Englewood Cliffs, N.J.: Prentice-Hall.

Shimp, C.P., and K.L. Wheatley. 1971. Matching to relative reinforcement frequency in multiple schedules with a short component duration. *Journal of the Experimental Analysis of Behavior 15*: 205–10.

Staddon, J.E.R. 1977. On Herrnstein's equation and related forms. *Journal of the Experimental Analysis of Behavior 28*: 163–70.

_____ . 1978. Theory of behavioral power functions. *Psychological Review 85*: 305–20.

_____ . 1979. Operant behavior as adaptation to constraint. *Journal of Experimental Psychology: General 108*: 48–67.

_____ . 1980. Optimality analyses of operant behavior and their relation to optimal foraging. In J.E.R. Staddon, ed., *Limits to Action: The Allocation of Individual Behavior*, New York: Academic Press.

Staddon, J.E.R., and S. Motheral. 1978. On matching and maximizing in operant choice experiments. *Psychological Review 85*: 436–44.

Staddon, J.E.R., and V.L. Simmelhag. 1971. The "superstition" experiment: A reexamination of its implications for the principles of adaptive behavior. *Psychological Review 78*: 3–43.

Todorov, J.C. 1972. Component duration and relative response rates in multiple schedules. *Journal of the Experimental Analysis of Behavior 17*: 45–49.

White, K.G. 1978. Behavioral contrast as differential time allocation. *Journal of the Experimental Analysis of Behavior 29*: 151–60.

12 CHOICE AND THE RESCORLA-WAGNER MODEL

William Vaughan, Jr.

Within the controversy regarding maximization versus matching accounts of behavior, a distinction that would prove fruitful is that between a heuristic and an algorithm. Given a certain end to be achieved, a heuristic is an approach that generally has a good chance of working but is not guaranteed to do so; an algorithm invariably works. It may be asked whether the process governing animals' behavior is best viewed as a heuristic or an algorithm. The following sort of process is probably best construed as an algorithm whose end is maximization of rate of food intake: Under a choice situation the distribution of behavior is varied, either systematically or randomly. Distributions that are correlated with higher rates of food intake tend to be maintained (reinforced by that higher rate of food intake), while other distributions tend to drop out. Distributions similar to those correlated with higher rates of food tend to be sampled more than those similar to distributions correlated with lower rates of food. Although such a process is in a strict sense a heuristic, in that it does not guarantee the absolute maximum of rate of food intake (the distribution of behavior may end up at a *local* maximum), under most environmental conditions it would probably give rise to an outcome not far from optimal. I assume that this model is consistent with the positions of both Baum (1973) and Rachlin, Green, Kagel, and Battalio (1976).

The argument to be made here is that behavior is governed by a process that may be unequivocally described as heuristic with respect to reinforcement maximization. The present approach is a particular elaboration of what has been

The work reported here was supported by Grant MH-15494 from the National Institute of Mental Health to Harvard University. I thank S.L. Greene, D.H. Loveland, and H.L. Miller, Jr., for collaboration in the design and running of the experiments.

termed *melioration* (Herrnstein and Vaughan, 1980). Although it may sound reactionary, the process is one of strengthening behaviors followed by an increase in value and weakening behaviors followed by a decrease in value. This model implies that aggregate experimental variables count for very little; rather, it is their specific local application that has the greatest effect on behavior.

A corollary of this approach is that the specific environmental contingencies are of great importance in governing behavior. It thus makes little sense to say, for example, that behavior is such as to maximize rate of reinforcement. Rather, one must say that under specific conditions rate of reinforcement is maximized, since the outcome will be a joint function of the environmental contingencies and the behavioral process.

The general approach to be pursued here is not necessarily inconsistent with the observation that animals often appear to maximize rate of food intake under naturalistic conditions. If the natural environment is structured in such a way that a fairly simple heuristic would lead to maximization of food intake, then by natural selection such a heuristic should come to be encoded in the germ plasm. However, if one wants to understand behavior per se, the behavioral process must be dissected out of its naturalistic context. Only then will it be possible to say what outcome can be expected under an arbitrary change in that environment.

The model discussed here is motivated in part by the Rescorla-Wagner (1972) model. It predicts the process of melioration (Herrnstein, Chapter 19 of this volume), along with several other experimental results to be discussed here. Let us begin by considering a pigeon facing a pair of keys, A and B. At any point in time, responding on key A may have one of two effects: Reinforcement C occurs, or it does not. Let V_A represent the value of key A at some point in time, and V_C the value of reinforcement. If reinforcement is produced, assume the value of A changes by

$$\Delta V_A = k_1 (V_C - V_A), \tag{12.1}$$

where $0 < k_1 < 1$. That is, the value of A changes by an amount proportional to the difference between its current value and the value of reinforcement.

Let V_A' represent the unconditioned value of A (V_A' would be taken to be zero, and V_C' greater than zero). During responding on key A but in the absence of transitions to reinforcement, the value of A is assumed to change continuously toward its unconditioned value, at a rate proportional to the difference between its current value and its unconditioned value:

$$dV_A/dt = k_2 (V_A' - V_A). \tag{12.2}$$

Again, $0 < k_2 < 1$.

Figure 12-1 shows qualitatively how V_A varies over time as a function of V_A' and the presentation of C (upward arrows), the value of which is V_C. Equilibrium is reached when the increase in V_A upon presentation of reinforcement

Figure 12-1. Value of key A (V_A) as a function of time at delivery of reinforcement (upward arrows). V_A is incremented by an amount proportional to $V_C - V_A$. In the absence of reinforcement, V_A changes continuously at a rate proportional to $V_A' - V_A$.

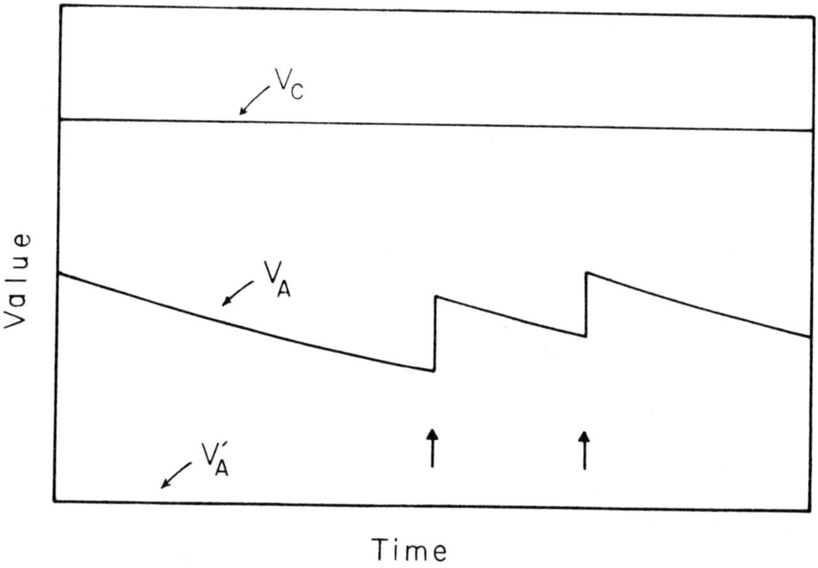

equals the descrease in V_A between reinforcements, on average. Together, Eqs. (12.1) and (12.2) allow one to say that at equilibrium the value of A, $V_{A\text{eq}}$, is a strictly monotonically increasing function of rate of reinforcement in its presence. That is, if R_A represents reinforcements for responding on key A and T_A time spent responding on A, and B represents a second key, then

$$\frac{R_A}{T_A} > \frac{R_B}{T_B} \rightarrow V_{A\text{eq}} > B_{B\text{eq}}, \qquad (12.3)$$

and, by symmetry, conversely. This relation is shown in Figure 12-2.

Suppose that keys A and B each have a variable-interval (VI) schedule associated with them. Given a session length of, say, 1 hour, assume for the moment that R_A, the number of reinforcements received on key A, is a constant, given that the time spent on key A, T_A, is greater than zero; similarly for key B. Under these conditions, R_A/T_A and R_B/T_B, the local rates of reinforcement, will be a function of T_A/T_B, or the ratio of times spent on side A to side B. Specifically,

$$\frac{T_A}{T_B} < \frac{R_A}{R_B} \rightarrow \frac{R_A}{T_A} > \frac{R_B}{T_B} \qquad (12.4)$$

Figure 12–2. The equilibrium values of keys A and B ($V_{A\text{eq}}$ and $V_{B\text{eq}}$) are a monotonically strictly increasing function of the local reinforcement rates on those keys, R_A/T_A and R_B/T_B. V_C and V_D refer to the values of reinforcement on keys A and B, respectively.

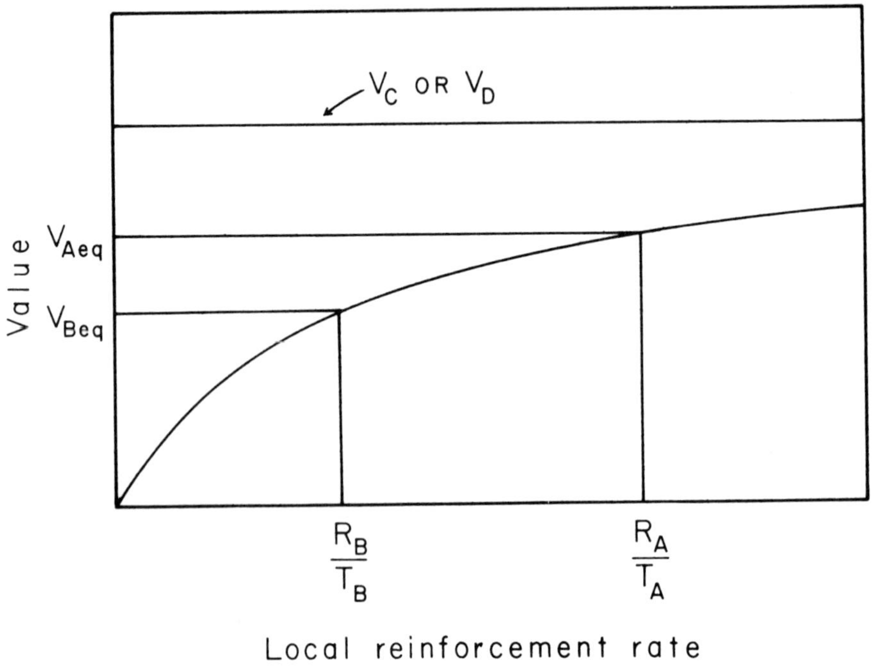

and conversely. By transitivity, then,

$$\frac{T_A}{T_B} < \frac{R_A}{R_B} \rightarrow V_{A\text{eq}} > V_{B\text{eq}} \,. \tag{12.5}$$

In other words, the assumptions made so far allow one to say that if the ratio of times on keys A and B is less than the ratio of reinforcements received on those keys, at equilibrium the value of A will be greater than the value of B.

In order to close the loop T_A/T_B must be shown to be some function of $V_{A\text{eq}}$ and $V_{B\text{eq}}$. This is accomplished as follows: Let V_{AB} represent the value of switching from A to B, and V_{BA} the value of switching from B to A. Along the lines of Eq. (12.1), assume

$$\Delta V_{AB} = k_3 (V_{B\text{eq}} - V_{A\text{eq}}) \,, \tag{12.6}$$

and conversely for ΔV_{BA}. That is to say, if $V_{A\text{eq}} > V_{B\text{eq}}$, then changeovers from B to A will increase the value of making the changeover in that direction, whereas changeovers from A to B will decrease the value of changing in that direction. Thus, eventually,

$$V_{A\text{eq}} > V_{B\text{eq}} \rightarrow \Delta V_{BA} > 0 > \Delta V_{AB} \quad . \tag{12.7}$$

That is, if at equilibrium key A is greater in value than key B, the value of changing to A will increase, whereas the value of changing to B will decrease. Combining implications (12.5) and (12.6), yields

$$\frac{T_A}{T_B} < \frac{R_A}{R_B} \rightarrow \Delta V_{BA} > 0 > \Delta V_{AB} \quad . \tag{12.8}$$

If the value of changing from A to B equals that of changing from B to A, assume that an equal amount of relative time is spent on each side. Further, given a constant value of changing from A to B, assume an increase in the value of changing from B to A will result in more relative time in general being spent on side A than before, with the increase in relative time occurring as a continuous function of changes in V_{BA}. Applying this assumption under all symmetrical cases means assuming a function f,

$$\frac{T_A}{T_A + T_B} = f(V_{BA}, V_{AB}) \tag{12.9}$$

such that $V_{BA} = V_{AB} \rightarrow T_A/(T_A + T_B) = 0.5$; $T_A/(T_A + T_B)$ increases continuously as V_{BA} increases or V_{AB} decreases, and it decreases as V_{BA} decreases or V_{AB} increases. That is, as the value of changing from B to A grows larger, and/or the value of changing from A to B grows smaller, relatively more time will be spent on key A.

$$\Delta V_{BA} > 0 > \Delta V_{AB} \rightarrow \Delta \frac{T_A}{T_A + T_B} > 0 \rightarrow \Delta \frac{T_A}{T_B} > 0 \quad . \tag{12.10}$$

Combining implications (12.8) and (12.10) yields

$$\frac{T_A}{T_B} < \frac{R_A}{R_B} \rightarrow \Delta \frac{T_A}{T_B} > 0 \quad . \tag{12.11}$$

That is, if the ratio of times on keys A and B is less than the ratio of reinforcements, relatively more time will come to be spent on key A. A point of stable equilibrium is reached at

$$\frac{T_A}{T_B} = \frac{R_A}{R_B} \quad . \tag{12.12}$$

There may or may not be oscillations about this point, depending on the value of k_3 and the shape of f, but let us assume that on average the system satisfies Eq. (12.12). We have thus shown that matching of the ratio of times to the ratio of obtained reinforcements (Eq. (12.12)) may be deduced from a set of assumptions specifying changes in various hypothesized dimensions of value.

The model as presented so far pictures a pigeon as evaluating *local* rates of reinforcement on each of a number of simultaneously available keys. Changing from a key with lower local rate of reinforcement to one with a higher rate increases the value of making that transition, leading to a redistribution of relative time among the keys. At equilibrium the local rates of reinforcement on all keys being sampled must be equal. Overall rate of reinforcement does not enter the computation.

Figure 12-3 shows the general structure of the model. Local rates of reinforcement on the various keys give rise to keys having associated with them certain values. Changeovers involving a change in value are either strengthened or weakened, leading to a change in the distribution of behavior. Local rates of reinforcement are in turn a function of the distribution of behavior.

In order to contrast the present model with a maximization analysis, four experiments will be discussed. Two of those experiments simultaneously favor the present model and go against a maximization analysis, the third simply goes against a maximization analysis, and the fourth simply favors the present model. Consider first single-key VI schedules of reinforcement, which, following Baum (1973), may be viewed as functions relating reinforcement rate to response rate. Figure 12-4 (top) shows a VI 2 minutes as a function of this form. The nonlinearity derives from the convention of stopping the tape reader whenever reinforcement sets up, until that reinforcement has been delivered.

It is possible to remove the nonlinearity by running the VI tape continuously, except during reinforcement, and storing a reinforcement (so to speak) whenever one sets up. Under these conditions, which were employed by Anger (1956), if responses occur at least as often as reinforcement sets up, obtained rate of reinforcement will come to equal the programmed rate, in this case the constant 30 reinforcements per hour. Figure 12-4 (bottom) shows the function obtained under these conditions.

Figure 12-3. Flow diagram indicating the basic variables assumed by the present model.

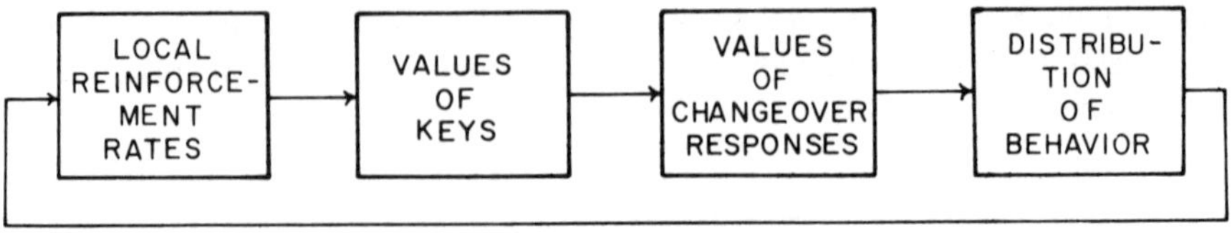

Figure 12–4. Nonlinear variable-interval 2-minute schedule (*top*). Linear variable-interval 2-minute schedule (*bottom*).

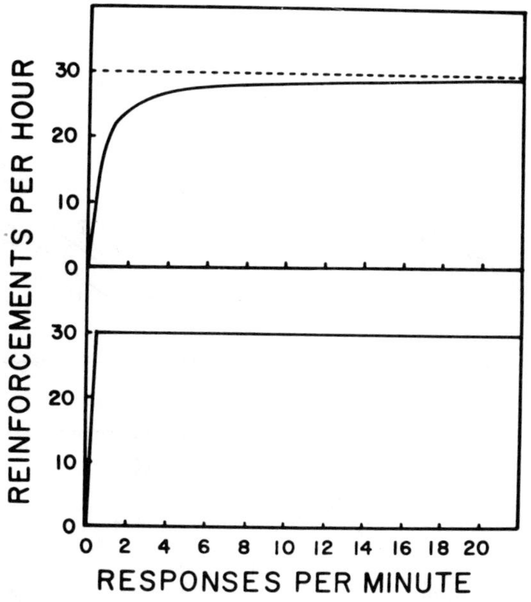

It has been suggested that on concurrent VI schedules, matching results because pigeons behave in a way that maximizes rate of reinforcement. Such an analysis relies upon the nonlinearity of VI schedules as usually programmed. If a pigeon were presented with concurrently available *linear* VI schedules, all distributions of behavior except extremes would give rise to the same overall rate of reinforcement. Under such conditions, if the pigeon were governed by a maximization algorithm, its behavior should become random within the extremes. This implies that the function showing logged response or time ratios against the logged reinforcement ratio should have a flat slope over a broad range.

On the other hand, the present analysis implies that under such conditions the distribution of time should match the distribution of reinforcements. In collaboration with H.L. Miller and D.H. Loveland, an experiment has been run in which pigeons were presented with pairs of concurrently available linear VI schedules. The overall rate of reinforcement was programmed to be 90 per hour, while the ratios of reinforcements on the two keys were 1:2, 2:1, 5:1, 1:1, and 1:5. In addition to a 3-second COD, immediately following reinforcement at least 3 seconds had to elapse before reinforcement could be obtained. Sessions ended after 40 reinforcements.

Figure 12-5. Log ratio of behavior on left and right keys as a function of log ratio of obtained reinforcements. Open circles represent time *T*; closed circles, pecks *P*. Least-square-linear functions are also shown.

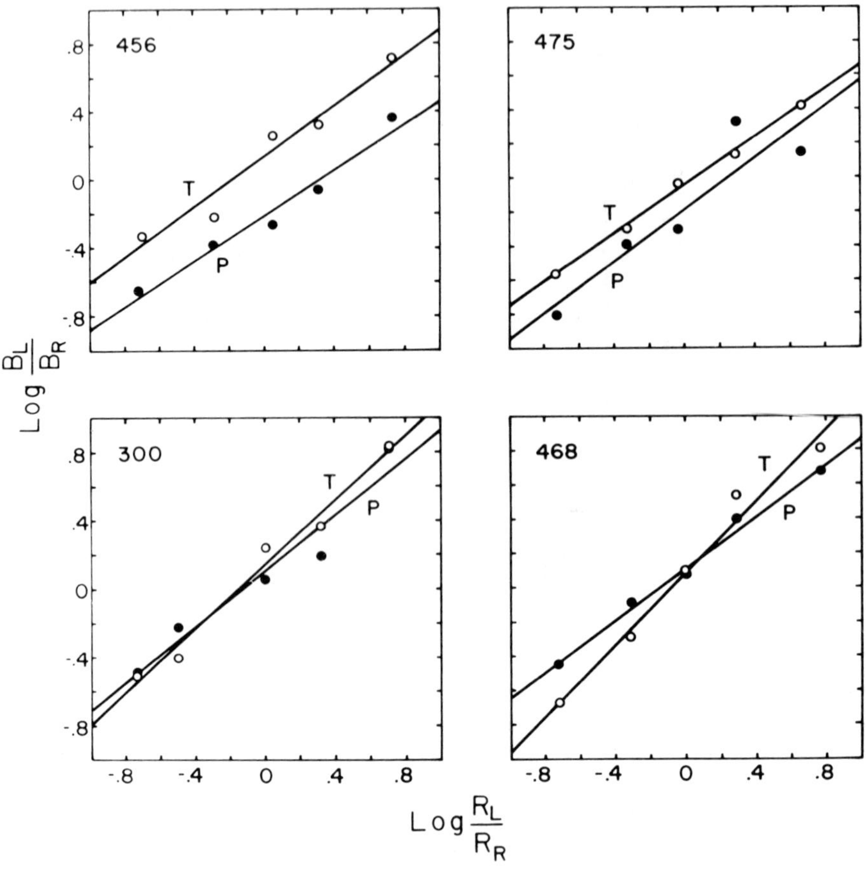

Figure 12-5 shows for each of the four birds the logged time ratios (open circles) and logged response ratios (closed circles) as a function of logged reinforcement ratios. For time, the slopes of the least-squares linear functions are 0.74, 0.71, 0.82, and 1.05; for responses they are 0.63, 0.77, 0.82, and 0.76. Although there is undermatching, this outcome favors the present model over a maximization analysis. By that analysis, for example, about half the slopes should be *negative*, leading to an average slope of zero.

In Figure 12-6 are shown a (linear) interval schedule *I* and a ratio schedule *R*. The latter is a linear function through the origin, since it specifies a propor-

Figure 12-6. By adding a linear interval *I* and a ratio *R* schedule, a function corresponding to their sum $(I + R)$ is produced.

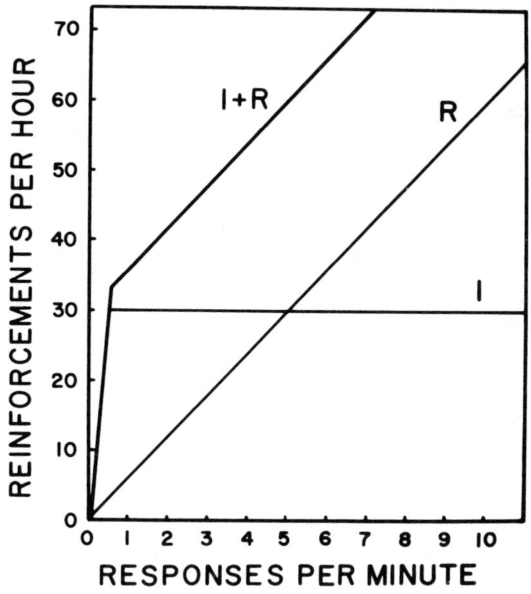

tionality between response rate and reinforcement rate. Given I and R, it is possible to construct their sum $I + R$, as follows. Both an interval and a ratio reader run in parallel, the former driven by time and the latter by responses. Whenever either sets up reinforcement, it is stored. Whenever a response occurs with one or more reinforcements so stored, reinforcement is delivered and the store is decremented by 1. In this way any function with a slope that is nonnegative and no steeper than the continuous reinforcement function (CRF) may be generated. This family provides a way of going continuously between (linear) VI and VR schedules.

It is also possible to subtract one schedule from another. Figure 12-7 (top) shows an interval and ratio schedule, and their difference, $I - R$. This may be programmed as follows. As before, an interval and ratio reader run in parallel and reinforcements from the interval reader are stored. If the ratio reader sets up, however, it subtracts a reinforcement from the storage device. If the device can store both surpluses and deficits, the function $I - R$ is generated. Figure 12-7 (bottom) shows $R - I$, which is generated by having a ratio reader add to the store and an interval reader subtract from it. Altogether, these procedures allow for the programming of *any* linear functions not steeper than continuous

Figure 12-7. Subtracting a ratio from a linear interval schedule produces a schedule with negative slope (*I* - *R*) (*top*). Subtracting a linear interval from a ratio schedule produces a schedule with what would be a negative intercept (*bottom*).

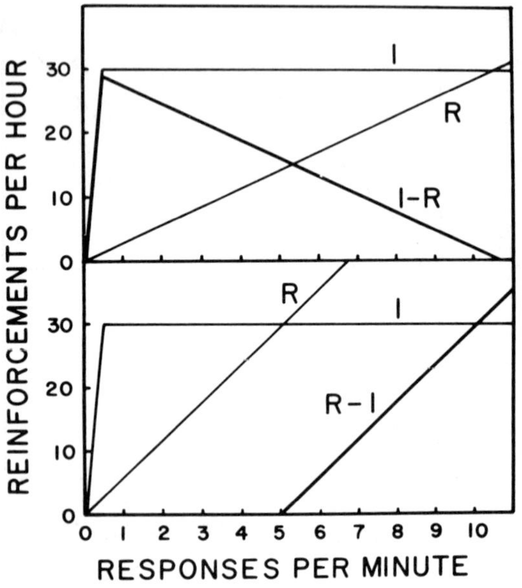

reinforcement (CRF): The slope and the intercept may be either less than, equal to, or greater than zero. Not only is it possible to go continuously between (linear) VI and VR schedules, but it is possible to go outside their range by these means.

It is the function $I - R$ that is of concern at this point. For each of these functions a particular response rate maximizes rate of reinforcement. Further, as that response rate is approached, from above or below, reinforcement rate grows monotonically. Such schedules thus provide an unambiguous test regarding the sufficiency of a maximization analysis of the sort discussed earlier.

In an experiment described in (Vaughan, 1976), birds were first tested on three schedules with negative slopes and then on three schedules with flat slopes. The latter schedules were such that each bird obtained approximately the same rate of reinforcement on each schedule that it had obtained on one of the negative-sloped schedules. Since what effect the negative-sloped schedules would have was unclear, nine birds were run so as to sample a moderately large range of conditions. The VI schedules that entered the negative slopes were VI 30, VI 45, and VI 90 seconds for all birds, in that order. One group of three birds had FR 20 subtracted from each of the VI schedules in turn; a second group had

Figure 12-8. Response rate and reinforcement rate (closed circles) on schedules with negative slopes (solid lines) as well as on schedules with zero slopes (open circles and dashed lines).

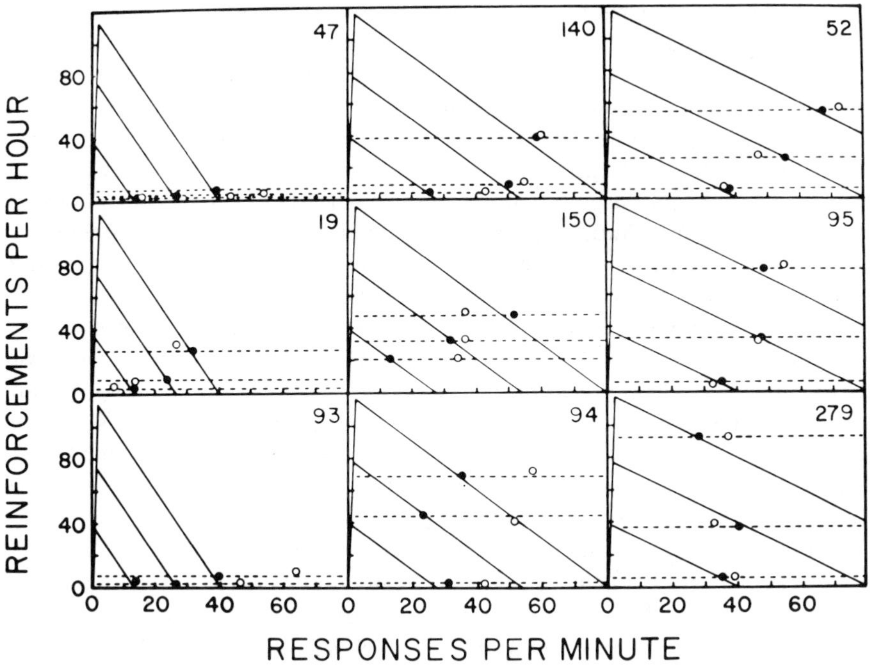

FR 40, and a third FR 60. It was assumed that FR, rather than VR, could be subtracted without the birds detecting periodicities in reinforcement rate.

In the first condition it became apparent that the birds were responding faster than optimal in terms of reinforcement rate. For that reason they were continued on that condition for a minimum of 60 sessions to ensure sufficient time for detection of the contingency in case that was all that was required.

Five-day averages are shown for each bird in Figure 12-8. Closed circles show behavior on negative slopes (solid lines), while open circles show behavior on flat slopes (dashed lines). Bird 93 was run on only two flat slopes, since obtained reinforcement rates on two of the negative slopes were very similar. Lumping all conditions together, response rate increased significantly going from negative slopes to flat slopes (Wilcoxon matched-pairs signed-ranks test (Siegel, 1956); $n = 26$, 1 tail, $p < 0.025$; 2 tail, $p < 0.05$). This implies that, upon going from negative to flat slopes, the birds detected the difference in correlation between response rate and reinforcement rate. Although sensitive to the differing correlations, they failed to behave in an optimal manner. Failure to optimize in this case is presumably not a discrimination problem, then, but derives from more basic principles governing response strength.

The results of the first experiment suggest that on concurrent schedules, some process other than or in addition to maximization of overall reinforcement rate must be operative, whereas the second experiment suggests that on single-key schedules response rate is not governed by maximization. This leaves open at least the possibility that on concurrent schedules maximization of reinforcement rate may operate in conjunction with some other process that in and of itself will give rise to matching.

Testing such a possibility requires a concurrent schedule such that all distributions of time produce matching while overall rate of reinforcement is some function of the distribution of time. If matching always occurs, the effect such a process could contribute to the distribution of behavior would then be negated, leaving unconstrained any maximization tendency.

The third experiment employed what has been termed an interdependent schedule (Herrnstein and Vaughan, 1980). For any distribution of behavior between the two keys, R_A/T_A equaled R_B/T_B but R_A/T_A and R_B/T_A both varied as a function of the distribution of time between the two keys. R_A/T_A and R_B/T_B were maximized at 120 reinforcements per hour, given that relative time on the right key equaled 0.25; deviations from that value reduced the reinforcement rates continuously. As exclusive preference was approached, the rates of reinforcement approached 60 per hour.

The schedule on which the birds were run is shown in each of the panels of Figure 12-9. The five data points show 5-day averages for each of the three birds (that is, point 1 is from sessions 1-5, and so on). Point 5 is averaged of the last 6 sessions of the 26 sessions. The last panel shows the average across birds of that last point.

The basic conclusion that may be drawn from this experiment is that the distribution of behavior between two keys is not influenced by overall rate of reinforcement, given that local reinforcement rates on the two keys are equal. This outcome is inconsistent with a maximization analysis, though it is consistent with the analysis being suggested here, since that analysis assumes that it is differences in the local reinforcement rates that govern the distribution of behavior.

Given the negataive implications with regard to maximization discussed so far, one may wonder whether the present analysis makes contact with those experiments in the literature generally thought to support a maximization analysis. Perhaps one of the clearest of such experiments is that of Herrnstein and Loveland (1975), who ran birds on concurrent VR schedules. When the two schedules differed in VR requirement, they found that the birds tended to show nearly exclusive preference for the better of the two schedules.

Herrnstein and Loveland discussed two possible interpretations of these data. The birds might have been minimizing deviations from matching (albeit advanta-

Figure 12-9. Relative time on right and rate of reinforcement for three birds on an interdependent schedule. Each data point is the average of five sessions (that labeled 1 is from sessions 1-5) except that labeled 5, which is from sessions 21-26. The lower right panel shows the average of the last point for each bird.

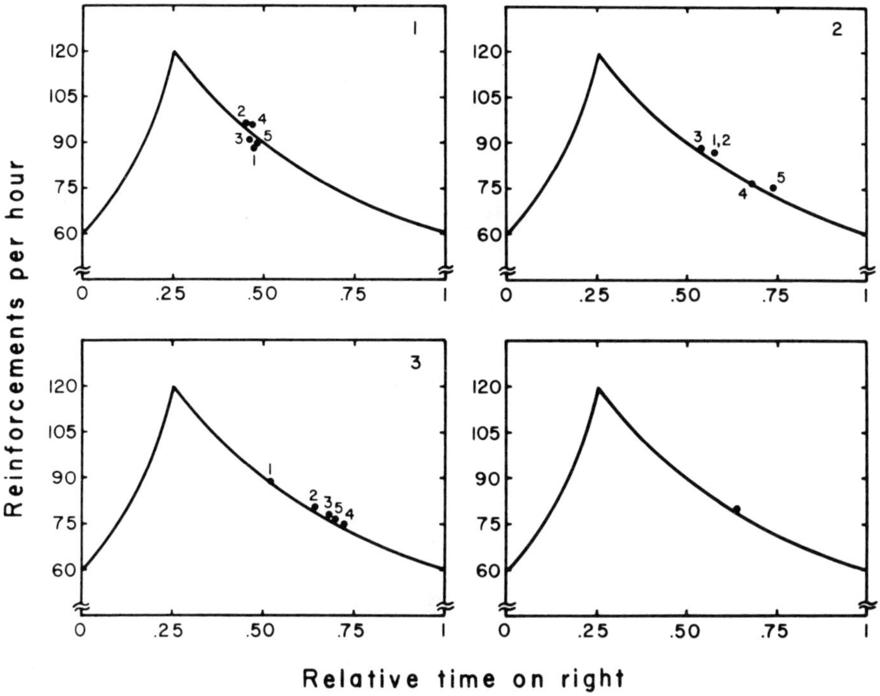

geously), or they might have been minimizing the time required to obtain a given number of reinforcements. The latter interpretation, which is equivalent to a maximization of reinforcement rate analysis, has found the greatest number of adherents.

According to the present analysis, however, a third process was actually at work. Given equal local rates of responding on the two keys, the key with the smaller requirement would deliver a higher local rate of reinforcement than the other. This difference in local reinforcement rates should, by the present model, drive the distribution of behavior toward the better key. The fact that overall rate of reinforcement is thereby maximized is incidental, from the present point of view.

The last experiment to be discussed, run in collaboration with S.L. Greene, explored a modification of the present model that again derives from the Rescorla-Wagner model. Suppose two stimuli, $A1$ and $A2$, are present on a key on which a pigeon is responding. Given reinforcement, assume:

$$\Delta V_{A1} = k_1 (V_C - \bar{V}) \; ,$$
$$\Delta V_{A2} = k_1 (V_C - \bar{V}) \; , \qquad (12.13)$$

where $\bar{V} = V_{A1} + V_{A2}$. In the absence of reinforcement, assume:

$$dV_{A1}/dt = k_2 (\bar{V}' - \bar{V}) \; ,$$
$$dV_{A2}/dt = k_2 (\bar{V}' - \bar{V}) \; , \qquad (12.14)$$

where $\bar{V}' = V'_{A1} + V'_{A2}$.

The application of these equations will be illustrated in terms of the present experiment. Four birds were run in a chamber equipped to project slides onto a response key. In the first condition each slide consisted of one geometrical shape, randomly placed. The birds were divided into two groups and run on the following multiple schedule:

		Birds 223, 226	Birds 157, 134
Red parallelogram	$(A1)$	VI 30 seconds	EXT
Blue hexagon	$(A2)$	VI 2 minutes	VI 2 minutes
Purple square	$(B1)$	EXT	VI 30 seconds
Brown circle	$(B2)$	VI 2 minutes	VI 2 minutes

Let $A1$ represent the red stimulus, $A2$ blue, $B1$ purple, and $B2$ brown. For the first group of birds, the following should hold: $V_{A1} > V_{A2}$; $V_{A2} = V_{B2}$; $V_{B2} > V_{B1}$; $V_{B1} = 0$. For the second group: $V_{B1} > V_{A2}$; $V_{A2} = V_{B2}$; $V_{B2} > V_{A1}$; $V_{A1} = 0$. For each group, then, following training on the first multiple schedule, we should have $V_{A2} = V_{B2}$. In order to verify this the birds were run in a single concurrent condition. Only $A2$ and $B2$ were presented. Responses to a second key advanced the slide tray, so that the procedure is similar to that employed by Findley (1958). Relative time in the presence of the two classes of slides was measured.

Next, some slides were made having two stimuli on them: red and blue, and purple and brown. For the slides with red and blue, it was assumed $\bar{V} = V_{A1} + V_{A2}$, while for those with purple and brown, $\bar{V} = V_{B1} + V_{B2}$. The following multiple schedule was run next:

		Birds 223, 226	Birds 157, 134
Red	$(A1)$	VI 30 seconds	EXT
Red + blue	$(A1 + A2)$	VI 2 minutes	VI 2 minutes
Purple	$(B1)$	EXT	VI 30 seconds
Purple + brown	$(B1 + B2)$	VI 2 minutes	VI 2 minutes

The first group, then, should have $V_{A1} > (V_{A1} + V_{A2})$; $(V_{A1} + V_{A2}) = (V_{B1} + V_{B2})$; $(V_{B1} + V_{B2}) > V_{B1}$; $V_{B1} = 0$. For the second group, $V_{B1} > (V_{B1} + V_{B2})$; $(V_{B1} + V_{B2}) = (V_{A1} + V_{A2})$; $(V_{A1} + V_{A2}) > V_{A1}$; $V_{A1} = 0$. These inequalities imply that the first group should have $V_{A2} < 0 < V_{B2}$, in spite of the fact that $A2$ and $B2$ have always received identical rates of reinforcement in their presence. This is because the context of $A2$ has been high in value, while the context of $B2$ has been low in value. The second group should have $V_{B2} < 0 < V_{A2}$, for analogous reasons.

In order to test this implication the birds were finally run on the same concurrent schedule as before. In both cases the local rates of reinforcement were the same in the presence of each of the two stimuli. Relative time in the presence of blue, $A2$, is shown in Figure 12-10 for both concurrent tests. The first group spent less time in the presence of blue during the second test than it had during the first test, whereas the second group spent more time. Qualitatively, then, these results indicate that the value a stimulus gains when reinforcement is presented depends on the context within which it is presented.

These four experiments, then, suggest that a process similar to that proposed by Rescorla and Wagner may be operating with regard to the values that pigeons associate with the response keys available to them. A major point that must be addressed, however, concerns the evolution of such a process. The general argument that has been made is that evolution should have produced organisms that are maximally fit within their environments. By this logic organisms should apparently be expected to behave so as to maximize rate of food, or energy, intake, at least within a wide range of only moderately difficult contingencies.

It is recognized that maximization of energy intake refers to net energy intake, that is, gross intake minus costs involved in securing that intake. Two other forms of cost should be considered, however. One is the biological cost of creating and maintaining the neural mechanisms of a complex behavioral strategy, as opposed to a simple one. The second form of cost relates to the speed with which a behavioral process will track changing environmental contingencies.

Consider an organism making use of an optimization process, spending some of its time in the presence of each of two alternatives, one paying off better than the other. The question is, Should behavior be redistributed and, if so, in which direction? The answer depends on how overall payoff varies as a function of changes in the distribution of behavior. Without making a priori assumptions, this can only be determined by in fact varying behavior and seeing how overall payoff in fact changes. Similarly, even if the payoff increases for changes in a particular direction, nothing says it would not increase more for going in the opposite direction or that it would continue to increase for continuing to change in the same direction.

What is being suggested here is that in the absence of prior assumptions an optimization process will track changing contingencies slowly, especially if feed-

Figure 12-10. Relative time spent in blue for birds in the fourth experiment, following the first and second training conditions.

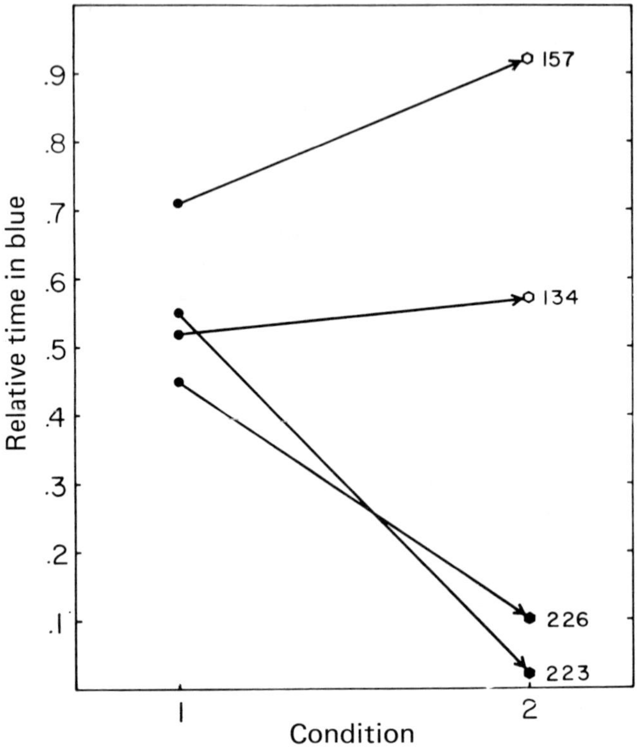

back is intermittent. In contrast, the process being suggested here need not wait for feedback with regard to overall payoff: Given differential local rates of payoff, behavior will shift in the direction of the better alternative. The speed of responding appears to be purchased at the cost of insensitivity to overall rate of payoff, as illustrated by the third experiment.

The question with regard to complexity of neural machinery must be more hypothetical. By either model, an organism must in some sense remember how much time it is distributing to each of a number of alternatives. Suppose there are N alternatives, and for simplicity assume that time spent in each may either increase, decrease, or remain constant. These three possibilities imply that, according to an optimization model, an organism must be able to correlate overall rate of return with each 3^N possibilities, picking that possibility showing the greatest increase in overall rate of return. According to the present model it is necessary for the organism to know the value of each of the N possibilities and distribute more time to the better and less time to the worse of those alternatives. Although certainly not conclusive, this argument suggests that an optimization process may require a more complex neural basis than the model being suggested here.

The general argument, then, is that an optimization process may have greater costs associated with it than the present model. Consider now the steady-state payoffs. On concurrent ratio schedules with any number of different alternatives, each with different requirements, both processes will settle down at exclusive preference for the best alternative. On concurrent interval schedules overall payoff appears to be relatively insensitive to the distribution of behavior. To the extent that the environment is structured in such a way that both behavioral processes pay off approximately equally at equilibrium, the lower costs of the model being suggested here would favor its propagation within evolution. Thus, given certain environmental constraints, a heuristic may provide for a better net payoff than a more strictly optimizing process.

REFERENCES

Anger, D. 1956. The dependence of interresponse times upon the relative reinforcement of different interresponse times. *Journal of Experimental Psychology 52*:145–61.

Baum, W.M. 1973. The correlation-based law of effect. *Journal of the Experimental Analysis of Behavior 20*:137–53.

Findley, J.D. 1958. Preference and switching under concurrent scheduling. *Journal of the Experimental Analysis of Behavior 1*:123.

Herrnstein, R.J., and D.H. Loveland. 1975. Maximizing and matching on concurrent ratio schedules. *Journal of the Experimental Analysis of Behavior 24*:107–16.

Herrnstein, R.J., and W. Vaughan, Jr. 1980. Melioration and behavioral allocation. In J.E.R. Staddon, ed., *Limits to Action: The Allocation of Individual Behavior.* New York: Academic Press.

Rachlin, H.; L. Green; J.H. Kagel; and R.C. Battalio. 1976. Economic demand theory and psychological studies of choice. In G. Bower, ed., *The Psychology of Learning and Motivation*, vol. 10, New York: Academic Press.

Rescorla, R.A., and A.R. Wagner. 1972. A theory of Pavlovian conditioning: Variations in the effectiveness of reinforcement and nonreinforcement. In A.W. Black and W.F. Prokasy, eds., *Classical Conditioning*, vol. 2, *Current Research and Theory*, New York: Appleton-Century-Crofts.

Siegel, S. 1956. *Nonparametric Statistics for the Behavioral Sciences.* New York: McGraw-Hill.

Vaughan, W., Jr. 1976. Optimization and reinforcement. Unpublished doctoral thesis, Harvard University.

13 DO INTERACTIONS IN MULTIPLE AND CONCURRENT SCHEDULES HAVE A COMMON BASIS?

Ben A. Williams

Much of the systematic investigation of schedule interactions has been inspired by Herrnstein's (1970) analysis, which focused on the "context of reinforcement." According to Herrnstein, the effects of simple schedules, multiple schedules, and concurrent schedules all are to be understood as a function of a single independent variable—relative rate of reinforcement—and can be described by a single equation:

$$P_1 = k \frac{R_1}{R_1 + mR_2 + R_0}, \qquad (13.1)$$

where P_1 represents the rate of behavior, R_1 the rate of reinforcement for that behavior, R_2 the rate of reinforcement for behavior under some other schedule, R_0 the rate of reinforcement for behavior not under control of the experimenter, and m the degree of interaction between the schedule components (going from 0 to 1.0. The denominator of the equation thus represents the total context of reward in which a reinforcement event occurs, and the response-strengthening effects of that event are inversely related to the value of the context.

Despite the conceptual simplicity offered by Herrnstein's analysis and the considerable evidence in its favor (cf. de Villiers and Herrnstein, 1976), it has been severely challenged by subsequent theoretical developments. In particular, several investigators have questioned whether effects of concurrent and multiple schedules have a common basis, claiming instead that interactions found with multiple schedules are less general and depend upon the development of elicited behavior controlled by Pavlovian (respondent), not operant, processes (Gamzu and Schwartz, 1973; Rachlin, 1973; Hearst and Jenkins, 1974; Schwartz and Gamzu, 1977). Moreover, relative rate of reinforcement is widely disputed as the controlling variable in concurrent schedules as well. As witnessed by several chapters in this volume, "matching" in concurrent schedules is not believed to result from relative rate of reinforcement, per se, but instead can be deduced from other assumptions. Notably, Herrnstein (Chapter 19) apparently has adopted this position.

The impetus for the research described in this chapter was that the rejection of relative rate of reinforcement as a primitive variable has been premature. With respect to concurrent schedules, impressive evidence still remains (Catania, 1962, 1963; Rachlin and Baum, 1972; Nevin, 1974) that relative rate of reinforcement has its effects independent of effects on response rate or response competition (that is, the effects of R_2 in Eq. (13.1) are independent of the behavior controlled by R_2). Moreover, attempts to provide alternative formulations of concurrent-schedule effects (Rachlin, Chapter 16; Herrnstein, Chapter 19; Staddon and Motheral, 1978) continue to provide no satisfactory treatment of such evidence. Also, the occurrence of matching in short-component multiple schedules (Shimp and Wheatley, 1971; Todorov, 1972; Killeen, 1972) provides prima facie evidence that multiple and concurrent schedule effects are fundamentally similar. Yet none of the alternative formulations of concurrent-schedule effects have offered any suggestion as to how matching in multiple schedules is to be explained. At the very least, these empirical facts demand a more thorough assessment of how multiple and concurrent schedules are related.

The research described in this chapter deals primarily with interactions between response rates in the components of multiple schedules. At issue is the identification of the independent variable. First, the evidence pertaining to the major alternatives to relative rate of reinforcement will be examined and then relative rate of reinforcement itself looked at more closely. Presaging this discussion, it will be argued that the alternative accounts of multiple-schedule interactions are inadequate and the empirical literature continues to provide strong evidence for relative rate of reinforcement as the controlling variable. It will also be argued, however, that the effects of relative rate of reinforcement cannot be understood adequately at a molar level (e.g., Herrnstein, 1970), because such effects have temporal properties not encompassed by a molar description.

TWO ALTERNATIVES TO THE RELATIVE LAW OF EFFECT

Response Competition Theory

Although response competition (and response substitution) effects have received wide attention in the treatment of both simple and concurrent schedules (Rachlin and Burkhard, 1978; Staddon, 1979; Allison, Miller, and Wozny, 1979), only recently has the notion been applied to multiple schedules. As developed by

Staddon (Chapter 11), the basic notion is that contrast in multiple schedules occurs because of competition between interim and terminal behaviors. Staddon argues that any interval schedule of reinforcement produces both types of behavior (cf. Staddon and Simmelhag, 1971). Thus, during both components of a multiple VI VI schedule, both interim and terminal behaviors occur, so that the time available for either is reduced. When the schedule is changed to multiple VI EXT, the terminal behaviors are eliminated in the EXT component. The interim behavior that did occur in the unchanged VI can then move into the EXT component, leaving more time for terminal responding in the unchanged VI, so that an increase in response rate (that is, contrast) occurs. The results of an experiment with rats, where the availability of interim behavior was manipulated explicitly (Hinson and Staddon, 1978) seems to provide strong evidence for this hypothesis.

Staddon's analysis is attractive because it offers the opportunity for a description of effects of multiple and concurrent schedules in a single theoretical framework. Moreover, it appears to explain the apparent species differences reported for multiple schedules (contrast being found with pigeons more easily than with rats), on the assumption that rats have fewer interim behaviors available in the usual operant training situation (cf. Hinson and Staddon, 1978). However, his account also faces the fundamental problem that the critical variable is apparently not the behavior maintained in the variable component, but the relative rate of reinforcement, per se (cf. Schwartz and Gamzu, 1977).

The first experiment tests Staddon's response-competition notion by investigating the effects of signaled reinforcement. Such effects are relevant because he assumes the critical variable to be the amount of time available for interim activity. Thus, if interim activity that occurred previously in the unchanged VI component can move to the EXT component when a multiple VI VI is changed to multiple VI EXT, then a similar shift should occur when the schedule is changed to multiple VI signaled VI. This is true because the signaled VI schedule is essentially similar to EXT with respect to the total amount of time that is discriminated EXT, so that little competition between interim and terminal behaviors should occur during the signaled component. Consequently the interactions found with signaled reinforcement should be more similar to those found with EXT than to those found with equivalent rates of unsignaled reinforcement.

Figure 13-1 depicts the mean results of four subjects that all received the following conditions. A two-component multiple schedule was used in which one component was always a green key light associated with a VI 3-minute schedule. The schedule in the second component was then varied among EXT, VI 1-minute, or signaled VI 1-minute. A vertical white line was presented during EXT; an oblique red line was presented during the VI 1-minute. With the signaled VI 1-minute, the stimulus was the vertical white line until the reinforcer was available and then changed to the oblique red line until the next peck produced the reinforcer. For each schedule condition all subjects were trained with component durations of 15, 60, and 300 seconds. The order of training for all

Figure 13-1. Responses per minute in the constant VI 3-minute component of a multiple schedule, averaged over four subjects, as a function of component duration. Different functions represent the effects of different schedules in the variable component.

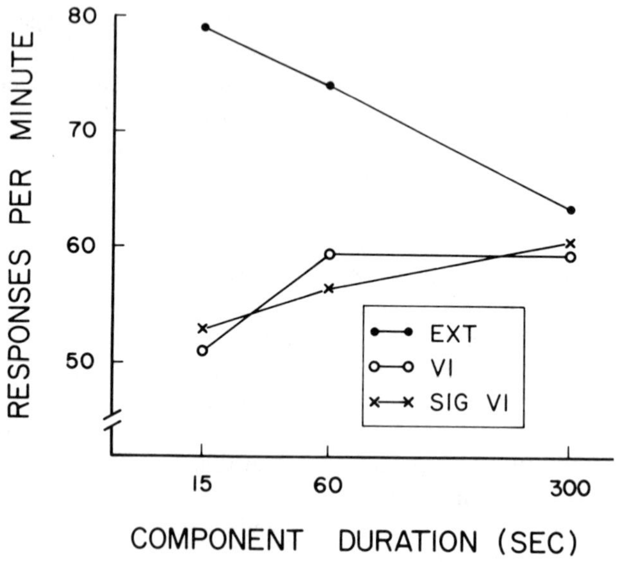

subjects is shown in Table 13-1. Throughout training, the two components of the schedule strictly alternated.

The results in Figure 13-1 depict the behavior in the unchanged VI 3-minute component during the last four sessions of each condition. In the case of 60-second component durations, where each condition was run twice, the results of both exposures to each condition were averaged together. For all component durations, responding in the variable component was low (0-3 responses/minute) for the EXT and signaled VI conditions, and at a high rate during the VI 1-minute condition. But the response rates in the variable component did not predict the behavior during the constant component. With EXT in the variable component, response rate was highest with the shortest component duration and then decreased as component duration was lengthened. With VI 1-minute in the variable component, response rate was lowest with the shortest component duration and then increased. Both of these findings replicate previous research (Shimp and Wheatley, 1971; Todorov, 1972). Of critical interest were the results with the signaled VI in the variable component. Despite little responding in the signaled VI component itself, responding in the constant component was nearly identical to that with the VI 1-minute condition. The results thus provide strong

Table 13-1. Sequence of conditions for subjects trained in multiple VI 3 VI 1, multiple VI 3 EXT, and multiple VI 3 signaled VI 1. The data shown in Figure 13-1 were taken from the last four sessions of each condition.

Order	*Variable Schedule*	*Component Duration (seconds)*
1	EXT	60
2	VI 1	60
3	sigVI 1	60
4	sigVI 1	300
5	VI 1	300
6	EXT	300
7	EXT	15
8	VI 1	15
9	sigVI 1	15
10	sigVI 1	60
11	VI 1	60
12	EXT	60

support for the view that relative rate of reinforcement in the two components, not the degree of responding, is the controlling variable.

Staddon has suggested that these results do not necessarily disprove his notion that response competition is the mechanism behind multiple schedule interactions, if it is assumed that the signaled VI schedule has the same effects as a variable-time (VT) schedule. This is not implausible because the only difference between a VT schedule and a signaled VI schedule is that primary reinforcers are response-independent for the former and the onset of the signal (presumably a conditioned reinforcer) is response-independent for the latter. It is known that VT schedules produce pecking, as do VI schedules (but not necessarily at the response key: Staddon and Simmelhag, 1971). The implication is that observation of the subejcts during the signaled VI component should show behavior more similar to that found with the VI 1-minute component than would be expected on the basis of the response rates themselves. Hence subjects were observed during all conditions. One of the four subjects did exhibit considerable wall pecking during the signaled VI condition, but the remaining three exhibited no pecking of any type except when the signal was present. For these three the behavior during the signaled VI was essentially similar to that during EXT, although the animals were considerably more active. Two of the subjects exhibited a great deal of turning behavior, some amount of preening, and some floor searching during both conditions. The fourth subject remained essentially stationary directly in front of the response key for the entire time. Despite this diversity in behavior patterns all subjects showed similar patterns of contrast. Apparently, therefore, observation of the subjects provides no more evidence for Staddon's response-competition notion than do the response rate data themselves. Consequently the similarity of the VI and signaled-VI conditions seen in Figure 13-1 must be taken as strong evidence against response competition as the source of interactions in multiple schedules.

Additivity Theory

Unlike response-competition theory, for several years additivity theory has had numerous adherents as an explanation of multiple schedule interactions. Its basic notion is that contrast occurs because of the addition of classically conditioned elicited responses to the key light signal. The major evidence in its favor is that contrast usually does not occur in situations in which the signal is located off the response key (cf. Schwartz and Gamzu, 1977) and that topographical tagging of "elicited" versus "operant" pecks shows the usual contrast manipulations to affect only elicited behavior, not operant behavior (Keller, 1974; Spealman, 1976; Schwartz, 1978).

The evidence in favor of additivity theory appears formidable. However, it will be argued here that it is inadequate for both logical and empirical reasons. Not only can it not provide a complete account for contrast (for which there is increasing agreement: see Schwartz, 1978), but in addition there is strong reason to doubt whether it is relevant *in any way* to steady-state interactions susceptible to quantitative description.

The best arena for demonstrating the failure of additivity theory is the analysis of component duration. As is well documented, contrast effects are larger with shorter components. Additivity theory explains this finding by assuming that elicited responding is at its maximum at the onset of the positive discriminative stimulus and then declines (Rachlin, 1973; Spealman, 1976), producing a gradient of local response rate like that shown in Figure 13-2. The effect of

Figure 13-2. Hypothetical pattern of local response rate postulated by additivity theory for the VI component of a multiple VI EXT.

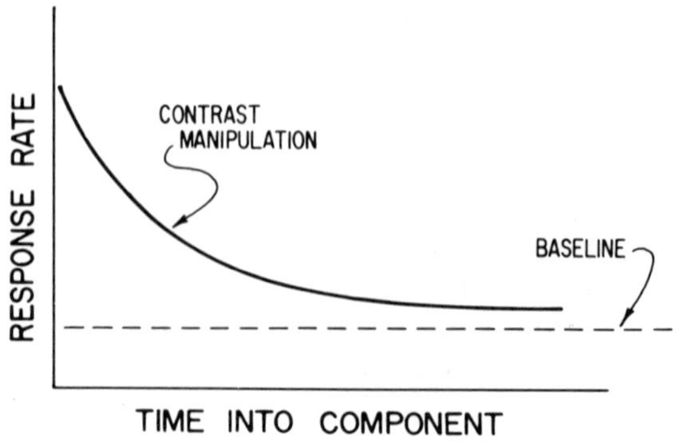

component duration on average rate within a component is then explained by the assumption that short components are functionally similar to the initial portions of the longer components. Thus, the average rate for a short component comes entirely from the initial portion of Figure 13-2, where elicited responding is high. Evidence in favor of this account appears substantial (Rachlin, 1973; Spealman, 1979; Schwartz, 1978).

Despite the evidence in its favor, the additivity account is challenged by the lack of generality of the local rate pattern seen in Figure 13-2. Although such a pattern has been obtained in some studies (cf. Schwartz and Gamzu, 1977), others have found such gradients to be only transitory, despite the persistance of contrast with respect to the average rate throughout the component (e.g., Nevin and Shettleworth, 1966). Other studies have shown overall contrast to occur when an *S*+ is alternated with a blackout, yet the use of blackout typically produces a flat local-rate gradient throughout training (Vieth and Rilling, 1972). Still other studies (Buck, Rothstein, and Williams, 1975; Hamilton and Silberberg, 1978; Williams, 1974) have shown a different type of local-rate gradient, with the highest local rate at the end of the VI component (prior to the transition *to* extinction). Because of the diversity of local-rate effects, the generality of the additivity account becomes questionable.

Figure 13-4 presents a very different perspective on the relation between local-response-rate gradients and the effect of component duration on overall contrast. The data were taken from the last 4 of 20 60-minute sessions of training on a multiple VI 1-minute EXT schedule. Training occurred first with the 30-second components and then with the 150-second components, so that an order effect could be confounded with component duration. This is unlikely, however, because all subjects had extensive experimental histories, including training in a previous contrast study (Williams, 1976), and over 200 sessions on various multiple schedules in the present study prior to the conditions depicted in Figure 13-3.

Figure 13-3 shows that all subjects had higher average response rates when component duration was 30 seconds, demonstrating the effect of component duration that has been obtained in previous studies. Yet the local-rate gradients were quite different from that predicted by additivity theory. Some subjects did show a gradient like that in Figure 13-2 when 150-second components were used (e.g., B-3), but no subject showed such a gradient with 30-second components. Instead, all subjects *increased* their response rate as the end of the 30-second components was approached. Consequently, the higher average rate during the shorter components was primarily a reflection of the high local rate at the end of the component, whereas with the longer components the high local rate at the end of the component did not generally occur (although it did to some extent for subjects R-11 and Y-22).

Also of interest from Figure 13-3 is the behavior during the initial portion of the 30-second components. For some subjects (e.g., R-11) an elevation in local

Figure 13-3. Local response rates for individual subjects trained on a multiple VI 1-minute EXT. Separate functions for each subject represent the local rates obtained with each component duration. Data are taken from the last four sessions of each condition.

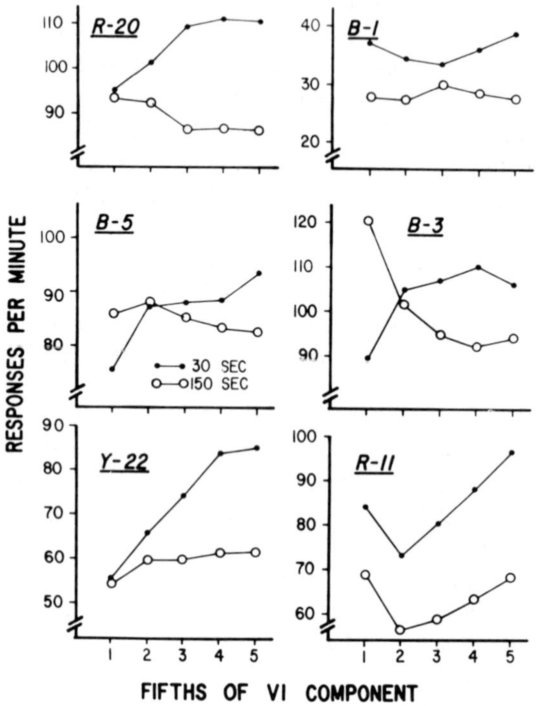

rate occurred at the beginning of the component, *in addition* to the increasing rate gradient at the end of the component. The joint occurrence of both effects argues against alternative interpretations of the increasing gradients that have been offered by other investigators (cf. Malone, 1976).

The local contrast effect at the beginning of the 30-second components was variable over subjects and often transitory over sessions. The relation between amount of training and changes in the local-rate gradient is seen in Figure 13-4, which shows the local rate averaged over subjects as a function of sessions of training. It should be noted that the exact from of the averaged functions is not representative of any individual subject; however the direction of the changes in the function across sessions was representative. In general Figure 13-4 shows that the small local contrast effect at the beginning of the component (which occurred for four of the six subjects early in training) decreased with continued training with the 30-second components, whereas the increasing local rate to-

Figure 13-4. Local rate of responding within the VI component of a multiple VI 1-minute EXT schedule, averaged over subjects. The different functions for each component duration correspond to different sessions of training. The baseline function comes from the last four sessions of training on a multiple VI 1-minute VI 1-minute schedule.

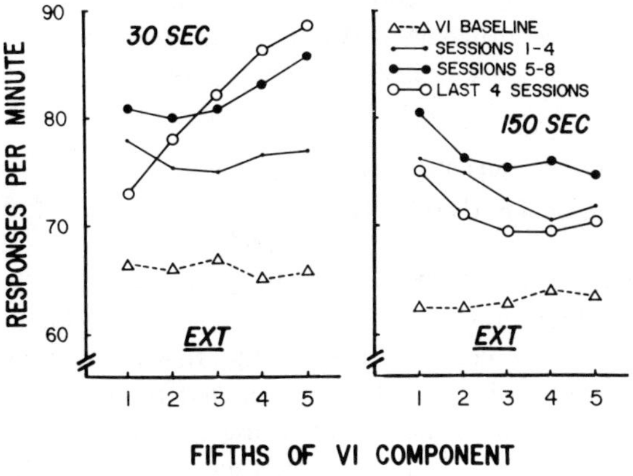

ward the end of the component increased with continued training. On the other hand, the local-rate gradient did not change systematically as a function of sessions with the 150-second components (right panel), as a decreasing local-rate gradient continued throughout. As can be seen by comparing Figures 13-3 and 13-4, however, the decreasing local-rate gradients with the 150-second components did not occur for all subjects.

The local-rate gradients seen in Figures 13-3 and 13-4 (and those of Buck et al., 1975) provide evidence against additivity theory as an account of contrast, at least as initially formulated (e.g., Rachlin, 1973). Recently, however, the significance of the increasing local-rate gradients has been challenged by Hamilton and Silberberg (1978) based on their finding that increasing local-rate functions also occur in situations where behavior is maintained entirely by response-independent reinforcement (multiple VT EXT). Because of such data they argue that the decreasing local-rate gradient seen in Figure 13-2 need not be a necessary feature of additivity theory. We have no quarrel with this position, if it is realized that the ability of additivity theory to explain the effect of component duration depends upon assumptions about the local-rate gradients as depicted in Figure 13-1. It is of course possible that additivity theory could be reformulated to explain the effects of component duration without consideration of local response rates, but at present no such reformulation has been forthcoming.

It should not be surprising that additivity theory has difficulty encompassing the effects of component duration, because it is incapable, in principle, of explaining the limiting form of the interaction—the matching of relative rate of responding to relative rate of reinforcement that occurs with short components (cf. Killeen, 1972). The problem is that additivity theory views the rate of pecking on a multiple schedule to come from two *separate* sources: the baseline rate maintained by the response contingency and the addition of autoshaped pecks elicited by the signal value of the key light stimulus. Yet matching involves both types of behavior, since relative response rate is a linear function of a single variable: relative rate of reinforcement. Additivity theory provides no rationale for why the sum of two separate sources of pecking should vary so lawfully.

CONTRAST AND THE ROLE OF SCHEDULE TRANSITIONS

Local-rate gradients like those seen in Figures 13-2 – 13-4 are of interest for reasons other than their implications for additivity theory. Namely, they provide important clues as to the nature of the variables that produce contrast. Decreasing gradients like that in Figure 13-2 suggest that the transition from the preceding period of extinction is the primary cause of contrast. Increasing gradients like those in Figures 13-3 and 13-4 suggest an alternative interpretation, that the controlling variable is the transition to the *following* period of extinction. Effects of the following schedule have been known for some time (Pliskoff, 1961; 1963; Wilton and Gay, 1969) but have been ignored in theoretical discussions of contrast. Still a third alternative is that the local effects of neither type of schedule transition is critical, as contrast may depend solely on the overall relative rate of reinforcement. To the extent these variables can be separated, the requirements for an adequate theory of contrast would be delineated more clearly.

In Williams (1976) the various possible controlling variables were partially separated by directly opposing the effects of the preceding and following schedule. Pigeons were trained initially on a multiple VI 90-second VI 90-second schedule with 90-second components that regularly alternated. Then the schedule was held constant during one component, while the schedule in the other was changed to a variant of a fixed-interval schedule (actually a mixed FI EXT). During all of the FI conditions the rate of reinforcement in the second component remained the same as during the baseline, since one reinforcer was obtained during *each* 90-second component. The critical variable was the temporal location of the reinforcer within that component. For one condition it occurred early (5 seconds after the start of the component); for a second condition it occurred in the middle (45 seconds after the start); for a third condition it occurred near the end (80 seconds after the start). For all three conditions extinction was in effect during the remainder of the component after the one reinforcer had been obtained. The results were that response rate in the unchanged VI 90-second component was monotonically related to the temporal location of the reinforcer, with the rates higher the *later* the FI reinforcer occurred. Moreover, relative to the baseline condition of multiple VI 90 seconds VI 90-seconds, responding in the unchanged component was suppressed by the FI 5-second condition but was enhanced by the FI 80-second condition. In other words, responding in the unchanged component was inversely related to the local rate of reinforcement in the following component. Of greatest interest to the present discussion is the fact that the FI 5-second condition, which produced response suppression, was the condition in which the VI 90-second component was preceded by the longest period of EXT (80-85 seconds). Thus, when the preceding- and following-schedule effects were pitted against each other, the following schedule was the more important.

One difficulty with the procedure of Williams (1976) was that the confounding of the preceding- and following-schedule effects prevented the isolated study of either. Thus, while the following-schedule effect was found to be dominant, it was not possible to determine how the effect of the preceding schedule was otherwise related to overall contrast. To overcome this problem a new procedure was used in Williams (1979). Figure 13-5 shows two sequences of a four-component multiple schedule. For two of the components (cued by the circle and the lines), the schedules were identical (VI 3 minutes), so that differences in response rate during those two components should be due only to their location in the component sequence. Subjects were presented one of the sequences for an extended period, so that the effects of the preceding schedule or the following schedule could be examined without confounding from the alternative effect. For example, when the Following sequence was used, the circle and lines were each preceded half the time by EXT and half the time by VI 1-minute, so that any effect of the preceding schedule should be equivalent for both components. The schedules following the circle and lines consistently differed, however, so any difference in response rate in the two components would be due to the following schedules. A similar argument holds for the Preceding sequence, where the circle and lines had consistently different preceding schedules but the same following schedules, so that any difference in response rate would be due to the preceding schedule. At issue was how the effects of the two different sequences varied with component duration.

All subjects received both sequences of presentation. Three component durations (15, 60, 180 seconds) were used for each sequence. Within the training period with each component duration the positions of the circle and lines in the sequences were as shown in Figure 13-5 for one series of sessions, and then their positions in the sequences were reversed, in order to control for stimulus preferences. Figure 13-6 summarizes the results for the entire set of conditions by showing data averaged over subjects for the last four sessions of training on each condition. The data for individual subjects were generally consistent with the mean functions that are shown (for further details see Williams, 1979). In the top half of Figure 13-6, the results are plotted in terms of the absolute measure

Figure 13-5. The sequences of component presentation used in Williams (1979). Each column corresponds to a different sequence. After the completion of the last component shown in a column, the sequence recycled to the first component shown in the column. The two sequences that are shown differed only with respect to the location of the lines and the circle in the sequence. The sequence labeled *Following Schedule* has the lines always followed by VI 1-minute (blue) and the circle always followed by EXT (red). The sequence labeled *Preceding Schedule* has the lines always preceded by VI 1-minute (blue) and the circle always preceded by EXT (red).

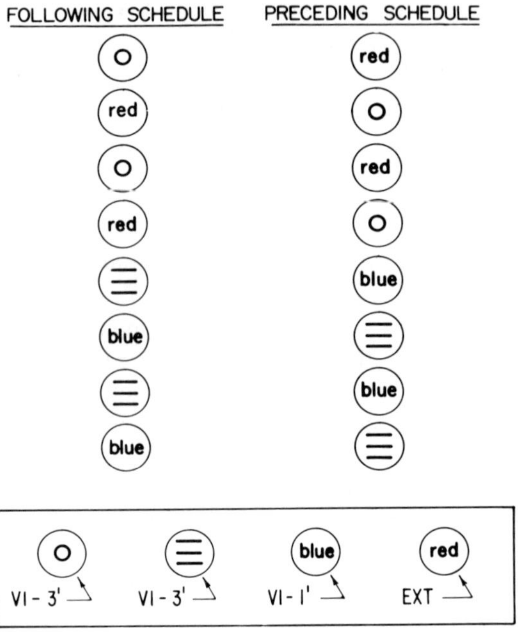

of responses per minute. In the bottom half, the data are shown in terms of the proportion of the total responses that occurred in the two VI 3-minute components. The preceding schedule produced no meaningful effect for either measure. The small difference shown was not consistent across subjects. It should be noted that the stimuli were chosen for their high discriminability, and the subjects had large amounts of previous training. Apparently, therefore, the preceding schedule was not an important determinant of schedule interactions in steady-state procedures and cannot explain the general effect of component duration. Theories that rely on the effects of the preceding schedule (Rachlin, 1973) should therefore no longer be considered viable.

Figure 13-6. Response rates, averaged over subjects, during the circle and lines components, which had identical schedules (see Figure 13-6). *Preceding Schedule* sequence (*left*); *Following Schedule* sequence (*right*); averages of the absolute response rates (*top*); averages of the relative rates (*bottom*).

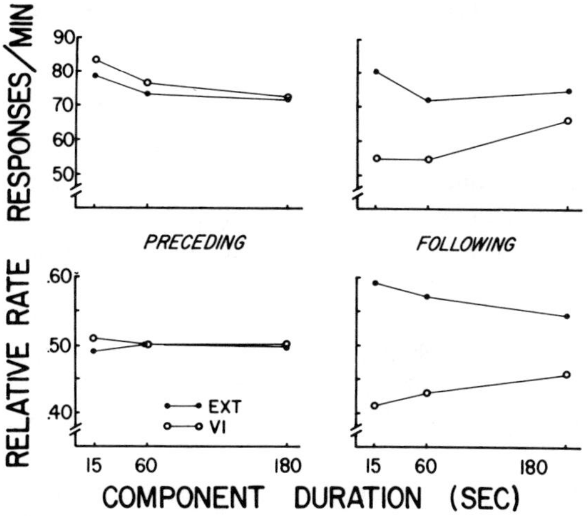

Figure 13-6 shows quite a different picture for the results of the Following sequence. Rate of responding was consistently higher in the VI 3-minute component that was followed by EXT, and the size of that effect decreased as component duration increased. The results thus support the findings of previous studies (Buck et al., 1975; Williams, 1976) where two-component schedules were used and suggest strongly that the following-schedule effect must be an important feature of any adequate theory of schedule interactions. Whether any extant theory can explain the asymmetry of the interactions seen in Figure 13-6 is questionable.

As noted earlier, three variables have been implicated as causes of contrast in multiple shcedules: the effect of the preceding schedule, the effect of the following schedule, and changes in relative rate of reinforcement independent of local transitions between schedules. Figure 13-6 demonstrated that the first of these is not an important variable, at least with steady-state procedures with highly discriminable stimuli (but see Malone, 1976, for situations where it is important). However, the results leave unresolved the relation between the second and third variables. Still at issue is whether the following-schedule effect accounts for *all* of steady-state interactions. If so, any formulation of the effect of "context of reinforcement" would become much more specific, with the possibility that such effects could be reduced to some more elementary mechanism.

Alternatively, the effect of the following schedule may be a local effect only, with there being a general effect of context of reinforcement apart from it. If such a separation were possible, an important issue would still remain. Namely, how do the two effects interact such that their sum obeys the matching relation previously found with short-component schedules? As discussed before, the occurrence of matching seems to imply that a single variable controls schedule interactions, so that any independent effects of the second and third variables should be unlikely.

In an attempt to resolve these issues, the subjects of Williams (1979) were exposed to a modification of the Following sequence for 30 additional sessions, using 15-second components. The sequence was the same as shown in Figure 13-5 except the positions of the circle and the lines were reversed. Thus, the lines were always followed by EXT and the circle was always followed by VI 1-minute, while both were preceded equally often by the VI 1-minute and EXT. Thirty additional sessions were then given with the only change being that the schedule during Blue was changed from VI 1-minutes to EXT. Consequently, both the lines and the circle were now followed by EXT. Finally, 25 additional sessions were given in which the schedule during Blue was returned to VI 1-minute.

At issue was whether the change in reinforcement during Blue would affect responding only during the circle component, which Blue always followed, or whether it would affect responding during both the circle and lines components. Any effect on responding during the lines presumably would not be due to the effect of the following schedule and thus would argue for an effect of relative rate of reinforcement independent of the following-schedule effect.

Table 13-2 shows the results for individual subjects for the last five sessions of each of the three phases of training. If only the mean results are considered, the data support the view that there is little effect of relative rate of reinforcement independent of the following-schedule effect, as responding during the lines varied only slightly (3-4 responses/minute), whereas that during the circle varied widely (30-35 responses/minute). Unfortunately, the mean results are not indicative of the results for individual subjects. Two subjects (Y-22, B-3) showed clear increases in response rate to the lines when Blue was changed to EXT, with this effect reversing when the schedule during Blue was returned to VI 1 minute. A third subject (R-11) showed a similar increase during phase 2, but the effect did not reverse. The means do not reflect these trends, however, because they were overshadowed by the large changes seen for the remaining subject (R-20) in the opposite direction. Its response rate during the lines *decreased* when the schedule during Blue was changed to EXT.

The variability in behavior for individual subjects poses interpretative difficulties. Nevertheless, the data in Table 13-2 do suggest that there was some effect of relative rate of reinforcement for two of the subjects that was independent of the following-schedule effect. The issue that must be confronted is

Table 13-2. Rates of responding during components with VI 3-minute schedules as a function of the schedule during Blue. The circle was always followed by Blue; the lines were always followed by Red, which was always associated with EXT. Data are taken from the last four sessions of each phase of training.

Subject	*Stimulus*	*VI 1-minute*	*EXT*	*VI 1-minute*
Y-22	Lines	57.7	66.0	60.5
	Circle	40.1	58.4	43.0
R-11	Lines	53.1	65.0	64.4
	Circle	41.8	65.3	51.0
B-3	Lines	135.0	147.0	134.3
	Circle	118.6	155.4	120.6
R-20	Lines	109.5	90.3	89.5
	Circle	86.6	128.0	80.6
Mean	Lines	88.8	92.1	87.2
	Circle	71.8	101.8	73.8

how this apparent independence of effects can be reconciled with the matching phenomenon in short-component schedules, which argues strongly for a *single* determinant of steady-state schedule interactions.

One resolution is to assume that the following-schedule effect extends beyond the immediately antecedent component. Component duration was always 15 seconds for the conditions shown in Table 13-2, which meant that the Blue component followed the lines by 30-60 seconds, although other components intervened. It is possible that the effect of changing reinforcement during Blue could extend over such a period, so that the changes in response rate to the lines still could be ascribed to the following-schedule variable.

A second alternative is to consider sources of control independent of the following-schedule effect. One possibility is the effect of the discrimination contingency on response topography. Stimulus generalization gradients are known to be steeper after differential reinforcement in the different components of a multiple schedule than after nondifferential reinforcement (cf. Lyons and Thomas, 1967). It is likely that different response topographies are associated with the different gradients, if, as commonly assumed, the flat gradients obtained after nondifferential reinforcement are due to stimulus control by features of the environment off the response key. Attention to such features should be expected to produce a response topography that is less efficient than when stimulus control is specific to the cues localized on the response key. Such changes in response topography might also have resulted from the variations in reinforcement during the conditions shown in Table 13-2. Therefore, a general increase in response rate should have occurred in phase 2, because the unit of behavior was more efficient at that time.

Clearly effects of differential reinforcement on response topography are quite close to the explanation of contrast provided by additivity theory, which I have argued against. There are two critical differences, however, Whereas additivity theory assumes that the increase in response rate during the positive discriminative stimulus is due to elicited contact with that stimulus alone, the present notion is that changes in the response unit may affect responding during *all* components, because the change is due to the loss of stimulus control by environmental features not on the response key. Consequently the change in response unit should have the effect of multiplying *all* of the responding that would have occurred otherwise. Predictions regarding relative rate of responding (e.g., matching) would be unaffected by such a transformation.

A second aspect of the present notion that differs from additivity theory lies in the identification of the sources of variability. Whereas additivity theory claims that changes in response topography are a function of the strength of the contingency between a stimulus and a reinforcer and should vary whenever that contingency is changed, the present notion suggests that changes in response topography may be at least partially irreversible, because the conditions that cause a stimulus to regain stimulus control after a return to nondifferential reinforcement are by no means clear. It is likely that the degree of reversibility will vary with several factors, including the nature of the discriminative stimuli, the nature of the response key and the conditioning environment that surrounds it, and the experience of the subject.

Whatever the explanation of the rate increases to the lines seen in Table 13-2, their occurrence should not overshadow the importance of the following-schedule effect itself. Inspection of Table 13-2 shows clearly that the following-schedule effect was the most robust effect that occurred, for all subjects, including those that showed some rate increase during the lines.

The discussion to this point bears recapitulation. It has been argued here that neither response-competition theory (Staddon, Chapter 11) nor additivity theory provides an adequate explanation of interactions seen in multiple schedules. Moreover, the occurrence of matching in short-component schedules (cf. Killeen, 1972) provides prima facie evidence that such interactions are under the control of a unitary variable. Unfortunately, simple characterization of that unitary variable is not now possible. Relative rate of reinforcement, per se, does not appear adequate, because of the temporal asymmetries in the way reinforcement in one component of the schedule affects behavior in other components (Williams, 1976; 1979). It is possible but by no means proved at this point that all steady-state interactions in multiple schedules can be reduced to the effects of the following schedule of reinforcement, although the mechanism underlying the following-schedule effect itself remains in question (see Williams, 1979, for further discussion of several possible mechanisms). It is clear, however, that any adequate theoretical characterization of the effects of context of reinforcement must encompass the following-schedule effect because, at the very least, it is responsible for a major component of contrast in multiple schedules. None of the present theoretical alternatives offer much hope of providing such a characterization.

CONCURRENT SCHEDULES AND THE FOLLOWING-SCHEDULE EFFECT

The last issue to be considered in this chapter is how the following schedule of reinforcement is related to contrast (and matching) in concurrent schedules, where the temporal order of components has no clear meaning. One answer is that it is not. Such a conclusion would be compatible with the theoretical positions taken in several chapters in this book (e.g., Herrnstein, Chapter 19) where mechanisms of concurrent-schedule interactions have been proposed that do not suggest any basis for comparable effects in multiple schedules. Arrayed against this conclusion is the large amount of data showing functional similarity between concurrent and multiple schedules. Most notable are the data of Killeen (1972), which indicates that multiple schedules produced relative distributions of behavior virtually identical to similar concurrent schedules, regardless of whether they were yoked to the concurrent schedules or presented with fixed component durations that were independent of the concurrent schedules. As yet, no one has provided an explanation of this similarity, other than that the two schedules involve the same independent variables. The implication is that the following-schedule effect that operates in multiple schedules somehow must be applicable to concurrent schedules as well.

The last experiment to be described asked whether the effects of the following schedule can be demonstrated in a choice situation. The experiments already described show that the following schedule can affect the rate of behavior. At issue is whether it also affects the "value" of the schedule component. Rate and value usually are correlated but need not be, as shown by choice studies where response rate has been enhanced by various second-order schedules (e.g., Arnett, 1972). Rate increases associated with contrast generally are though to reflect an increase in value, however, as shown by the correlation between the variables that produce contrast and those that produce preferences in concurrent schedules (cf. Schwartz and Gamzu, 1977). Moreover, Hinson, Malone, McNally, and Rowe (1978) have provided evidence that the effect of component duration on contrast is also associated with a change in value. In their study the same VI schedule was used for both a long and short component of a multicomponent schedule (the remaining components were all EXT). Response rate during the short component was higher than during the long component. But most important, response rate during the long component was suppressed whenever it was preceded by the short component. If local contrast effects can be taken to reflect different values of the schedule components (cf. Malone, 1976), the implication is that the short component had higher value than the long component, despite their being associated with the same density of reinforcement. In view of such evidence the following schedule effect, which determines the effect of component duration (Williams, 1979) also should be correlated with changes in stimulus value.

The present study used a concurrent-chain procedure (cf. Fantino, 1977). Two response keys were available in the first link of the chain, each of which was associated with an independent VI 60-second schedule where the reinforcer was access to the one of the terminal links of the chain. Choice proportions during the initial link should thus reflect the value of the different terminal links associated with each choice response. Entry into either terminal link terminated the initial-link choice phase, which again resumed only after the terminal-link exposure for that cycle had been completed. This meant that exposure to only one of the two terminal links could occur during any one cycle.

The two terminal links consisted of identical 30-second exposures to the same VI 1-minute food schedule but differed with respect to the events that occurred after the 30-second exposures had ended. For both terminal links, 90 seconds intervened between the exposure to the VI food schedule and the return to the initial-link choice phase. After one terminal-link stimulus the 90 seconds consisted of a TO condition in which the chamber was darkened and food was not available. After the other terminal-link stimulus a houselight was illuminated and a VT 30-second schedule of response-independent food delivery operated (the key lights were extinguished). It was expected that a higher response rate would occur in the terminal link followed by the TO. The issue would then be whether the higher rate would be accompanied by an increase in preference for that terminal link, as assessed by the choice proportions during the initial link.

The first row in Table 13-3 shows the choice proportions during the initial link; the second row shows the ratio of the response rates in the terminal links.

Table 13-3. Following-Schedule effects on behavior in a concurrent-chain schedule. The top number for each condition represents the preference during the initial link of the schedule (0.50 = indifference); the bottom number represents the ratio of absolute response rates in the terminal links (1.0 represents equal rates).

	90 Seconds of TO versus 90 Seconds of VT 30-seconds				
	B-1	*R-3*	*R-7*	*R-9*	*Mean*
Initial Choice	.28	.40	.30	.20	.29
relative rate	1.17	1.09	1.09	1.12	1.12
Reversal Choice	.42	.34	.36	.30	.35
relative rate	.95	1.14	1.06	1.14	1.07

Choice proportions greater than 0.50 indicate preference for the terminal link followed by TO. Relative rate measures greater than 1.00 indicate higher response rates in the terminal link followed by TO. Rows 1 and 2 represent the last 5 of 25 sessions of the initial determination; rows 3 and 4 represent the data for the last 5 of 25 sessions of the reversal, in which the relation between initial-link choice and terminal-link schedule was reversed.

Rows 2 and 4 show that higher response rates did generally occur in the terminal link followed by the TO, as 7 of the 8 entries were larger than 1.0. Nonetheless, the effect was considerably smaller than the effects described earlier for the multiple schedule studies (see Figure 13-6). Rows 1 and 3 showed larger effects on choice, however, as all subjects preferred the terminal link followed by the VT schedule, for both the initial determination and reversal, despite higher response rates in the terminal link followed by TO. Thus, preference and response rate were dissociated. The apparent implication is that the following-schedule effect does *not* play a role in interactions involving choice between schedules.

Unfortunately, the choice data shown in rows 1 and 3 are ambiguous in interpretation. It is possible that preference was controlled not by the value of the terminal-link stimuli but, rather, by the direct relation between choice response and the schedule that occurred 30 seconds later. This possibility is enhanced by the results of a continuation of the study not shown in Table 13-2. The only modification was that reinforcement during the terminal-link stimuli was abolished. Thus, choice responses were followed by 30-seconds access to one or the other terminal-link stimulus, both of which were now associated with EXT, followed in turn by either 90 seconds of TO or 90 seconds of a VT-30 second schedule. The result of this change was that preference for the terminal link followed by the VT schedule not only continued but was considerably enhanced. Apparently, therefore, the dissociation between absolute rate in the terminal links and preference in the initial link may mean very little about the value of the terminal-link stimuli themselves. The results shown in Table 13-2 may imply only that preference can be controlled by events that occur 30 seconds after the choice response has occurred.

CONCLUSIONS

The experiments that have been described pose more questions than they answer. The following-schedule effects seen in Williams (1976; 1979) and the subsequent data reported here do show that several widely held interpretations of contrast in multiple schedules (e.g., additivity theory) are invalid, at least applied to interactions found in steady-state procedures. The results also challenge any molar notion of relative rate of reinforcement as the controlling variable (e.g., Herrnstein, 1970), although some more specific variation of that notion may prove applicable. On the negative side of the ledger, it has not been possible to reduce the following-schedule effect to any more fundamental basis (see Williams, 1979: experiment 2-4) and the attribute of the following schedule that controls behavior remains unknown. Nevertheless, the results do allow the conclusion that any adequate theory of schedule interactions must have an account of the following-schedule effect as a central focus.

The issue most germane to the theme of the book is how multiple and concurrent schedule interactions are related. If there exists a common basis to both types of interaction, then "maximizing" interpretations of concurrent schedule effects become extremely unlikely. This is true because interactions in multiple schedules do not affect the distribution of reinforcement received in the two components of the schedule, so it is not clear how the subject can improve its total reward by changing its distribution of behavior. The occurrence of matching in multiple schedules makes a strong case that the two schedules do have a common basis. Nevertheless, recent theories of concurrent schedule behavior (Silberberg and Ziriax, Chapter 5; Rachlin, Chapter 16) are predicated on the assumption that this apparent commonality is illusory. The present data are germane to this issue but do not resolve it. On the one hand, the data challenge previous explanations of contrast in multiple schedules that have postulated mechanisms specific to multiple schedules (e.g., additivity theory). On the other hand, the following-schedule effect has been shown to account for at least a large portion of contrast in multiple schedules, and the relation of the following-schedule effect to concurrent schedules is uncertain. Until this uncertainty is reduced, it seems unlikely that a general theory of schedule interactions will be forthcoming.

REFERENCES

Allison, J.; M. Miller; and M. Wozny. 1979, Conservation in behavior. *Journal of Experimental Psychology: General 108*: 4-34.

Arnett, F.B. 1972. Inverse relation between choice and local response rate with a schedule of response-produced blackouts. *Journal of the Experimental Analysis of Behavior 17*: 37-43.

Buck, S.; B. Rothstein; and B.A. Williams. 1975. A re-examination of local contrast in multiple schedules. *Journal of the Experimental Analysis of Behavior 24*: 291-301.

Catania, A.C. 1962. Independence of concurrent responding maintained by interval schedules of reinforcement. *Journal of the Experimental Analysis of Behavior 5*: 175-84.

_____. 1963. Concurrent performances: Reinforcement interaction and response independence. *Journal of the Experimental Analysis of Behavior 6*: 253-63.

de Villiers, P., and R.J. Herrnstein. 1976. Toward a law of response strength. *Psychological Bulletin 83*: 1131-53.

Fantino, E.J. 1977. Conditioned reinforcement: Choice and information. In W.K. Honig and J.E.R. Staddon, eds., *Handbook of Operant Behavior*, Englewood Cliffs, N.J.: Prentice-Hall.

Gamzu, E., and B. Schwartz. 1973. The maintenance of key pecking by stimulus-contingent and response-independent food presentation. *Journal of the Experimental Analysis of Behavior 19*: 65–72.

Hamilton, B.E., and A. Silberberg. 1978. Contrast and autoshaping in multiple schedules varying reinforcer rate and duration. *Journal of the Experimental Analysis of Behavior 30*: 107–22.

Hearst, E., and H.M. Jenkins. 1974. *Sign-tracking: The Stimulus-Reinforcer Relation and Directed Action.* Austin, Texas: Psychonomic Society.

Herrnstein, R.J. 1970. On the law of effect. *Journal of the Experimental Analysis of Behavior 13*: 243–66.

Hinson, J.M., and J.E.R. Staddon. 1978. Behavioral competition: A mechanism for schedule interactions. *Science 202*: 432–34.

Hinson, J.M.; J. Malone; K. McNally; and D. Rowe. 1978. Effects of component length and of the transitions among components in multiple schedules. *Journal of the Experimental Analysis of Behavior 29*: 3–16.

Keller, K. 1974. The role of elicited responding in behavioral contrast. *Journal of the Experimental Analysis of Behavior 21*: 249–57.

Killeen, P. 1972. A yoked-chamber comparison of concurrent and multiple schedules. *Journal of the Experimental Analysis of Behavior 18*: 13–22.

Lyons, J., and D.R. Thomas. 1967. Effects of interdimensional training on stimulus generalizations. II. Within subjects design. *Journal of Experimental Psychology 75*: 572–74.

Malone, J.C. 1976. Local contrast and Pavlovian induction. *Journal of the Experimental Analysis of Behavior 26*: 425–40.

Nevin, J.A. 1974. Response strength in multiple schedules. *Journal of the Experimental Analysis of Behavior 21*: 389–408.

Nevin, J.A., and S.J. Shettleworth. 1966. An analysis of contrast effects in multiple schedules. *Journal of the Experimental Analysis of Behavior 9*: 305–15.

Pliskoff, S.S. 1961. Rate-change effects with a pre-schedule-change stimulus. *Journal of the Experimental Analysis of Behavior 4*: 383–86.

____. 1963. Rate-change effects with equal potential reinforcements during the "warning stimulus." *Journal of the Experimental Analysis of Behavior 6*: 557–62.

Rachlin, H.C. 1973. Contrast and matching. *Psychological Review 80*: 217–34.

Rachlin, H.C., and W. Baum. 1972. Effects of alternative reinforcement: Does the source matter? *Journal of the Experimental Analysis of Behavior 18*: 231–41.

Rachlin, H., and B. Burkhard. 1978. The temporal triangle: Response substitution in instrumental conditioning. *Psychological Review 85*: 22–47.

Schwartz, B. 1978. Stimulus-reinforcer contingencies and local behavioral contrast. *Journal of the Experimental Analysis of Behavior 29*: 297–308.

Schwartz, B., and E. Gamzu. 1977. Pavlovian control of operant behavior: An analysis of autoshaping and its implications for operant conditioning. In

W.K. Honig and J.E.R. Staddon, eds., *Handbook of Operant Behavior*, New York: Prentice-Hall.

Shimp, C., and K. Wheatley. 1971. Matching to relative reinforcement frequency in multiple schedules with a short component duration. *Journal of the Experimental Analysis of Behavior 15*: 205-10.

Spealman, R.D. 1976. Interactions in multiple schedules: The role of the stimulus-reinforcer contingency. *Journal of the Experimental Analysis of Behavior 26*: 79-93.

Staddon, J.E.R. 1979. Operant behavior as adaptation to constraint. *Journal of Experimental Psychology: General 108*: 48-67.

Staddon, J.E.R., and S. Motheral. 1978. On matching and maximizing in operant choice experiments. *Psychological Review 85*: 436-44.

Staddon, J.E.R., and V.L. Simmelhag. 1971. The "superstition" experiment: A re-examination of its implications for the principles of adaptive behavior. *Psychological Review 78*: 3-43.

Todorov, J.C. 1972. Component duration and relative response rates in multiple schedules. *Journal of the Experimental Analysis of Behavior 17*: 45-49.

Vieth, H., and M. Rilling. 1972. Comparison of time-out and extinction as determinants of behavioral contrast: An analysis of sequential effects. *Psychonomic Science 27*: 281-82.

Williams, B.A. 1974. The role of local interactions in behavioral contrast. *Bulletin of the Psychonomic Society 4*: 543-45.

_____. 1976. Behavioral contrast as a function of the temporal location of reinforcement. *Journal of the Experimental Analysis of Behavior 26*: 77-85.

_____. 1979. Contrast, component duration, and the following schedule of reinforcement. *Journal of Experimental Psychology: Animal Behavior Processes 5*: 379-96.

Wilton, R., and R. Gay. 1969. Behavioral contrast in one component of a multiple schedule as a function of the reinforcement conditions operating in the following schedule. *Journal of the Experimental Analysis of Behavior 12*: 239-46.

IV PUNISHMENT

14 MAXIMIZING, MINIMIZING, AND MATCHING BETWEEN REINFORCING AND PUNISHING SITUATIONS

Marvin Z. Deluty

How do rewards affect an animal's distribution of choices to various alternatives? What is the nature of the relation between choice and reinforcement? The matching law represents an attempt to answer these questions: It is a quantitative relation between an organism's actions and the reinforcements of those actions.

When animals are responding on concurrent variable interval, variable interval (conc VI VI) schedules of reinforcement, their proportion of responses or time to one key or lever approximately equals or matches the proportion of obtained reinforcers associated with that key or lever (Baum, 1973; de Villiers, 1977; Herrnstein, 1970, 1971; Myers and Myers, 1977). That is,

$$\frac{R_1}{R_1 + R_2} = \frac{r_1}{r_1 + r_2} \;; \qquad (14.1)$$

$$\frac{T_1}{T_1 + T_2} = \frac{r_1}{r_1 + r_2} \;. \qquad (14.2)$$

R_1 and T_1 represent, respectively, the number of responses and the amount of time allocated to one key or lever; R_2 and T_2 represent, respectively, the number of responses and the amount of time allocated to the second key or lever; and r_1 and r_2 denote, respectively, the number of reinforcers for the first and second manipulandum.

This research was done in collaboration with Russell M. Church and was supported by Public Health Service Grant 23247 to Brown University.

The question then arises as to how punishers affect an animal's distribution of choices to various alternatives. What is the quantitative relation between choice and punishment? This chapter describes an attempt to answer this question.

CHOICE ALLOCATION BETWEEN PUNISHING SITUATIONS

Experiment 1

In this study three male, naive albino Norway rats of the Charles River CD strain served as subjects. The procedure was very simple: There was a single lever in a standard operant chamber. In the presence and absence of white noise, electric shocks were delivered to the animals according to various variable-time (VT) schedules. There was a noise component and a no-noise component. By making a single lever-press response, the animal could switch from one component to the other component. [(The white noise was 70 decibels (re 0.0002 dyne/square centimeter)]. Throughout the study, a changeover delay (COD) of 2 seconds was also used; that is, shocks were not delivered for 2 seconds following the animal's switch from one component to the other component.

The amount of time the animals spent in each component was measured, as the shock schedules in the components were varied. The schedule in a component was either VT 30 seconds, VT 2 minutes, VT 4 minutes, or no punishment. Shocks were delivered according to an exponential waiting-time distribution, with 0.1 second the minimum intershock time. The probability of shock presentation was equal to 0.1 divided by the mean interval. Shocks were of 0.5-second duration and 0.8 milliampere intensity. In this experiment, the VT schedule for a component was not in operation when the rat was in the other component: The schedule timer in a component ceased to advance when the animal was not in that component (Deluty and Church, 1978: experiment 1).

Table 14-1 shows the mean proportion of time spent in component 1, $T_1/T_1 + T_2$, and the obtained number of punishers per session for each component in every condition. Each value is the mean of the last five sessions of each condition.

The top panel of Figure 14-1 shows the proportion of time allocated to component 1 as a function of the proportion of obtained shocks in component 2. p_1 is the obtained number of shocks in component 1, and p_2 is the obtained number of shocks in component 2. The different geometric shapes represent the different subjects. The solid diagonal line represents the matching relation between the proportion of time spent in component 1 and the proportion of shocks in component 2. The broken line is the least squares regression line; the equation of the regression line is indicated.

Figure 14-1 presents a time-allocation matching relation for punishment. The proportion of time allocated to one component equals or matches the propor-

Figure 14-1. Experiment 1. The proportion of time spent in component 1 as a function of the proportion of obtained punishers in component 2 (*left*). The proportion of time allocated to component 1 as a function of the square roots of the relative delays until shock (*right*). The broken lines are the least squares regression lines. (From M.Z. Deluty and R.M. Church, Time-allocation matching between punishing situations, *Journal of the Experimental Analysis of Behavior*, 1978, *29*:191–98. Copyright 1978 by the Society for the Experimental Analysis of Behavior, Inc. Reprinted by permission.)

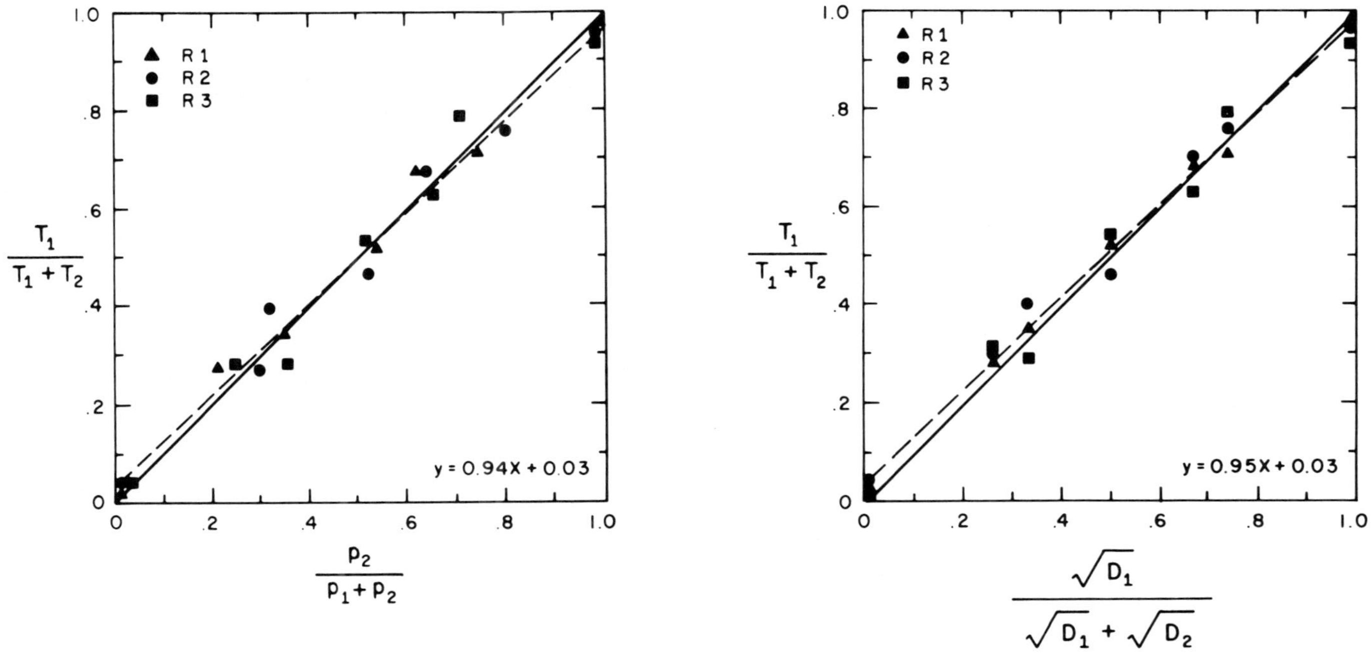

Table 14-1. Sequence of conditions in Experiments 1 and 2 and the results.

Condition	*Scheduled Punishment Rate (per minute)* Comp. 1	Comp. 2	Rat	$\frac{T_1}{T_1 + T_2}$	*Obtained Punishers per Session* Comp. 1 (p_1)	Comp. 2 (p_2)
			Experiment 1			
1	2.0	0.0	1	0.02	1.0	0.0
			2	0.02	1.2	0.0
			3	0.02	1.0	0.0
2	0.0	2.0	1	0.98	0.0	0.6
			2	0.98	0.0	1.0
			3	0.93	0.0	2.8
3	2.0	2.0	1	0.52	26.0	29.8
			2	0.46	24.0	26.0
			3	0.53	22.2	24.0
4	0.5	2.0	1	0.68	7.8	12.6
			2	0.68	9.0	15.0
			3	0.63	7.8	14.4
5	2.0	0.5	1	0.35	13.8	7.4
			2	0.40	17.8	8.2
			3	0.29	13.4	7.6
6	0.25	2.0	1	0.71	5.2	15.8
			2	0.76	3.0	12.6
			3	0.79	4.2	10.4
7	2.0	0.25	1	0.28	13.4	3.6
			2	0.28	11.6	5.0
			3	0.29	13.8	4.6
			Experiment 2			
1	2.0	0.0	1	0.07	2.8	0.0
			2	0.09	5.8	0.0
			3	0.00	0.8	0.0
2	0.0	2.0	1	0.97	0.0	1.4
			2	0.97	0.0	2.8
			3	0.93	0.0	4.6
3	2.0	2.0	1	0.35	18.2	35.6
			2	0.49	23.8	24.8
			3	0.59	29.0	17.2
4	0.5	2.0	2	0.92	10.8	4.6
			3	0.92	10.8	5.0
5	1.0	2.0	1b	0.77	24.4	12.4
			2	0.70	23.0	15.2
			3	0.78	17.2	14.6

Table 14-1. continued

Condition	Scheduled Punishment Rate (per minute) Comp. 1	Comp. 2	Rat	$\frac{T_1}{T_1 + T_2}$	Obtained Punishers per Session Comp. 1 (p_1)	Comp. 2 (p_2)
		Experiment 2 (continued)				
6	2.0	0.5	1b	0.09	5.4	12.8
			2	0.07	3.4	10.4
			3	0.09	4.0	6.8
7	1.0	2.0	1b	0.69	19.8	18.4
			2	0.79	21.0	10.6
			3	0.97	21.2	2.6
8	2.0	1.0	1b	0.17	9.2	21.2
			2	0.16	8.0	21.6
			3	0.11	9.7	21.8

tion of punishers in the *other* component. For example, if 80 percent of all shocks occur in component 2, then the animal will allocate 80 percent of its time to component 1.

In Deluty (1976), it was suggested that the effect of reinforcement and punishment on the proportion of responses in concurrent schedules may be expressed by the following equation:

$$\frac{R_1}{R_1 + R_2} = \frac{r_1 + p_2}{r_1 + r_2 + p_1 + p_2} . \qquad (14.3)$$

Assuming that the time-allocation matching relation for reinforcement and punishment is the same as the response-allocation matching relation for reinforcement and punishment, then the proportion of time will be allocated as follows:

$$\frac{T_1}{T_1 + T_2} = \frac{r_1 + p_2}{r_1 + r_2 + p_1 + p_2} . \qquad (14.4)$$

Experiment 1 investigated a special case of Eq. (14.4): namely, the situation where there is no reinforcement and only punishment is present. Under these conditions ($r_1 = 0$ and $r_2 = 0$), the result is the following:

$$\frac{T_1}{T_1 + T_2} = \frac{p_2}{p_1 + p_2} . \qquad (14.5)$$

Figure 14-1 (left) examines the adequacy of this formula: Equation (14.5) accounts for 98 percent of the variance over all animals.

It should be pointed out that a different model of punishment has been proposed by de Villiers (1977, 1980) and also by Farley and Fantino (1978). They have suggested that the effect of reinforcement and punishment on the proportion of choices allocated to various alternatives may be expressed by the following equation:

$$\frac{T_1}{T_1 + T_2} = \frac{r_1 - p_1}{r_1 + r_2 - (p_1 + p_2)} \quad . \tag{14.6}$$

Problems exist however, with this formulation. First, let us assume that the intensity of punishment remains constant and the rates of reinforcement remain constant. Then as the rates of punishment increase, the value of the denominator of this equation can be zero or negative. When the denominator is zero, the equation is indeterminate; when the denominator is negative, it predicts a negative proportion of choices.

In addition there is a second problem with this model. What does the model predict will occur in the situation that was studied in Experiment 1? How will animals allocate their time when there is no reinforcement and only punishment is present? In this special case, where $r_1 = 0$ and $r_2 = 0$, the result is the following:

$$\frac{T_1}{T_1 + T_2} = \frac{p_1}{p_1 + p_2} \quad . \tag{14.7}$$

This formula states that the proportion of choices to an alternative will equal the proportion of punishments for that alternative: An animal will prefer the alternative with the greater proportion of shocks. If, for example, 70 percent of all shocks occur to alternative 1, then the animal will allocate 70 percent of its time to alternative 1. But the results of Experiment 1, as shown in Figure 14-1 (left), show that just the opposite is true: If 70 percent of all shocks occur to alternative 1, then the animal will allocate 70 percent of its time to alternative 2. (It should be noted, however, that recent work has shown that Eq. (14.3) does not always correctly predict how reinforcement and punishment together affect choice in some situations (de Villiers, 1980 and Chapter 15 in this volume; Farley, 1980).

Until now the topic has been the correlation of two obtained variables—the proportion of time spent in a component and the proportion of received shocks. It is possible, though, to predict the proportion of time allocated to a component based on the expected times until the aversive events. If D_1 and D_2 represent the expected delays until shock in components 1 and 2, respectively, then the expected number of shocks in the components is

$$E(p_1) = T_1/D_1 \quad ; \tag{14.8}$$

$$E(p_2) = T_2/D_2 \quad . \tag{14.9}$$

If these formulas are substituted into Eq. (14.5), the result is the following:

$$\frac{T_1}{T_1 + T_2} = \frac{\sqrt{D_1}}{\sqrt{D_1} + \sqrt{D_2}} \quad . \tag{14.10}$$

Figure 14-1 (right) shows for each animal the proportion of time allocated to component 1 as a function of the square roots of the relative delays until shock. The fit is very good, with the least squares regression line accounting for 99 percent of the variance over all subjects.

Experiment 2

Experiment 1 differed from most choice-allocation reinforcement studies in one important respect. In the punishment experiment the schedule in a component was in effect only when an animal was in that component; in the reinforcement experiments it is typically the case that the schedules for the components are always in effect. But does this procedural factor affect the manner in which animals allocate their time between punishing situations?

In Experiment 2 the three animals that began the experiment were those used in the first study. During one of the conditions one of the subjects died and was replaced by an experimentally naive rat (R1b). The procedure was exactly the same as the procedure in Experiment 1 except that the shock schedules for both components were always in operation; the "tapes" for both components were always in effect. The schedule in a component was either VT 30 seconds, VT 1 minute, VT 2 minutes, or no punishment. As in Experiment 1, a 2-second COD was also used (Deluty and Church, 1978: experiment 2). Table 14-1 shows the mean proportion of time spent in component 1 and the mean obtained number of punishers per session for each condition.

The first question to be asked is whether the proportion of time in a component equals the proportion of shocks in the other component as was the case in Experiment 1. Figure 14-2 (left) shows the proportion of time allocated to component 1 as a function of the proportion of obtained shocks in component 2. The diagonal represents the punishment matching relation (Eq. (14.5)). It is quite apparent that except for the extremes, the points do not fall along the diagonal

The second question to be asked is whether Eq. (14.10) holds under the conditions of Experiment 2. Figure 14-2 (right) shows the proportion of time spent in component 1 as a function of the square roots of the relative delays until shock. Minimizing is indicated by the step function. Minimizing indicates

Figure 14-2. Experiment 2. The proportion of time spent in component 1 as a function of the proportion of obtained punishers in component 2 (*left*). The proportion of time allocated to component 1 as a function of the square roots of the relative delays until shock (*right*). The step function indicates minimizing. (From M.Z. Deluty and R.M. Church, Time-allocation matching between punishing situations, *Journal of the Experimental Analysis of Behavior*, 1978, *29*:191-98. Copyright 1978 by the Society for the Experimental Analysis of Behavior, Inc. Reprinted by permission.)

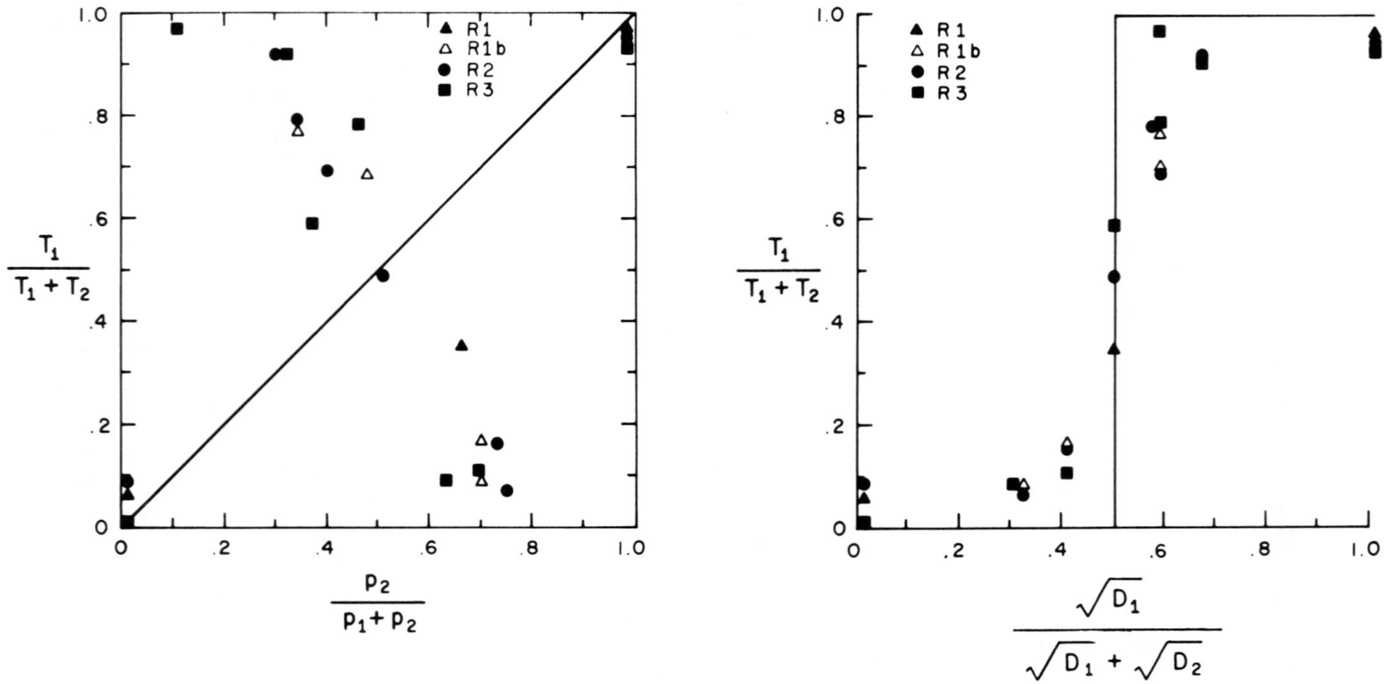

that the animal allocates all of its time to the component where the expected time until shock is longer. To the left of the step at 0.5, the expected delay until shock in component 1 is less than that in component 2: Minimizing will therefore occur if the subject spends all of its time in component 2. To the right of the step at 0.5, the expected delay until shock in component 1 is greater than that in component 2: Minimizing will therefore occur if the subject spends all of its time in component 1. Figure 14-2 indicates that the points fall close to the step function.

The results of the first two experiments reveal that time allocation between punishing situations depends upon whether the shock schedules for both components are always in effect, or whether the shock schedule in a component is in effect only when the animal is in that component. The punishment matching relations (Eqs. (14.5) and (14.10)) resulted when only one schedule was in operation at any time. If both schedules were always in operation, however, then the longer the animal spent in one component, the greater the probability that a shock had been set up in the other component. A changeover from one component to the other would result in the delivery of shock, thereby forcing the subject to allocate virtually all of its time to the component with the smaller rate of shock.

CHOICE ALLOCATION BETWEEN REINFORCING SITUATIONS

Is this procedural factor also of importance in the manner in which animals allocate their choices between reinforcing situations? The purpose of the next series of three experiments was to see if response and time allocation between reinforcing situations are dependent upon whether both schedules are always in effect or whether only one schedule is in effect. In each of these previously unpublished experiments, the responses of rats to one lever were reinforced on one random-interval (RI) schedule of reinforcement, and the responses to the other lever were reinforced on a second RI schedule, In Experiment 3 the schedule in one component was not in operation when the animal was in the other component. In Experiment 4 the typical concurrent schedules situation was used: The schedules for both alternatives were always in operation. In Experiment 5 the procedure of Experiment 4 was again employed but with the addition of a COD.

Experiment 3

In Experiment 3 the tape for a lever stopped when the rat was not working on that lever (the schedule for one lever was not in operation when the rat was responding on the other lever). Such a procedure has occasionally been used in the past (Fallon, Thompson, and Schild, 1965; Findley, 1958: fig. 5; see also Killeen, 1972: experiment 2, parts 3 and 4).

Four naive rats were given a daily ration of 14 grams of ground Purina Chow mixed with 25 cubic centimeters of water. Water was always available in each rat's cage. Following magazine training and the shaping of both lever-press responses, the animals were exposed to the conc RI RI schedules shown in Table 14-2. Under the RI reinforcement schedules, responses on a lever were reinforced with a specific probability every 0.1 second. The probability was equal to 0.1 divided by the mean interval in seconds. For response 1 (left lever), the reinforcement schedule was always a RI 2-minute schedule. For response 2 (right lever), the reinforcement schedule was either a RI 8-minute, RI 4-minute, RI 2-minute, RI 1-minute, or RI 0.5-minute schedule. Following condition 1, the sequence of the RI reinforcement schedules for response 2 was different for each subject (Latin square design). Each session lasted 50 minutes. The rat was recorded as spending time on a lever from the occurrence of a response on that lever until the occurrence of a response on the other lever. Data analysis is based upon the last five sessions of each condition.

Table 14-3 gives the response rates and the obtained reinforcement rates for each lever at each of the scheduled proportion of reinforcers. Figure 14-3 shows for each subject the proportion of responses and time allocated to lever 1 as a function of the scheduled proportion of reinforcements to lever 1. The step function indicates maximizing. To the left of the step at 0.5, the scheduled proportion of reinforcements for lever 1 is less than that for lever 2: Maximizing will therefore occur if the animal allocates all of its responses and time to lever 2 (0 percent for lever 1). To the right of the step at 0.5, the scheduled proportion of reinforcements for lever 1 is greater than that for lever 2: Maximizing will therefore occur if the animal allocates all of its responses and time to lever 1. Although exact maximizing did not occur, the data points fall somewhat near the step function.

Figure 14-4 shows for each rat the proportion of responses and time allocated to lever 1 as a function of the obtained proportion of reinforcers to lever 1.

Table 14-2. Sequence of RI x-minutes schedules for lever 2 for each subject in Experiment 3. For lever 1 a RI 2-minute schedule was always in effect. The same sequence of conditions was used in Experiments 4 and 5.

Animal	*1 (10 sessions)*	*2 (15 sessions)*	*3 (20 sessions)*	*4 (25 sessions)*	*5 (30 sessions)*
R1	2	4	1	8	0.5
R2	2	8	0.5	4	1
R3	2	0.5	8	1	4
R4	2	1	4	0.5	8

Figure 14-3. Experiment 3. The proportion of responses (*top*) and time (*bottom*) allocated to lever 1 as a function of the scheduled proportion of reinforcements to lever 1. The step function indicates maximizing.

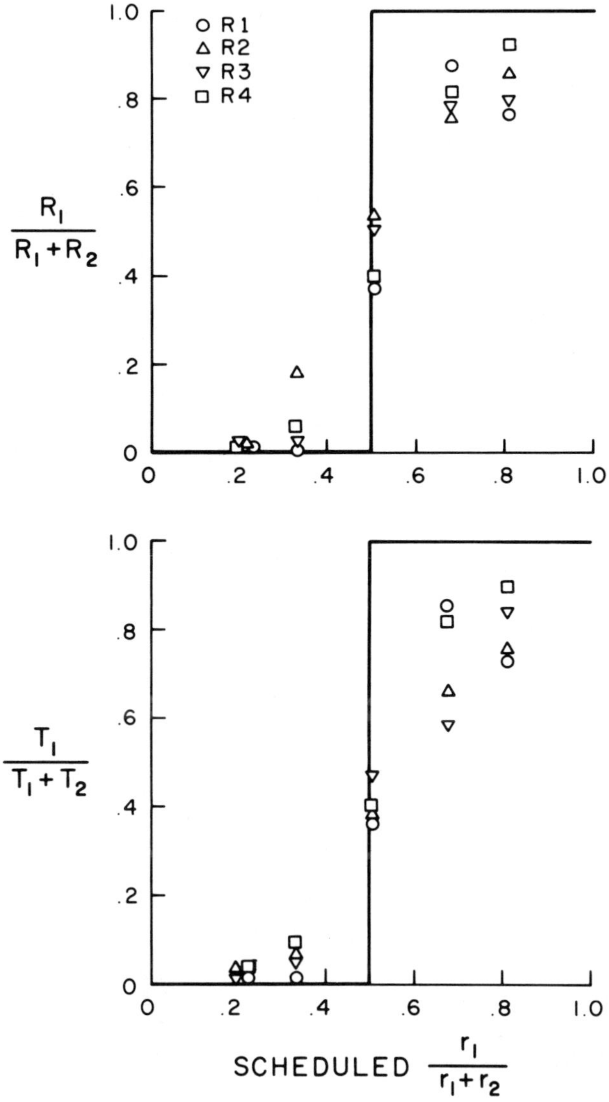

Table 14-3. Responses per minute on lever 1 $[R_1/(T_1 + T_2)]$ and lever 2 $[R_2/(T_1 + T_2)]$ and obtained reinforcements per minute on lever 1 $[r_1/(T_1 + T_2)]$ and lever 2 $[r_2/(T_1 + T_2)]$ at each of the scheduled proportion of reinforcers in Experiments 3-5.

Scheduled		*Responses*		*Reinforcements*	
$\frac{r_1}{r_1 + r_2}$	*Rat*	*Lever 1* $R_1/(T_1 + T_2)$	*Lever 2* $R_2/(T_1 + T_2)$	*Lever 1* $r_1/(T_1 + T_2)$	*Lever 2* $r_2/(T_1 + T_2)$
			Experiment 3		
0.20	1	0.2	51.1	0.00	1.82
	2	0.4	25.4	0.00	1.67
	3	0.2	13.2	0.00	1.49
	4	0.8	61.2	0.00	1.60
0.33	1	0.2	40.5	0.00	0.90
	2	3.1	13.3	0.04	0.82
	3	0.3	10.4	0.01	0.84
	4	2.5	37.9	0.06	0.86
0.50	1	10.4	17.2	0.28	0.33
	2	8.0	6.8	0.15	0.27
	3	3.2	3.0	0.21	0.23
	4	7.8	12.4	0.14	0.32
0.67	1	21.2	2.9	0.27	0.04
	2	11.3	3.5	0.34	0.06
	3	2.9	0.9	0.18	0.08
	4	29.1	6.3	0.42	0.05
0.80	1	13.9	4.0	0.32	0.03
	2	6.2	1.0	0.31	0.04
	3	4.5	1.1	0.29	0.03
	4	55.3	4.5	0.41	0.01
			Experiment 4		
0.20	5	12.7	34.3	0.40	1.66
	6	9.1	10.6	0.46	1.54
	7	11.6	25.8	0.45	1.73
	8	31.2	42.5	0.46	1.80
0.33	5	18.6	32.3	0.49	0.95
	6	30.8	41.3	0.44	0.89
	7	25.5	34.0	0.51	0.90
	8	27.1	19.3	0.48	0.96
0.50	5	11.4	19.2	0.43	0.41
	6	11.0	9.0	0.53	0.44
	7	18.3	14.0	0.56	0.45
	8	30.4	11.5	0.51	0.52
0.67	5	18.7	19.5	0.57	0.23
	6	19.4	10.7	0.51	0.28
	7	18.2	7.9	0.50	0.20
	8	40.5	11.3	0.51	0.22

Table 14-3. continued

Scheduled $\frac{r_1}{r_1 + r_2}$	Rat	Responses: Lever 1 $R_1/(T_1 + T_2)$	Responses: Lever 2 $R_2/(T_1 + T_2)$	Reinforcements: Lever 1 $r_1/(T_1 + T_2)$	Reinforcements: Lever 2 $r_2/(T_1 + T_2)$
		Experiment 4 (continued)			
0.80	5	42.2	13.0	0.48	0.13
	6	16.4	7.8	0.41	0.10
	7	25.4	11.8	0.53	0.09
	8	49.6	14.1	0.50	0.12
		Experiment 5			
0.20	9	17.6	52.5	0.41	1.78
	10	12.6	27.8	0.36	1.57
	11	8.7	41.2	0.35	1.87
	12	6.3	44.1	0.28	1.64
0.33	9	25.1	36.0	0.40	0.90
	10	5.3	6.9	0.26	0.39
	11	18.5	54.3	0.44	0.88
	12	21.5	29.4	0.42	0.88
0.50	9	26.6	17.6	0.49	0.42
	10	8.8	6.0	0.39	0.33
	11	17.0	17.5	0.42	0.42
	12	16.2	11.2	0.35	0.43
0.67	9	30.1	22.5	0.44	0.22
	10	13.3	5.0	0.37	0.16
	11	24.6	27.5	0.44	0.19
	12	20.9	7.5	0.43	0.17
0.80	9	39.6	11.1	0.44	0.10
	10	10.4	3.0	0.38	0.09
	11	18.3	11.3	0.41	0.10
	12	18.6	4.3	0.41	0.07

The solid diagonal lines represent the matching of the proportion of lever 1 responses and time to the obtained proportion of lever 1 reinforcements. The points fall close to the diagonal with the least squares regression lines indicated by the broken lines. The response-allocation regression line accounts for 96 percent of the variance; the time-allocation regression line accounts for 98 percent of the variance.

The results of Experiment 3, in which rats were responding on conc RI RI schedules, with the schedule for one alternative not in operation when the animal chose the other alternative, may be summarized as follows: There was approximate maximizing of the scheduled reinforcement proportion, and there was approximate matching of the obtained reinforcement proportion. The maximizing outcome has not been reported when animals are responding on concurrent

Figure 14-4. Experiment 3. The proportion of responses (*top*) and time (*bottom*) allocated to lever 1 as a function of the obtained proportion of reinforcements to lever 1. The broken lines are the least-squares regression lines.

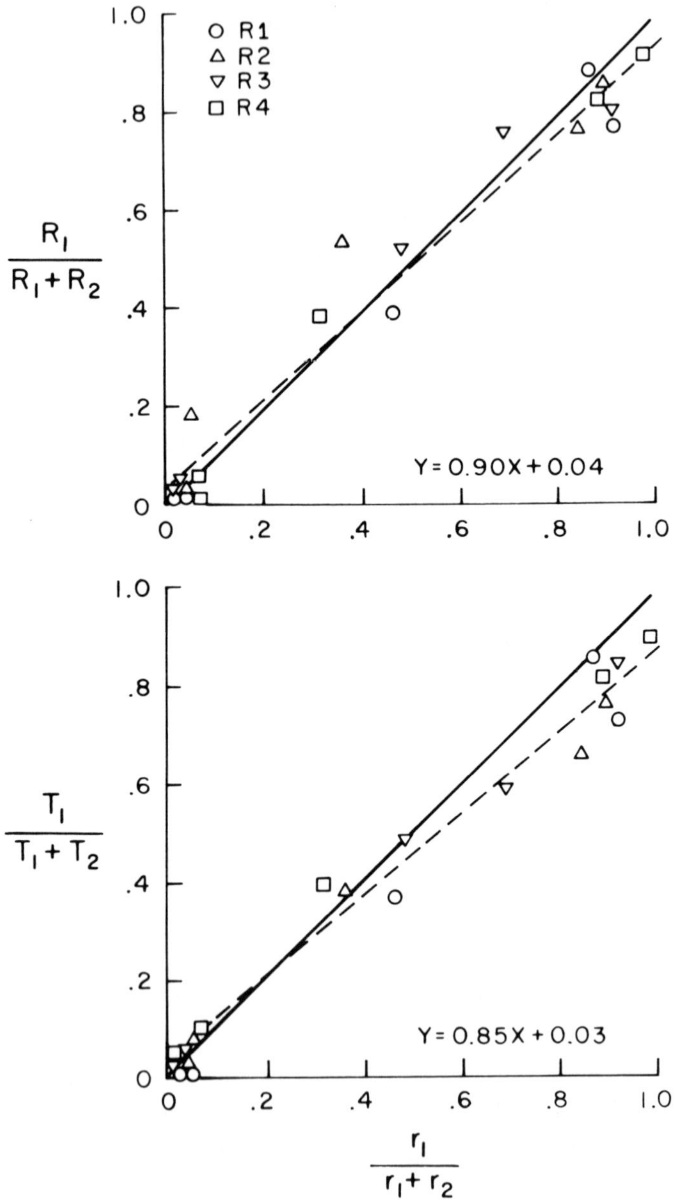

interval schedules with both schedules always in operation but only when they are responding on concurrent ratio schedules (Herrnstein, 1958; Herrnstein and Loveland, 1975; Rider, 1979). It should also be noted that under the present conditions the matching of the obtained reinforcement and choice proportions was obtained without the use of a COD. It was not necessary that a delay elapse following the animal's switch from one lever to the other lever before reinforcers be delivered. The use of a COD has often been necessary for matching to occur, however, when the schedules for both alternatives are always in operation (Catania, 1966; de Villiers, 1977).

Experiment 4

In Experiment 4 the procedure was exactly the same as in Experiment 3 except that the RI schedules for both levers were always in operation; that is, the tape for a lever ran whether or not the rat was working on that lever. Four naive rats served as subjects. Following magazine training and the shaping of both lever-press responses, the animals were exposed to the conc RI RI schedules shown in Table 14-2. The sequence of conditions for animals R5, R6, R7, and R8 corresponded, respectively, to the sequence of conditions for animals R1, R2, R3, and R4.

Table 14-3 gives the response rates and the obtained reinforcement rates at each of the scheduled proportion of reinforcers. Figure 14-5 shows for each rat the proportion of responses and time allocated to lever 1 as a function of the obtained proportion of reinforcers to lever 1. Large deviations from matching occurred, although less so with respect to time allocation. The time-allocation linear regression line accounts for more of the variance than the response-allocation linear regression line (87 percent vs. 66 percent).

Experiment 4 differed from Experiment 3 in only one respect: In the former the schedules were always in operation, but in the latter only one could be in effect at any time. Yet, whereas both matching and approximate maximizing occurred in Experiment 3, neither was obtained in Experiment 4.

Experiment 5

In Experiment 4 the schedules for both levers were in effect at all times, regardless of the behavior of the animal. Experiment 5 replicated these conditions except for the addition of a COD of 5 seconds. Four naive rats served as subjects. Following magazine training and the shaping of both lever-press responses, the animals were exposed to the conc RI RI schedules shown in Table 14-2. The sequence of conditions for animals R9, R10, R11, and R12 corresponded, respectively, to the sequence of conditions for animals R1, R2, R3, and R4.

Figure 14-5. Experiment 4. The proportion of responses (*top*) and time (*bottom*) allocated to lever 1 as a function of the obtained proportion of reinforcements to lever 1. The broken lines are the least-squares regression lines.

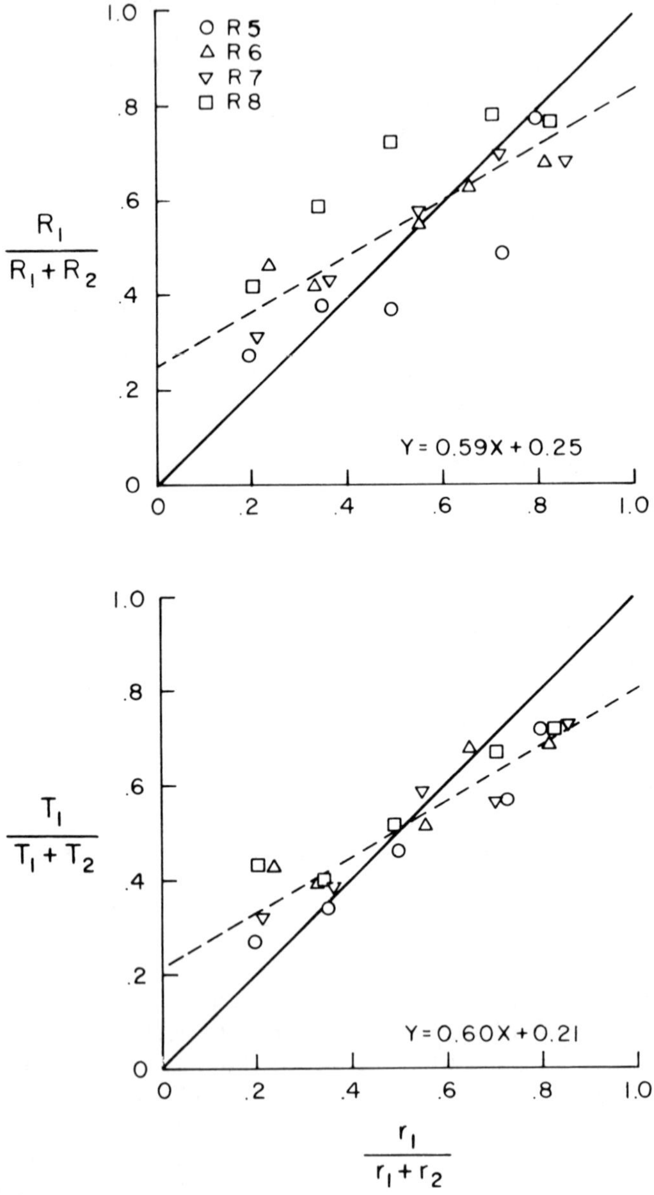

Figure 14-6. Experiment 5. The proportion of responses (*top*) and time (*bottom*) allocated to lever 1 as a function of the obtained proportion of reinforcements to lever 1. The broken lines are the least-squares regression lines.

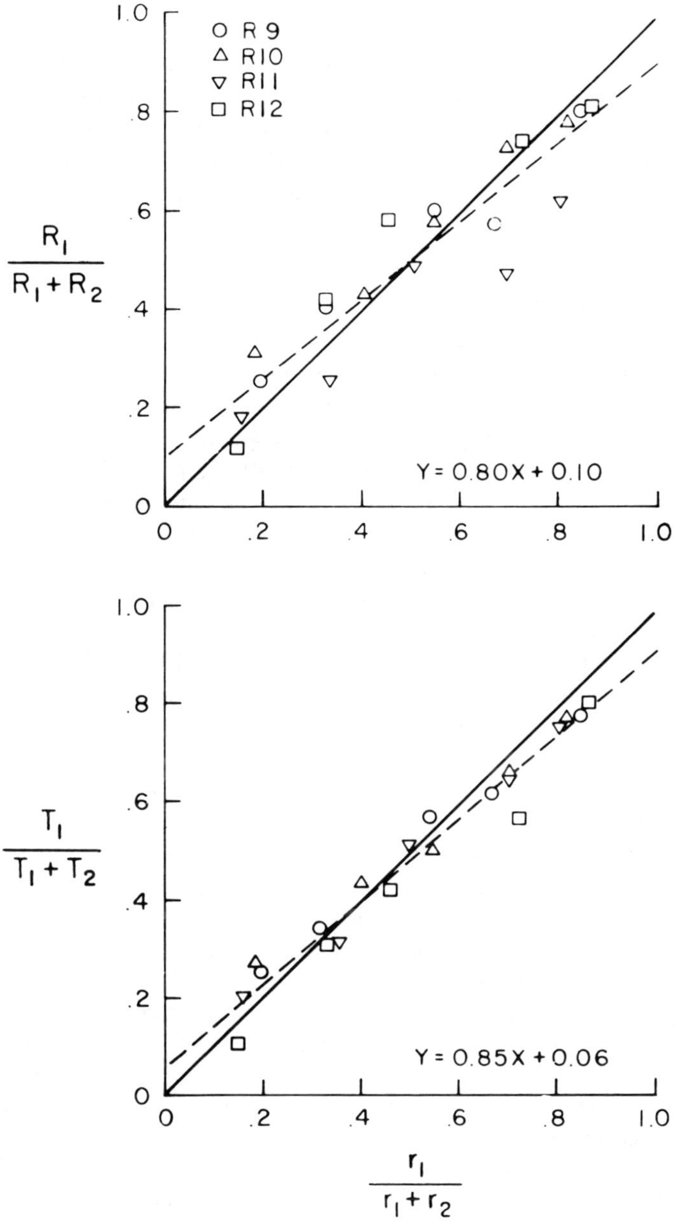

Table 14–3 provides the response rates and the reinforcement rates at each of the scheduled proportion of reinforcers. Figure 14–6 shows for each rat the proportion of responses and time allocated to lever 1 as a function of the obtained proportion of reinforcers to lever 1. The response-allocation linear regression line accounts for 85 percent of the variance; the time-allocation linear regression line accounts for 96 percent of the variance. As in Experiment 4, there were smaller deviations from matching for time allocation than for response allocation (cf. Baum, 1976; Norman and McSweeney, 1978, Shull and Pliskoff, 1967).

The present experiment differed from Experiment 4 in only one respect: In Experiment 5 a COD of 5 seconds was used, while no COD was employed in Experiment 4. Under these conditions, there was approximate response- and time-allocation matching.

DISCUSSION

It is apparent from these experiments that choice allocation between reinforcing and punishing situations is dependent upon whether the schedules for the alternatives are always in effect or whether the schedule for one alternative is not in effect when the animal has chosen the other alternative. If the schedule for one alternative is not in operation when the animals has chosen the other alternative, then both matching (Eqs. (14.1) and (14.2)) and approximate maximizing will occur between reinforcing situations (Experiment 3); between punishing situations, the matching relations of Eqs. (14.5) and (14.10) will occur (Experiment 1). If the schedules for both alternatives are always in operation, maximizing will not result and matching (Eqs. (14.1) and (14.2)) will occur between reinforcing situations only if a COD is present (Experiments 4 and 5); between punishing situations only minimizing will result (Experiment 2).

Although the presence of the COD affected choice allocation between reinforcing situations, its role in punishing situations was not determined. The matching relations that were discovered in Experiment 1 may be dependent upon the presence of the COD. Because the COD is a shock-free period of time, it is possible for an animal to avoid all of the shocks in both components if its changeover rate is large enough. Reinforcement studies have shown that as the value of the COD increases, the number of changeovers decreases (e.g., Shull and Pliskoff, 1967). Although it remains to be determined, it may be that as the value of the COD increases in punishing situations, the number of changeovers increases. The punishment matching relations were found when a COD of 2 seconds was used, but perhaps an animal would allocate all of its time to the component having the smaller punishment rate if no COD were used. If the COD does indeed affect choice allocation between punishing situations, its role would have to be represented in Eqs. (14.5) and (14.10).

It should be noted that in one respect the matching relations for reinforcement (Eqs. (14.1) and (14.2)) and one of the matching relations for punishment

(Eq. (14.10)) are comparable. If Eqs. (14.1) and (14.2) are expressed in ratio form, the general case is the following (de Villiers, 1977):

$$\frac{R_1}{R_2} \text{ or } \frac{T_1}{T_2} = a\left(\frac{r_1}{r_2}\right)^b , \tag{14.11}$$

where a and b are constants. The parameter a signifies factors that result in a bias to one of the alternatives. The parameter b represents the sensitivity of choice allocation to changes in the ratio of the reinforcers. Eqs. (14.1) and (14.2) are special cases of Eq. (14.11) where $a = 1$ and $b = 1$. Similarly, if Eq. (14.10) is expressed in ratio form, the general case is the following:

$$\frac{T_1}{T_2} = a\left(\frac{D_1}{D_2}\right)^{.5} . \tag{14.12}$$

Although the value of b is 0.5 in the case of punishment, the value of b is typically in the range of 0.8 to 1.0 (Baum, 1979).

One approach in accounting for all of the above results involving reinforcing and punishing situations might be based on some sort of optimizing process (e.g., Rachlin, Green, Kagel, and Battalio, 1976; Shimp, 1969; Staddon and Motheral, 1978). Between reinforcing situations, if the animals respond to the alternative with the shorter expected time to reinforcement, at least the qualitative nature of the results observed would follow. Let us consider a specific case in which the random interval to food on the left lever is 2 minutes and the random interval to food on the right lever is 1 minute. At the beginning of a session the rat would respond predominantly on the right lever. In Experiment 3, where the expected time to reinforcement remained constant throughout the session, the rat would continue to respond predominantly on the right lever throughout the session. This could result in maximization, as observed. In Experiment 4, however, the expected time to reinforcement on the left lever decreases as the rat spends more time on the right lever. This would produce less overmatching; in fact, undermatching was observed. A COD, of course, would reduce this tendency; this was observed in Experiment 5.

In aversive situations, like appetitive situations, if the animals respond to the alternative with the longer expected time to shock, again the qualitative nature of the results observed would follow. Let us consider a specific case in which the random interval to shock in one component is 2 minutes and the random interval to shock in the other component is 1 minute. At the beginning of a session under both the "tape stops" and "tape moves" procedures, the rat would remain predominantly in the 2-minute component. In the tape-stops procedure, where the expected time to shock remains constant throughout the session, the rat would continue to remain predominantly in the 2-minute component throughout the session. This could result in the observed matching (Experiment 1). In the tape-moves procedure, however, the expected time to shock in the 1-minute component decreases as the rat spends more time in the 2-minute component. This would lead to the allocating of time almost exclusively to the 2-minute component (that is, the observed minimization, Experiment 2). Thus, the same process of time estimation and choice of the better time (short for food and long for shock) would lead to more exclusive responding on one lever for the tape-stops reinforcement procedure than for the tape-moves reinforcement procedure, but more exclusive time allocation to one component for the tape-moves shock procedure than for the tape-stops shock procedure. As in the case of positive reinforcement, the approach described may be more promising than a list of special response- and time-allocation rules for different procedures.

REFERENCES

Baum, W.M. 1973. The correlation-based law of effect. *Journal of the Experimental Analysis of Behavior 20*: 137–53.

_____. 1976. Time-based and count-based measurement of preference. *Journal of the Experimental Analysis of Behavior 26*: 27–35.

_____. 1979. Matching, undermatching, and overmatching in studies of choice. *Journal of the Experimental Analysis of Behavior 32*: 269–81.

Catania, A.C. 1966. Concurrent operants. In W.K. Honig, ed., *Operant Behavior: Areas of Research and Application*, New York: Appleton–Century–Crofts.

Deluty, M.Z. 1976. Choice and the rate of punishment in concurrent schedules. *Journal of the Experimental Analysis of Behavior 25*: 75–80.

Deluty, M.Z., and R.M. Church. 1978. Time-allocation matching between punishing situations. *Journal of the Experimental Analysis of Behavior 29*: 191–98.

de Villiers, P.A. 1977. Choice in concurrent schedules and a quantitative formulation of the law of effect. In W.K. Honig and J.E.R. Staddon, eds., *Handbook of Operant Behavior*, Englewood Cliffs, N.J.: Prentice–Hall.

_____. 1980. Toward a quantitative theory of punishment. *Journal of the Experimental Analysis of Behavior 33*: 15–25.

Fallon, D.; D.M. Thompson; and M.E. Schild. 1965. Concurrent food- and water-reinforced responding under food, water, and food-plus-water deprivation. *Psychological Reports 16*: 1305–11.

Farley, J. 1980. Reinforcement and punishment effects in concurrent schedules: A test of two models. *Journal of the Experimental Analysis of Behavior 33*: 311–26.

Farley, J., and E. Fantino. 1978. The symmetrical law of effect and the matching relation in choice behavior. *Journal of the Experimental Analysis of Behavior 29*: 37–60.

Findley, J.D. 1958. Preference and switching under concurrent scheduling. *Journal of the Experimental Analysis of Behavior 1*: 123–44.

Herrnstein, R.J. 1958. Some factors influencing behavior in a two-response situation. *Transactions of the New York Academy of Sciences 21*: 25–45.

——. 1970. On the law of effect. *Journal of the Experimental Analysis of Behavior 13*; 243–66.

——. 1971. Quantitative hedonism. *Journal of Psychiatric Research 8*: 399–412.

Herrnstein, R.J., and D.H. Loveland. 1975. Maximizing and matching on concurrent ratio schedules. *Journal of the Experimental Analysis of Behavior 24*: 107–16.

Killeen, P. 1972. A yoked-chamber comparison of concurrent and multiple schedules. *Journal of the Experimental Analysis of Behavior 18*: 13–22.

Myers, D.L., and L.E. Myers. 1977. Undermatching: A reappraisal of performance on concurrent variable-interval schedules of reinforcement. *Journal of the Experimental Analysis of Behavior 27*: 203–14.

Norman, W.D., and F.K. McSweeney. 1978. Matching, contrast, and equalizing in the concurrent lever-press responding of rats. *Journal of the Experimental Analysis of Behavior 29*: 453–62.

Rachlin, H.; L. Green; J.H. Kagel; and R.C. Battalio. 1976. Economic demand theory and psychological studies of choice. In G.H. Bower, ed., *The Psychology of Learning and Motivation*, vol. 10, New York: Academic Press.

Rider, D.P. 1979. Concurrent ratio schedules: Fixed vs. variable response requirements. *Journal of the Experimental Analysis of Behavior 31*: 225–37.

Shimp, C.P. 1969. Optimal behavior in free-operant experiments. *Psychological Review 76*: 97–112.

Shull, R.L., and S.S. Pliskoff. 1967. Changeover delay and concurrent schedules: Some effects on relative performance measures. *Journal of the Experimental Analysis of Behavior 10*: 517–27.

Staddon, J.E.R., and S. Motheral. 1978. On matching and maximizing in operant choice experiments. *Psychological Review 85*: 436–44.

15 TOWARD A QUANTITATIVE THEORY OF PUNISHMENT

Peter A. de Villiers

Despite the importance of aversive contingencies in the control of behavior and the prevalence of punishment procedures in behavior therapy and behavioral pharmacology, progress in specifying the interaction between punishment and positive reinforcement in quantitative terms has lagged behind the development of quantitative models of positive reinforcement alone. The theoretical issue that dominated much of the research on punishment in the 1960s and early 1970s concerned the relative merits of accounting for response suppression in terms of competing avoidance or escape responses, or in terms of a negative version of the law of effect. Avoidance theories of punishment suggest that response suppression is a by-product of the strengthening of competing behaviors. In a punishment procedure proprioceptive or contextual stimuli associated with responding are paired with the punisher and become conditioned aversive (or fear-provoking) stimuli. Any behavior that terminates these aversive stimuli is negatively reinforced and increases in frequency. The punished response is suppressed because the escape responses interfere with its occurrence (Dinsmoor, 1954; Mowrer, 1960; Solomon, 1964). An alternative competing response theory suggests that the avoidance or escape responses are species-specific defense reactions elicited by stimuli associated with aversive stimulation, and there need be little or no instrumental conditioning involved, but the suppression of the punished response is still the result of the increase in other behaviors (Bolles, 1972). In contrast, the negative law of effect holds that aversive stimuli directly inhibit the responses on which they are contingent, a process analogous, but opposite in effect, to the direct strengthening of responding by positive reinforcement (Thorndike, 1911; Estes, 1969; Rachlin and Herrnstein, 1969).

Two different lines of research have attempted to test predictions made by the opposing theories. The first of these compared the effects of contingent and noncontingent aversive stimulation on responding and choice. Several studies established the importance of the contingency between responding and the punisher in producing response suppression (Azrin, 1956; Rachlin, 1967; Schuster and Rachlin, 1968; Williams, 1969), suggesting that incompatible behaviors elicited by the aversive stimulus or by contextual stimuli associated with it were not sufficient to account for the suppression of the punished response. For example, Schuster and Rachlin (1968) investigated pigeons' choice in concurrent chain schedules with terminal links that contained either response-contingent or noncontingent electric shock as well as equal variable-interval (VI) schedules of food reinforcement. When the frequency of shock was equal in the two terminal links, key pecking was substantially more suppressed in the link with contingent shock than in the link with noncontingent shock. On the other hand, choice in the initial link of the chain was determined solely by the relative frequency of shock in the terminal links, not by the difference in contingency. As long as shocks were equally frequent in the response-contingent and noncontingent conditions the pigeons were indifferent between them. These results are in keeping with a negative law of effect that stresses the importance of the contingency between responding and the aversive stimulus and the symmetry between positive reinforcement and punishment. Choice in concurrent and concurrent chain schedules of food reinforcement also seems to be determined primarily by the relative frequency of food presentations and not by whether it is contingent on responding or not, and contingent food maintains a rate of responding far higher than that maintained by free food (Autor, 1969; Bauman, Shull, and Brownstein, 1975).

Schuster and Rachlin argue that their findings contradict an apparent prediction of the avoidance theory of punishment. The greater suppression of responding with contingent rather than noncontingent shock suggests that the crucial conditioned aversive stimuli postulated by the theory must be proprioceptive stimuli associated with responding. If this is so the pigeons would be expected to prefer the noncontingent shock condition in the choice between terminal links. First, there would be conditioned aversive feedback from every response in the contingent shock condition that should add to the aversiveness of that condition as a whole, making it more aversive than the noncontingent condition when the shock frequency in each link was equal. Second, the responses that produced shock in the terminal link and chose that alternative in the initial link were pecks on the same key, so it might be expected that aversive proprioceptive feedback would suppress responding on that key in both links of the chain. In fact, terminal link response rates and initial link choices varied independently in the experiment, and contingent and noncontingent shocks had equal effects on choice. Nevertheless, as Schuster and Rachlin themselves point out, an avoidance theory of punishment can account for the results if it makes certain assumptions.

In particular, the theory must assume that the pigeons discriminate clearly between pecking at the white key in the choice link and pecking at the red (or green) key in the terminal link. Then any aversiveness associated with pecking in the terminal link would not be expected to generalize to the choice link. This is not an implausible assumption to make. Second, response-produced feedback would acquire more conditioned aversiveness in the contingent-shock condition because it would be a more valid predictor of shock than situational cues. In the noncontingent condition, however, the situational cues would be the better predictors of shock and they would acquire more associative strength than response feedback. For the pigeons to be indifferent between the two conditions, the greater aversiveness of response-produced stimuli must be balanced by the lower frequency of responding in the contingent condition and the greater aversiveness of contextual stimuli in the noncontingent condition. For this to occur across the whole range of shock frequencies examined by Schuster and Rachlin would be serendipitous to say the least. Alternatively the theory must argue that the value of the discriminative stimuli for each terminal link depends solely on the frequency of food and shock associated with them. But then conditioned aversive stimuli do not add to the aversiveness of a situation in the way that primary aversive stimuli do, and the parallel between conditioned and unconditioned aversive stimuli assumed by the theory breaks down. In short, an avoidance theory of punishment can accommodate itself to these results in a somewhat tortuous and ad hoc fashion but not without some cost.

The second line of research was directed at a prediction that both proponents and opponents agree is made by the avoidance theory of punishment. The theory regards the behavior leading to punishment as a chain of several responses, each one serving as or providing a discriminative stimulus for the next response in the chain. In these terms a simple lever press might be decomposed into such responses as approaching the lever, lifting a paw, touching the lever, pressing it down, and so on. Under conditions of punishment stimuli associated with components of the chain furthest from the final, punished response would acquire the least aversiveness. Subjects would continue to initiate the chain of behavior leading to reinforcement but would tend to terminate the chain at the point where the acquired aversiveness of that component outweighed its conditioned positive reinforcing properties. If only response-produced stimuli become conditioned aversive stimuli in a punishment procedure, then response initiations should not decrease in frequency no matter how intense the punishment. The initial response in the chain cannot be displaced by an escape response, because there is no aversive stimulus to be terminated before the first response in the chain is made. In practice, however, it is difficult to determine exactly where the chain of behavior begins, so the first response that the experimenter measures could also be suppressed at high intensities of punishment. Nevertheless, it should be possible to suppress substantially completions of the response chain with little if any suppression of the initial response, and punishment should produce a great many abortive responses in which the chain is begun but not completed.

The results of experiments that set out to test this prediction have been somewhat equivocal. Mansfield and Rachlin (1970) investigated the effects of different punishment intensities, extinction, and satiation on a response chain that consisted of a peck at a green key followed within 10 seconds by a peck at a red key. Illumination of the red key required an initial peck at the green key, and green-red peck sequence were both reinforced and punished. The frequency of sequence initiations and completions declined together under all three experimental manipulations, and the proportion of initiations to completions did not change as responding decreased. There was therefore no evidence of an increase in aborted sequences when the final response was punished.

In contrast, Carman (cited by Dinsmoor, 1977) and Millenson and McMillan (1975) observed increases in the percentage of aborted response sequences when they punished a lever-press chain. Carman measured lever touches and presses on a standard VI 1-minute schedule of positive reinforcement. When lever presses were punished on a VI 1-minute schedule, both touches and criterion presses decreased in frequency, but the proportion of touches to presses increased. Millenson and McMillan shaped rats to hold a lever down for a minimum of 10 seconds in order to receive a food pellet. When lever releases following a 10-second press were both reinforced by food and punished by brief 0.3 milliampere shocks, the number of criterion lever holds declined but the rate of initial lever-presses remained unchanged. During extinction both the frequency of response initiations and the holding times declined gradually and simultaneously, and partial satiation by prefeeding reduced the rate of response initiations without having any effect on holding times. Thus the percentage of aborted response chains increased during punishment but not during extinction or satiation. Millenson and McMillan account for the discrepancy between their results and those of Mansfield and Rachlin on two grounds. In the Mansfield and Rachlin experiment both the initiation and maintenance of the chain of responses was under the control of exteroceptive rather than proprioceptive stimuli, since particular configurations of the key lights cued the appropriate times for pecks at the green and red keys. Furthermore, the extra green-key pecks that Mansfield and Rachlin counted as abortive sequences did not set the animal back at an earlier point in the sequence and thus cancel earlier members of the chain in the way early releases did in the Millenson and McMillan procedure.

In an evaluation of the status of the avoidance theory of punishment. Dinsmoor reviewed the studies on abortive response chains (but not Schuster and Rachlin's findings on choice), and concluded: "Thus a certain amount of empirical support is now available supporting the avoidance theory of punishment, and after two decades it still appears quite viable" (1977:94). However, as Millenson and McMillan noted, although their results are predicted by the avoidance theory, they are equally compatible with the negative law of effect if it includes a delay of punishment gradient analogous to the delay of reinforcement gradient for positive reinforcement (Renner, 1966). Criterion releases are closest to the punisher in time, so they would be the most affected elements in the chain; response components further and further from the punisher would be less suppressed. Suppression of criterion holds, but little or no suppression of press initiations would lead to the increase in abortive responses observed by Millenson and McMillan. A similar explanation could be given for Carman's findings.

In short, with a bit of stretching each theory could account for the results of both lines of research and it became difficult to distinguish between them on empirical grounds. Choice between the theories became more an issue of theoretical preference, and arguments between the theorists tended to stress theoretical purity and conceptual simplicity rather than the confirmation or disconfirmation of empirical predictions. The avoidance theory has the apparent parsimony of appealing to just one strengthening process of instrumental conditioning but at the cost of introducing unobserved classically conditioned aversive stimuli. The negative law of effect does not appeal to any such unobserved intervening process, but it invokes two primary principles of reinforcement, one concerned with the strengthening of behavior, the other with its inhibition.

Since the 1960s, however, three quantitative formulations of the effects of punishment have emerged, two of which in particular can be considered the lineal descendants of competing response theories and the negative law of effect. In certain situations these models make quite distinct predictions. It therefore becomes possible to examine the implications of the opposing theories of punishment in quantitative form and so distinguish between them more clearly on empirical grounds.

THREE MODELS

All three formulations attempt to incorporate the effects of punishment within the set of equations generalized by Herrnstein (1970, 1974) from the matching relation. The matching equation states that

$$\frac{R_1}{R_1 + R_2} = \frac{r_1}{r_1 + r_2}, \qquad (15.1)$$

where R_1 and R_2 are response frequencies for two alternatives, and r_1 and r_2 are their respective reinforcement rates. To account for absolute response rates in concurrent VI schedules, Herrnstein proposed the equation

$$R_1 = \frac{kr_1}{r_1 + r_2 + r_e}, \qquad (15.2)$$

where k represents the maximal response rate when there is no reinforcement from sources other than r_1, and r_e is the sum of reinforcement from all sources other than those specified by the experimenter (that is, r_1 and r_2). In the case of a single VI schedule, the equation becomes

$$R_1 = \frac{kr_1}{r_1 + r_e} \tag{15.3}$$

since there is only one specified reinforcer, r_1. These equations capture the relation between several measures of response strength and a wide range of reinforcement parameters (de Villiers and Herrnstein, 1976; de Villiers, 1977).

Bradshaw, Szabadi, and Bevan (1977, 1978) worked with Eq. (15.3). They studied human subjects pressing a button for monetary reinforcers on five different VI schedules. Punishing responding by subtracting money according to a VR 37 schedule decreased the value of the k parameter but increased the size of r_e. Punishing responding according to a VI 170-second schedule, on the other hand, only increased r_e; k remained roughly constant. If r_e is interpreted as a measure of the reinforcement for competing behaviors, as Herrnstein (1974) suggests, an increase in r_e captures the essential aspects of competing response theories of punishment.

Deluty (1976) based his theory on the finding that when one response is punished, alternative unpunished responses increase in frequency (Azrin and Holz, 1966). For concurrent VI schedules involving both reinforcement and punishment, he proposed the following equation:

$$\frac{R_1}{R_1 + R_2} = \frac{r_1 + p_2}{r_1 + p_2 + r_2 + p_1}, \tag{15.4}$$

where p_1 and p_2 are frequencies of punishment for the two alternatives. In an experiment with rats Deluty maintained the relative frequency of food reinforcement constant at 0.50 but punished responding on the two levers according to different VI schedules of shock presentation. Over a range of shock frequencies Eq. (15.4) provided an adequate fit to the observed changes in relative response rate. To the extent that the concurrent responses compete with each other, this too can be considered a formalization of a competing response theory (Farley, 1980): The punished response is suppressed because an alternative response increases in value. But the two competing response theories described in the introduction are more appropriately represented by an increase in the r_e parameters of Eq. (15.2) and (15.3) than by Deluty's equation. The competing responses postulated by the theories are not normally conceived of as being alternative ways of obtaining the same reinforcer as the punished response.

Third, Estes (1969) suggested a simple subtractive model of the interaction between reinforcement and punishment. It was adopted in Millenson and de Villiers (1972) to account for conditioned suppression with signaled noncontingent shock and in de Villiers, (1977) to explain the effects of punishment on concurrent VI schedules of positive reinforcement. In the latter the following equation was proposed for the case in which concurrent responses are both reinforced and punished.

$$\frac{R_1}{R_1 + R_2} = \frac{r_1 - p_1}{r_1 - p_1 + r_2 - p_2} \quad . \tag{15.5}$$

Farley and Fantino (1978) put forward a ratio version of the same equation but refined it by adding a scaling parameter c that equates the value of one positive reinforcer and one punisher. Thus,

$$\frac{R_1}{R_2} = \frac{r_1 - cp_1}{r_2 - cp_2} \quad . \tag{15.6}$$

They suggest that the subtractive model quantifies the negative law of effect since punishers act directly on the response on which they are contingent, just like positive reinforcers but with the opposite effect. The response is suppressed because its net reinforcing value is directly reduced.

These three models of the effects of VI punishment make different predictions about choice in concurrent schedules with unequal rates of positive reinforcement but equal rates of VI punishment for the two alternatives. Since r_e is usually assumed to be the same for both alternatives in a concurrent schedule (Herrnstein, 1970; de Villiers, 1977), increases in r_e alone result in a decrease in the absolute response rates but no change in the relative response rates, which should continue to match the obtained relative reinforcement frequency. The denominators in Eq. (15.2) get larger, but they remain equal, so they still cancel out to produce the matching equation. Deluty's equation predicts a shift in preference toward indifference between the two alternatives, that is, toward undermatching between relative response rates and relative reinforcement frequencies. When $p_1 = p_2$ and $r_1 > r_2$,

$$\frac{r_1}{r_1 + r_2} > \frac{r_1 + p_2}{r_1 + p_2 + r_2 + p_1} \quad . \tag{15.7}$$

In contrast, the subtractive model predicts overmatching, a shift in preference toward the schedule with the higher reinforcement rate, since

$$\frac{r_1}{r_1 + r_2} < \frac{r_1 - p_1}{r_1 - p_1 + r_2 - p_2} \tag{15.8}$$

when $p_1 = p_2$, $r_1 > r_2$ and $r_2 \neq 0$. The ratio versions of Eqs. (15.4) and (15.5), whether or not they include the exchange parameter c, make the same contrasting predictions as long as $c > 0$.

EMPIRICAL SUPPORT FOR THE SUBTRACTIVE MODEL

Two papers have presented results that support the subtractive model. Two experiments with pigeons are described in de Villiers (1980). In the first of these, three pigeons responded on a concurrent VI 1-minute VI 3-minute schedule for grain. Intermittent punishment by brief electric shocks was then arranged by a single VI 30-second schedule and assigned with an equal probability to each key. A 3-second changeover delay (COD) was in effect for the food reinforcement throughout the experiment, but there was no COD for the punishment. In the absence of punishment the pigeons closely matched relative response rates and time allocation (cumulative interchangeover time) to the relative reinforcement frequency of 0.75. With increasing punishment intensity, however, the relative response and time distributions deviated more and more from matching toward the key associated with the higher frequency of food. At most intensities of shock response rate on the VI 3-minute schedule was substantially more suppressed than responding on the VI 1-minute schedule.

The second experiment was similar to the first except that the baseline concurrent schedule used a single-VI scheduling procedure (Stubbs and Pliskoff, 1969). A single VI 40-second schedule arranged food reinforcers that were then assigned with a probability of 0.75 to the left key and 0.25 to the right key. A VI 30-second schedule programmed the punishers with equal probability of the assignment to each key. Again a 3-second COD was in effect for the food schedule but not for the punishment schedule. As in the first experiment, the pigeons' preference shifted toward the key with the greater probability of reinforcement as the punishment intensity increased, and all three birds overmatched their relative response rate to the relative frequency of food. The overmatching is apparent in Figure 15–1, which also shows the differential suppression of the two responses with increasing shock intensity. Two of the pigeons were then exposed to a concurrent VI schedule in which the grain reinforcement as well as the punishment was assigned with an equal probability to the two alternatives. Despite substantial suppression of responding, no change in the baseline preference was observed during the punishment conditions of this procedure. The shift in preference when reinforcement assignments were unequal can therefore not be explained by the introduction of a response bias toward one key during the punishment conditions.

In a test of the same predictions of the subtractive model and Deluty's equation, Farley (1980: experiment 1) obtained shifts in preference comparable to those reported in de Villiers (1977, 1980). Three pigeons were exposed to five pairs of concurrent VI reinforcement schedules. In the punishment condition a VI 2-minute schedule of brief 6-milliampere shocks was superimposed on each concurrent schedule pair. Punishers and reinforcers could not occur within 3 seconds of each other, and there was a 3-second COD on both the reinforcement

Figure 15-1. Deviation of relative response rates (*top*) and relative time distribution (*middle*) from the obtained relative reinforcement frequencies as a function of punishment intensity. Positive deviations represent overmatching; negative deviations represent undermatching. Relative reinforcement frequencies were unequal, but relative rates of punishment were equal on the two keys. (*Bottom*) The suppression ratios for responding on each key calculated in terms of $B/(A+B)$, where B is response rate in a particular punishment condition and A is response on that key in the baseline condition. (From de Villiers, 1980.)

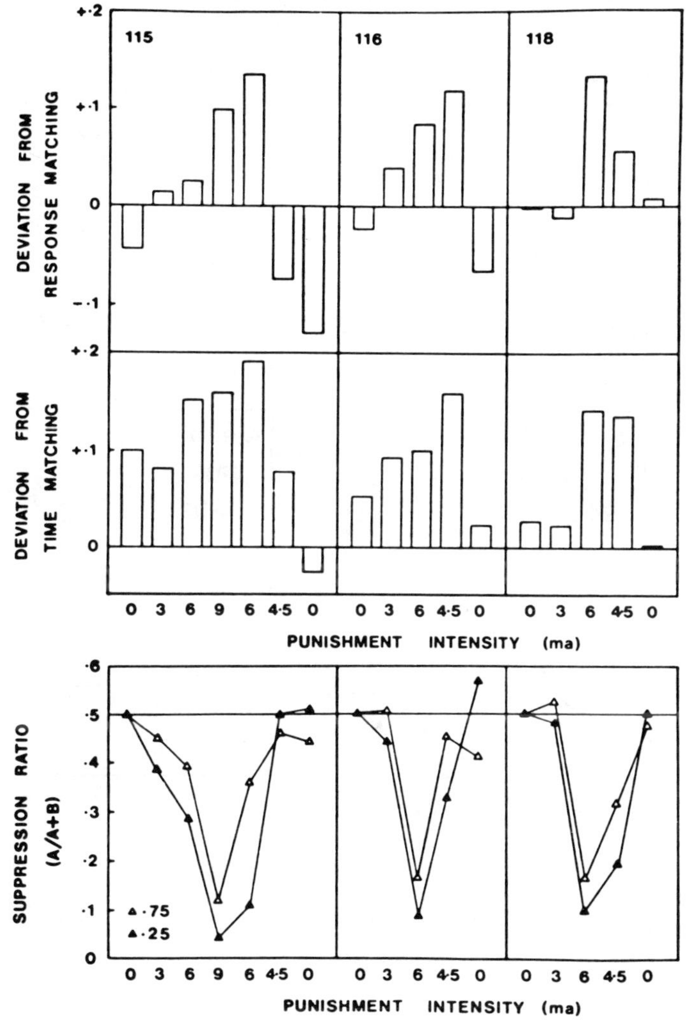

Figure 15-2. Analysis of baseline (filled circles, solid lines) and punishment (open circles, broken lines) data from Farley (1980: experiment 1) in terms of least-squares fit linear functions relating log response ratios to log reinforcement ratios. The percentage of the data variance accounted for is given below each function.

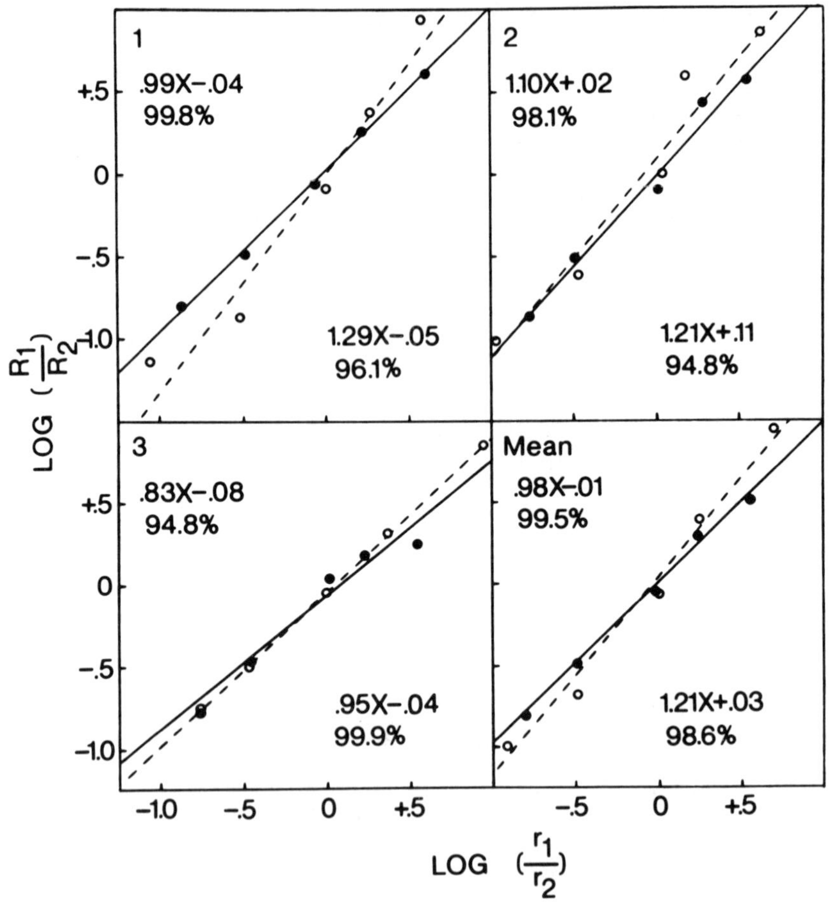

and the punishment schedule. Figure 15-2 plots the logarithm of the response ratios against the logarithm of the obtained positive reinforcement ratios from the baseline and punishment conditions. When the animal is matching, the slope of this function is 1.0. Overmatching is represented by a slope > 1.0; undermatching by a slope < 1.0. An intercept greater than or less than zero indicates a bias toward one or another of the schedules (Baum, 1974; de Villiers, 1977). All of the pigeons shifted toward more extreme preferences or overmatching, in keeping with the prediction made by the subtractive model embodied in Eqs. (15.5) and (15.6). For P1 and for the mean of the three birds the slope of the linear regression function in the punishment condition was significantly higher than the baseline slope ($p < 0.025$ in each case), while for the other two birds the change in slope was in the same direction though not statistically significant.

The absolute response rate equation derived from Eq. (15.5) (including a scaling parameter on the punisher) is

$$R_1 = \frac{k(r_1 - cp_1)}{(r_1 - cp_1) + (r_2 - cp_2) + r_e} . \qquad (15.9)$$

The corresponding equation derived from Deluty's model (Eq. (15.4)) would be

$$R_1 = \frac{k(r_1 + cp_2)}{(r_1 + cp_2) + (r_2 + cp_1) + r_e} . \qquad (15.10)$$

By examining the directional derivatives of these equations, Farley (1980) found several conditions under which they made contrasting predictions. For instance, if the rate of reinforcement for the two alternatives is kept equal and constant but the absolute rate of punishment increases, Eq. (15.9) predicts that the absolute response rate for each alternative should decline; Eq. (15.10) predicts that it should increase. Similarly, if r_1 and r_2 are unequal but are maintained in the same proportion to each other while the absolute frequency of reinforcement is increased, and the punishment frequencies remain constant, Eq. (15.9) predicts that the subject's preference for the denser reinforcement schedule should decrease. Equation (15.10) predicts that the preference for the denser schedule should become more extreme. For each of these predictions Farley demonstrated that the change in the subject's response rate and preference is in keeping with Eq. (15.9), the subtractive model, and directly counter to the predictions of Deluty's model.

All of these results suggest that the subtractive model is the most appropriate account of the interaction between reinforcement and punishment. In de Villiers (1980) I demonstrated that Eq. (15.5) also fits Deluty's (1976) data, but Deluty's equation (Eq. (15.10)) cannot account for the shifts in preference observed by both myself and Farley, nor for the results of Farley's other two experiments. A change in the value of the k and r_e parameters of Eq. (15.2) cannot account for the overmatching found in my experiments and in Farley's experiments either, unless the parameter values are allowed to differ between the two alternatives in the concurrent schedule. But unless there is a clear difference in the context in which each response is emitted (as, for example, when the manipulandums are very different), there is no good reason for assuming that the two parameters differ across the responses and several reasons for assuming that they are the same. First, if k and r_e are allowed to differ across concurrent responses, there are four instead of two free parameters when Eq. (15.2) is fitted to absolute response rates from concurrent schedules; two for R_1 and two for R_2. But as shown in Herrnstein, 1970, and de Villiers, 1977, the equation provides an excellent fit to the data from concurrent VI schedules when a single value is calculated for each of the parameters. Second, in fitting Eq. (15.3) to the response rates from several single VI schedules the experimenter must assume that k and r_e are the same for the different VI schedules, or the equation becomes too awkward to apply. Similarly, the parameter values are plausibly assumed to be the same for both responses in a two-response concurrent schedule when the two responses are the same and there are different VI schedules for each alternative. In my experiments the pigeons responded on a two-key concurrent VI schedule and the punishment frequency was equal for the two alternatives, so there is no principled reason for assuming that k or r_e differed between the two responses, yet substantial overmatching was observed in the punishment conditions.

Nevertheless, even if k and r_e are the same for each alternative, the question remains whether the value of r_e increases during the punishment condition, as predicted by a competing response theory. Equation (15.9) could not sensibly be tested by the absolute response rates from my experiments since too few data points were collected from any one intensity and frequency of punishment, but Farley studied the effects of a fairly constant relative and absolute frequency of punishment on five different concurrent VI schedules. I therefore analyzed the response rates from the baseline and punishment conditions of Farley's first experiment in terms of Eqs. (15.2) (baseline) and (15.9) (punishment). For the punishment condition Eq. (15.9) was fitted three times; once with k and r_e fixed at their best fit values from the baseline, once with k kept constant but with r_e free to vary to its best fit value for the punishment data, and once with both parameters varying until a least-squares fit was obtained. For each pigeon the scaling parameter c was set at the value that returned the slope of the linear function relating log response ratios to log reinforcement ratios to its baseline value shown in Figure 15-2. Since this represented the pigeons' baseline preference between the reinforcement schedules in the absence of punishment, it was used as the standard on which to equate reinforcers and punishers. The analysis was performed by a computer program using an iterative least-squares fit procedure (de Villiers and Herrnstein, 1976); the results are given in Table 15-1.

For P_2 the data was also analyzed with c set to a value that produced a slope of 1.0, assuming matching to be the norm for scaling the reinforcers and punish-

Table 15-1. Parameter values and percentage of the data variance accounted for by least-squares fits of Eqs. (15.3) and (15.9) to absolute response rates from Farley (1980: experiment 1). For the punishment condition the equation was fitted three times, once with the parameters kept constant at their baseline values, once with k kept constant but r_e free to vary, and once with both parameters free to vary. The exchange constant c was set at a value that produced the baseline slope in the function relating log response ratio to log reinforcement ratio. k is given in responses per minute; r_e is scaled as reinforcements per hour.

			Punishment		
Subject	*Parameter*	*Baseline*	*Both Parameters Constant*	*k Constant*	*Both Parameters Varying*
P 1	k	155.9	155.9	155.9	155.9
(c = .11)	r_e	34.8	34.8	48.0	11.4
		99.5 %	91.6 %	96.7 %	97.1 %
P 2	k	60.5	60.5	60.5	49.0
(c = .06)	r_e	0.006	0.006	15.0	0.09
		96.7 %	69.7 %	90.9 %	91.4 %
(c = .10)[a]	k	60.5	60.5	60.5	48.0
	r_e	0.006	0.006	14.4	0.08
		96.7 %	70.2 %	91.2 %	92.0 %
P 3	k	54.6	54.6	54.6	40.3
(c = .07)	r_e	5.4	5.4	20.4	0.06
		91.3 %	72.6 %	96.6 %	98.4 %
Mean	k	84.1	84.1	84.1	61.0
(c = .11)	r_e	12.0	12.0	24.6	2.4
		99.4 %	84.9 %	97.5 %	98.6 %

a. The c value required to produce a slope of 1.0.

ers. Since P_3 was undermatching even in the punishment conditions, it made no sense to scale the punishers to produce matching. For P_1 and the mean data, the baseline slope was so close to 1.0 that the c value was not changed when it was set to produce matching. As can be seen from the table, the analysis produced essentially the same outcome regardless of which scaling assumption was made. Equation (15.2) provided an excellent fit to the baseline data with the same k and r_e value used for both responses. The only change in the punishment condition was that the same VI 2-minute punishment schedule was superimposed on both alternatives, so a single k and r_e value was used for that condition too. When k and r_e were both allowed to vary in fitting Eq. (15.9) to the punishment data, the best fit value of k was consistently smaller than it was in baseline, but changes in r_e were not consistent from one pigeon to another. For one pigeon r_e decreased, but for two pigeons it increased in value. The other two applications of the equation to the punishment data are perhaps more intreresting. One represents the ability of the subtractive model to account for the suppression of the absolute response rates without any change in the parameters of the equation; the other represents a combination of the subtractive model with a change in the value of competing responses. The subtractive model alone does not account for Farley's absolute response rate data very well. The percentage of data variance accounted for when k and r_e are fixed at their baseline values is substantially smaller than the percentage accounted for by the least-squares fit in which both parameters could vary. On the other hand, when the subtractive model is supplemented by an increase in r_e it accounts for practically the same percentage of the variance as the best fit with both parameters varying.

To summarize, the choice and absolute response rate data from de Villiers (1980) and Farley (1980) suggest that the most appropriate quantitative model of punishment combines crucial aspects of both the negative law of effect and competing response theories. A subtractive model that quantifies the negative law of effect accounts for the obtained shifts in preference in concurrent VI schedules when VI punishment is added, but in order to account for the changes in absolute response rates it must be supplemented by an increase in r_e, the reinforcement for competing responses (Herrnstein, 1974). Of course this does not establish which competing response theory of punishment is the appropriate one, that which appeals to behaviors elicited by the punisher or signals for its occurrence or that which appeals to classically conditioned aversive proprioceptive stimuli and escape responses that are strengthened by negative reinforcement. These theories must be evaluated on other grounds.

LIMITATIONS AND LOOSE ENDS

Despite the empirical support for the subtractive model, two findings indicate that the range of situations to which it readily applies is limited. First, Eqs. (15.5) and (15.9) predict that as punishment frequency or intensity increases or positive reinforcement decreases, the value of one or both alternatives goes to zero or becomes negative. The animal should show exclusive preference for one alternative if the net reinforcing value of the other goes to zero or becomes negative, and it should cease responding altogether if the value of both goes to zero. In fact pigeons do stop responding when the concurrent VI food schedules are removed but the punishment schedules remain in effect. On the other hand, Deluty and Church (1978) found that rats would work in a particular kind of concurrent schedule, in which they chose between different frequencies of noncontingent shock in the absence of any food reinforcement for responding. In keeping with Deluty's (1976) equation (Eq. (15.4), the rats matched the relative time spent in one component to the relative frequency of shock in the other component. However, a number of considerations argue that this finding may be limited to a particularly narrow range of situations and question whether it represents evidence against the subtractive model of punishment. The inverse matching in Deluty and Church's study only held if the two variable-time (VT) shock schedules did not run concurrently, as they do in the normal concurrent scheduling procedure. If both schedules ran simultaneously, the rats tended to minimize the shock frequency by exclusive preference for the component with the lower rate of shock. The rats were first trained to respond on the changeover lever to escape from a component with VT shock to one with no shock. This initial escape training could be essential for responding to be maintained in the final procedure. Finally, a 2-second COD was in effect for each VT schedule, and since the COD provided a shock-free period following every changeover response, it may have maintained responding. The matching that was observed is in keeping with other studies showing time matching to relative shock-free time (Baum, 1973).

Second, a recent experiment by Bradshaw, Szabadi, and Bevan (1979) found results that are not in keeping with the subtractive model as embodied in Eqs. (15.5) and (15.6). In one phase of their experiment human subjects who were responding on concurrent VI schedules of monetary reinforcement were punished for one alternative according to a VR schedule of monetary subtraction. In the concurrent schedule one alternative was kept constant at VI 51 seconds while the other varied from VI 8 seconds to VI 720 seconds. The VR 34 punishment schedule was superimposed on the constant VI 51-second alternative on alternate days. Without the punishment schedule all three subjects matched relative response rates to the relative frequency of payoff on each schedule, and the response rate on the constant VI schedule varied inversely with the reinforcement rate for the alternative response, as predicted by Eq. (15.2). In the punishment condition, however, response rate on the invariant VI schedule was reduced to a fairly constant low rate and did not vary with the frequency of reinforcement for the other alternative. This produced a large bias toward the unpunished response, but it also led to pronounced undermatching, in that the slope of the linear function relating log response ratios to log reinforcement ratios for each subject was practically flat. Many factors need to be explored to account for the apparent discrepancy between this result and the findings of the animal experiments by de Villiers and Farley. The discrepancy may arise from species differences; similar procedures involving VR punishment of concurrent VI schedules have not been studied with both animal and human subjects. In addition Bradshaw et al. punished only the alternative with the constant rate of reinforcement. Since the rate of reinforcement and punishment on that alternative did not vary across conditions, the human subjects may have tended to dismiss that schedule and respond only very rarely on it in the punishment condition. Finally the feedback functions relating response rates to reinforcement rates in concurrent schedules with both VI and VR components are complex, and it is not clear yet how they should be incorporated into the subtractive model of punishment. Prelec and Herrnstein (1978) demonstrated that interactions between responding and reinforcement rates on concurrent VI schedules were quite different when a VR alternative was added. The two VI schedules seemed to be decoupled from one another so that the rate of responding and reinforcement on one did not seem to affect the other. Prelec and Herrnstein showed that this followed from the nature of the feedback functions for VI and VR schedules that they had derived. A similar decoupling of the concurrent VI schedules may have occurred when the VR punishment schedule was added in Bradshaw et al.'s experiment.

To summarize, the subtractive model of punishment embodied in Eqs. (15.5), (15.6), and (15.9) applies well to situations in which both positive reinforcement and punishment are arranged according to VI schedules. Progress in extending it to other situations will depend on advances in specifying the feedback functions that relate the frequency of responses to their punishing and reinforcing consequences on different schedules.

REFERENCES

Autor, S.M. 1969. The strength of conditioned reinforcers as a function of frequency and probability of reinforcement. In D.P. Hendry, ed., *Conditioned Reinforcement*, Homewood, Ill.: Dorsey Press, pp. 127–62.

Azrin, N.H. 1956. Some effects of two intermittent schedules of immediate and nonimmediate punishment. *Journal of Psychology 42*: 3–21.

Azrin, N.H., and W.C. Holz. 1966. Punishment. In W.K. Honig, ed., *Operant Behavior: Areas of Research and Application*, New York: Appleton-Century-Crofts, pp. 380–447.

Baum, W.M. 1973. Time allocation and negative reinforcement. *Journal of the Experimental Analysis of Behavior 20*: 313–22.

____. 1974. On two types of deviation from the matching law: Bias and undermatching. *Journal of the Experimental Analysis of Behavior 22*: 231–42.

Bauman, R.A.; R.L. Shull; and A.J. Brownstein. 1975. Time allocation on concurrent schedules with asymmetrical response requirements. *Journal of the Experimental Analysis of Behavior 24*: 53–57.

Bolles, R.C. 1972. *Theory of Motivation.* 2nd ed. New York: Harper & Row.

Bradshaw, C.M.; E. Szabadi; and P. Bevan. 1977. Effects of punishment on human variable-interval performance. *Journal of the Experimental Analysis of Behavior 27*: 275–79.

____. 1978. Effect of variable-interval punishment on the behavior of humans in variable-interval schedules of monetary reinforcement. *Journal of the Experimental Analysis of Behavior 29*: 161–66.

____. 1979. The effect of punishment on free-operant choice behavior in humans. *Journal of the Experimental Analysis of Behavior 31*: 71–81.

Deluty, M.Z. 1976. Choice and the rate of punishment in concurrent schedules. *Journal of the Experimental Analysis of Behavior 25*: 75–80.

Deluty, M.Z., and R.M. Church. 1978. Time-allocation matching between punishing situations. *Journal of the Experimental Analysis of Behavior 29*: 191–98.

de Villiers, P.A. 1977. Choice in concurrent schedules and a quantitative formulation of the law of effect. In W.K, Honig and J.E.R. Staddon, eds., *Handbook of Operant Behavior*, Englewood Cliffs, N.J.: Prentice-Hall, pp. 233-87.

____. 1980. Toward a quantitative theory of punishment. *Journal of the Experimental Analysis of Behavior 33*: 15-25.

de Villiers, P.A., and R.J. Herrnstein. 1976. Toward a law of response strength. *Psychological Bulletin 83*: 1131-53.

Dinsmoor, J.A. 1954. Punishment. I. The avoidance hypothesis. *Psychological Review 61*: 34-46.

____. 1977. Escape, avoidance, punishment: Where do we stand? *Journal of the Experimental Analysis of Behavior 28*: 83-85.

Estes, W.K. 1969. Outline of a theory of punishment. In B. Campbell and R.M. Church, eds., *Punishment and Aversive Behavior*, New York: Appleton-Century-Crofts, pp. 57-82.

Farley, J. 1980. Reinforcement and punishment effects in concurrent schedules: A test of two models. *Journal of the Experimental Analysis of Behavior 33*: 311-26.

Farley, J., and E. Fantino. 1978. The symmetrical law of effect and the matching relation in choice behavior. *Journal of the Experimental Analysis of Behavior 29*: 37-60.

Herrnstein, R.J. 1970. On the law of effect. *Journal of the Experimental Analysis of Behavior 13*: 243-66.

____. 1974. Formal properties of the matching law. *Journal of the Experimental Analysis of Behavior 21*: 159-64.

Mansfield, R.J.W., and H.C. Rachlin. 1970. The effect of punishment, extinction, and satiation on response chains. *Learning and Motivation 1*: 27-35.

Millenson, J.R., and P.A. de Villiers. 1972. Motivational properties of conditioned anxiety. In R.M. Gilbert and J.R. Millenson, eds., *Reinforcement: Behavioral Analyses*, New York: Academic Press, pp. 97-128.

Millenson, J.R., and A. St. C. McMillan. 1975. Abortive responding during punishment of bar holding. *Learning and Motivation 6*: 279-88.

Mowrer, O.H. 1960. *Learning Theory and Behavior.* New York: Wiley.

Prelec, D., and R.J. Herrnstein. 1978. Feedback functions for reinforcement: A paradigmatic experiment. *Animal Learning and Behavior 6*: 181-86.

Rachlin, H.C. 1967. The effect of shock intensity on concurrent and single-key responding in concurrent-chain schedules. *Journal of the Experimental Analysis of Behavior 10*: 87-93.

Rachlin, H.C., and R.J. Herrnstein. 1969. Hedonism revisited: On the negative law of effect. In B. Campbell and R.M. Church, eds., *Punishment and Aversive Behavior*, New York: Appleton-Century-Crofts, pp. 83-110.

Renner, K.E. 1966. Temporal integration: Relative value of reward and punishment as a function of their temporal distance from the response. *Journal of Experimental Psychology 71*: 902-907.

Schuster, R., and H.C. Rachlin. 1968. Indifference between punishment and free shock: Evidence for the negative law of effect. *Journal of the Experimental Analysis of Behavior 11*: 777-86.

Solomon, R.L. 1964. Punishment. *American Psychologist 19*;239–53.

Stubbs, D.A., and S.S. Pliskoff. 1969. Concurrent responding with fixed relative rate of reinforcement. *Journal of the Experimental Analysis of Behavior 12*:887–95.

Thorndike, E.L. 1911. *Animal Intelligence: Experimental Studies.* New York: MacMillan.

Williams, J.L. 1969. Response contingency and effects of punishment: Changes in autonomic and skeletal responses. *Journal of Comparative and Physiological Psychology 68*:118–25.

V ECONOMIC MAXIMIZATION THEORY

16 ECONOMICS OF THE MATCHING LAW

Howard Rachlin

Choice situations in the laboratory and everyday life may be described by two molar theories. One theory, the matching law, (Herrnstein, 1961, 1970) says that organisms allocate behavior among alternatives in proportion to reinforcement from those alternatives. The other theory, economic maximizing, says that organisms adjust behavioral allocation among alternatives so as to maximize the utility of reinforcements from the alternatives (Rachlin, Green, Kagel, and Battalio, 1976). In a wide variety of choice situations both of these theories describe behavior well. But in some situations where choice is between qualitatively different commodities, the economic maximizing theory provides meaningful interpretation of parameters in terms of preference and substitutability of the commodities, whereas the matching law, although also describing the data, does not by itself provide a meaningful account of parameters. This chapter shows how the matching law may be derived from economic maximizing theory and how economic maximizing theory gives meaning to the matching-law parameters.

The two theories that describe how organisms allocate time to various activities are molar because they ignore the occurrence of particular responses and are silent on the mechanism by which responses are integrated; they focus only on the allocation of behavior to one or another activity over the course of time. The degree of temporal integration of various subjects in various situations differs, of

The research was supported by grants from the National Science Foundation. I am grateful to Gary Becker for comments on an earlier version of this chapter. Another version based on the same presentation was published previously (Rachlin, Kagel, and Battalio, 1980).

course. A pigeon may integrate reinforcements and responses over a few minutes whereas a human being may integrate those variables over much longer periods. Molar theories apply only over periods during which integration does occur. For instance, a bird may spend so much of an hour sitting at a nest and so much of the hour foraging for food. The observed allocation may depend directly on reinforcement over the course of the hour for each of the two activities. Neither the activities nor the reinforcement need occur at any particular time within the period of temporal integration so long as they occur somewhere within that period. A molar theory would not, by itself, differentiate between a bird that alternated nest-sitting and foraging every 10 minutes for 60-minutes and a bird that spent the first half-hour nest-sitting and the second half-hour foraging. Both birds would have allocated half of the hour to nest-sitting and half of the hour to foraging.[a]

Although only temporal allocations will be considered formally, much of this analysis applies directly to allocation of discrete responses. Baum (1976) has shown that operant conditioning experiments with pigeons and rats, as well as nonoperant maze and straight-alley experiments (where running rather than pecking or lever pressing is the observed response), may be converted to temporal allocations. (A pigeon, in a 1-hour experimental session, may be assumed to be pecking for a certain fraction of the hour, proportional to its overall rate of pecking.) Henceforth it is assumed that such conversions can be made.

Of the two molar theories of choice that have been proposed, the matching law says that organisms allocate time to various activities so as to *match* with their behavior the reinforcements contingent on those activities, whereas the economic theory says that organisms allocate time to activities so as to *maximize* overall utility. Most choice situations are analyzable in terms of both of these molar theories. That is, matching and maximizing have both been shown to be reasonable interpretations of behavior in those situations. Proponents of each theory hold that theirs is the fundamental process whereas the other is an incidental result (see Herrnstein, Chapter 19; Vaughan, Chapter 12). This chapter argues in favor of the economic theory of maximization as the fundamental process (hence, that matching is incidental).

Figure 16-1 diagrams a simple choice situation in molar terms. An animal A must choose between two environments, E_1 and E_2. For instance, a bird A might choose between sitting on a nest, E_1, and looking for food E_2. E_1 and E_2

a. Psychologists (and economists) often assume that organisms cannot automatically integrate events over time. Although economic theories are all molar theories, economists assume that temporal integration is a function of the rational mind, hence, that economics applies only to human behavior. There is a great deal of evidence, however, that not only human beings but all animals (as well as inanimate objects) integrate over greater or lesser intervals. Economics is often said to be a science of rational behavior, but "rationality" means no more to economists than conformance to economic theory. Economics is properly a science of maximization rather than rationality (Rachlin, 1980).

Figure 16–1. Flow diagram of a molar choice situation. An animal A in a period of time $(T_1 + T_2)$ allocates part of the period T_1 to an environmental situation E_1 and part of the period T_2 to another environmental situation E_2. The animal receives commodity C_1 from E_1 and C_2 from E_2.

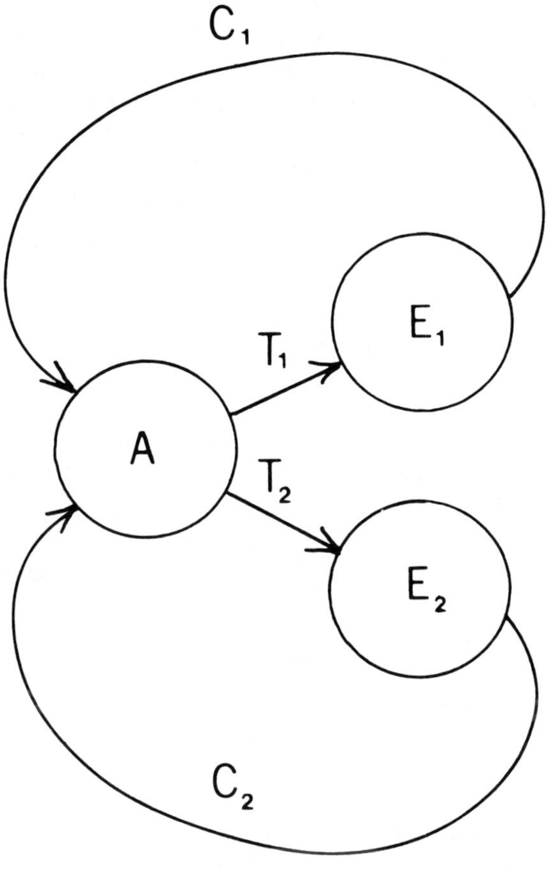

are assumed to be mutually exclusive and exhaustive. The time of observation is also assumed to be fixed at unit value, so that

$$T_1 + T_2 = 1 \ . \tag{16.1}$$

The consequences of choosing E_1 and E_2 are represented as C_1 and C_2. Thus T_1 and T_2 are outputs of A whereas C_1 and C_2 are inputs to A.

For the sake of simplicity choice between only two alternatives and only the positive case (where C_1 and C_2 are rewards) will be considered.[b] The theories we will discuss are attempts to express choice (the ratio T_1/T_2) as a function of reward (C_1 and C_2). In that sense they are theories of incentive (Logan, 1965).

No distinction is made here between situations in which rate of a single response is measured and situations in which choice between two equivalent responses is measured. In the former T_1 corresponds to time spent responding and T_2, to time spent not responding. In the latter T_1 corresponds to time spent at one alternative and T_2, to time spent at the other alternative. Molar theory allows both situations to be considered alike (as choice between E_1 and E_2). In Herrnstein (1970) and Rachlin (1978) it was shown how the matching law and economic maximization theory may be extended from symmetrical choice (concurrent responses) to asymmetrical choice (responding versus nonresponding).

The Feedback Function

There are three functional transformations implied in Figure 16-1. One is the transformation made by the subject, A, in which C_1 and C_2 are converted to T_1/T_2. This transformation is the ultimate object of interest. The other two transformations are those that take place in the environments E_1 and E_2. These transformations are feedback functions (Baum, 1973). Different environments will be represented by different feedback functions. In operant conditioning experiments concurrent schedules of reinforcement are represented by the feedback functions for E_1 and E_2. Let us examine the properties of one particular feedback function:

$$C = aT^m \quad . \tag{16.2}$$

There are several reasons for considering this function. First, it is simple. Second, it can represent ratio schedules. A ratio schedule (Ferster and Skinner, 1957) provides reinforcement proportional to responding. Normally in ratio schedules number of reinforcements is set proportional to number of responses. But, assuming that rate of responding is usually constant, such schedules may be characterized by a proportionality of C to T (Baum, 1976). Thus, where $m = 1.0$ in Eq. (16.2), the feedback function is that of a ratio schedule. Ratio schedules are found in everyday life in working conditions where the payoff C is proportional to the time spent working, T.

Figure 16-2 shows feedback functions of Eq. (16.2) with various values of m. The ratio-schedule feedback function ($m = 1$) is the diagonal straight line. The curved line and the horizontal line correspond to interval and time schedules.

b. The empirical observations and the theoretical framework on which this chapter is based have been extended to more than two alternatives. See Newman (1965) for an extension of economic demand theory to more than two alternatives. See Rachlin and Burkhard (1978) for a discussion of the three-alternative case in general terms.

Figure 16–2. The feedback function, $C = aT^m$ where $a = 1$ and $m = 0$ (variable-time schedule), $m = 0.2$ (variable-interval schedule), and $m = 1.0$ (variable-ratio schedule).

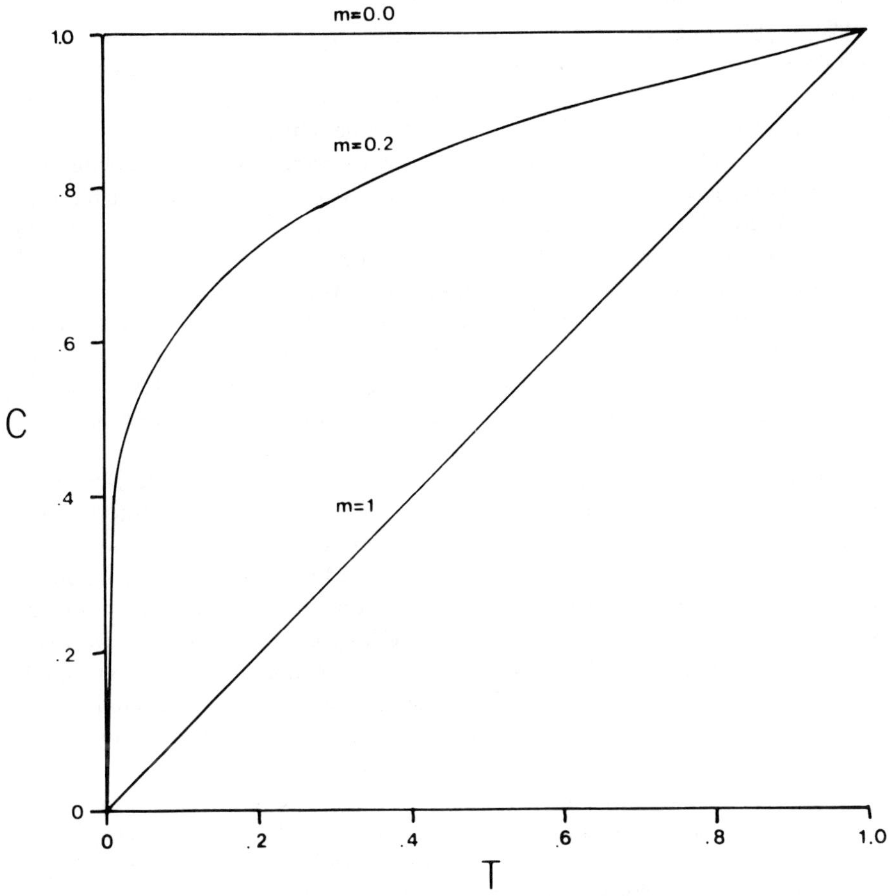

A variable-interval schedule (Ferster and Skinner, 1957) programs reinforcement in temporal terms. The program sets a maximum rate of reinforcement a. If the subject does not respond (for example, if the pigeon does not peck the key) no reinforcement is delivered. As responding increases from zero, rate of reinforcement increases rapidly and approaches a. If the subject spent all its time responding ($T = 1.0$) it would receive reinforcement at the maximum rate a. Everyday feedback functions often have this characteristic. For instance, to be quite prosaic, consider the response of collecting eggs from a chicken. (Suppose that the chicken starts the process leading to laying a new batch of eggs only when the previous batch is collected.) If the farmer never collects the eggs he gets no eggs. If the farmer continuously stands by the chicken he gets eggs at the maximum rate. But the farmer can get eggs at *almost* the maximum rate by collecting them at reasonable intervals. A similar feedback function governs many temporally defined everyday processes such as people collecting mail from the mailbox or checking the oven to see when food is ready.

A variable-time schedule is one where reinforcement is delivered freely at a certain rate a regardless of the subject's behavior. In Eq. (16.2), when $m = 0$, reinforcement rate C would equal the maximum rate a regardless of responding T. In Eq. (16.2), $m = 0$ characterizes variable-time schedules.

The feedback function of Eq. (16.2) treats variable-interval schedules as fundamental. It characterizes other schedules (variable-time and variable-ratio schedules) as extreme forms of variable-interval schedules (at $m = 0$ or $m = 1$). At $m \simeq 0$, the variable-interval schedule would be timelike. At $m \simeq 1$, the variable-interval schedule would be ratiolike. Thus, this feedback function models the molar properties of various schedules (usually considered as discrete and unrelated operations) in terms of the continuous variation of a single parameter.

Laboratory procedure in programming variable-interval schedules varies considerably. Most often the variable-interval timer is stopped after a reinforcement is scheduled but before it is collected. Sometimes, however, reinforcements are accumulated (Vaughan, Cahpter 12). With concurrent variable-interval schedules, sometimes the two timers are stopped together (Stubbs and Pliskoff, 1969) and sometimes they are completely independent (Herrnstein, 1961). Sometimes only one timer is used and the reinforcers assigned randomly or quasi-randomly to one or the other alternative (Shimp, 1969). Sometimes the reinforcer intervals are distributed uniformly over their range (Herrnstein, 1961), sometimes, exponentially (Fleshler and Hoffman, 1961). No single feedback function can exactly model all laboratory procedures. Thus, a simple function with the required general properties is better than a complicated function that exactly models one procedure and (necessarily) fails to model the others.

With regard to standard operant conditioning procedures, two questions need to be asked of a proposed feedback function: Does it account for results obtained with typical variable-interval schedules? And, when it is deliberately programmed, does it produce typical variable-interval behavior?

McSweeny (1975) varied the body-weight of pigeons by food deprivation of various degrees. The rate of response of the pigeons with a given variable-interval schedule of reinforcement varied inversely with body-weight. Equation (16.2) was a good predictor of rate of reinforcement as a function of rate of response in that experiment (Rachlin, 1978).

It remains to be shown that the simple function, Eq. (16.2) is not so far off the mark that, were it deliberately programmed, "typical" variable-interval behavior would not be observed. Although the required experiments have not yet been performed, the range of procedures that yield typical variable-interval behavior with single alternatives and matching with multiple alternatives is so wide (de Villiers, 1977) that there is little doubt that typical properties would be found.

The form of Eq. (16.2) is like one that has been used in economics (Becker, 1965) to express income as a function of time devoted to activities. The curve of Figure 16-2 corresponding to $m = 0.2$ demonstrates diminishing returns. The more responding (the greater T) the less the rate of increase of reinforcement.

The case where $m > 1$ (not shown in Figure 16-2) would correspond to the opposite of diminishing returns, increasing returns. It conforms to no typical laboratory contingency of reinforcement, but it does conform approximately to some everyday contingencies of reinforcement. With $m > 1$ the more time spent in an activity the greater the payoff. Obviously this situation could not go on forever or we would all be continuously doing that activity. But it does conform intuitively to some activities (up to a point) such as listening to music or reading novels. Equation (16.2) provides a meaningful link between such activities and ordinary activities where diminishing returns is the rule.

Combing Eqs. (16.1) and (16.2):

$$\left(\frac{C_1}{a_1}\right)^{\frac{1}{m_1}} + \left(\frac{C_2}{a_2}\right)^{\frac{1}{m_2}} = 1 \ . \qquad (16.3)$$

This is the restriction on C_1 and C_2 set by the feedback equation and the constraint in time of observation. In economic terms the feedback equation sets the *price* (how much T must be spent for a unit of C) and the time-constraint sets the subject's *income* (how much T is available). Price and income together determine a *budget*, the equation for which is Eq. (16.3).

Economic Theory

Figure 16-3 is a geometric representation of a microeconomic system with two commodities. The two axes represent amounts of the commodities, C_1 and C_2. Each point in the positive quadrant represents a package of the commodities. A point high and to the right in the quadrant represents a package with a lot of both commodities whereas a point near the origin represents a package with little of the two commodities. Points on either axis represent packages of only one commodity. These "packages" correspond to amounts of C_1 and C_2 that would be obtained by the subject A in Figure 16-1 in a unit session. A series of fundamental axioms about choice (for instance, transitivity, diminishing marginal rates of substitution) imposes restrictions that define a class of indifference contours in this space. The contours are characterized by the function

$$U = f(C_1, C_2) \ , \qquad (16.4)$$

Figure 16-3. Amount of C_1 versus amount of C_2. Each point in the space represents a package of C_1 and C_2. The line $a_1 a_2$ represents a budget line. The subject *A* can obtain packages of C_1 and C_2 only on (or below) this line. The thin lines represent presumed indifference curves. The dark point represents the maximum indifference curve reachable within the constraints of the budget.

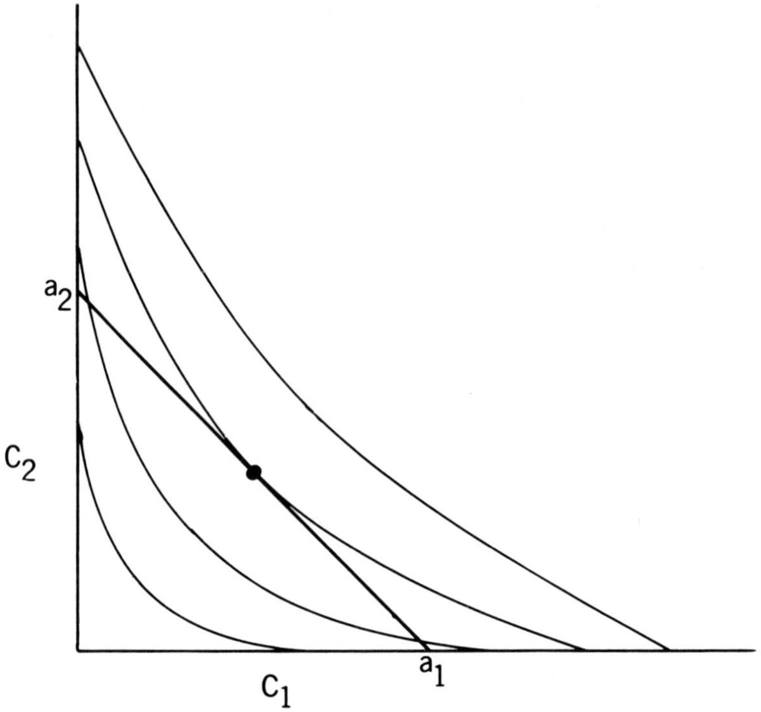

where *U* stands for utility. The indifference contours are constant-utility functions. Within the economists' theory of commodity choice behavior (demand theory) the indifference contours are assumed to fill the quadrant, not to cross, and to be everywhere convex with respect to the origin. The path of reasoning from the fundamental assumptions about choice to the shape of the contours is straightforward and can be found in any standard microeconomics text (for instance, Newman, 1965). In Figure 16-3, the thin lines show a sample of a set of contours consistent with demand theory. All points on a given contour represent packages of C_1 and C_2 of equal utility; they could be exchanged for each other by a consumer without gain or loss. The slope of a contour at any point represents the rate at which C_2 would be exchanged for C_1 as a consumer moves along a contour. This slope is the consumer's "marginal rate of substitution" of C_2 for C_1. Packages on higher contours are assumed to be more valuable than packages on lower ones. A consumer, according to demand theory would willingly exchange any package on a lower contour for a package on a higher contour.

Choice among packages is restricted in theory by the consumer's budget, however. The budget restrictions are imposed by Eq. (16.1), which limits time available and Eq. (16.2), which expresses reinforcement as a function of time. Their combination, Eq. (16.3), with $m_1 = m_2 = 1$ is represented by the heavy line ($a_1 a_2$) in Figure 16-3. If the consumer spent his entire budget on C_1 ($C_2 =$ 0) he would buy the package at point a_1. If he spent his entire budget on C_2 ($C_1 = 0$) he would buy the package at point a_2. If he distributed his purchases among C_1 and C_2, he could buy any package within the triangle, $0\, a_1\, a_2$. Points *within* the triangle would not exhaust the budget. Points on line $a_1 a_2$ would exhaust the budget.

The assumption of demand theory, that a consumer would exchange any package of lower value for a package of higher value, implies that the consumer will purchase the particular package of goods at the point (of tangency, here) where the budget line $a_1 a_2$ touches the highest reachable contour. This is the package of highest attainable value. It is represented in Figure 16-3 by the solid dot.

As applied to the behavior of an individual, demand theory assumes a constant set of indifference contours that represent his underlying preferences. If choice is found to vary, the variations are assumed to be due to changes in the *budget* (which represents contingencies imposed from without) rather than to variations in the indifference contours (the consumer's basic preference structure). Economics provides some techniques for determining what indifference contours look like but, within economic theory, there are few mechanisms for *change* of indifference contours. The assumption of fixed preference does not limit economic theory as much as might be thought. Stigler and Becker (1977) show how budget changes alone may account for changes in fashion, development of tastes for music or literature, susceptibility to advertising, and various addictions such as alcohol addiction or heroin addiction.

The Matching Law

A version of the matching law for which there is considerable empirical evidence with human and other animal subjects (de Villiers, 1977; and various chapters in this volume), says

$$\frac{T_1}{T_2} = b\left(\frac{C_1}{C_2}\right)^u . \qquad (16.5)$$

This relation is obtained in operant conditioning experiments where various schedules of reinforcement serve as E_1 and E_2. For instance, a pigeon's peck on the left of two keys might be reinforced according to a variable-interval schedule with a maximum rate of reinforcement, a_1, of 1 per minute while pecks on the right key might be reinforced according to a variable-interval schedule with a maximum rate of reinforcement of 2 per minute (a_2) until stable values for T_1, T_2, and C_1 and C_2 were obtained.

The pair of ratios (T_1/T_2 and C_1/C_2) determines a single point on a plot of log T_1/T_2 versus log C_1/C_2. Then a_1 and a_2 are systematically varied, behavior with each pair determining another point. If the points are in a straight line the matching law is confirmed; the slope of the line is the constant u; the antilog of the y intercept is the constant b.[c]

When the constant b differs from 1.0, the subject is considered to be biased toward E_1 or E_2. If C_1 and C_2 as measured do not represent the entire set of consequences, unmeasured consequences may produce bias ($T_1 \neq T_2$ even when $C_1 = C_2$). Examples of unmeasured consequences in pigeon choice experiments are different key colors, differential resistance of keys to pecking, differential fatigue, preferences for place and direction of movement. These unmeasured factors, when they make a difference, are assumed to act multiplicatively (that is, b is multiplied rather than, say, added to the right-hand term of Eq. (16.5). Some evidence that this assumption is valid is the empirical constancy of b as C_1 and C_2 are varied and the fact that various *measured* consequences such as rate and amount of food reinforcement have been shown to combine multiplicatively (Baum, 1974; Baum and Rachlin, 1969).

The exponent u has often been found to be equal to unity. However, in some experiments u has deviated from unity, mostly downward (de Villiers, 1977). One possible cause of less-than-unitary values of u is a failure of the subject to discriminate the source of C_1 in E_1 from the source of C_2 in E_2 (Baum, 1974). At the extreme, E_1 and E_2 would be completely confused and T_1/T_2 would remain constant even while C_1 and C_2 varied. (Thus the value of u obtained would be zero.) This explanation, however, cannot account for all cases in which u deviates from unity. First, u has sometimes been found to be greater than unity. Confusion of E_1 and E_2 would not explain $u > 1$. Second, in some experiments in which every effort was made to enhance discriminability of E_1

c. An equation trivially resembling the matching law emerges from the feedback equation (Eq. (16.2)) alone: $C_1/C_2 = (a_1/a_2)\,(T_1/T_2)^m$. But this equation does no more than repeat the feedback equation. The feedback equation expresses the relation of C to T for given values of a_1 and a_2 but not what the observed values of C and T will be. Furthermore, in an actual experiment, a_1 and a_2 would be independent variables. Eq. (16.5) says that C and T will vary so as to keep b constant even while a_1 and a_2 are changed (and it will be shown that u remains constant even for different values of m). The matching law applies across conditions (across variations of a_1 and a_2), whereas the feedback function only specifies relations between C and T with a given value of a_1 and a_2.

and E_2, u has stubbornly refused to approach unity (Leigland, 1979; Trevitt, Davison, and Williams, 1973). Furthermore, as shown in Rachlin, 1978, and as will be shown here, subjects in choice experiments may ignore the differential source of C_1 and C_2 and still obey Eq. (16.5) (with $u = 1$) by altering T_1/T_2 so as to maximize $C_1 + C_2$. There is another explanation for $u \neq 1$ in Eq. (16.5).

Substitutability and Matching

Implicit in economic theory is the concept of an animal with a set of needs and an environment capable of supplying a set of commodities that can satisfy those needs. But a given commodity may satisfy more than one need (as milk may satisfy both hunger and thirst); two commodities may both partially satisfy a single need and partially satisfy different ones. To the extent that two commodities satisfy the same set of needs they are substitutable. To the extent that they satisfy different needs they are not substitutable. Thus to a thirsty organism milk would be largely substitutable for water whereas dry food would be largely unsubstitutable for water.

Substitutability would depend on the use to which the commodities were to be put. For instance, if foods were to be used only to fatten livestock they might be valued by a farmer only for their caloric content. The farmer buying them would care only how many calories they had. Whatever part of C_1 or C_2 that did not contain calories would be useless. This implies straightline indifference contours in Figure 16-3, (the thin lines would be straight instead of curved). The indifference contours would be nothing but constant calorie contours. If C_1 and C_2 had the same number of calories per unit the contours would be straight lines angling 45° downward to the right. If C_1 and C_2 had different calories per unit the contours would still be straight but at another angle (and the straight budget line would touch the highest possible straight indifference contour at an axis).

In most operant studies of concurrent schedules C_1 and C_2 are the same commodity. A pigeon in these studies may vary its allocation of pecks between two keys but receives deliveries of the same grain mixture as a consequence of pecking either key. Referring to Figure 16-1 the choice in concurrent operant experiments is usually between different schedules (feedback functions with different parameters) of food reinforcement. The reinforcements in these studies may be assumed to be completely (or almost completely) substitutable. The matching law (Eq. (16.5)) has been shown to apply very well to this class of choice experiments (de Villiers, 1977). But in what way does the matching law apply (or does it apply at all) when C_1 and C_2 are not completely substitutable? As long as C_1 and C_2 are made to be completely substitutable, the matching law, while supported, is of limited usefulness. If the matching law could be extended to commodities of different kinds (not completely substitutable) it might be used to scale both preference and substitutability of those commodities and serve as an instrument for classifying drives and incentives. This chapter presents one way in which the matching law may be extended to account for incompletely substitutable commodities. There may be others. For instance, Herrnstein (1981) has suggested a method based on different rates of satiation. But Herrnstein's method has not been specified in any detail. It is not clear how different rates of satiation would affect the exponent u of Eq. (16.5). As will be seen, the theory presented here treats this exponent as a measure of substitutability.

As the substitutability between two commodities decreases from complete to partial substitutability to complete nonsubstitutability, the indifference contours change from straight lines to curves like those in Figure 16-3 to sharp-cornered angles. At the extreme of nonsubstitutability a pair of commodities will only be accepted in a fixed proportion (as the ingredients for a precise recipe might be). The "leftover" commodity in a given package beyond that proportion would be useless. For instance, if food and water *had* to be consumed in the ratio of 3 to 1 units, a package consisting of three units of food plus one unit of water would be worth just as much as a package consisting of three units of food plus two units of water. The extra unit of water would be useless.

Substitutability is indexed by the curvature of the indifference contours. A function will now be specified for Eq. (16.4) that can express curvature of indifference contours as a single parameter.

The Utility Function

Consider the following utility function:

$$U = x_1 C_1{}^y + x_2 C_2{}^y \quad . \tag{16.6}$$

Like the feedback function, Eq. (16.2), the utility function is assumed to be a power function.[d]

When $y = 1$ the utility function is a straight line and indifference contours (constant U) would slope downward at an angle depending on the fraction, x_1/x_2. This is the case of complete substitutability. The ratio x_1/x_2 represents the degree to which C_1 and C_2 contain the desired, completely substitutable, commodity. If, for example, calories were the only valued aspect of two foods, C_1 and C_2, then x_1 and x_2 would represent the calories per unit of C_1 and C_2.

d. If, in Eq. (16.6), $x_i = z_i/y$, $z_i > 0$ and $y \leq 1$, the utility function would be one referred to in the economics literature as a CES utility function. In cases where $y = 0$, Eq. (16.6) reduces to $U = z_1 \log C_1 + z_2 \log C_2$ (see Barten, 1977: 38-39 for example). Allen (Note 2) has also suggested their use in psychology.

In cases where $y < 0$, all utilities are negative. Maximization is construed as choice of the least negative of available utilities. The sign of the utility is not meaningful in any other way (negative utilities are not assumed to be aversive, for instance).

In situations where C_1 and C_2 are the very same commodity (most concurrent-operant experiments) it can be argued that $y = 1$ and $x_1 = x_2$ (approximately). Then, according to Eq. (16.6), $U = C_1 + C_2$. Utility would simply be the sum of the reinforcers from the two concurrent sources. In Rachlin, 1978, and Staddon and Motheral, 1978, it was argued that maximization of this sum (maximization of utility) occurs when the subject matches T_1/T_2 to C_1/C_2. Rachlin used the feedback function presented here, Eq. (16.2). Staddon and Motheral used an exponential feedback function. In both cases (with completely substitutable commodities) matching was shown to result from maximizing overall rate of reinforcement. This implies that a subject simply adds the reinforcers together and adjusts T_1/T_2 until this sum is as big as it can possibly get. At that point the subject would be matching T_1/T_2 to C_1/C_2 (In Eq. (16.5), $u = b = 1.0$).

When $y < 1$ the indifference contours depart from straightness and bend inward as they do in Figure 16-3. Just as fractional exponents of Eq. (16.2) represent diminishing marginal returns from spending time at an activity, less-than-unity exponents of Eq. (16.6) represent diminishing marginal rates of substitution as more and more of a given commodity is accumulated. With utility held constant a subject could "move along" a single constant-utility contour by trading one commodity for another. In Figure 16-3, consider a subject with zero amount of C_1 and some finite amount of C_2 (a package represented by a point on the ordinate above zero). At the ordinate the contours all slope sharply downward. As the subject moved along a contour he would give up large amounts of the commodity he had (C_2) for small amounts of the commodity he did not have (C_1). But as the person moved further along the contour, losing C_2 and accumulating C_1, he would be willing to give up less and less of C_2 for a unit of C_1. Of course, in cases where the pattern of reinforcers is vastly different between alternatives (all the reinforcers in a lump versus the same number of reinforcers spread out in time) the assumptions of molarity stated at the beginning of the chapter would be violated. In such cases more molecular considerations would govern choice. Such molecular choices may still be governed by economic considerations.

The constant y, as it causes the curvature of the contours to vary, is thus an index of substitutability.

What is the consequence of maximizing utility as expressed by Eq. (16.6), with $y \neq 1$? Substituting Eqs. (16.1) and (16.2) into Eq. (16.6):

$$U = x_1 a_1^y T_1^{my} + x_2 a_2^y (1 - T_1)^{my} .$$

Differentiating with respect to T_1 and setting the derivative equal to zero yields

$$x_1 a_1^y T_1^{my-1} = x_2 a_2^y (1 - T_1)^{my-1} .$$

Substituting T_2 for $(1 - T_1)$ and C_i/T_i^m for a_i $(i = 1, 2)$ and rearranging terms;

$$\frac{T_1}{T_2} = \frac{x_1}{x_2} \left(\frac{C_1}{C_2} \right)^y . \tag{16.7}$$

In cases where $m_1 \neq m_2$ this procedure yields

$$\frac{T_1}{T_2} = \frac{m_1}{m_2} \cdot \frac{x_1}{x_2} \left(\frac{C_1}{C_2} \right)^y . \tag{16.8}$$

When the second derivative of the utility function is negative, Eqs. (16.7) and (16.8) are maxima. This condition will hold when $my < 1$. The only cases considered where the second derivative could be nonnegative would be with concurrent ratio schedules ($m_1 = m_2 = 1$) together with completely substitutable commodities ($y = 1$). Then $U = x_1 a_1 T_1 + x_2 a_2 T_2$. When $x_1 a_1 \neq x_2 a_2$ this is obviously a maximum at either $T_1 = 1$ or $T_2 = 1$, depending on x and a. In all other cases, Eqs. (16.7) and (16.8) represent unique global maxima.

Both Eqs. (16.7) and (16.8) are forms of the matching law, Eq. (16.5). Equations (16.1), (16.2), and (16.5) thus imply that *the matching law is a consequence of maximization even when the reinforcers are not completely substitutable.* The ratio x_1/x_2 represents preference for one alternative over the other independent of degree of substitutability and the feedback function (schedule of reinforcement). This corresponds to the notion of bias b in matching in Eq. (16.5). But bias may arise in the matching law from another source, $m_1 \neq m_2$, to be discussed shortly.

The exponent y in Eqs. (16.7) and (16.8) corresponds to the exponent u in the matching equation. This correspondence is what enables us to extend the matching law from a virtually tautologous description of the properties of the schedules of reinforcement, useful mostly in the development of better operant-conditioning techniques (Rachlin, 1971), to a true measuring instrument. It allows us to use the precision of operant procedures to measure the substitutability among commodities for an individual organism. Henceforth the constants x and y of Eqs. (16.7) and (16.8) will refer to presumed properties of C_1 and C_2. The equivalent constants of Eq. (16.5) (b and u) will refer to experimental results.

The particular utility function used here, Eq. (16.6), together with the particular feedback function, Eq. (16.2), results in matching. Obviously choice depends on the interaction of utility and feedback functions.

Heyman and Luce (1979) have shown that matching is inconsistent with maximization of $C_1 + C_2$ assuming certain feedback functions and random temporal distribution of discrete responses. Equation (16.6) would be inconsistent with any set of indifference contours that were not homothetic. A homothetic set of indifference contours will have equal slopes along any single ray from the origin. Imagine a contour of any shape and a set of rays from the origin going through the contour. A contour homothetic to the first would cross each ray at the same angle that the first contour crossed that ray.

In economics the slope of a budget line (such as $a_1 a_2$ in Figure 16-3) is a measure of relative price whereas the height of the line is a measure of income. With homothetic contours T_1/T_2 at the point of maximization would be dependent on relative price and independent of income. If a straight budget line were tangent to a contour at one point (as in Figure 16-3) all parallel budget lines would be tangent to homothetic contours along the same ray (the same value of C_1/C_2). The matching law indicates this fact by its assertion that choice depends only on the ratio C_1/C_2 and not in any way on the sum $C_1 + C_2$. This is already a fairly strict limitation on the indifference contours. The finding of matching under various conditions of deprivation, noncontingent electric shocks, and provision of reinforcement from other sources (de Villiers, 1977) is evidence that contours in operant situations (with $y = 1$) are homothetic under a variety of conditions.

Equation (16.6) defines a set of homothetic contours because $\partial C_1/\partial C_2$ (slope of utility function) is constant for a given C_1/C_2 (ray from the origin):

$$\frac{\partial U}{\partial C_1} = y\, x_1\, C_1{}^{y-1} \quad ;$$

$$\frac{\partial U}{\partial C_2} = y\, x_2\, C_2{}^{y-1} \quad ;$$

$$\frac{\partial C_1}{\partial C_2} = \frac{x_2}{x_1}\left(\frac{C_2}{C_1}\right)^{y-1} .$$

Thus, Eq. (16.6) implies both the matching law and homothetic indifference contours. If contours were found to be nonhomothetic it would disconfirm Eq. (16.6) and, necessarily, the matching law. There is not yet a general set of rules for what class of utility functions together with what class of feedback functions would be consistent with matching (Eq. (16.5)) and what set of utility functions together with what sets of feedback functions would be inconsistent with matching.

Concurrent Ratio Schedules

With concurrent ratio schedules, $m_1 = m_2 = 1$. Equation (16.3) becomes

$$\frac{1}{a_1} C_1 + \frac{1}{a_2} C_2 = 1 \quad .$$

This function is a straight budget line such as that of Figure 16–3. Budget lines for concurrent ratio schedules will always be straight lines with slope dependent on a_1 and a_2.

When identical reinforcers are used for the two ratio schedules ($y = 1; m = 1$) animals come to choose the better alternative (higher value of a) exclusively (Herrnstein and Loveland, 1975). This serves to maximize overall reinforcement. In Figure 16–3, a straight budget line would touch the highest of a set of straightline indifference contours at one or the other axis. The dot would be on the axis corresponding to $C_1 = a_1$ or $C_2 = a_2$.

Equations (16.1) and (16.2) imply that at $T_1 = 1$ (and $T_2 = 0$) $C_1 = a_1$. Alternatively, at $T_2 = 1$ (and $T_1 = 0$) $C_2 = a_2$. One of these conditions would apply when the straight budget line and straight indifference contours intersected. If the budget line were parallel to the indifference contours ($a_1/a_2 = x_1/x_2$) there would be tangency at all points and T_1/T_2 could not be predicted. Such congruency is nominally predicted with $a_1 = a_2$ and identical reinforcers (because then $x_1 = x_2$) but the slightest unprogrammed side preference could alter the angle of one of the lines and shift the maximum to exclusive preference for one side or another.

When C_1 and C_2 are not the same commodity the matching law (Eq. (16.5)) by itself says nothing about choice. One might suppose either $b \neq 1$ or $u \neq 1$ but there is no a priori reason to suspect how one or the other constant will vary Economic theory, on the other hand, asserts that b is a measure of preference while u is a measure of substitutability of the two commodities. With concurrent ratio schedules, because $m_1 = m_2 = 1$, Eq. (16.7) applies. That equation is somewhat awkward to use because both T and C are dependent variables. But combining Eq. (16.7) with Eq. (16.2) and rearranging terms:

$$\frac{C_1}{C_2} = \left(\frac{x_1}{x_2}\right)^s \left(\frac{a_1}{a_2}\right)^s \quad , \tag{16.9}$$

where $s = 1/(1 - y)$. The fraction a_1/a_2 (in economics, relative price) is an independent variable. Let us now consider how Eq. (16.9) (which is just a transformed version of the matching law) describes results of concurrent ratio schedules with different reinforcers.

Rachlin et al. (1976) studied rats' choices between different foods and liquids by means of concurrent ratio schedules. Rats could press either of two levers in a Skinner box. Pressing one lever a certain number of times (ratio schedule) produced one reinforcer (C_1). Pressing the other lever similarly produced the other reinforcer (C_2). The budget line was varied by varying the number of presses required by the ratio schedules and the amount of reinforcement per delivery (which together determine economic price) and also varying the total allowed number of presses on both levers combined (defined as economic income). The session was terminated when a rat used up its alloted presses (its income).

Figure 16-4 shows the results with one rat choosing between rootbeer and (nonalcoholic) tom collins mix. Line aa was the first budget line imposed (prices on each lever = one lever press per 0.05 cubic centimeters fluid; income = 300 presses). The point on that line shows how the rat allocated its presses. Then the price of tom collins was decreased, the price of rootbeer was increased, and income was compensated to allow (but not compel) the rat to maintain its consumption as before. The new budget line is shown as bb (price on lever 1 = 1 press per 0.1 cc; price on lever 2 = 1 press per 0.025 cc; income = 483 presses). Budget line cc shows the effects of an increase in the price of tom collins and decrease in the price of rootbeer (price on lever 1 = 1 press per 0.025 cc; price on lever 2 = 1 press per 0.10 cc; income = 267 presses). Note that the point (representing the obtained allocation) on bb is to the right of the point on aa (and the point on cc is to the left of the point on aa) showing a reduction in con-

Figure 16-4. Amount of rootbeer C_1 versus amount of tom collins C_2 for a rat in the experiment of Rachlin, Green, Kage, and Battalio, 1976. The solid lines are budget lines. The point on each line represents the actual consumption of rootbeer and tom collins by the rat with that budget.

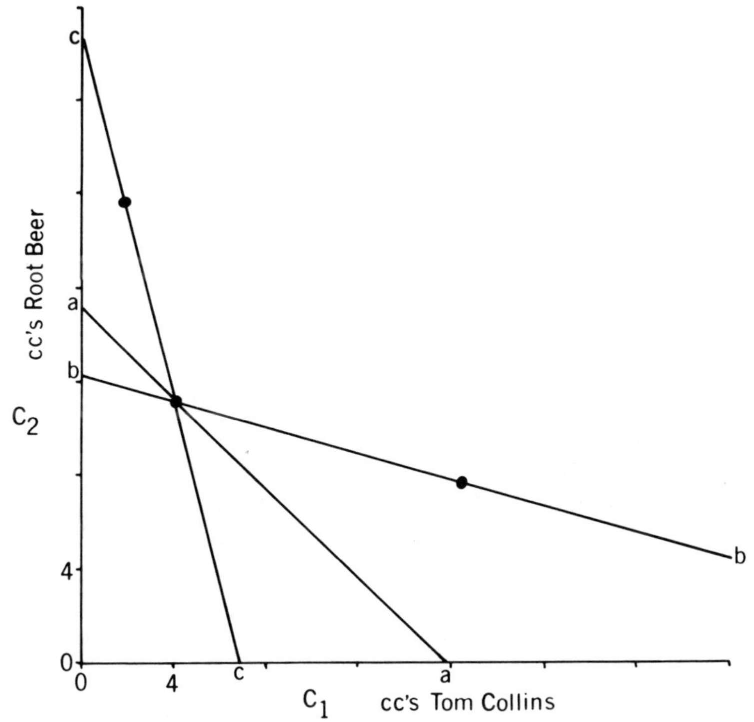

Figure 16-5. The slope of the budget line of Figure 16-4 (a_1/a_2) versus the ratio of consumption (C_1/C_2) by the rat with that budget. Log-log plot.

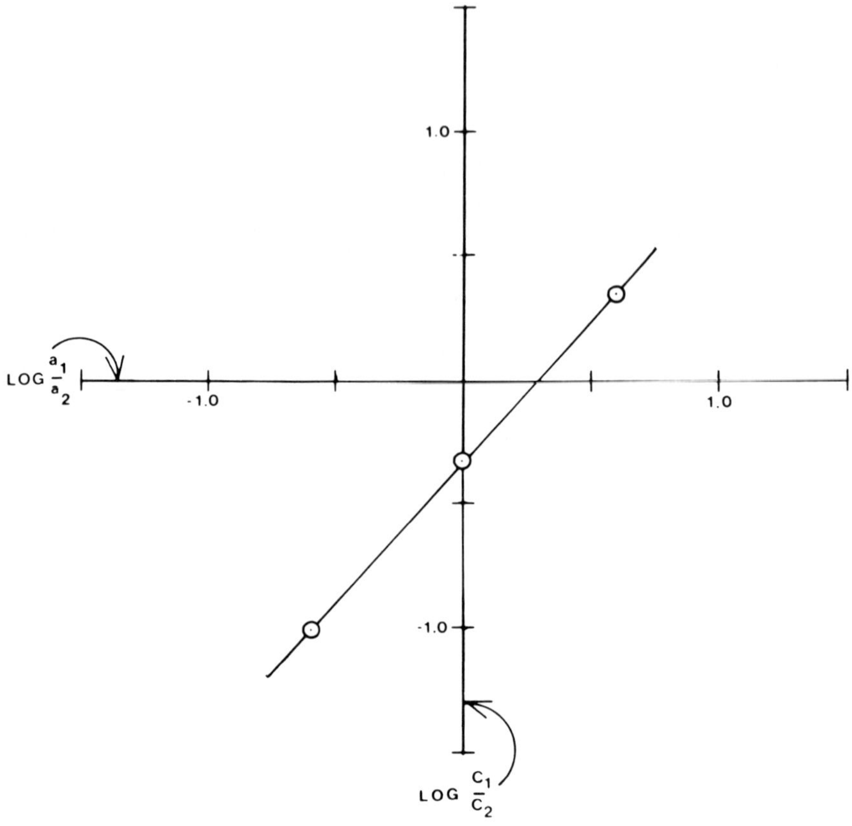

sumption of the more expensive commodity and an increase in consumption of the less expensive commodity in each case.[e]

For each of the three points in Figure 16-4 relative price (a_1/a_2) has been set and relative consumption (C_1/C_2) can be calculated. For the points on aa, bb, and cc, relative prices are 1.0, 4.0, and 0.25; relative consumptions are 0.364, 2.097, and 0.091. Figure 16-5 is a plot of the log of relative consump-

e. The budget lines were varied so as to test economic demand theory that asserts that indifference contours are everywhere convex to the origin. Indifference contours would have this curvature, in the present analysis, when $0 < y < 1$. But, more generally, the fact that the points on bb and cc are outside of the traingle Oaa ensures that demand theory is satisfied (Newman, 1965).

tion versus the log of relative price. The three points fall very close to a straight line. According to Eq. (16.9), the slope of this line is equal to the exponent s. The slope of the line in Figure 16-5 is 1.143. This value is the s from Eq. (16.9). Thus, $y = (s - 1)/s = 0.125$.

As noted previously, Eq. (16.7) is the same as Eq. (16.5). Equation (16.7) was derived from maximization of utility. The constant y in Eq. (16.7) is a measure of substitutability. The same, then, must be true of the constant, u, of Eq. (16.5) (the matching function). When this constant is determined directly in a choice experiment it is called u. When the constant is determined as a result of maximization it is called y. But it is only a single constant. The less-than-unity value of y thus strictly implies (in Eq. (16.5)) $u < 1$. When the exponent u is less than 1, the animal is said to be undermatching (Baum, 1974). The very low value of y in the experiment reported here indicates extreme undermatching.

The intercept on the ordinate of Figure 16-5 is log $(x_1/x_2)^s$. This value is equal to -0.439. Hence $x_1/x_2 = 0.413$.

In other words, rootbeer and tom collins mix are far from completely substitutable ($y \ll 1$) and preferences are biased toward rootbeer in a ratio of about 2.5 to 1 for the rat whose behavior is plotted in Figures 16-4 and 16-5. Economic theory thus allows the matching law to be applied in a straightforward way to concurrent ratio schedules with different reinforcers for the two alternatives.

In another experiment with rats and one with pigeons, the two commodities chosen between with concurrent ratio schedules were food and water. These experiments will not be described here in any detail. See Battalio, Kagel, Rachlin, and Green (1981); Kagel, Battalio, Green, and Rachlin, (1980); Kagel, Battalio, Rachlin, Green, Basmann, and Klemm (1975); and Rachlin, Green, Kagel, and Battalio (1976) for more detailed descriptions of these and similar experiments. But a critical difference between the food/water experiments and the rootbeer/tom collins experiments is worth noting. For all the pigeons and rats tested, food and water were even less substitutable than were rootbeer and tom collins mix for the rat of Figures 16-4 and 16-5. Complete nonsubstitutability between two commodities indicates, as mentioned previously, that the animal tends (literally, at all costs) to keep the two commodities within a small range of proportions. Thus, when prices of food and water were varied in the same way as the prices of rootbeer and tom collins mix (as in Figure 16-4), adjusting income so as to allow the animals to obtain their initial package) much smaller swings in consumption ratios were found. Both pigeons and rats kept C_1/C_2 at about the same value throughout. Consequently, in plots such as that of Figure 16-5, food/water lines were more nearly horizontal than that of Figure 16-5. Slopes s of 0.1 to 0.3 were typical (Battalio, Kagel, Rachlin, and Green, 1981). This corresponds to values of y of -9 to -2.3. In the matching equation negative values of y would imply a sort of antimatching; as C_1/C_2 increased, T_1/T_2 would decrease. Negative values for y (and u) make no sense if the matching law is considered in a noneconomic context. Why should increasing rate of reinforcement for an alternative cause a *decrease* in rate of response? It makes perfect sense considered in an economic context. Referring to Figure 16-1, when the subject, A acts always to keep the food/water rati) C_1/C_2 at or near its desired level, the higher the rate of food reinforcement C_1 from E_1 (relative to the rate of C_2 from E_2) the less time T_1 the animal has to spend in E_1 to maintain C_1/C_2 at a fixed level.

Concurrent Interval Schedules

With concurrent interval schedules ($m_1 \simeq m_2 < 1$). Equation (16.3) becomes

$$\left(\frac{C_1}{a_1}\right)^{\frac{1}{m}} + \left(\frac{C_2}{a_2}\right)^{\frac{1}{m}} = 1 \ .$$

This is the budget line for interval schedules. With $m < 1$ the budget line is a curved line.[f] Figure 16-6 shows this budget line with $a_1 = a_2 = 1.0$ and $m = 0.2$. The indifference contours are drawn as straight lines, which would be the case when $y = 1$. This approximates the indifference contours for the preponderance of experiments with concurrent variable-interval schedules because, in those experiments, C_1 and C_2 have almost always been the same commodity (most often grain for pigeons). The matching law was first formulated to describe the results of such experiments (Herrnstein, 1961). Reinforcers have been varied from one experiment to the next and the matching law (Eq. (16.5)) has been verified with various organisms under a wide variety of experimental conditions (de Villiers, 1977). But variation of experimental conditions has rarely included making the commodity C_1 different from C_2. Thus, most of the information on concurrent variable-interval schedules comes from the restricted case (shown in Figure 16-6) where $y = 1$. The curvature of the budget line ($m < 1$) represents diminishing marginal rate of return (considering time at an alternative as an investment). If the indifference contours were curved ($y < 1$) their curvature would represent diminishing marginal rate of substitution. With concurrent ratio schedules and different commodities (Figure 16-3) the curvature of the *indifference contours* provides unique, nonedge, maximization solutions. With concur-

f. The constant, m, could be deliberately varied in practice in the following way: In Figure 6-1 consider the variable-interval schedule imposed by E_1 as consisting of two subschedules, subschedule A running during T_1 and delivering reinforcements immediately and subschedule B running during T_2 but holding reinforcements until the subject returns to T_1. A corresponding pair of schedules would apply for E_2. With typical variable-interval schedules, subschedule A and subschedule B would both be the same schedule. With typical variable-ratio schedules, subschedule B would be extinction. As subschedule B increased in value from extinction to the equivalent of subschedule A, (for both E_1 and E_2) the constant m would vary from 1.0 downward.

Figure 16-6. A plot of C_1 versus C_2 such as that of Figure 16-3. In this figure the indifference contours are straight (representing completely substitutable commodities rather than incompletely substitutable commodities as in Figure 16-3) and the budget line is curved (representing concurrent variable-interval schedules rather than concurrent variable-ratio schedules as in Figure 16-3.

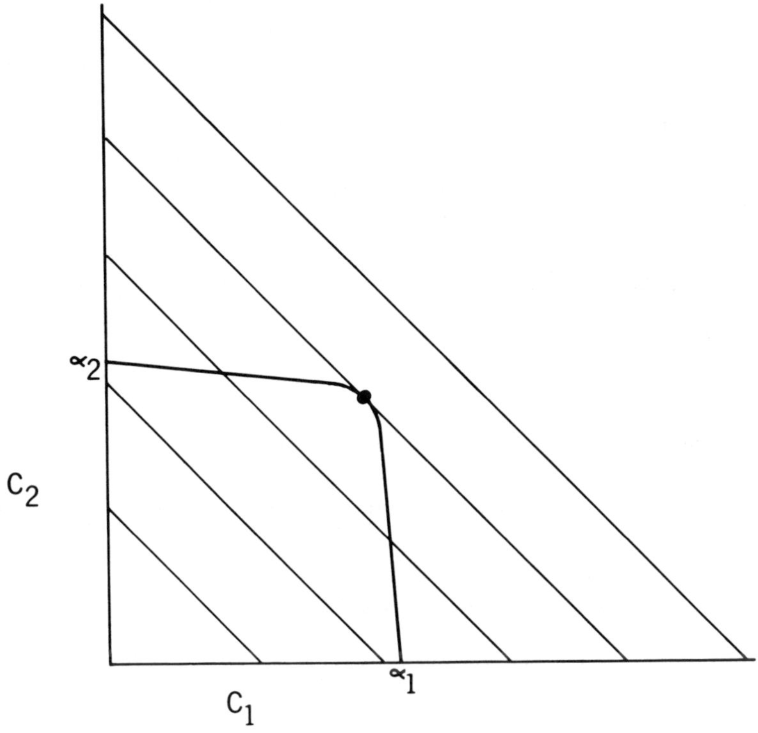

rent interval schedules (Figure 16-6) the curvature of the *budget line* provides unique, nonedge, maximization solutions, In both cases Eq. (16.5) (or, in other terms, Eq. (16.7) describes the results.

Often, with concurrent variable-interval schedules, some slight undermatching is found ($u \simeq 0.8$) even when C_1 and C_2 are the same reinforcer, most often grain delivered to pigeons from the same hopper (Trevitt, Davison, and Williams, 1973). According to the present analysis $u \neq 1$ could not arise from inadvertent inequality of m_1 and m_2 because Eq. (16.8) asserts that such inequality should cause "bias" ($b \neq 1$ in terms of Eq. (16.5)) rather than "undermatching" ($u <$ 1 in terms of Eq. (16.5)). This analysis also asserts that undermatching cannot arise from a side preference because preference for one side or another ($x_1/x_2 \neq 1$ in terms of Eq. (16.7) implies $b \neq 1$ (in terms of Eq. (16.5)) and again leaves the exponent unaffected.

Undermatching, despite C_1 and C_2 consisting of the same reinforcer, must arise because of some residual nonsubstitutability. This could come about not because the animal prefers one side to another but because the animal prefers to keep responses in a certain proportion, say to alternate them. A pigeon or rat might get tired of leaning to one side to peck the key or press the lever, or a pigeon might get tired of pecking a key of a constant color and prefer color to change, or a person might prefer to press two buttons to one button. Catania (1975) has shown that pigeons prefer to peck two keys to pecking a single key even where the consequences in the two cases are identical. This "preference for proportion" is what is meant by nonsubstitutability. To the extent that it exists in choice experiments it will produce undermatching ($u < 1$).

When the tendency to alternate is severely restricted, as when a barrier that the subject must cross to switch from E_1 to E_2 is placed between the concurrently available manipulandums, undermatching should not be obtained. Baum (Note 1) has found that with such a barrier, with pigeons, overmatching ($u > 1$) is in fact the result.

Occasionally results have been reported of concurrent-interval experiments where C_1 and C_2 were different commodities. In two of these studies (Hollard and Davison, 1971; Miller, 1976) undermatching was *not* found ($u \simeq 1$). From the point of view of this analysis this could only occur because, despite the apparent difference in quality and despite preference one way or the other, C_1 and C_2 were completely substitutable.

In Miller's experiment the commodities were different types of grain comprising the normal diet of the pigeon. Since taste is a minor factor in preference by pigeons and birds in general (Gustavson, 1977) various grains may be expected to vary only in a single dimension (perhaps size or color or some multiplicative combination of attributes expressible as a single dimension). Thus despite preferences different grains may be substitutable for each other for the pigeons. Hollard and Davison compared food and reinforcing brain stimulation. Because food deprivation has a strong effect on brain stimulation responding, it has been suspected that reinforcing brain stimulation acts somehow on the same internal mechanism that regulates food intake (Trowill, Panksepp, and Gandelman, 1969). If such speculations were valid the two commodities would be completely substitutable. The Hollard-Davison result is strong evidence that the speculations are valid.

Two concurrent variable-interval studies have been reported in which C_1 and C_2 are clearly nonsubstitutable. A study by Hamblin and Miller (1977) used milk and a sucrose solution as C_1 and C_2 for rats. Experiments in my laboratory with concurrent *ratio* schedules indicate that milk and sucrose solution are not completely substitutable for rats. With these commodities as reinforcers with concurrent *interval* schedules Miller found consistent undermatching ($u < 1$) as predicted here.

Another study (Hursh, 1978) reported results with food and water as C_1 and C_2 with rhesus monkeys responding on concurrent interval schedules. As indicated previously, food and water have a low degree of substitutability. As predicted in the previous section, Hursh found antimatching ($u \simeq -10$, a somewhat more negative slope than the values of -9 to -2.3 found with pigeons using concurrent ratio schedules of food and water). The higher the rate of reinforcement of a given commodity, the less time the monkeys spent at the response by which it was obtained. A similar result has been found with pigeons (Baum, Note 1) with food and water as reinforcers on concurrent variable-interval schedules.

Another study was reported (Matthews and Temple, 1979), in which cows chose between concurrent variable-interval schedules of hay versus meal. Since cows, unlike pigeons, do taste their food some degree of nonsubstitutability in the two commodities might be expected. The cows choices obeyed Eq. (16.5) and, as expected, the exponent u was far less than unity, averaging 0.41 for responses and 0.63 for times.

These findings are powerful evidence that matching, when it is found, is a consequence of maximization. If, as Staddon and Motheral (1978) speculated, the reverse were true (that organisms had evolved so as to maximize reinforcement only through a more fundamental matching mechanism) one must ask why organisms match with food versus food and water versus water but actually do the reverse of matching with food versus water. With the matching law as expressed in Eq. (16.5) there is only one relevant dimension of reinforcement, its "value" to the organism (Baum and Rachlin, 1969; Rachlin, 1971). But a dimension in addition to simple value of a reinforcer must affect choice with different reinforcers. It has been argued here that this dimension is their substitutability. It is possible, of course, that organisms match fundamentally when choosing between schedules of the same reinforcer and maximize fundamentally when choosing between schedules of different reinforcers, but this seems an unparsimoneous way to explain results that are also explainable in terms of maximizing alone.

Concurrent Ratio-Interval Schedules

These schedules are part of the general calss of schedules where $m_1 \neq m_2$. The budget line is Eq. (16.3). Equation (16.8) predicts that, other things equal, organisms should prefer the schedule with the higher value of m but that matching should still occur. According to Eq. (16.8), where one schedule is a ratio schedule ($m = 1$) and the other is an interval schedule ($m < 1$) and C_1 and C_2 are the same commodity ($y = 1$, $x_1 = x_2$), subjects should match (obey Eq. (16.5)) with $u = 1$: no overmatching or undermatching) but should show a bias toward the ratio schedule.

Behavior does not conform to this prediction in all respects. Results with concurrent ratio-interval schedules obtained by Herrnstein and Heyman (1979) show matching (with $u = 1$) but do not reveal the expected bias toward the ratio schedule. One possible reason for the apparent failure of subjects to maximize reinforcement is that the choice situation as expressed in Figure 16-1 is oversimplified. So far C_1 and C_2 have been assumed to be the reinforcers delivered by one schedule or another—such commodities as grain or pellets of food or dippers of water or milk or (in experiments with human subjects) points on a counter, or money. But these "goods" are not the only ones available to a subject in an experiment. As Herrnstein (1970) has pointed out, other reinforcers are available to an animal in an experimental chamber. In the case of pigeons, preening, pecking at the floor, wing flapping, and other manifestations of what Staddon and Simmelhag (1971) classify as interim behavior may be positively valued.

Battalio, Kagel, and Green (1979), Rachlin and Burkhard (1978), Rachlin (1978); and Green, Kagel, and Battalio (Chapter 18) have shown that "lesure," time spent not pecking the key (or pressing the lever) and therefore available for interim responses, may be treated as a good, partially substitutable for other goods such as food and water.

As Figure 16-2 shows, schedules with high values of m require more time spent responding (higher values of T) than do schedules with low values of m, in order to obtain equivalent reinforcement (equivalent values of C). This implies that ratio schedules (with high values of m) allow less leisure than interval schedules (with low values of m). The higher is m, the lower is the lesire allowed by the schedule. Thus although Eq. (16.8), ignoring leisure, asserts a bias directly proportional to m, consideration of leisure implies a bias inversely proportional to m.

The foregoing analysis implies that if leisure activity were somehow removed or made unavailable (say, by restraining the subject) bias should shift toward ratio schedules, whereas if leisure activity were somehow enhanced or increased (say, by providing a mirror to pigeons or wood chips to rats or reading material to human subjects), bias should shift toward interval schedules. Such tests have not, to my knowledge, been made.

Another subclass of concurrent schedules with differences in m are single (nominally, nonconcurrent) schedules. Performance on such schedules may be considered as choice between responding and not responding. Choice between responding and not responding is described by the matching law (Eq. (16.5)) by substitution of time spent responding and not responding for T_1 and T_2 and substitution of reinforcement for C_1. Reinforcement assumed to arise from leisure (Herrnstein, 1970) corresponds to C_2. In practice C_2 is estimated from the data (de Villiers, 1977).

Performance with various single schedules can be described in terms of matching and maximization just as concurrent ratio-interval schedules can. Since corresponding issues arise in concurrent and nominally nonconcurrent cases single schedules will not be discussed further here. See Herrnstein (1970) and de Villiers (1977) for a description in terms of matching and Rachlin (1978) for a description in terms of maximizing.

DISCUSSION

Economics has developed largely as a nonexperimental science. Through the years scattered attempts have been made to study individual choice behavior in economics experimentally (see MacCrimmon and Toda, 1969, for a review of the work), but these experiments have no coherent set of operations. Operant conditioning techniques provide coherent operations that can constitute an experimental paradigm for the study of individual choice behavior (and a number of related areas) in economics. Perhaps the best of these operant procedures for this purpose are concurrent schedules of reinforcement. Such schedules have straightforward application to everyday choices and, by expressing variables in relative terms, they avoid many of the complications of those choices.

It has elsewhere been shown how economic demand theory and labor supply may be modeled in operant paradigms (Rachlin et al., 1976; Battalio et al., 1979; Green, Kagel, and Battalio (Chapter 18). Here it has been shown that the matching law, the best available empirical description of choice behavior with concurrent schedules (and, by extension, of choice behavior in describable situations outside of the laboratory), is an economic law. It provides measures (the consistency of which can be tested) of the preference and substitutability of commodities.

REFERENCE NOTES

1. Baum, W. Unpublished data. (University of New Hampshire).
2. Allen, C.M. *A general economic model of choice.* Unpublished manuscript, 1979. (Arizona State University).

REFERENCES

Barten, A.P. 1977. The systems of consumer demand functions approach: A review. *Econometrica 45*: 23–51.

Battalio, R.C.; J.H. Kagel; and L. Green. 1979. Labor supply behavior of animal workers: Towards an experimental analysis. In V.L. Smith, ed., *Research in Experimental Economics*, vol. 1, Greenwich, Conn.: JAI Press.

Battalio, R.C.; J.H. Kagel; H. Rachlin; and L. Green. 1981. Consumer demand behavior with pigeons as subjects. *Journal of Political Economy 89*: 67–92.

Baum, W.M. 1973. The correlation based law of effect. *Journal of the Experimental Analysis of Behavior 20*: 137–53.

_____. 1974. On two types of deviation from the matching law: Bias and undermatching. *Journal of the Experimental Analysis of Behavior 22*: 231–42.

_____. 1976. Time-based and count-based measurement of preference. *Journal of the Experimental Analysis of Behavior 26*: 27–37.

Baum, W.M., and H.C. Rachlin. 1969. Choice as time allocation. *Journal of the Experimental Analysis of Behavior 12*: 861–74.

Becker, G.S. 1965. A theory of the allocation of time. *The Economic Journal 85*: 495–518.

Catania, A.C. 1975. Freedom and knowledge: An experimental analysis of preference in pigeons. *Journal of the Experimental Analysis of Behavior 24*: 89–106.

de Villiers, P.A. 1977. Choice in concurrent schedules and a quantitative formulation of law of effect. In W.K. Honig and J.E.R. Staddon, eds., *Handbook of Operant Behavior*, Englewood Cliffs, N.J.: Prentice-Hall.

Ferster, C.B., and B.F. Skinner. 1957. *Schedules of Reinforcement.* New York: Appleton-Century-Crofts.

Fleshler, M., and H.S. Hoffman. 1962. A progression for generating variable-interval schedules. *Journal of the Experimental Analysis of Behavior 5*: 529–30.

Gustavson, C.R. 1977. Comparative and field aspects of learned food aversions. In L.M. Barker, M.R. Best, and M. Domjan, eds., *Learning Mechanisms in Food Selection*, Waco: Baylor University Press.

Hamblin, R.L., and H.L. Miller, Jr. 1977. Matching as a multivariate power law: Frequency of behavior versus frequency and magnitude of reinforcement. *Learning and Motivation 8*: 113–26.

Herrnstein, R.J. 1961. Relative and absolute strength of response as a function of frequency of reinforcement. *Journal of the Experimental Analysis of Behavior 4*: 267–72.

_____. 1970. On the law of effect. *Journal of the Experimental Analysis of Behavior 13*: 243–66.

_____. 1981. A first law for behavioral analysis. *Behavioral and Brain Sciences 4*: 392–95.

Herrnstein, R.J., and G.M. Heyman. 1979. Is matching compatible with reinforcement maximization on concurrent variable-interval variable-ratio? *Journal of the Experimental Analysis of Behavior 31*: 209–23.

Herrnstein, R.J., and D.H. Loveland. 1975. Maximizing and matching on concurrent ratio schedules. *Journal of the Experimental Analysis of Behavior 24*: 107–16.

Heyman, G.M., and R.D. Luce. 1979. Operant matching does not result from maximizing reinforcement rate. *Animal Learning and Behavior 1*: 133–40.

Hollard, V., and M.C. Davison. 1971. Preference for qualitatively different reinforcers. *Journal of the Experimental Analysis of Behavior 16*: 375–80.

Hursh, S.R. 1978. The economics of daily consumption controlling food- and water-reinforced responding. *Journal of the Experimental Analysis of Behavior 29*: 475–91.

Kagel, J.H.; R.C. Battalio; L. Green; and H. Rachlin. 1980. Consumer demand theory applied to choice behavior of rats. In J.E.R. Staddon, ed., *Adaptation to Constraint: The Biology, Economics and Psychology of Individual Behavior*, New York: Academic Press.

Kagel, J.H.; R.C. Battalio; H. Rachlin; L. Green; R.L. Basmann; and W.R. Klemm. 1975. Experimental studies of consumer demand behavior using laboratory animals. *Economic Inquiry 13*: 22–38.

Leigland, S. 1979. Deviations from matching as a measure of preference for alternatives in pigeons. *Journal of the Experimental Analysis of Behavior 32*: 1–13.

Logan, F.A. 1965. Decision making by rats: Delay versus amount of reward. *Journal of Comparative and Physiological Psychology 59*: 246–51.

MacCrimmon, K.R., and M. Toda. 1969. The experimental determination of curves. *Review of Economic Studies 36*: 433–451.

McSweeny, F.K. 1975. Concurrent schedule responding as a function of body weight. *Animal Learning and Behavior 3*: 264–70.

Matthews, L.R., and W. Temple. 1979. Concurrent schedule assessment of food preference in cows. *Journal of the Experimental Analysis of Behavior 32*: 245–54.

Miller, H.L. 1976. Matching-based hedonic scaling in the pigeon. *Journal of the Experimental Analysis of Behavior 26*: 335–47.

Newman, P. 1965. *The Theory of Exchange*. Englewood Cliffs, N.J.: Prentice-Hall.

Rachlin, H. 1971. On the tautology of the matching law. *Journal of the Experimental Analysis of Behavior 15*: 249–51.

_____ . 1978. A molar theory of reinforcement schedules. *Journal of the Experimental Analysis of Behavior 30*: 345–60.

_____ . 1980. Economics and behavioral psychology. In J.E.R. Staddon, ed., *Adaptation to Constraint: The Biology, Economics and Psychology of Individual Behavior*, New York: Academic Press.

Rachlin, H., and B. Burkhard. 1978. The temporal triangle: Response substitution in instrumental conditioning. *Psychological Review 85*: 22–48.

Rachlin, H.; L. Green; J.H. Kagel; and R.C. Battalio. 1976. Economic demand theory and psychological studies of choice. In G. Bower, ed., *The Psychology of Learning and Motivation*, vol. 10, New York: Academic Press.

Rachlin, H.C.; J.H. Kagel; and R.C. Battalio. 1980. Substitutability in time allocation. *Psychological Review 87*: 355–74.

Shimp, C.P. 1969. Optimal behavior in free-operant experiments. *Psychological Review 76*: 97–112.

Staddon, J.E.R., and S. Motheral. 1978. On matching and maximizing in operant choice experiments. *Psychological Review 85*: 436–45.

Staddon, J.E.R., and V.L. Simmelhag. 1971. The "superstitution" experiment: A reexamination of its implications for the principles of adaptive behavior. *Psychological Review 78*: 15–43.

Stigler, G.J., and G.S. Becker. 1977. De gustibus non est disputandum. *The American Economic Review 67*: 76–90.

Stubbs, D.A., and S.S. Pliskoff. 1969. Concurrent responding with fixed relative rate of reinforcement. *Journal of the Experimental Analysis of Behavior 12*: 887–95.

Trevitt, A.J.; M.C. Davison; and R.J. Williams. 1973. Performance in concurrent interval schedules. *Journal of the Experimental Analysis of Behavior 19*: 451–58.

Trowill, J.A.; J. Panksepp; and R. Gandelman. 1969. An incentive model of rewarding brain stimulation. *Psychological Review 76*: 264–81.

17 PREFERENCE AND RESPONSE SUBSTITUTABILITY IN THE MAXIMIZATION OF BEHAVIORAL VALUE

Barbara Burkhard

The work described here concerns response substitutability. This work is part of a larger project to develop a general theory of behavior using concepts borrowed from microeconomics. Some success has been achieved in describing behavior maintained by schedules of reinforcement in these terms (Rachlin and Burkhard, 1978; Rachlin, 1978; Rachlin, Chapter 16 of this volume; Green, and Kagel, and Battalio, Chapter 18). In general, the approach has been to show that behavior observed under constraint, such as a constraint imposed by a contingency, is the result of a maximization process. That is, the value of the total behavior output is maximized given the constraints on the behavior. This parallels the economic notion that consumers each maximize the utility of their total purchases given the constraints of their individual budgets.

In microeconomics the analysis of the maximizing process subsumes a preference structure for the commodities in question. That is, the maximizing process proceeds according to individual preference. The structure of preference in microeconomics is based on commodity substitutability. In this work, it is similarly proposed that response substitutability is a viable parameter for describing preferences, especially when used in this context of value maximization. In as much as substitutability is a relatively unexplored parameter in psychological theories of preference or choice, the goal here is to show how response substitutability can be measured and how it can be used to analyze behavior change.

The research in this chapter was supported by a grant from the National Science Foundation. The author wishes to thank David G. Ebin for his extensive help with the mathematical and statistical analyses.

Before launching into a discussion of substitutability per se consider the question, How does a theory of preference, whether based on substitutability or not, aid in the experimental analysis of behavior?

Premack was perhaps the first to make use of a preference model for the analysis of instrumental reinforcement effects. The model of preference he proposed (Premack, 1971) involves a simple scale of response value. That is, the organism orders the response alternatives available to it. The relative values of the various response alternatives can be measured by the relative durations of the times spent on the activities in an unrestricted or "baseline" condition. Premack used the relative values of responses to predict reinforcement and punishment effects. That is, responses of higher value reinforce responses of lower value whereas lower valued responses punish higher valued responses when the responses are restricted by a contingency. Various parametric studies by Premack and his colleagues indicated more quantitative predictions may be possible using the value scale. The studies show that the level of instrumental responding is directly proportional to the baseline levels of the instrumental and contingent responses (Brownstein, 1962; Landford, Benson, and Weisman, 1968; Premack, 1961, 1963; Schaeffer, 1965).

Premack's work has laid the foundation for more recent models of reinforcement. The deprivation model (Timberlake and Allison, 1974), the conservation model (Allison, 1976), and Mazur's (1979) value-averaging model all retain Premack's notion that preference structure can be described by the value scale. These three models are similar to Premack's model in that reinforcement and punishment effects result from an interaction of the preferences and the constraints imposed by the contingency. In essence they offer various rules for how behavior is rearranged between baseline and contingency conditions.

The models differ from Premack's and each other in how response value and contingency restrictions are represented and evaluated. Evaluating the relative merits of these models, is beyond the scope of this chapter, but a problem with all models using Premack's value scale must be pointed out. All of the models so far mentioned have been tested primarily with eating, drinking, and running responses with rats. The values of these responses are often stable, predictable, and independent of fluctuations of other response alternatives. As Mazur (1979) has indicated, this stability or constancy is a prerequisite. If the contingency restrictions differentially change the value of a given response or in some way induce an interaction among the alternatives (for example, if increased running induces thirst), Premack's model may not hold.

Problems of response interactions are even more clearly illustrated in studies of adjunctive behavior and interim (cf. Staddon, 1977) behavior. When responses other than the usual instrumental and contingent responses are available or measured in such studies, the "rearrangement" of behaviors between baseline and contingency conditions is far more complicated. The value scale is of limited utility. The new distribution of behaviors under a contingency may depend on not only the individual response values but a host of interactions among responses as well. This point is simply illustrated in a study involving children.

Two children were observed individually over several sessions in a playroom with eight toys. The children were allowed free access to all the toys for the first four sessions. It was possible to obtain a stable rank order of preference for the toys for each child based on the time spent with each item. The more time spent with the toy, the higher the rank order of preference. In a final session, play with the most preferred toy was made contingent on play with the least preferred toy. This is a direct test of Premack's differential probability hypothesis and, as expected, the time on the instrumental toy was found to increase and the time on the contingent toy decreased. However, this is an inadequate way of describing what actually happened. Both children became noticeably agitated. Both verbalized spontaneously that they did not wish to play with either toy involved in the contingency. Rather, they substituted one of the other toys available. This play was noticeably different in character. During baseline sessions both children played quietly with toys, going from one to another. They did not spend a long period of time with any given toy. During the contingency session, spontaneous verbalizations increased and the children became more involved with the single toy that was substituted for the pair of toys in the contingency. One child spent the entire session on this substitute toy after an initial short period with the pair involved in the contingency. The results indicate that the values of the toys changed dramatically under the contingency. A toy that did not rank particularly high initially increased dramatically when the contingency between two other toys was imposed. It is also noteworthy here that none of the available models cited based on the simple scale of value could account for the allocations of time to the instrumental and contingent toys in this study or in a more involved study along the same lines done with a larger group of children (Burkhard, 1976). A different type of preference structure is needed which allows the representation of response interactions.

Tversky (1972) has suggested a model of choice applicable to response alternatives that are not independent. Tversky's model is specifically applicable to the substitutable choices. Suppose a boy is presented with a red bicycle, a green bicycle, and a coloring book. If the boy prefers the red bicycle to the green bicycle and any bicycle to the coloring book he mignt nevertheless spend 90 percent of his time playing with the red bicycle, 10 percent playing with the coloring book, and 0 percent with the green bicycle. But, if the red bicycle is removed he may spend 90 percent of his time with the green bicycle and 10 percent still on the coloring book. Tversky's model accounts for this seemingly paradoxical choice pattern where a less preferred toy (the coloring book) was chosen over a more preferred toy (the green bicycle). The rank order of preference for the three toys was marked by the substitutability of the green and red bicycles as evidenced by the complete substitution of the green for the red bicycle when the red was removed.

Tversky's model deals with what we might call static choices. It explains why such paradoxical choices as in the bicycle example exist. However, as it has been presented, it does not readily provide a rule for the allocation of time to various responses in operant situations, where allocations are likely to change over the course of time. The model to be described here, like Tversky's, provides a representation for nonindependent response alternatives. However, this model, based on the concept of substitutability, allows the representation of these alternatives over time.

Assume the existence of the *value function*, which orders temporal distributions of responses on a scale of value. Distributions having higher values are more preferred to those having lower values. The function is written: $V = f(R_1, R_2, \ldots, R_n)$ where R_i is the duration of the ith response. Two properties are assumed for this function: (1) It is increasing, and (2) it is convex. The first assumption requires that an increase in any one response given all others constant results in an increase in V. The second assumption has to do with substitution. Consider, first, the two-response case.

The level sets of a function V are the sets of distributions having equal values. Figure 17-1 shows three such sets graphically. Each point on the graph represents a particular combination of R_1 and R_2. All combinations of R_1 and R_2

Figure 17-1. Indifference curves for combinations of two responses R_1 and R_2. All points on a single curve have equal value of the value function V as indicated by the constants K_1, K_2, and K_3 ($K_3 > K_1$). The arrow from the origin shows the direction of preference.

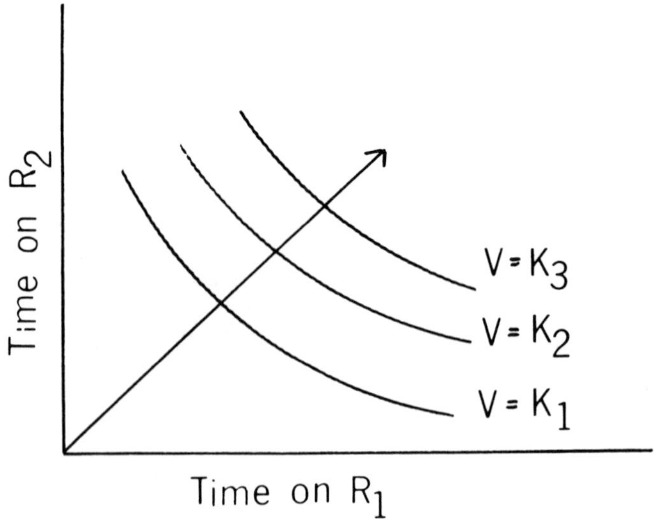

on a single curve have equal value, either K_1, K_2, or K_3. Economists call these curves "indifference curves," since the subject shows no preference for any point on a given curve over any other point on the same curve. However, any combination of R_1 and R_2 represented by a point on a higher curve is preferred to any combination represented as a point on a lower curve.

The rate of substitution is defined as the negative of the slope of the curve. The convexity property implies that the curve will be convex, that the slope will increase as R_1 increases. The marginal rate of substitution (MRS) is the negative of the slope at a given point. A limiting case (of complete substitutability) is a curve that has a constant slope of -1. In this case the curve is a straight line, indicating that R_1 and R_2 are completely substitutable for each other over the entire range of values. As the curve departs from this limit, substitutability decreases.

The two-dimensional representation of Figure 17-1 is limited in application to psychological problems because contingencies such as reinforcement or punishment involve at least three responses, the instrumental response, the contingent response (consumption of the reinforcer), and some other behavior that the subject could perform. Without this other behavior the contingency would completely restrict the instrumental response. (The other response may also be considered the behavior of not responding, or not making the instrumental response. Obviously this behavior must be permitted or the instrumental response would not be free to vary). Consideration must thus be widened from two to three responses.

Figure 17-2 shows three responses in a three-dimensional representation. Indifference surfaces rather than indifference curves are shown. Again, any point on a higher surface is preferred to any point on a lower surface.

The three-dimensional plot can be reduced to two dimensions by considering only distributions of responses in a constant time period $T (R_1 + R_2 + R_3 = T)$. This restricts the possible combinations of R_1, R_2, and R_3 to a triangular plane with vertices at $R_1 = R_2 = R_3$, as shown in Figure 17-3. The intersection of the triangular plane with the indifference surfaces yields a set of closed curves, indifference contours. One possible set of contours is shown in Figure 17-4, where the triangle is shown in the plane of the paper. The convexity property of the value function implies that the contours will bound convex sets. That is, they will have no "dips" in them. (A more detailed discussion of the assumptions of this model may be found in Rachlin and Burkhard, 1978). Contour a in Figure 17-5 bounds a convex set, whereas contour b does not.

The shape and position of the contours reflect the substitutabilities of the responses as they change over the various distributions contained in the triangle. Figure 17-6 contains a contour that shows different substitutabilities among the three responses. The linearity of the contour in the region parallel to the $R_1 - R_2$ side of triangle indicates that R_1 and R_2 are very substitutable for each other. This is much more so than in the case of either R_1 or R_2 for R_3. The part

Figure 17-2. Indifference surfaces shown in three-dimensional space. The arrow indicates the direction of preference. All points on a single surface have equal values. ($K_2 > K_1$).

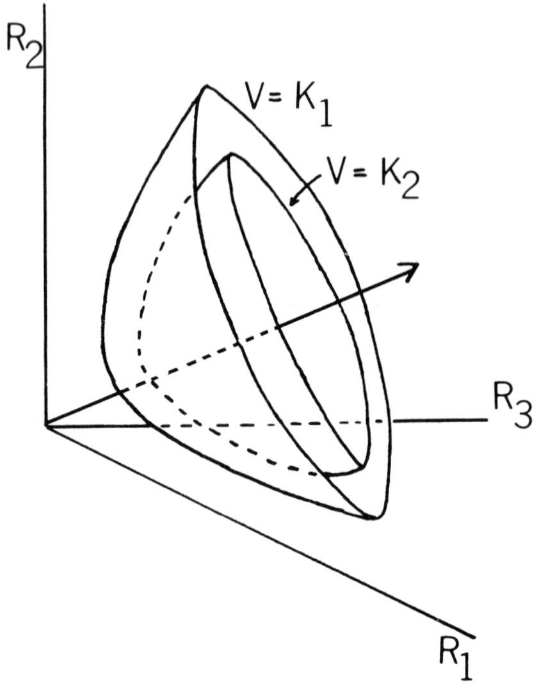

of the curve parallel to the side of the triangle reflects a rate of substitution approaching -1.

Indifference contours can be illustrated with data. Figure 17-7 shows contours from two subjects. Each subject was asked to select three enjoyable activities that could be done in combination over a particular period of time. An appropriate time period was selected. An example might be playing the piano, reading a newspaper, and going for a walk in a two-hour period. Each subject was asked to indicate a maximally preferred distribution. This point is marked *X* on the triangles of Figure 17-7. Each subject was also given a list of 65 time distributions of the three responses and asked to rate each distribution as to how enjoyable that particular combination would be. A scale of 1 to 5 was used, 5 being the most preferred.

The triangles of Figure 17-7 show half the points for two subjects. As can be seen, each triangle contains distinct regions of equal value. Contours are drawn by hand. The contours for the two subjects show differing substitutabilities.

Figure 17-3. The locus of distributions of R_1, R_2, and R_3 contained in a constant time interval T.

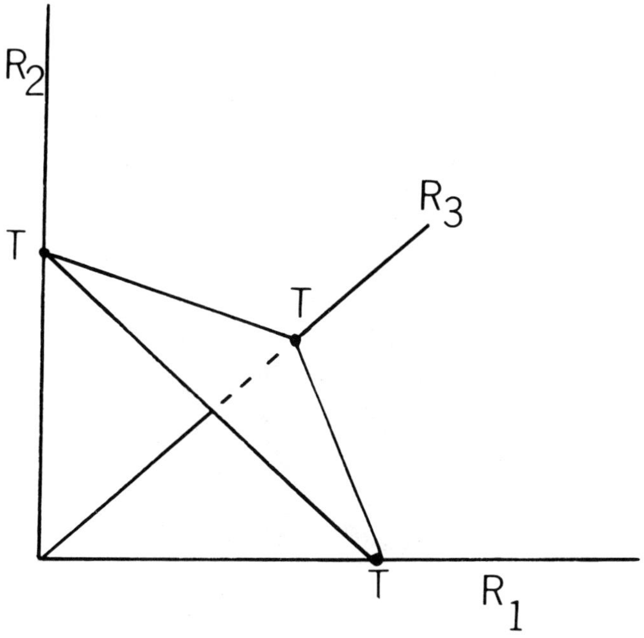

Those for S_2 show that R_1 and R_2 (selected for this similarity) were generally more substitutable for each other than either was for R_3.

Do these contours actually bound convex sets? To test this all points of equal value were connected with straight lines. The outer lines of the figures thus constructed constitute a polygon. The number of points inside the polygon having lower value were then counted. If the contours bound convex sets there should be no points of lower value within a polygon in the perfect case. Using the binomial theorem one can calculate the probabilities that the specific number of interior points of lower value that were obtained occurred by chance. Of a total of 32 contours tested (4 for each of 8 subjects) only 6 were greater than 0.05. One-third was less than 10^{-4}. It can be concluded that the contours probably bound convex sets.

The second test was to see if maximum value points could be predicted if behavior were restricted to a schedule-like situation. Figure 17-8 illustrates the geometrical representation of a ratio-like (part a) and an interval-like (part b) schedule. A ratio-like schedule restricts responding to the set of distributions falling on a line from one vertex of the triangle to the opposite side of the triangle. In the figure the line goes from vertex R_3 to the $R_1 - R_2$ side. In this

Figure 17-4. The intersection of indifference surfaces such as those of Figure 17-2 with the triangular plane of Figure 17-3 as shown in two dimensions. The intersection yields indifference contours on the triangle. Inner contours have higher value.

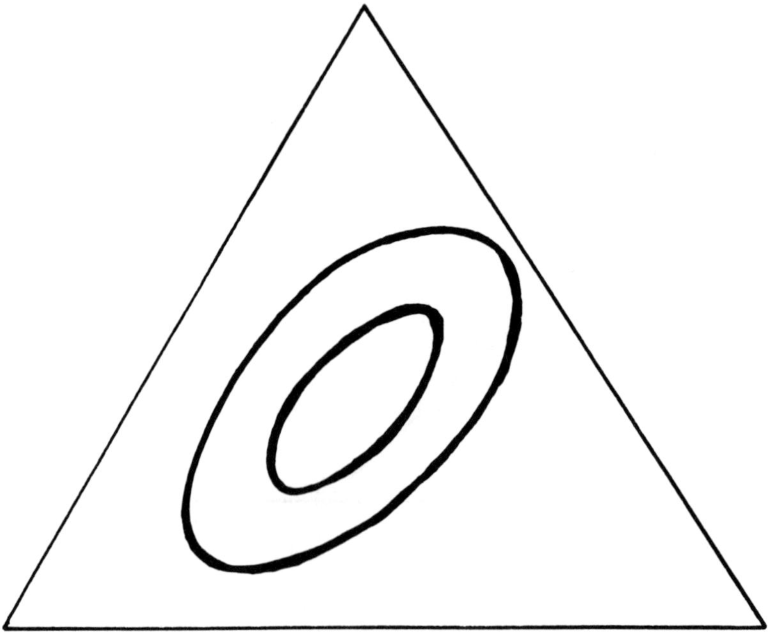

case R_3 is free to vary from 0 to 100 percent while R_1 and R_2 occur in a constant ratio. An interval-like schedule is represented by the restriction of one response (in the figure R_2) to a constant level.

In the study subjects were given 25 different restrictions. For ratio restrictions the subjects were asked to choose the best possible combination of their three responses given that two of them must be done in a specific ratio such as 2:1. That is, a subject might be asked how much time she would spend reading the newspaper, playing the piano, and going for a walk given that twice as much time must be spent reading the newspaper as playing the piano? For interval restrictions the subjects were asked to choose the most preferred combination given that one response, say playing the piano, must be done for a fixed proportion of the time, say 20 minutes.

It was possible to predict with a fair degree of accuracy the combinations the subjects selected. (See the appendix to this chapter for statistical analysis.) These predictions were based on the point at which the indifference contour having the

Figure 17-5. (a) Indifference contour that bounds a convex set. (b) Indifference contour that does not bound a convex set.

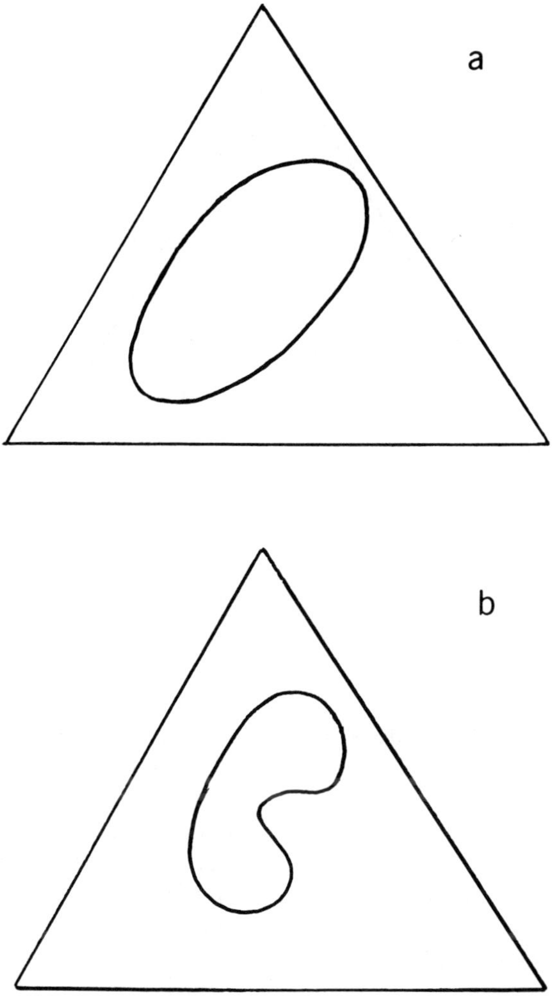

highest possible value is tangent to the restriction line. The present theory implies that this point of tangency is the point at which the value function is maximized given the particular restriction. Specifically, it was found that the predicted points based on the contours were in nearly all cases closer to the points selected by the subjects than would be expected if the two points were simply randomly distributed along the line.

Figure 17-6. Indifference contour showing nearly complete substitutability between R_1 and R_2.

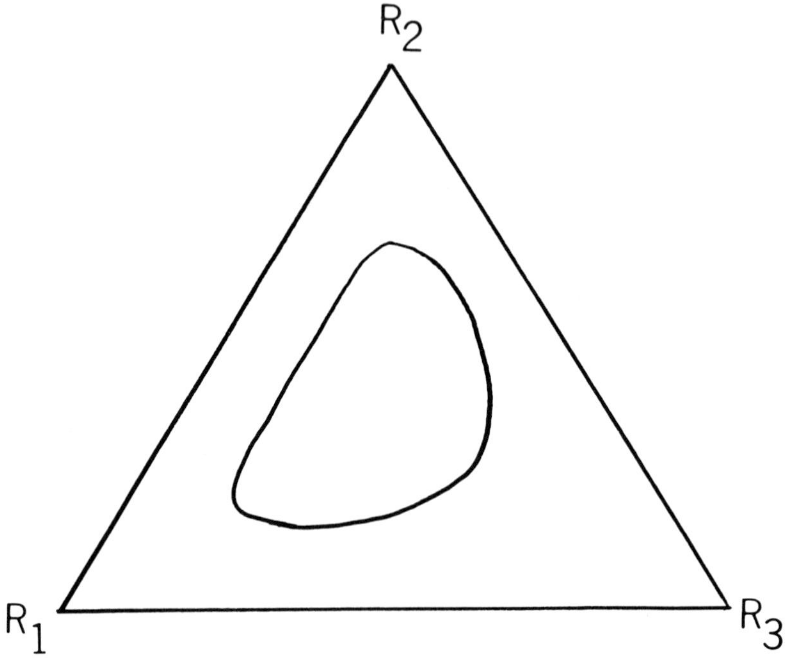

Pilot data for a larger study provides illustration of how the theory of preference based on substitutability can be used to study choices. The situation studied is similar to the problem of choice between bicycle and coloring book. Previous approaches such as Premack's have suggested that the value of individual responses is a relevant parameter for behavioral predictions. The model here suggests that substitutability as represented by the indifference map is the relevant parameter. This model incorporates the earlier approaches into a broader representation of preference, which has more flexibility to deal with problems of nonindependent choices.

Three children ages 4, 5, and 7 were used as subjects. The experimenter worked with each child individually in a separate room in the child's home. The experimenter presented many sets of toys and snacks for the child to select. Each set was made up of three items such that two items were similar and the third item dissimilar. One set, for example, was made up of M&M candies, candy corn, and marbles. The child was told he or she could "buy" 12 items from each set. That is the child might select 7 M&M candies, 3 candy corns, and 2 marbles. For several sessions the child was permitted to select any combination

Figure 17-7. Values of distributions for two subjects. Contours are drawn by hand.

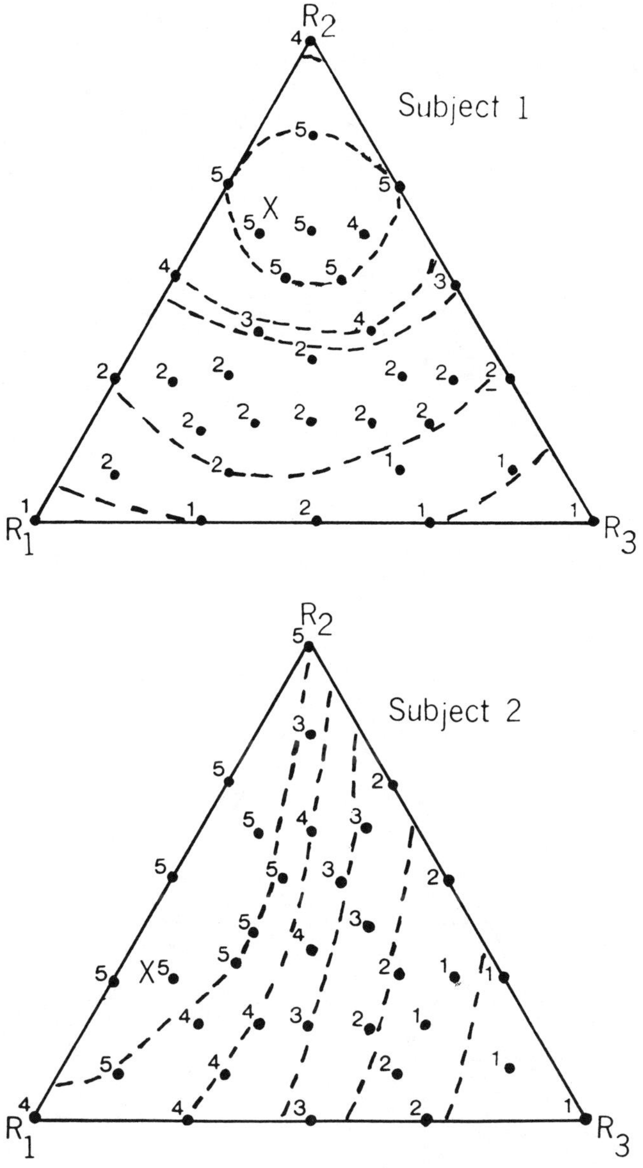

Figure 17–8. (a) Geometrical representation of a ratio-like schedule. R_3 is free to vary along the line from R_3 to $R_1 - R_2$ side. R_1 and R_2 are restricted to a constant proportion. (b) Geometrical representation of an interval-like schedule. R_2 is restricted to a constant level.

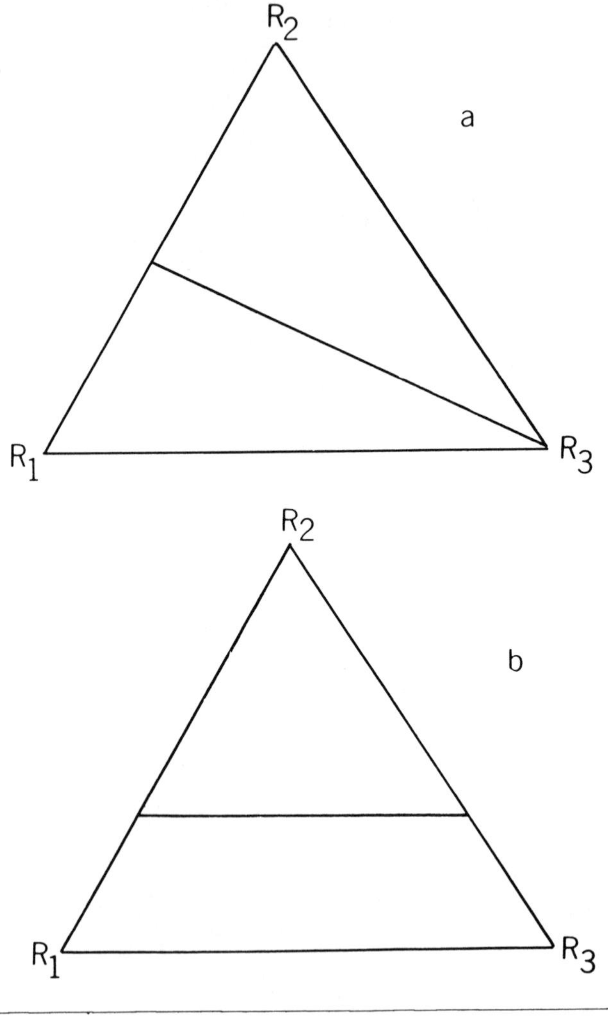

Figure 17-9. Representation of the boy-snack experiment in triangular coordinates. *A* and *a* represent the two similar items. Item *B* is dissimilar. Line *ab* is a ratio line representing a constant proportion between *A* and *B*. Line *cd* is the locus of possible choice sets with a constant level of item *A*.

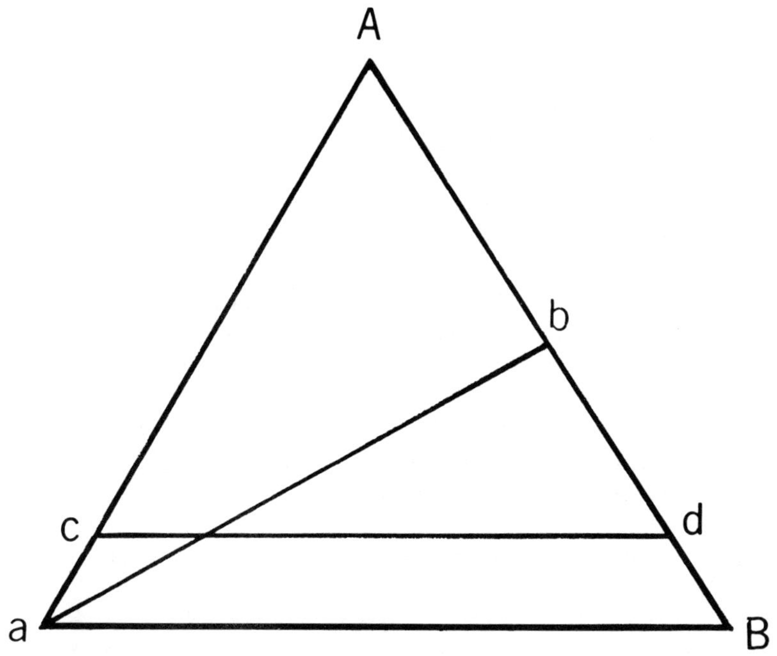

of 12 items from each set. If the choice became fairly stable, the more preferred of the similar pair was identified as item *A*, the less preferred as item *a*, and the dissimilar item as *B*. Preference in this case was the simple rank order of the items selected. If more candy corns were selected than M & M's, candy corn became *A* and M & M's became *a*. Marbles, the dissimilar item were *B*.

In subsequent sessions the subject was again allowed to choose 12 items, but with constraints. Two types of constraints that are similar to those created by variable-time and fixed-ratio schedules of reinforcement were examined. Figure 17-9 shows the possible purchases of the 3 items in a given set in triangular coordinates. Instead of time, the coordinates represent numbers of items purchased. (This, of course, is a discrete rather than continuous measure. However, this does not alter the result that will be analyzed. Each vertex represents a choice set containing 12 of the items identified and zero amounts of the remaining 2 items.

Lines cd and ab represent the constraints used in the study. Line cd (similar to a variable-time schedule) sets a limit on the number of one of the items (A) that could be taken by the child. For example, the child was told that he could select 12 items as long as only 2 of them were candy corns. The question was how the child would allocate the remaining 10 selections to the M & M's and marbles.

Line ab represents a ratio between one of the similar items and the dissimilar item. In this case the child was told that he could choose 12 items as long as he chose an equal number of candy corns and marbles. Here we looked to see how far along the restriction line the choice set appeared.

As this was a pilot experiment the restrictions were varied widely across many sets. The results are too extensive to describe in detail here. However, two basic results are important for the thesis of this chapter. First, substitution was observed between the similar items (A and a) when one or the other was reduced to a low level, as represented by line cd. That is, if candy corns were constrained to a low level, M & M's increased relative to the baseline sessions while marbles remained approximately the same. This is the same result as the bicycle-coloring book example. This result occurred more or less irrespective of whether the substitute of the similar pair had a higher or lower rank than the dissimilar item. If the choices, were in fact completely independent under this constraint they might be expected to follow Luce's (1959) axiom. That is, M & M's and marbles should both increase in the proportion in which they were originally chosen. In fact, the substitutability overshadowed this effect. Only the similar item increased.

This is not to say that constrained choice in no way reflects the rank order of preference. In the ratio contraint (where subjects were asked to select one of the similar pair and the dissimilar item in a set proportion) the rank order of the dissimilar item was very important. If item B, the dissimilar item, was less preferred to items A and a to the extent that it was rarely chosen, then the constrained choice was at point a. Only a was selected. If B was more preferred in that it was frequently selected during baseline, the constrained choice moved toward point b. Less of a and more of the constrained pair was chosen.

These results can be analyzed using the theory of preference by specifying the relations among the three choice alternatives with an indifference map. For simplicity consider the extreme examples, which can be represented by straight-line indifference maps. This will allow the use of simple analytic procedures to predict results.

Figure 17-10 shows the three cross-sections for a set of three-dimensional indifference surfaces whose sides are parallel planes. The figure shows the three pairwise comparisons for the three choice situation in the experiment. The cross-sections reveal the basic preference structure of the situation. Cross-section (a) shows that A and a are highly similar choice alternatives. The rate of substitution of A and a is -1, indicating the highest degree of substitutability. The in-

Figure 17-10. Cross-sections of three-dimensional indifference surfaces whose sides are parallel planes. (a) Indifference lines for *A* and *a* with *B* held constant. (b) Indifference lines for *a* and *B* with *A* constant. (c) Indifference lines for *A* and *B* with *a* held constant.

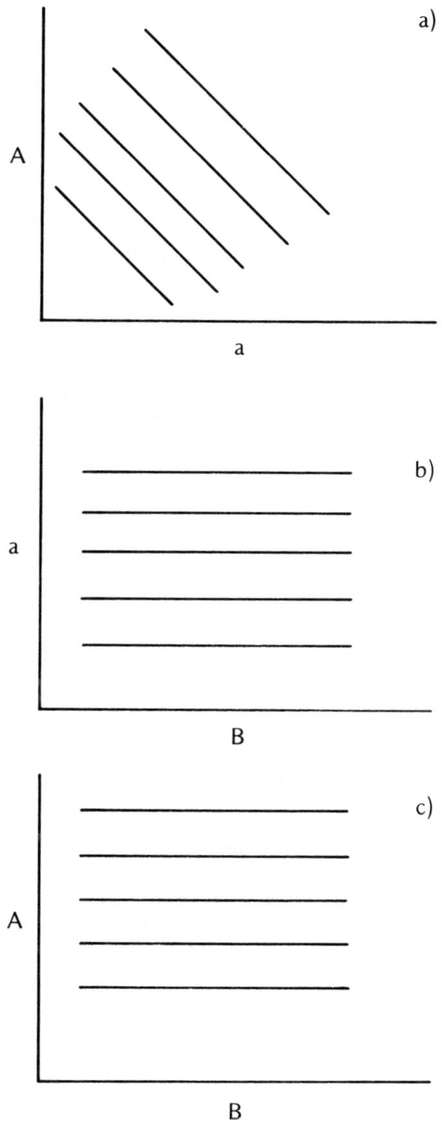

difference lines in (b) and (c) show that a and B and A and B are not substitutes. Furthermore, the straight lines indicate that B is rarely chosen over a or A. The rate of substitution is close to zero, indicating that the subject is unwilling to give up even the smallest amounts of a and A in exchange for an increase in B.

In this case the value function is given by

$$V = A + a + \epsilon B \quad , \tag{17.1}$$

where A, a, and B represent the number of each item. ϵ is a number that will be varied depending on the rank order of B. If the slope of the indifference lines in Figure 17-10 b and c is in fact zero, the value of ϵ is 0 and the value function is given as

$$V = A + a \quad . \tag{17.2}$$

Since the experiment requires that 12 and only 12 items be bought, the budget is

$$A + a + B = 12 \quad . \tag{17.3}$$

This budget generates the triangular plane in three-space shown in Figure 17-9 in the plane of the paper.

Constraining A to a low level, say 2, as is represented by the line cd in the Figure 17-9, yields

$$A = 2 \quad . \tag{17.4}$$

Thus there are three equations for three unknowns (Eqs. (17.2)–(17.4)):

$$\begin{aligned} V &= A + a \quad ; \\ 12 &= A + a + B \quad ; \\ A &= 2 \quad ; \end{aligned}$$

To maximize V it is quite clear that $A = 2$ and $a = 10$ and $B = 0$—one result found in the study.

With a ratio restriction such as line ab,

$$A = B \quad . \tag{17.5}$$

The three equations are now Eqs. (17.2), (17.3), and (17.5),

$$\begin{aligned} V &= A + a \quad ; \\ 12 &= A + a + B \quad ; \\ A &= B \quad . \end{aligned}$$

Clearly to maximize V, $A = 0, B = 0$, and $a = 12$. This means that the choice set is at point a. No items involved in the ratio are selected as a is completely substituted.

One can similarly deal with the situation where B is not zero and is in fact preferred to a. The slopes of the indifference curves of Figure 17-10 b and c change from flat to negatively sloped lines. Again, the value function, Eq. (17.1), is

$$A + a + \epsilon B = V$$

where $\epsilon > 1$. Again, the budget is Eq. (17.3),

$$A + a + B = 12 \quad .$$

With the same ratio line ab,

$$A = B \quad ,$$

(Eq. (17.5). Solving these three equations yields the expression

$$V = 12 - (1 - \epsilon)A \quad . \tag{17.6}$$

To maximize V, A must be as large as possible. Within the budget $A = B = 6$, which is point b.

It is interesting to note what happens if all three indifference map cross-sections have slopes of -1. This situation reflects complete substitutability among the three items. If choice is constrained to a ratio, it is impossible to predict the location of the choice set on the line. Any point along the line is completely substitutable for any other as the line falls wholly on a single indifference surface.

The representations of the indifference maps here are all surfaces with flat sides. In fact, the theory specifies that the surfaces are curved. It is this curvature that allows the constrained choice set to fall between the extreme points on the ratio line. The value function is represented by a more complicated function of A, a, and B than in (17.1). The procedure for finding the choice point is the nonetheless same. The function is maximized subject to the constraints of the budget triangle and the ratio.

The theory of preference presented here is different from previous approaches in that the "rule" for the rearrangement of behaviors from baseline to contingency situations is simply the maximization rule. Different results covering experiments using different types of responses are handled via the preference structure of the particular responses involved. It is the value function by which the differences between one behavioral situation and another are represented. A single function would therefore not be expected to describe many situations containing different responses. Rather, different functions would be expected to represent different situations. Different functions, however, would share the parameters they represent. That is, each value function reveals the preference map. The preference map provides a representation of the substitutabilities of the various responses through the first derivative of the value function. It also tells how these substitutabilities change as time is expended on various responses via the second derivative. This model thus allows a great deal of flexibility in representing and predicting behavior change.

APPENDIX 17A

The statistical inference in this case was based on the distance between the predicted points Y and the chosen points Y' along the various contingency lines. If the predictions using the contours were perfect, the distance between the points would be zero. If the predictions using the contours were completely useless, the distance between Y and Y' would be the expected distance between two points independently and uniformly positioned along the line.

The expected value of the distance between two points uniformly distributed along a line segment of length ℓ is $\ell/3$. This is computed as follows:

Let Y and Y' be two uniformly distributed random variables on the interval $(0, \ell)$. Then each has probability density $1/\ell$. Hence the pair (Y, Y') is distributed on the square $(0, \ell) \times (0, \ell)$ with density $1/\ell^2$. Hence the distance between Y and Y' is the absolute value of $Y - Y'$ and the expected value of the distance is

$$\int_0^{\ell} \int_0^{\ell} |Y - Y'| 1/\ell^2 \; dY dY' \; .$$

Evaluating this integral yields $\ell/3$.
Thus, the expected distance between two points on a line of length ℓ is one-third the length of the line.

Table 17A-1 presents the expected values ($\ell/3$) of the distance between Y and Y' for both types of restriction lines used in the experiment. The mean distances between the predicted and chosen points are also shown. As can be seen from the table, the distances between chosen and predicted points is less than the expected value of this distance if the points were uniformly distributed.

The distance between the predicted and chosen points is less than $\ell/8$ for 14 of the 24 predictions. The probability that the distance is $\ell/8$ or less (for uniformaly distributed points) can be determined by the following:

Consider a square whose sides are distance ℓ in length. The diagonal of the square represents the points where $Y = Y'$ as the distance between the two points is zero. Lines qr and $q'r'$ can be drawn parallel to the diagonal at a distance say a. The probability that two points are this distance a or less apart is given by the ratio of the area of the sum of the two trapezoids created by the lines qr and $q'r'$ and the total area of the square (ℓ^2).

Hence,

$$P|(Y - Y'| \leq a) = 1/\ell 2 \left[(\ell - a) \sqrt{2} + \ell\sqrt{2} \right] \frac{a\sqrt{2}}{2} .$$

$$= \frac{2a}{\ell} - \frac{a^2}{\ell^2} .$$

The probability of the distance between the two points being $\ell/8$ or less is 0.23. The probability that at least 14 of 24 predictions would be $\ell/8$ or less is

$$\sum_{k=14}^{24} \binom{24}{k} (.23)^k \, (.77)^{24-k}$$

Table 17A-1. Observed mean distance ($n = 4$) between Y and Y' on all restriction lines. Also given are the lengths of the restriction lines ℓ the expected distance (Expected $|Y - Y'|$) and the proportion of the line spanned by the observed distance.

Restriction	ℓ	*Expected* ($\lvert Y - Y'\rvert$)	*Observed* ($\lvert Y - Y'\rvert$)	*Proportion*
1. $R_1 = 10$	127.3	42.4	15.9	1/8
2. $R_1 = 20$	113.1	37.7	4.4	1/26
3. $R_1 = 40$	84.9	28.3	8.8	1/10
4. $R_1 = 80$	28.3	9.4	5.3	1/5
5. $R_2 = 10$	127.3	42.4	23.0	1/6
6. $R_2 = 20$	113.1	37.7	24.8	1/5
7. $R_2 = 40$	84.9	28.3	17.7	1/5
8. $R_2 = 80$	28.3	9.4	5.3	1/5
9. $R_3 = 10$	127.3	42.4	12.4	1/10
10. $R_3 = 20$	113.1	37.7	14.7	1/8
11. $R_3 = 40$	84.9	28.3	5.3	1/16
12. $R_3 = 80$	28.3	9.4	5.3	1/5
13. $R_2 = 2R_1$	124.8	41.6	12.6	1/10
14. $R_3 = 2R_2$	124.8	41.6	32.7	1/4
15. $R_1 = 2R_3$	124.8	41.6	9.4	1/13
16. $R_3 = 2R_1$	124.8	41.6	13.1	1/10
17. $R_1 = 2R_2$	124.8	41.6	14.0	1/9
18. $R_2 = 2R_3$	124.8	41.6	19.6	1/6
19. $R_1 = R_2$	122.5	40.8	14.3	1/9
20. $R_2 = 3R_1$	127.5	42.5	15.3	1/8
21. $R_2 = 4R_1$	129.6	43.2	15.4	1/8
22. $R_2 = 5R_1$	131.2	43.7	21.7	1/6
23. $R_2 = 10R_1$	135.5	45.2	16.8	1/5
24. $R_2 = 20R_1$	138.2	46.1	20.0	1/7

Note: $p\left(\binom{24}{14} \leq \ell/8\right) = 0.0002$

or approximately 0.0002. We can thus infer that our prediction is better than chance.

REFERENCES

Allison, J. 1976. Contrast, induction, facilitation, suppression and conservation. *Journal of the Experimental Analysis of Behavior 25*: 185–98.

Brownstein, A.J. 1962. Predicting instrumental performance from independent rates of contingent response in a choice situation. *Journal of Experimental Psychology 63*: 29–31.

Burkhard, B. 1976. Contingent behavior in children: A test of Premack's theory of reinforcement. Unpublished doctoral dissertation. State University of New York at Stony Brook.

Landford, A.; L. Benson; and R.G. Weisman. 1968. Operant drinking and prediction of instrumental performance. *Psychonomic Society 3*: 571–73.

Luce, R.D. 1959. *Individual Choice Behavior: A Theoretical Analysis.* New York: Wiley.

Mazur, J.E. 1979. The value-averaging model: An alternative to "conservation in behavior." *Journal of Experimental Psychology: General 108*: 43–47.

Premack, D. 1961. Predicting instrumental performance from the independent rate of the contingent response. *Journal of Experimental Psychology 61*: 163–71.

____. 1963. Predicting of the comparative reinforcement values of running and drinking. *Science 139*: 1062–63.

____. 1971. Catching up with common sense or two sides of a generalization: Reinforcement and punishment. In R. Glaser, ed., *The Nature of Reinforcement*, New York: Academic Press.

Rachlin, H. 1978. A molar theory of reinforcement schedules. *Journal of the Experimental Analysis of Behavior 30*: 345–60.

Rachlin, H., and B. Burkhard. 1978. The temporal triangle: Response substitution in instrumental conditioning. *Psychological Review 85*: 22–48.

Schaeffer, R.W. 1965. The reinforcement relation as a function of instrumental response base rate. *Journal of Experimental Psychology 69*: 419–25.

Staddon, J.E.R. 1977. Schedule-induced behavior. In W.K. Honig and J.E.R. Staddon, eds., *Handbook of Operant Behavior*, Englewood Cliffs. N.J.: Prentice-Hall.

Timberlake, W., and J. Allison. 1972. Response deprivation: An empirical approach to instrumental performance. *Psychological Review 81*: 146–64.

Tversky, A. 1972. Elimination by aspects: A theory of choice. *Psychological Review 79*: 281–99.

18 RATIO SCHEDULES OF REINFORCEMENT AND THEIR RELATION TO ECONOMIC THEORIES OF LABOR SUPPLY

Leonard Green, John H. Kagel, and Raymond C. Battalio

As early as 1953 B.F. Skinner suggested that ratio schedules of reinforcement used to study the behavior of laboratory animals could be viewed as piecework wage rates like those paid to employees in human economic systems. Under a ratio schedule reinforcement is determined by the number of responses emitted by the subject. Skinner's proposition is the starting point for this chapter: The economist's theory of labor supply is adapted to characterize laboratory procedures used in studying ratio schedules. Experiments based on this adaptation were designed and conducted. The results of these experiments are used in several ways. First, they are used to test directly whether or not subjects will substitute leisure (not working) for consumption, as implied by the economic theory of labor supply, and to identify how substitutability varies with changes in random ratio (work) requirements. Second, the results test a number of molar quantitative models proposed in the psychology literature to account for performance under ratio schedules—Allison's conservation model (Allison, Miller, and Wozny, 1979), Herrnstein's (1970) matching law, Rachlin and Burkhard's (1978) temporal triangle, and Staddon's (1979) minimum-distance hypothesis. Finally, the experimental results and the model are extended to provide an account of some of the key characteristics of behavior under concurrent variable-interval, variable-ratio (VI VR) schedules.

In the next section the economic theory of labor supply is outlined in order to provide the rationale for the experiments conducted and an analytic framework within which to interpret the results. The experimental design and results

Research support from the National Science Foundation is gratefully acknowledged. We thank Gerald P. Dwyer, Jr., for helpful discussions and computational assistance.

follow. In the discussion section the results are evaluated in terms of the economic model offered, the relation of the model to psychological theories of instrumental responding is noted, and the models drawn from psychology are evaluated in terms of the data.

ECONOMIC THEORY OF LABOR SUPPLY: APPLICATION TO INSTRUMENTAL RESPONDING

The economist's theory of labor supply is a direct extension of the framework used to analyze choices between consumption goods (see Rachlin, Green, Kagel, and Battalio, 1976, for an application to choice among consumption goods). In using the theory to analyze instrumental responding, it is assumed, as is usually the case, that there is a single consumption good (a single contingent activity such as eating food) and a single job (a single instrumental response such as key pecking), the performance of which is required to obtain access to the consumption good. It is also assumed that the amount of time taken to perform an instrumental response is, for all practical purposes, constant throughout an experimental session. That is,

$$T_r = \alpha P \quad , \qquad (18.1)$$

where T_r is time spent responding, P is the number of pecks or responses made, and α is the constant time associated with each response. (Thus, time spent responding and number of responses can be used interchangeably in the analysis.) In the simple case of a time-constrained choice session in which consumption occurs outside the time constraint (say by stopping the session control clock during reinforcement[a]), Equation (18.1) implies that responding is directly competitive with other activities that obtain reinforcement (*leisure* in the economist's jargon) such as sleeping, preening, and so on, so that increases in response rate reduces the amount of time available for these other, unprogrammed, reinforcers.[b] More precisely, total session time T is made up of time spent responding T_r and time on other (leisure) activities, T_o. That is,

$$T = T_r + T_o \quad , \qquad (18.2)$$

a. Allowing consumption to occur within the time constraint has no effect on the analysis other than making Eq. (18.3) a bit more complicated provided all consumption opportunities are used and the consumption activity (eating) involves a constant time cost per unit of consumption.

b. As represented in Eq. (18.2) leisure refers to the time spent engaged in all activities other than the instrumental response. It is similar to r_o in Herrnstein's (1970) equation. Leisure is not a hypothetical construct. It encompasses all other activities that have their own reinforcers associated with them and that can to a great extent be measured and manipulated.

so that substituting (18.1) into (18.2) and rearranging yields

$$T_o = T - \alpha P \quad , \tag{18.3}$$

where total session time T is held constant.

Choices are over commodity bundles containing nonnegative amounts of the consumption good (in this case, food), instrumental responding (work), and time spent on other reinforcers (leisure). The individual is faced with the task of selecting the most preferred commodity bundle from those that can be obtained within the environmental constraints. Preferences over these commodity bundles are represented by a set of *indifference curves*, with an indifference curve defined as those commodity bundles equally preferred to each other. Figure 18-1a plots a hypothetical set of indifference curves. The vertical axis measures the amount of consumption (number of food reinforcements) or time spent on the contingent activity. The horizontal axis measures the amount of time spent on leisure (other) activities, as well as the number of responses or time spent on the instrumental response (work time). Both work and leisure can be represented on the horizontal axis since, as Eq. (18.3) shows, an increase in responding directly reduces the time available for other activities and vice versa. As the organism decreases its instrumental responding it receives fewer consumption goods but gains more leisure time. The dashed vertical line intersecting the horizontal axis on the right indicates that there is a physical limit on the amount of time spent on other (leisure) activities within a time-constrained choice session.

The indifference curves are drawn as declining over most values only becoming positively sloped as the level of other activities approaches its maximum value (when time spent working is close to zero). The positively sloped segment of the indifference curve implies that some amount of work (instrumental responding) is a *good*; that is, the individual would be willing to sacrifice some consumption in order to work. This would likely be the case when wheel running or licks on a water spout served as the instrumental activity with rats. Beyond some level of instrumental responding, however, additional responses are a *bad*, so that the individual is unwilling to spend additional time responding unless compensated with additional consumption. Commodity bundles on indifference curves further out from the origin are preferred to bundles on lower ones since with the same amount of responding the organism would receive more consumption on a higher curve.[c] Obviously the positively sloped segments of the indifference curves need not exist; work may be regarded as a bad relative to consumption no matter how little the amount of work. Nor are the slopes of the indifference curves invariant with respect to the other reinforcers available in the environment or to the relative value of the contingent activity. For example,

c. It is assumed here that consumption (the contingent activity) is a "good" at all points in the choice space. The model is readily extended to account for conditions where the consumption activity becomes a "bad" after a certain point (see Walsh, 1970: chap. 6).

Figure 18-1. (a) Indifference curves for consumption and leisure (not working) and the constraint on choices imposed by a ratio schedule of reinforcement (line 0A). Equilibrium choice point is where the constraint line is tangent to the highest attainable indifference curve, point A*. (b) Hypothetical effects of changes in ratio requirements on responding (work). Ratio-schedule values increase from 0C to 0A to 0B. Shown is the case of a bitonic response function where responding first increases and then decreases as the work requirement increases. (c) Effect of introducing free reinforcement while holding the ratio requirement constant. Straight line FA yields free reinforcement (of amount F) while earned reinforcement is obtained under the same ratio requirement as the original schedule, 0A. The indifference curves show that the addition of free reinforcement reduces responding (from A* to F*) and increases total consumption (free plus earned), as is usually assumed.

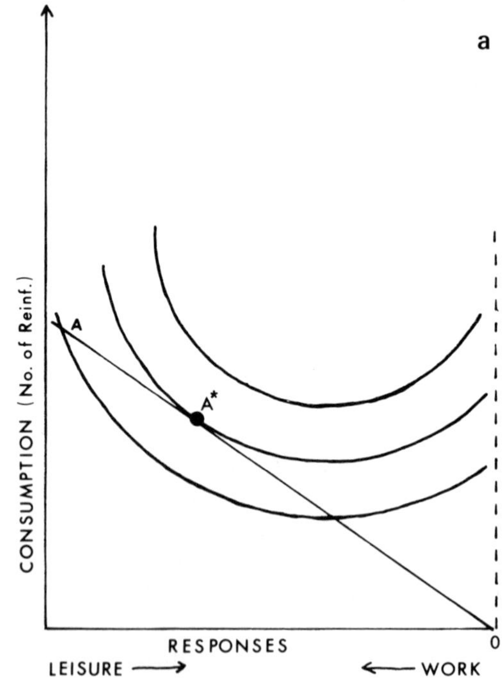

Figure 18-1. continued

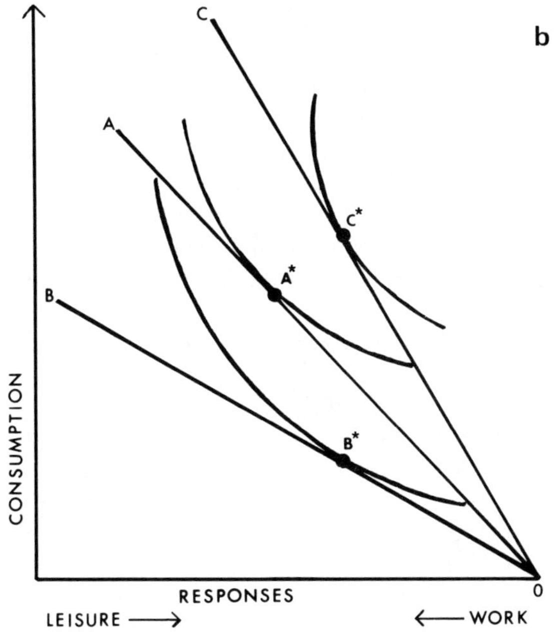

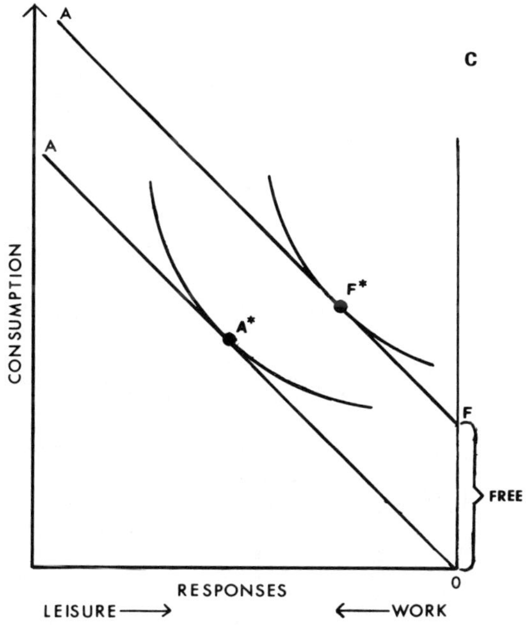

increasing the value of nonwork activities by introducing a running wheel into the experimental chamber for a rat should increase the slope of the declining segment of the indifference curves (as well as altering the point at which they become positively sloped); the increase in consumption required to compensate for an increase in instrumental responding rises as a result of enhanced alternative activities.[d] The actual shape of the indifference contour depends on the utility function and parameter values chosen. Several such utility functions have been proposed (we return to this at the end of the chapter). However, it is crucial to note that the shapes of indifference curves are not arbitrarily and capriciously chosen. Once the utility function and values for its parameters have been specified, then the indifference curve is precisely defined (see Rachlin, Battalio, Kagel, and Green, 1981).

The indifference curves describe workers' preferences over different commodity bundles. However, the parameters of the schedule of reinforcement imposed by the experimenter determine which commodity bundles can actually be obtained. Simple ratio schedules of reinforcement are characterized by straight lines (referred to as budget lines) such as 0A in Figure 18-1. Any point on 0A represents a work-consumption bundle attainable under this ratio schedule. At a zero response rate (all time devoted to nonwork activities) the consumption rate is zero. Increases in the response rate are rewarded with proportionate increases in consumption. The most preferred work-consumption package in the attainable set occurs at the point of tangency of an indifference curve to the ratio-schedule line, point A* in Figure 18-1a. Labor supply theory is primarily concerned with analyzing and predicting *changes* in work-consumption packages resulting from changes in the constraints.

An increase in the ratio requirement results in a flatter slope for the budget line characterizing the attainable work-consumption bundles (line 0B in Figure 18-1b), while a decrease in the ratio requirement results in an increased slope (line 0C in Figure 18-1b, for example). *Increases* in the ratio requirement correspond to *decreases* in real wages in the economist's terminology and vice versa. Economists generally hypothesize that, starting from a high ratio requirement (low wage rate), as the ratio requirement is decreased, responding will increase (the amount of labor supplied will increase). However, further decreases in the

d. Although the older, cardinal utility approach to choice in economics relied on arguments of absolute satiation for commodities to motivate the preference structures hypothesized, the more modern, indifference curve analysis relies on what might be called relative satiation; what the consumer requires in exchange for giving up a unit of commodity A, while staying on the indifference curve, depends upon the amount of A in hand *relative* to all other goods. Observationally this assumption is equivalent to Herrnstein's (1979) notion of relativity, which says that at any given moment an increment in reinforcement is quantified by the organism with respect to the context of total reinforcement. One may or may not choose to invoke arguments about shifts in motivation levels in order to argue persuasively the relativity hypothesis, but such arguments leave the observational implications of the hypothesis unaffected.

ratio requirement (increases in the wage rate) will eventually result in reduced responding (reduced amount of labor supplied). Figure 18-1b illustrates such a bitonic response function: As the ratio requirement decreases (from B to A to C), responding first increases (from B* to A*) and then decreases (to C*). Further, even when instrumental responding decreases, the individual still obtains more consumption. (Although C* involves less work than A*, the subject still receives more of the consumption good.) Data corresponding to such response functions are commonly reported in the literature (see Battalio, Kagel, and Green, 1979; Rachlin and Burkhard, 1978; Staddon, 1979, for summaries of these results).

Delivering a predetermined amount of noncontingent (free) consumption (like freely delivered food) while holding constant the ratio schedule for obtaining response-contingent consumption results in a straight line with the same slope as the ratio schedule but a positive intercept equal to the amount of noncontingent (free) reinforcement. Line FA in Figure 18-1c characterizes one such schedule obtained by adding F noncontingent consumption to ratio schedule 0A. Economists generally assume that the addition of free consumption without altering the wage rate (the ratio requirement) will reduce the response rate or the amount of work performed (say, from A* to F*) as shown in Figure 18-1c. However, the drop in response rate will not be so large as to reduce total consumption—the sum of contingent and noncontingent earnings.[e]

The key assumption underlying the indifference curves in Figure 18-1a, that beyond some level of instrumental responding the individual is unwilling to spend additional time responding unless compensated with additional consumption, is common to all particular economic theories of labor supply. The indifference curves must therefore be convex to the origin. This characteristic of the indifference curves may be tested as follows: Starting from some baseline schedule such as 0A in Figure 18-2, first observe the equilibrium work-consumption bundle, point X. Introduce a new schedule with noncontingent (free) consumption equal to F while simultaneously *increasing* the ratio requirement so that if subjects respond at the baseline rate they will obtain the original work-consumption package X. This yields constraint 0FXA in Figure 18-2. From the assumed convexity of the indifference curves, subjects *must* choose a work-consumption bundle such as Y, which lies along the FX portion of the constraint 0FXA. That is, subjects should *reduce responding and total consumption* (both contingent and noncontingent) in favor of increased levels of reinforcement from other activities in the environment. In economic terms, subjects should substitute leisure (not working) for consumption, with the magnitude of this substitution

e. In technical terms both consumption and leisure (not working) are assumed to be *normal goods* over the operating points in the choice space. That is, as real income is increased, *both* consumption and leisure will increase. If increases in real income result in a decrease in consumption, then that consumption good is referred to as an inferior good. (An example of an inferior good may be retread tires.)

Figure 18-2. Effects of income-compensated wage changes. Constraint on choices imposed by original ratio requirement is line 0A with equilibrium choice point X. Constraint 0FXA results from the income-compensated wage change in which free consumption equal to F is given while the ratio requirement for earned consumption increases so that responding at baseline rates would result in the original work-consumption package *X*. The new equilibrium point Y results in a work reduction (substitution effect) equal to $R_y - R_x$ and a reduction in consumption. Line 0D results from eliminating free reinforcement F while maintaining the higher ratio requirement. Eliminating the free reinforcement thus increases responding by $R_z - R_y$.

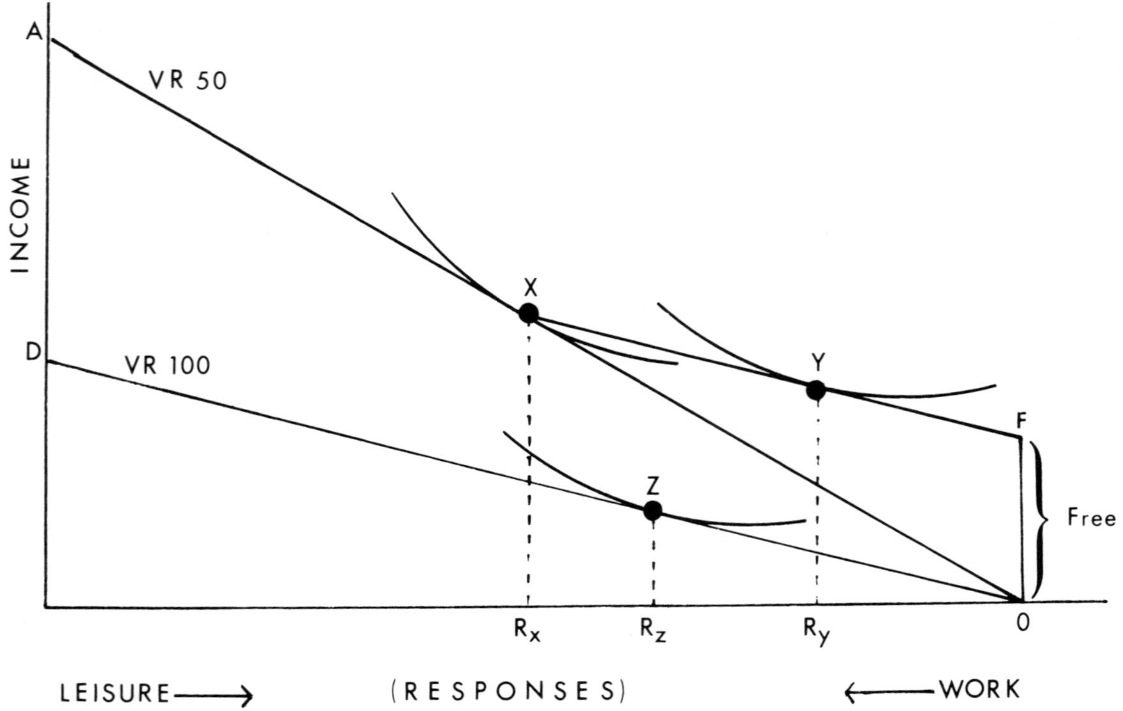

effect measured by the reduction in response rate. An alteration in the constraints such as illustrated in Figure 18-2 is referred to in the economics literature as an *income-compensated wage change.* It is to be emphasized that regardless of the actual utility function and parameter values chosen, *all* economic models predict that an income-compensated change *must* lead to a substitution effect (reduction in responding). If reduced responding were not to occur, then leisure would not be a *good.*

The effect on response rate of a simple change in just the ratio requirement alone is generally indeterminate, because the change always sets off two competing forces, one tending to increase the response rate, the other tending to reduce it, and the magnitude of the two effects is not determinate in the absence of a particular mathematical specification of the indifference curves. Figure 18-2 illustrates this phenomenon for an increase in the ratio requirement (change from budget line 0A to 0D). For the hypothetical case illustrated, the total effect of the increase in the ratio requirement (from VR 50 to VR 100) is to decrease responding from R_x to R_z: Total effect = $R_z - R_x$. As a matter of definition this increase in ratio requirements can be decomposed into (1) an income-compensated wage change (budget line 0FXA), which as a result of the convexity of the indifference curves must result in *decreased* work levels ($R_z - R_x$); and (2) a second change, which eliminates the free consumption F delivered under the compensated wage change while maintaining the higher ratio requirement in effect. This results in a movement from budget line 0FXA to 0D and results in an *increase* in the level of work from R_y to R_z. The response to the income-compensated component of the change is referred to as the (Slutsky) *substitution effect.* The response to the elimination of free consumption, holding the ratio requirement constant, is called the *income effect.* The total response to the increase in ratio requirements is the sum of these two effects.

Total change in work	≡	Slutsky substitution effect	+	Income effect	(18.4)
$(R_z - R_x)$		$(R_y - R_x)$		$(R_z - R_y)$	

Since the income and substitution effects have opposite effects on responding, the total effect of an increase in ratio requirements on response rate or work depends upon the relative strength of these two opposing forces.

In terms of income and substitution effects, the bitonic labor supply function results first from the income effect being larger than the substitution effect at the lower ratio requirements. Thus increases in ratio requirements increase responding. With further increases in ratio requirements the substitution effect must come to dominate the income effect and thus lead to reduced responding. However, this may occur in a number of different ways. The income effect may simply increase in (absolute) size with the substitution effect remaining approximately constant or vice versa, to name two possibilities. Particular patterns of behavior with respect to income and substitution effects characterize particular theories of labor supply, which are expressed as mathematical models characterizing the structure of the indifference curves. Finally, to the extent that behavior in the laboratory mimics behavior within national economic systems, this work and the models so developed have important public implications since the impact of alternative tax and welfare policies on labor supply are completely characterized in terms of their implied income and substitution effects.[f]

In the present experiments income-compensated wage changes (schedules such as 0FXA in Figure 18-2) were introduced starting from a number of different baseline random-ratio schedules—RR 12.5, RR 25, RR 50, and RR 100. Following each compensated wage change, the noncontingent consumption was eliminated while the increased ratio requirement was maintained, resulting in a line such as 0D in Figure 18-2. This enables us to sort out the effects of noncontingent consumption from the increase in the ratio requirement under schedule 0FXA. Since this was repeated under a number of different ratio requirements, we can also characterize how the income and substitution effects change with changes in the ratio requirements.

EXPERIMENTAL METHODS

Subjects. Six experimentally naive male White Carneaux pigeons were maintained at between 80 percent and 80 percent + 20 grams of their free-feeding body weights throughout the experiment. Water and grit were continuously available in their home cages.

Apparatus. The experiment was conducted in a sound-insulated pigeon test chamber that measured 27.9 centimeters long by 25 centimeters wide by 30.5 centimeters high. The response key was 2.5 centimeters in diameter and fabricated from clear acrylic plastic. It was located 17.8 centimeters above the floor and 10.5 centimeters from each side of the test chamber. The key was transilluminated with amber light and required a force of at least 15 grams to operate. Effective key pecks produced an auditory feedback click by operating a 28-volt relay. Reinforcement consisted of 3 seconds access to mixed pigeon grains, during which time the food hopper (located 3 centimeters from the floor and below the response key) was elevated and lighted by a 7-watt white bulb. General chamber illumination was provided by a 7-watt house light situated above

f. In this context welfare plans that involve loss of benefits as income increases have unambiguous disincentive effects on labor supply for the welfare recipient. The availability of free income has a disincentive effect as long as leisure is a normal good (this is the income effect of the welfare program), while the loss of benefits that results from working lowers the effective real wage rate, a further disincentive effect (this is the substitution effect of the program). Understanding the relative importance of these income and substitution effects is critical in designing an optimal (least cost) welfare program aimed at achieving certain minimum welfare levels (see Rees, 1977, for clarification of these points).

the response key with the light deflected upward. Both the house light and response key were darkened during food delivery. A fan provided ventilation and masked extraneous sounds. All scheduling and recording were performed automatically by solid-state programming equipment located in an adjacent room.

Procedure. The pigeons were first trained to eat from the food hopper and then trained to peck the response key using a modified autoshaping technique. Birds 47-50 were then given several days of training with increasing random-ratio schedules of reinforcement. After approximately 4 days of such training, these four birds were all placed on a random-ratio 50 schedule of reinforcement (RR 50), in which reinforcement occurred after a random number of responses had been emitted, with the arithmetic average number of responses per reinforcement equaling 50. Birds 29 and 30 were placed on a RR 12.5 schedule following initial key-peck training.

The effects of two classes of experimental conditions were studied in the present experiment (recall Figure 18-2): (1) *Baseline random-ratio schedules of reinforcement* (RR): Under these schedules, food reinforcement was dependent upon the animal's completing a required number of key-peck responses, the actual number of responses necessary for each reinforcement varying irregularly and unpredictably from reinforcement to reinforcement. Under ratio schedules rate of reinforcement is directly related to the rate of responding, since the faster the animal responds the sooner it will receive the food reinforcement. Random-ratio schedules used were 12.5, 25, 50, 100, 200, and 400. (2) The second class of conditions was *random-ratio schedules plus variable time (VT) free food deliveries (RR + VT free)*: Under these conditions reinforcement was still dependent upon the animal's emission of key pecks according to the ratio schedule in effect, as before. However, in addition, a VT schedule was in effect that delivered food completely independent of responding at variable intervals of time throughout the session. The rate at which the noncontingent food was delivered was determined by the rate of reinforcement obtained on the previous baseline random-ratio schedule as follows. The mean reinforcement rate from the last five sessions of the baseline condition was calculated for each bird. Half the number of these obtained reinforcements were now delivered independent of responding and at irregular times throughout the session, according to the distribution suggested by Fleshler and Hoffman (1962). In addition, the baseline random ratio was now doubled. For example, after completing the RR 50 schedule (line 0A in Figure 18-2) the bird was placed on a RR 100 schedule, with half the number of reinforcements previously obtained under the RR 50 now delivered freely and independent of responding according to the VT schedule (line 0FXA). In effect, if a bird responded at the same rate during the RR 100 plus VT free condition as it had under the baseline RR 50, it would obtain the same total number of food presentations under both conditions. However, half of the food deliveries were presented freely throughout the session whereas the other half were delivered dependent upon responding. Reduced responding during this condition would result in a decrease in the rate of contingent reinforcement but have no effect on the number or rate of delivery of the free foods. The RR + VT free schedules were run under RR values of 25, 50, 100, and 200.

Conditions were maintained until the following criteria for stability were attained: (1) The bird had been on the condition for at least 30 days; (2) on day 30 and every day thereafter until stability was achieved, the last 9 days were divided into three successive blocks of three sessions each and the medians of these blocks (Md_i) showed neither an upward nor a downward trend; that is, neither $Md_1 < Md_2 < Md_3$ nor $Md_1 > Md_2 > Md_3$; and (3) there was no trend during the final 5 days. Sessions were conducted 7 days a week, with each session lasting 40 minutes not counting the time during which the food hopper was presented.

RESULTS

Table 18-1 presents the order in which the experimental conditions were studied for each bird, the number of sessions at each condition, the number of free food deliveries programmed, the mean response rate over the last 5 days of each condition and the mean number of reinforcements obtained (contingent and noncontingent) over the last 5 days of each condition.

Periods with RR + VT free schedules were designed to test the convexity of the indifference curves as indicated in Figure 18-2. In each case these schedules immediately followed a baseline RR schedule. For example, bird 47 in period 1 was on an RR 50 schedule and obtained an average of 85.7 food reinforcements. In period 2 the number of free food deliveries was set at 43, 50 percent of the number obtained in period 1, and the random-ratio requirement was doubled (RR 100). If the bird responded in period 2 at its previous baseline rate, then the number of reinforcements obtained (contingent plus noncontingent) would just equal that obtained in the previous baseline condition. If the response rate dropped to zero, the number of reinforcements would equal the number of free foods delivered. Bird 47 showed a 95 percent reduction in response rate, from 107.2 responses per minute under the RR 50 schedule to 4.9 responses per minute under the RR 100 + VT free condition. This reduction in response rate provides the measure of the substitution effect. In addition, the total number of food deliveries obtained under the RR 100 + VT free schedule decreased approximately 47 percent, from 85.7 to 45.0, as almost all food obtained was free.

RR + VT free schedules were implemented in 22 different periods across 6 subjects (birds 47-50, and 29 and 30). In 19 cases the birds reduced their response rates and reinforcement rates, as predicted: Severely deprived birds gave up food to spend more time in nonwork (leisure) activities. Occasional replications of the baseline RR schedules showed that in these cases where the birds had reduced responding and reinforcement rates, the behavior was reversible.

That is, comparing response rates under the RR + VT free schedule with either the original RR schedule or the replication of that schedule, the reduction in response rates and consumption were of approximately the same magnitude.

In only three cases did responding increase under the RR + VT free schedule as compared to the baseline RR schedule. In one case the increase in responding was so small that the null hypothesis that there was no difference in response rate between the two conditions (bird 49, conditions 7 and 8) could not be rejected. In the second case replication of the baseline RR schedule showed a substantial increase in response rate (bird 50, conditions 6 and 9), suggesting that either a shift in preferences (the shape and location of the indifference curves) or some undetected change in experimental conditions had occurred that would explain the failure to substitute in period 7. Increasing the number of free food deliveries to equal 50 percent of those obtained during the replication of baseline (period 9) now produced consistent changes, the bird reducing its response rate and total consumption (bird 50, conditions 9 and 10). For the third inconsistency (bird 30, periods 1 and 2), eliminating free consumption under the RR + VT free schedule resulted in a marked increase in response rate and reinforcement rate relative to the baseline (bird 30, periods 1 and 3). Increases in the reinforcement rate following increases in ratio requirements are unusual (Hogan and Roper, 1978), suggesting that preferences might have shifted in this case also. However, one common characteristic of all three of these inconsistencies is that they occurred starting from the baseline RR 12.5 requirement (the lowest ratio requirement or highest wage rate).

RR + VT free schedules delivering 50 percent of the previous period's consumption on a noncontingent basis were implemented starting from each of four different baseline RR schedules—RR 12.5, 25, 50, and 100—for subjects 47-50. Reductions in response rates found under these schedules systematically increased (in absolute value) with increases in baseline random ratio requirements for all subjects. Table 18-2 shows these regularities. The size of the substitution effect (percentage decrease in responding when changed from baseline RR schedule to RR + VT free schedule) clearly increases with increases in the ratio requirement. This positive relation between the size of the substitution effect and the baseline RR requirement in cases of consistent choice helps to explain the fact that all of the inconsistent choices showed up exclusively at the lowest baseline RR requirement. The treatment effect is simply smaller there and it is easier for any inherent variability in choices to nullify tendencies to substitute.

The effects of free food deliveries on responding can be found by comparing those periods with the same RR requirement with and without free food (for example, comparison of period 2 with period 7 for bird 47). In virtually all cases, free food by itself reduces responding while at the same time increasing the overall reinforcement rate (free plus contingent). Table 18-3 draws together the data on the impact of free food deliveries on responding at the different RR requirements. The most striking result to note is that although the number of

Table 18-1. Summary of results: mean of last five sessions.

Period	*Number of Sessions*	*RR Requirement*	*Number of Free Rein-forcements*	*Response Rate (per minute)*	*Total Food Deliveries*
			Bird 47		
1	41	50	0	107.2	85.7
2	30	100	43	4.9	45.0
3	30	25	0	91.4	145.8
4	31	50	73	37.2	102.7
5[a]	48	100	108	11.5	112.5
6	30	100	43	15.1	47.5
7	30	100	0	117.0	47.2
8	30	200	24	1.1	24.0
9	37	200	0	45.5	9.4
10	47	12.5	0	113.5	362.9
11	30	25	181	71.4	300.3
12	30	25	0	102.4	163.6
13[b]	21	400	0	13.0	1.4
			Bird 49		
1	30	50	0	132.2	106.0
2	30	100	53	26.6	63.5
3	34	100	0	130.9	52.1
4	33	25	0	122.6	196.1
5	32	50	98	27.4	120.2
6	46	50	0	145.0	116.2
7	30	12.5	0	98.3	316.2
8	30	25	158	100.7	329.0
9	30	25	0	133.1	213.2
10	33	100	0	154.1	61.7
11	30	200	31	8.2	33.6
12	30	200	0	127.0	25.3
13	30	400	0	76.3	8.4
			Bird 29		
1	30	12.5	0	62.5	201.5
2	38	25	101	47.1	173.5
3	36	25	0	83.3	135.0

a. Under this condition the free reinforcement equaled 3/4 of those obtained under the RR 25 baseline condition (period 3) instead of half the number as in all other RR + VT free conditions. The RR requirement was increased to 100 so that if the subject responded at the baseline rate the obtained total reinforcement rate (contingent and noncontingent) would equal the baseline rate.

b. Minimum number of sessions for this period was reduced to 21 days. No other changes were made in the stability criteria.

Table 18-1. continued

Period	Number of Sessions	RR Requirement	Number of Free Rein-forcements	Response Rate (per minute)	Total Food Deliveries
			Bird 48		
1	30	50	0	172.7	137.3
2	32	100	69	106.2	112.3
3	31	100	0	140.8	56.2
4	32	25	0	166.2	266.0
5	31	50	133	104.3	215.1
6	34	50	0	161.9	129.5
7	33	100	0	146.0	58.1
8	30	200	29	72.2	43.4
9	55	200	0	123.0	24.5
10	30	12.5	0	103.6	331.2
11	30	25	166	92.2	316.9
12	30	25	0	144.6	231.3
13[b]	21	400	0	103.1	10.3
			Bird 50		
1	30	50	0	134.4	107.5
2	34	100	54	89.9	89.8
3	31	25	0	118.9	190.3
4	31	50	95	96.9	172.4
5[a]	30	100	143	70.4	171.4
6	32	12.5	0	86.1	274.7
7	31	25	137	91.9	301.9
8	30	25	0	120.1	192.4
9	30	12.5	0	121.2	387.0
10	30	25	193	100.2	354.0
11	30	100	0	133.6	53.7
12	33	200	27	72.8	41.7
13	30	200	0	113.3	22.6
14	31	100	0	106.1	42.7
15	31	200	0	125.2	25.7
16[b]	21	400	0	89.7	9.2
			Bird 30		
1	34	12.5	0	59.2	189.2
2	25	25	97	73.1	228.1
3	30	25	0	147.5	237.1

Table 18-2. Substitution effects (percentage decrease in responding) as a function of baseline random-ratio requirement.

Baseline RR Schedule		*Income-Compensated Schedule*	*Change in Response Rates (%)*[a]			
			47	*48*	*49*	*50*
RR 12.5	→	RR 25 + VT free	-37	-11	+2	-17
RR 25	→	RR 50 + VT free	-59	-37	-78	-18
RR 50	→	RR 100 + VT free	-95	-38	-80	-33
RR 100	→	RR 200 + VT free	-99	-51	-95	-45

a. Percentage decrease in responding from baseline random ratio schedule (RR) to a random ratio in which the ratio was doubled and half the baseline amount of food was delivered freely on a variable-time schedule (RR + VT free). For example, bird 47 decreased its response rate by 37 percent when changed from RR 12.5 to RR 25 + VT free.

Table 18-3. Income effects (percentage decrease in responding as a function of free food) as a function of random-ratio requirement.

	Bird 47		*Bird 48*	
RR Schedule	*Number of Free Food Deliveries on RR + VT Free*	*Change in Response Rate (%)*[a]	*Number of Free Food Deliveries on RR + VT Free*	*Change in Response Rate (%)*[a]
25	181	-22	166	-45
50	73	-65	133	-40
100	43	-96	69	-27
200	24	-98	29	-41
	Bird 49		*Bird 50*	
25	158	-18	193	-16
50	98	-79	95	-28
100	53	-83	54	-33
200	31	-94	27	-36

a. Percentage decrease in responding from baseline random-ratio schedule (RR) to the same random-ratio schedule but with the addition of freely delivered food (RR + VT free). For example, bird 47 decreased its response rate by 22 percent when changed from RR 25 to RR 25 + VT free.

free food deliveries consistently increased as the response requirements were reduced (more free food deliveries as the schedule changed from RR 200 to RR 25), none of the birds reduced their response rate in proportion to these increases in free food deliveries. In fact, for three of the four birds, percentage reductions in response rate consistently increased with increases in the RR requirements even though the number of free food deliveries had declined substantially. Thus, the impact of free food is clearly a function of the ratio requirement in effect. In theoretical terms, the size of the income effect increases with increases in RR requirements.

Figure 18-3. Mean rates of responding for individual birds as a function of the different random-ratio schedules of reinforcement with no free consumption. Open circles represent replications.

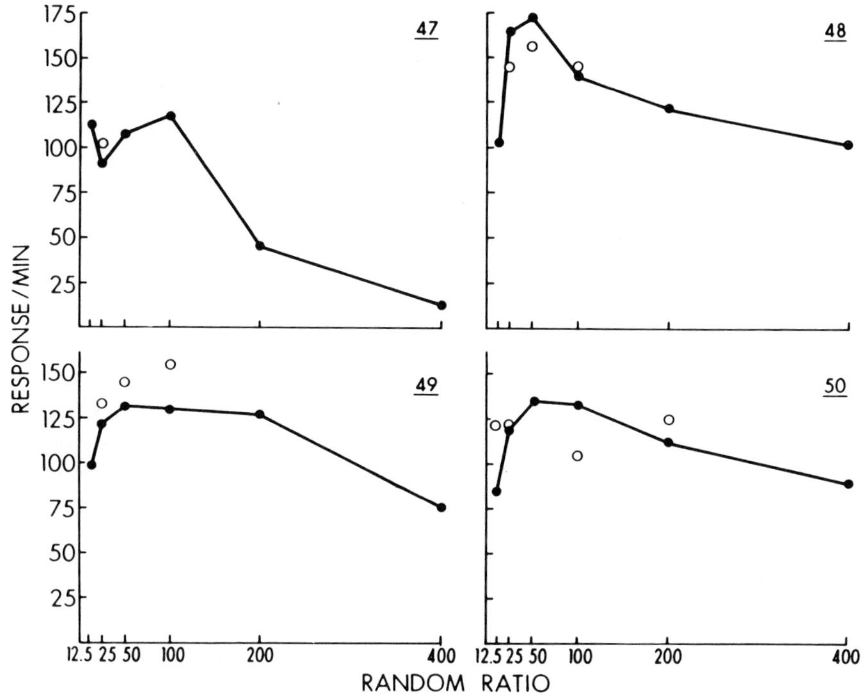

Figure 18-3 shows the relation between response rate and random ratio requirements during periods with no free consumption (baseline RR schedules) for each bird. The data display the familiar bitonic pattern with response rates increasing at first with increasing ratio requirements, reaching a maximum at around RR 50 to RR 100, and declining thereafter. In all cases, increases in RR requirements were accompanied by decreases in the number of food deliveries. That is, even though response rate increased at first with increases in RR requirements, the increase in response rate was insufficient to maintain reinforcement rate, consistent with results commonly found in the literature (Allison, 1979; Hogan and Roper, 1978). In economic terms, as the real wage decreased, or what is the same thing, as the effort price of food consumption increased (corresponding to increases in the ratio requirement), total food consumption decreased. This is referred to as the simple law of demand. Finally, the fact that response rate first increased with increases in RR requirements rules out an explanation of the reductions in response rate (the substitution effects) under the RR + VT free schedules as simply a response to the increase in ratio requirements. Under the RR + VT free schedules the increase in response requirements by themselves promote increased responding at least until RR 50 or RR 100 values are obtained.

DISCUSSION

Income-compensated wage changes consistently resulted in birds' reducing their response rate and reinforcement rate as predicted by the economic theory of labor supply. Thus, food-deprived birds gave up food to spend more time in nonwork (leisure) activities. These results clearly indicate that leisure is an economic good that must be taken into account in maximization models of animal behavior. The size of the substitution effect (reductions in response rate) consistently increased with increases in response requirements and were quite large, particularly at the higher ratio requirements. Free food deliveries consistently reduced response rate and increased total food intake when compared to schedules with the same ratio requirement and no free food. Here too, consistent empirical regularities appeared with increases in ratio requirements resulting in larger reductions in response rate even though the (absolute) number of free food deliveries was decreased. Finally, the data for responding in the absence of free food showed the bitonic response pattern, increasing at first with increases in ratio requirements and declining thereafter, while food intake declined continuously with increases in response requirements.

Relation of Economic Theories of Labor Supply to Psychological Theories of Instrumental Responding

A particular economic theory of labor supply consists of, among other things, a characterization of response patterns under changing ratio requirements (changing real wage rates) and response patterns under changing levels of noncontingent (free) consumption. To the extent that psychological theories of instrumental responding make predictions about molar response rates under changing ratio requirements and noncontingent earnings they may be viewed as particular economic theories of labor supply. Further, it is not necessary to identify the indifference curves that could give rise to a particular psychological theory of instrumental responding before viewing that theory as a particular model of labor supply or determining the consistency of the theory with the data.

The foregoing experiments provide data for testing a number of psychological theories of instrumental responding. The theories to be considered are Allison's conservation model (Allison, Miller, and Wozny, 1979), Herrnstein's (1970) matching law, Rachlin and Burkhard's (1978) temporal triangle, and Staddon's (1979) minimum-distance hypothesis. The implications of these theories are briefly outlined and related to the results of the experiments reported in the previous section.

Conservation theory implies a linear, positively sloped response function in the face of changing ratio requirements. That is, under all conditions, *increases* in the ratio requirement result in *increased* responding where the increase in response rates is characterized by the following function:

$$P = \frac{b}{\kappa + C/N} \ , \quad \kappa, b > 0 \ , \tag{18.5}$$

where P is the number of instrumental responses, N is the programmed ratio requirement, C measures the amount of the consumption good delivered per reinforcement, and κ and b are positive constants to be calculated from the data.[g] The bitonic nature of the response functions found here (recall Figure 18-3) and elsewhere in the literature argues against employing conservation theory with its prediction of a linear, positively sloped response rate as a parsimonious explanation of behavior.[h]

The matching law (Herrnstein, 1970) characterizes behavior under a single-ratio schedule of reinforcement as follows:

$$\frac{T_r}{T_o} = \frac{R_r}{R_o} \ , \tag{18.6}$$

where T_r is the proportion of time spent responding on the ratio schedule, R_r is the reinforcement rate on the ratio schedule, T_o is the proportion of time devoted to other activities, and R_o is the (unobserved) reinforcement from other activities. Timberlake (1977), following an article by Pear (1975), has shown that the matching law predicts that response rates will always *decrease* with *increases* in the ratio requirement and that this prediction appears contradicted by considerable data. In Appendix 18A it is shown that this prediction of the matching law even holds when R_o increases with increases in the ratio requirement, the plausible alternative to Pear's (1975) assumption that R_o is constant. Further, the inverse relation between responding and ratio values holds under the generalized matching law (Baum, 1974):

$$\frac{T_r}{T_o} = a\left[\frac{R_r}{R_o}\right]^b \ , \quad a, b > 0 \ , \tag{18.7}$$

g. Equation (18.5) follows directly from Allison et al.'s (1979) Eq. (3) where the constant b equals baseline performance characteristics: $b = \kappa 0_i + 0_c$, following the notation of Allison et al.

h. Allison (1982) has recently modified the conservation model so as to allow for the bitonic function. The newer version assumes that as the cost of the consumption good (such as food) rises through increases in the ratio requirement, the animal will eventually come to substitute its bodily stores of food for the increasingly costly external consumption good, thus leading to a decrease in responding. An evaluation of this latest addition to the model, however, awaits empirical testing.

where a and b are constant, provided $b \leq 1$, the typical parameter value found in applications of the model (Myers and Myers, 1977). This result follows whether R_o is constant or increases with increases in the ratio requirement (see Appendix 18A).

One defense offered for the matching law in the face of bitonic response functions reported in the literature has been that overall response rate is not an appropriate measure of responding since it includes time in which the animal is engaged in other behaviors. The extreme form of this argument holds that response rates should be calculated on a time base obtained by subtracting reward time and postreinforcement pause time from total session time (see Timberlake, 1977, for a statement of this position and citations of the relevant literature). In the present experiments, however, reward time was not included in total session time, the session control clock stopped whenever reinforcement occurred, while the use of RR as compared to FR schedules largely eliminated postreinforcement pauses. Nevertheless, we still find bitonic response functions. Under the circumstances it is difficult to reconcile these results with the matching law.[i]

The economic theory of labor supply developed here is essentially the same as Rachlin and Burkhard's (1978) temporal triangle model. Rachlin and Burkhard specify that the value of a set of responses is equal to their duration:

$$V = F(T_c, T_r, T_o) \quad , \tag{18.8}$$

where T_c, T_r, and T_o are times spent on the contingent activity (eating), instrumental responding, and all other activities, respectively. The formal assumptions made with respect to the value function are the same as those employed within economic theory to specify a util˙.y function, or what is the same thing, a consistent set of indifference curves in three-dimensional space: that the law of diminishing marginal rate of substitution holds, thus yielding convex indifference curves and that subjects allocate time so as to maximize V.

The three-dimensional value function of the temporal triangle may be reduced to two dimensions through the help of the time constraint:

$$T_c + T_o + T_r = 1 \tag{18.9}$$

Rearranging the terms of (18.9) and substituting for T_r in (18.8) yields

$$V = F(T_c, T_o, 1 - T_c - T_o) \quad , \tag{18.10}$$

i. Herrnstein (personal communication) notes that the value of k in his equation is likely to be smaller for small ratio requirements than for larger ratio requirements. This would be the case since the first response is topographically unique involving, as it does, approach to the response manipulandum. Such an assumption, along with others that may also be proposed, might permit the flexibility needed for the matching law to account for our results. More precise specification will be needed, of course.

which is a value function in two dimensions. This value function can be represented by indifference curves in two dimensions, like those shown in Figure 18-1a. Although for some applications the three-dimensional representation employed by Rachlin and Burkhard is preferable to the two-dimensional representation used here, the predictions of the two theories are nevertheless the same. In particular the result predicted to the compensated wage change of schedule 0FXA in Figure 18-2 is the same and indicates the degree of substitutability of the activities under investigation. More precise predictions for either the Rachlin-Burkhard model or the labor supply model require specifying a particular value function in place of Eq. (18-10).

Staddon's (1979) minimum-distance hypothesis is an example of just such a value function, one giving rise to elliptical indifference curves. According to the minimum-distance hypothesis, when a constraint is imposed, subjects minimize a weighted average of squared deviations from a free behavior ("bliss") point. The value function has the form

$$V = -a_c\,(T_c - \epsilon_c)^2 - a_r\,(T_r - \epsilon_r)^2 - a_o\,(T_o - \epsilon_o)^2 \quad , \qquad (18.11)$$

where T_c, T_r, and T_o refer to times spent on the contingent activity, instrumental responding, and all other activities, respectively. The terms $\epsilon_c, \epsilon_r, \epsilon_o$ refer to the unconstrained values of these activities—that is, the most preferred distribution among them in the absence of explicit contingencies. The terms a_c, a_r, a_o are positive constants. The minimum-distance hypothesis is an intuitively attractive, relatively simple value function that yields a bitonic response function in the face of changing ratio requirements, an essential characteristic of any value function designed to organize data from the present experiment as well as from a number of others reported in the literature.

A distinguishing characteristic of the minimum-distance hypothesis is its prediction concerning the impact of free reinforcement under different ratio requirements. It has embedded in it the assertion that, under a given ratio requirement with varying levels of free reinforcement, the scatter of points relating total consumption (free plus earned consumption) and response rates will be linear (such a scatter of points is referred to as an *expansion path* in the economics literature). Further and more specifically, expansion paths obtained under different ratio requirements are predicted to converge on the free behavior point. Figure 18-4 shows, in two-dimensional space, the relation between expansion paths implied by the theory under two different ratio requirements. Point A is the free behavior or bliss point involving little instrumental responding and a great deal of consumption. Line AB shows the relation between T_c (earned and free consumption) and T_r under differing levels of free reinforcement, holding the ratio requirement for earned reinforcers constant (at RR 25 in this case). Line AC represents a similar relation, except here the ratio requirement for earned reinforcers has been increased to RR 100. Note both the convergence of the expansion paths to the bliss point and the increased slope (in absolute

Figure 18-4. Expansion paths implied by the minimum-distance hypothesis under two different ratio schedule requirements. Point A is the free behavior ("bliss") point. Lines AB and AC show the relation between total consumption and responding as different amounts of *free* reinforcement are added to a low and a high baseline ratio schedule, respectively.

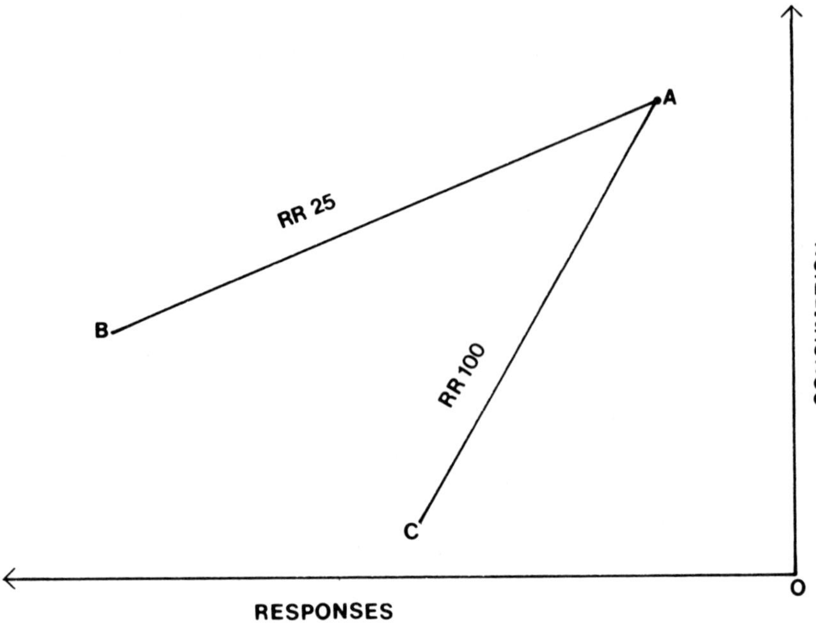

value) when the random-ratio requirement is increased (see Appendix 18B for proof of these propositions).[j]

These strong implications regarding the expansion paths are readily tested using data from the present experiment. Figure 18-5 shows the paths obtained under RR 25 and RR 100 for all four birds (similar results are found comparing expansion paths under the other ratio requirements). For three of the four birds the data clearly do not conform to the predictions of the minimum-distance hypothesis. Since there are a number of alternative methods for delivering free

j. Similar implications hold for a generalized version of the minimum distance hypothesis

$$V = -a_c(T_c - \epsilon_c)^Y - a_r(T_r - \epsilon_r)^Y - a_o(T_o - \epsilon_o)^Y \,,$$

where $Y > 1$.

Figure 18-5. Expansion paths obtained under RR 25 and RR 100 schedules of reinforcement for each bird as the number of *free* reinforcements delivered was varied. The least-squares line is presented for each data set.

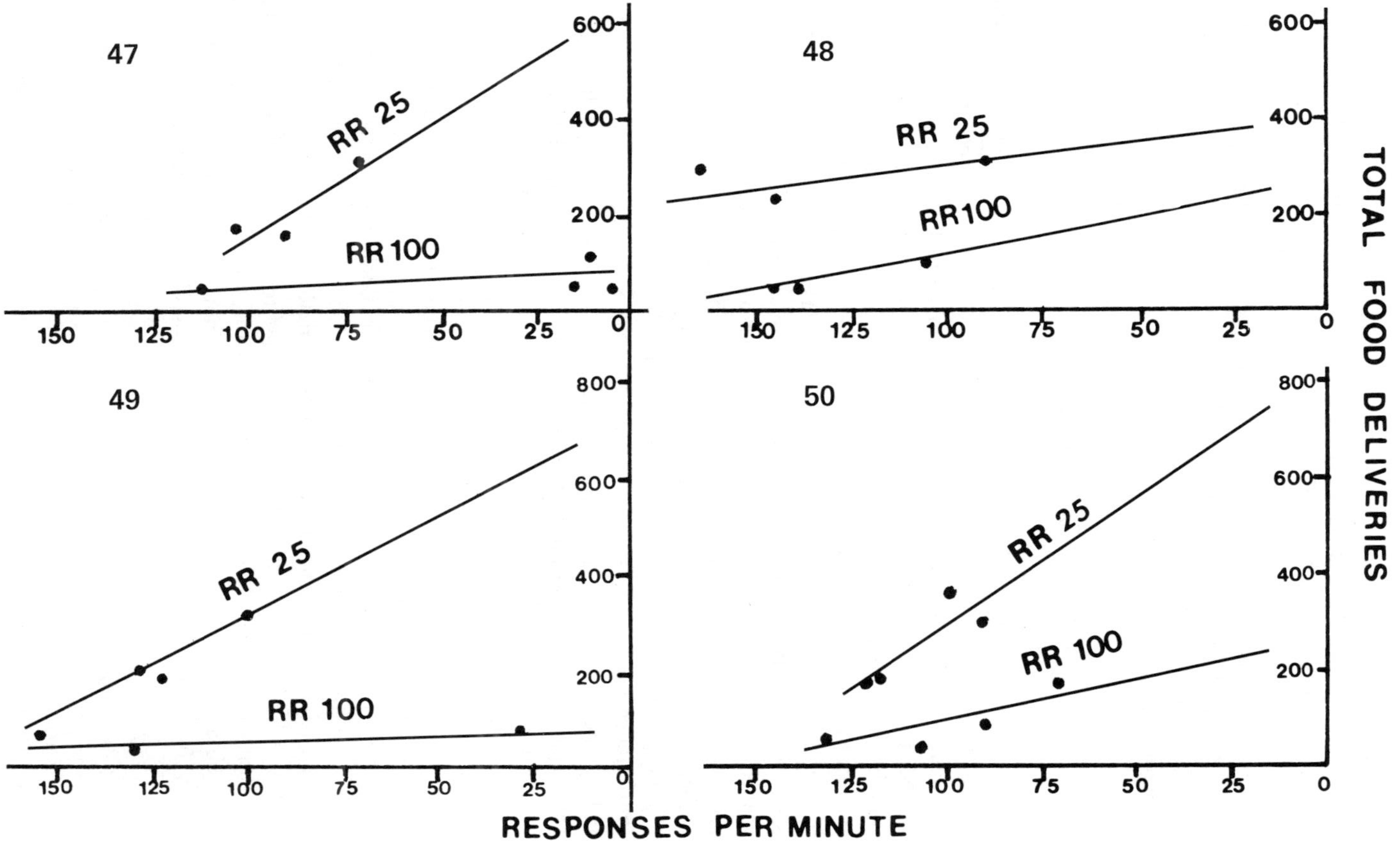

reinforcement it remains to be seen how sensitive the slope of the expansion paths is to these alternative procedures and whether the minimum-distance hypothesis might provide a better characterization under different experimental procedures.

A value function that both yields a bitonic response function and generates expansion paths compatible with the data in Figure 18-5 is a generalized constant elasticity of substitution (CES) utility function. It may be written (in two dimensions) as

$$V = \frac{a_c}{y}(T_c - \gamma_c)^y + \frac{a_o}{y}(T_o - \gamma_o)^y \ , \tag{18.12}$$

where $y \leq 1$, $a_c, a_o > 0$ and $(T_c - \gamma_c), (T_o - \gamma_o) \geq 0$, T_c and T_o are times spent on consumption and leisure (other) activities, respectively, and γ_c, γ_o are constants. The indifference contours of this utility function are homothetic with respect to the point represented by the γs. If the γs are positive the individual may be interpreted as spending time and energy to obtain "minimum necessary" quantities of consumption and leisure equal in magnitude to the size of the γs.[k] The indifference contours exhibit a constant elasticity of substitution with respect to amounts of consumption and leisure in excess of the minimum requirements, where $1/(1 - y)$ is the elasticity of substitution. That is, the individual is seen as establishing (protecting) minimum consumption levels first, with an altogether different rule for dealing with any income left over.

The generalized CES utility function still defines a logically consistent set of indifference contours when one or more of the γs is negative. In this case the indifference contours are homothetic to a point outside the quadrant of the value space representing positive amounts of consumption and leisure. Negative γs might be expected to be found in experiments employing nonessential commodities (flavored fluids with ad lib access to food and water) or in short-term experiments where subjects are maintained at constant body weights. If the γs are negative it makes no sense to describe the individual as purchasing a necessary (negative) collection of goods.

Efforts to fit the generalized CES response function to the data from this and other experiments have proven quite successful (see Battalio, Dwyer, and Kagel, Note 1). The model explains much of the variation in the data and provides a substantially better fit than a number of competing hypotheses. For example, in addition to the prediction of the expansion paths, comparisons of response functions obtained from the minimum-distance hypothesis and the generalized CES utility function show the CES model to fit the data far better than the mini-

k. If the γs are positive, the utility function is not defined under conditions where the time left over for working after meeting minimum leisure requirements is insufficient to meet minimum consumption requirements given the instrumental response requirement in effect. The model simply does not characterize behavior over this region of the choice space.

mum-distance hypothesis, both functions having essentially the same number of free parameters. The generalized CES function also reduces to the generalized matching relation between time spent in an activity and obtained reinforcement when the γs (the constants) are assumed to be zero and the power term y is assumed to be between 0 and 1. However, these restrictive assumptions necessary to produce generalized matching cannot be maintained in light of the present data. Thus, for this reason as well as for its ability to predict the bitonic labor supply function, the CES utility function is preferred to the matching law.

Extension to Concurrent VI VR Schedules

Under the RR + VT free conditions of the present experiment, hungry pigeons would have needed to maintain their response rates at baseline levels in order to maintain a constant food intake. Instead, they consistently reduced responding thereby reducing food consumption. This occurred even under conditions where the increase in RR requirements under the RR + VT free schedule would by itself have resulted in an increase in response rate. Consequently, the pigeons consistently failed to obtain the maximum food intake they were capable of under a given schedule. These reductions in food intake were far from trivial, at times amounting to close to 50 percent of the baseline consumption rate, the maximum reduction possible under the circumstances (see, for example, bird 47 in period 2 compared to period 1 in Table 18-1). Our interpretation of this result, indeed the whole rationale underlying the experiments, was that the value function that birds and other organisms maximize includes variables in addition to food, one of these variables being the other reinforcers available in the environment.

Herrnstein and Heyman (1979) reported that under concurrent VI VR schedules pigeons consistently failed to maximize food intake, giving up an average of 30 percent of the maximum food reinforcers available (about 60 food reinforcers per hour). This occurred because subjects spent "too much" time on the VI schedule. On the basis of this result, they argued that subjects do not maximize and rejected the use of maximization models to organize pigeons' behavior. However, a maximization approach does *not* posit that food alone is the commodity to be maximized. Leisure (not pecking) is also a commodity, one that is highly substitutable for food, as shown here. Indeed, viewing leisure as a commodity explains the bias toward the VI schedule observed under the concurrent VI VR schedules.

The failure to maximize food intake under concurrent VI VR schedules is associated with differential response rates on the two schedules. Herrnstein and Heyman's birds consistently pecked at a faster rate on the VR than on the VI schedule. Similar differences between response rates on VI as compared to VR schedules are well documented (e.g., Catania et al., 1977; Zuriff, 1970). Indeed, it is this tendency to peck at a substantially higher rate on VR as compared to VI schedules that in large measure accounts for the discrepancy between maximum and obtained consumption rates in their experiment. These differences in response rates give rise to consumption/leisure trade-offs, the net effect of which *must* be to bias time toward the VI schedule and consequently away from maximum food consumption.

Figure 18-6 illustrates this point. The functional relation between time spent on the interval schedule (T_i) and consumption rates obtained during the concurrent VI VR schedule from any of the feedback functions offered in the literature is shown by the humpback curve with maximum consumption rate at point A. Increases in time spent on the interval schedule beyond the maximum consumption rate *reduce* rates of food intake while *increasing* the amount of leisure time. Applying the indifference curve analysis presented previously to the present situation, we would postulate negatively sloped curves that are convex to the origin, like I_1 and I_2 in Figure 18-6. The attainable indifference curve furthest removed from the origin *must* be tangent to the feedback function to the *right* of the maximum consumption point A, so that the maximum for a value function that includes leisure as one of its arguments *must* occur to the right of the point of maximum food intake. The observed differences between maximum food intake and realized consumption rates depends upon the valuation of consumption relative to leisure in the utility function.

Herrnstein and Heyman attempted to correct for the differential response rate on the VI as compared to the VR schedule by weighting VI reinforcers more heavily. In terms of Figure 18-6 this would involve an upward displacement of the feedback function with a rightward shift in the maximum consumption point. The weight for VI reinforcers chosen by Herrnstein and Heyman consisted of the bias term obtained from fitting the matching equation to the pooled subject data. Even with this differential weighting the maximum consumption point still required subjects to spend substantially more time on the VR schedule than they did.

Correcting for the differential response rate on the VI schedule within the context of the model offered here requires comparing the maximum consumpton rate A with the utility maximizing point B. Implementing this correction procedure requires employing a specific utility function with definite parameter values and a parametrized feedback function. The generalized CES utility function (Eq. (18.12)) served as the valuation function in simulations, with parameters calculated over the pooled data from birds 47-50 who participated in the present experiment. The same feedback function as Herrnstein and Heyman employed was used with consumption rates for alternative time allocations calculated from the programmed values of the VI and VR schedules along with obtained performance parameters for the birds in their experiment. In the simulation the coefficients a_c and a_o of the utility function were allowed to vary within ± 2 standard errors of their mean values (the coefficients were thus per-

Figure 18-6. Feedback function for concurrent VI VR schedule and hypothetical indifference curves (I_1, I_2) for consumption and leisure. Leisure (not working) increases as relative time spent on the interval schedule (T_i) increases since the local response rate in the VI is 56 percent of that in the ratio schedule in the Herrnstein–Heyman (1979) experiment. Point of maximum food intake is A, whereas the equilibrium choice point B, *must* lie to the right of this if the indifference curves are convex and negatively sloped.

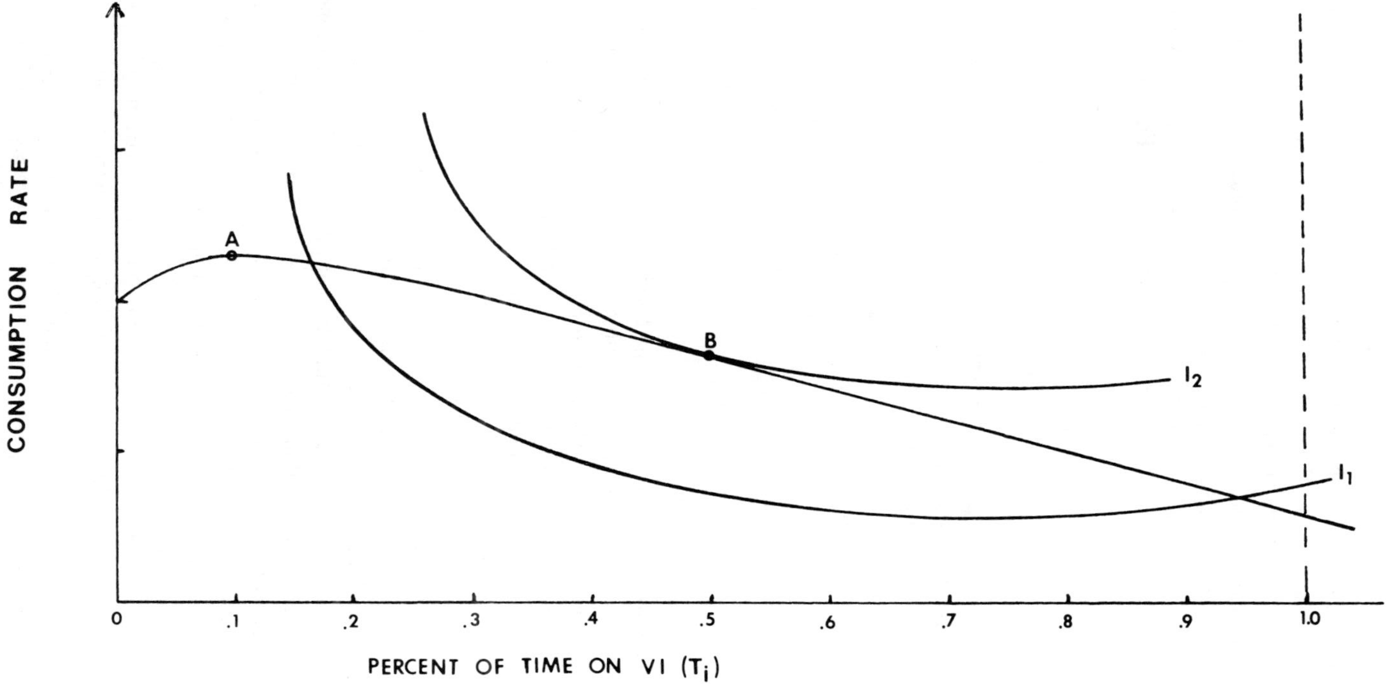

mitted to vary only within approximately the 95 percent confidence interval). Moreover, since the specification of one of the parameters completely determines the value of the other, only one of the parameters was, in effect, permitted to vary within its statistical limits. All other parameters were held at their average estimated values. A time cost of 0.1 second per peck was also assumed (each peck cost 0.1 second of leisure).

This simulation procedure produced biased matching over the pooled subject data. The resulting generalized matching relation (Eq. 18.7)) was

$$\frac{T_I}{T_R} = 1.156 \left(\frac{R_I}{R_R}\right)^{1.165} ,$$

where T_I and T_R refer, respectively, to times on the interval and ratio schedules, and R_I and R_R refer to obtained consumption rates on the two schedules. These values for T and I were chosen so as to maximize utility. The particular value function to which they are the maximum solution is Eq. (18.12) with the following parameter values[1]:

$$\frac{a_c}{y} = -.167 , \quad \gamma_c = -175.8 , \quad y = -.534 , \quad \frac{a_o}{y} = -.565 ,$$

$$\text{and} \quad \gamma_o = -1496.2 .$$

Maximizing utility rather than food consumption rates would, based on this value function, also result in subjects' giving up an average of 32 percent of the maximum available food reinforcers, or approximately 77 hopper presentations per hour. The bias toward the VI schedule and deviations from maximum consumption rates obtained are close to the values reported in Herrnstein and Heyman (1979).

The important issue here is whether there are valuations of leisure that predict the results obtained by Herrnstein and Heyman. Of course, several methods could be used to adjust for the differential work output in order to account for the data. How is the choice to be made between the competing explanations? We hasten to note that in our approach the valuations of consumption and leisure were taken from birds 47–50 of our experiments. These parameter values were then applied to the data reported by Herrnstein and Heyman. Consequently our simulation clearly demonstrates that there are plausible valuations of consumption and leisure that can give rise to the biased matching obtained under their procedure.

Although it stops short of a complete explanation of behavior under concurrent VI VR schedules, our simulation demonstrates that plausible valuations for

1. T_c is the total seconds of food hopper presentations per 40-minute session and leisure time (T_o) is also measured in seconds. T_c was based on a 40-minute time period since that was the session length in the present experiment.

leisure exist that can give rise to the biased matching reported under these schedules. The fact that pecking on the VR schedule exceeds responding on the VI can also be explained in terms of consumption/leisure trade-offs (see Rachlin and Burkhard, 1978: 39-40). Finally, we have not found it necessary to account for leisure in our analysis of concurrent VI schedules (Rachlin, Green, Kagel, and Battalio, 1976; Rachlin, Kagel, and Battalio, 1980) since in most such studies local response rates are found to be the same for each VI schedule component (de Villiers, 1977: 238), so that there are no important leisure differences in the schedule components to be accounted for. In cases where average interreinforcement times in the component VI schedules are of sufficiently different duration that they do give rise to substantially different local response rates, accounting for these differences should then become important.

SUMMARY

Behavior under ratio schedules of reinforcement was interpreted within the economist's theory of labor supply. Experiments were conducted comparing response rate and food consumption under simple random-ratio schedules (such as RR 50) to a higher random-ratio schedule that included free food delivery (such as RR 100 + VT free) but which was designed so that if response rate remained unchanged food intake would also remain unchanged. Consistent with predictions from the economic theory, pigeons maintained at 80 percent of normal body weight consistently reduced both response rate and food intake, often by large amounts, under the RR + VT free schedules as compared to the simple RR schedules. That is, severely deprived birds gave up food to spend more time in nonwork (leisure) activities. Moreover, the size of these substitution effects (reductions in response rate) consistently increased with increases in response requirements.

Free food deliveries reduced response rate but increased total food intake when compared to schedules with the same ratio requirement and no free food (such as RR 100 + VT free *versus* RR 100). Here too, consistent empirical regularities appeared. As the ratio requirement increased, the birds reduced their rate of responding more even though the absolute number of free food deliveries decreased. Finally, the data for responding in the absence of free food showed the familiar bitonic response pattern: As the ratio requirement was increased, responding first increased and then declined, whereas food intake declined continuously with increases in response requirements.

The results of these experiments clearly indicate that leisure (behaviors other than the target instrumental response) is an economic good that must be accounted for in maximizing models of animal behavior. Allison's conservation model, Herrnstein's matching law, and Staddon's minimum-distance hypothesis are all found to have important deficiencies in organizing these experimental data. The labor supply model can also account for empirical regularities reported under concurrent VI VR schedules.

A maximization model does not posit that food alone is the commodity to be maximized. Rather, leisure is also a commodity, one that, we showed, is highly substitutable for food. Subjects are viewed as choosing over commodity bundles containing amounts of the consumption good (such as food), instrumental responding (work, such as key pecking) and time spent on other activities (leisure). The subject selects the most preferred commodity bundle from those that can be obtained within the environmental constraints.

APPENDIX 18A
RATIO RESPONSE FUNCTIONS DERIVED FROM THE MATCHING LAW

For ratio schedules of reinforcement the matching law asserts

$$\frac{T_r}{T_o} = \frac{R_r}{R_o} , \qquad (18A.1)$$

where

T_r = time spent responding on the ratio schedule;

T_o = time spent on other activities;

R_r = reinforcement obtained from the ratio schedule;

R_o = other reinforcement.

By assumption, total session time T_t is

$$T_t = T_o + T_r \qquad (18A.2)$$

as time spent in consumption occurs outside the session time constraint.

Each response is assumed to take a fixed amount of time α, so that

$$T_r = \alpha P , \qquad (18A.3)$$

where P is number of responses. Substituting Eq. (18A.3) into (18A.2) gives

$$T_o = T_t - \alpha P . \qquad (18A.4)$$

The ratio schedule imposes the following relation between responding and reinforcement:

$$R_r = \frac{P}{N} , \qquad (18A.5)$$

where N is the average ratio requirement.

Substituting Eqs. (18A.3)–(18A.5) into (18A.1), with some manipulation, yields

$$P = \frac{T_t}{\alpha} - NR_o \quad . \tag{18A.6}$$

Taking the derivative of Eq. (18A.6) with respect to N,

$$\frac{dP}{dN} = -\left[N\frac{dR_o}{dN} + R_o\right] \quad ,$$

which is negative whether

$$\frac{dR_o}{dN} = 0 \quad \text{or} \quad \frac{dR_o}{dN} > 0 \quad ,$$

meaning that the response rate decreases with increases in the ratio requirement whether the value of other reinforcers is constant or increases with increases in N.

For the generalized matching law,

$$\frac{T_r}{T_o} = a\left[\frac{R_r}{R_o}\right]^b \quad \text{where } a > 0 \quad , \tag{18A.7}$$

and most data show $b \leqq 1$.

Substituting Eqs. (18A.3)–(18A.5) into (18A.7), with some manipulation, yields

$$aP^{b-1}\,T_t - \alpha P^b - \alpha_o\,R_o^b\,N^b = 0 \quad , \tag{18A.8}$$

which does not yield an explicit solution for P. However, we can still obtain the desired derivative in the neighborhood of points for which (18A.8) has a solution through total differentiation (see Chiang, 1974: chap. 8).

$$\frac{dP}{dN} = \frac{\left[N^b R_o^{b-1}\dfrac{dR_o}{dN} + R_o^b\;N^{b-1}\right]}{P^{b-1}\left[\left(\dfrac{b-1}{\alpha b}\right) a\,P^{-1} - 1\right]} \quad . \tag{18A.9}$$

The numerator of Eq. (18A.9) is positive when

$$\frac{dR_o}{dN} = 0 \quad \text{or} \quad \frac{dR_o}{dN} > 0 \quad ,$$

and the denominator is negative as long as $b \leqq 1$. Thus, the response rate decreases with increases in the ratio requirement under the generalized matching law, provided undermatching ($b \leqq 1$) holds.

APPENDIX 18B
EXPANSION PATHS IMPLIED BY THE MINIMUM-DISTANCE HYPOTHESIS

Staddon's (1979) minimum-distance hypothesis has subjects maximizing the following value function:

$$V = -a_c(T_c - \epsilon_c)^2 - a_r(T_r - \epsilon_r)^2 - a_o(T_o - \epsilon_o)^2 \quad , \tag{18B.1}$$

where T_r, T_c, and T_o represent time spent responding, consuming, and engaged in other reinforcing activities, respectively; ϵ_r, ϵ_c, ϵ_o represent the free behavior values of these activities; and a_c, a_r, and a_o are positive constants.

Under out experimental procedures T_t, total session duration, was made up of

$$T_t = T_r + T_o \tag{18B.2}$$

since food hopper presentations stopped the session control clock. It follows from this and the definition of the ϵs as free behavior points that

$$\epsilon_r + \epsilon_o = T_t \quad . \tag{18B.3}$$

Solving for T_o and ϵ_o in Eqs. (18B.2) and (18B.3), respectively, and substituting into (18B.1) gives

$$V = -a_c(T_c - \epsilon_c)^2 - \bar{a}_r(T_r - \epsilon_r)^2 \quad , \tag{18B.4}$$

where $\bar{a}_r = a_r + a_o$.

The ratio schedule imposes the following relation between consumption and responding:

$$T_c = \frac{P}{N} \quad , \tag{18B.5}$$

where N is the response requirement under the ratio schedule; we have omitted, for the sake of simplicity, the constant converting number of reinforcements into time spent consuming. Allowing for positive levels of free consumption C_F, gives

$$T_c = \frac{P}{N} + C_F \quad . \tag{18B.6}$$

Assuming that the time to make a peck α is constant,

$$T_r = \alpha P \ . \tag{18B.7}$$

To determine the maximum of Eq. (18B.4) subject to the constraint (18B.6), the Lagrangian was first set up as follows:

$$F(T_c, P, \lambda) = -a_c(T_c - \epsilon_c)^2 - \bar{a}_r(\alpha P - \epsilon_r)^2 - \lambda\left(\frac{P}{N} + C_F - T_c\right) ,$$

where (18B.7) was used to substitute for T_r. Then the partial derivatives were taken with respect to the three variables, T_c, P, λ, which were set equal to zero, which produced the necessary and sufficient conditions for a maximum.

$$\frac{\partial V}{\partial T_c} = -2a_c(T_c - \epsilon_c) + \lambda = 0 \ ; \tag{18B.8}$$

$$\frac{\partial V}{\partial P} = -2\bar{a}_r(\alpha P - \epsilon_r)\alpha - \frac{1}{N}\lambda = 0 \ ; \tag{18B.9}$$

$$\frac{\partial V}{\partial \lambda} = \frac{P}{N} + C_F - T_c = 0 \ . \tag{18B.10}$$

Solving Eq. (18B.8) and (18B.9) for λ and setting the expressions equal to each other, gives the equation of the expansion path,

$$T_c = \epsilon_c - \frac{\bar{a}_r}{a_c} \alpha N(\alpha P - \epsilon_r) \ . \tag{18B.11}$$

Defining the free behavior response rate as

$$\epsilon_p = \frac{\epsilon_r}{\alpha}$$

and substituting it for ϵ_r in (18B.11) gives

$$T_c = \epsilon_c - \frac{\bar{a}_r}{a_c} \alpha^2 N(P - \epsilon_p) \ . \tag{18B.12}$$

From (18B.12) we see that the expansion paths are linear and have a negative slope, $dT_c/dP < 0$, as $((\bar{a}_r)/(a_c)\ \alpha^2 N) > 0$. Further, when P equals ϵ_r, the free behavior point T_c equals ϵ_c, and this point is common to all the expansion paths, irrespective of the value of N, the ratio requirement. This common point corresponds to point A in Figure 18-4. Finally, as the ratio requirement N increases, the slope of the expansion path becomes steeper in absolute value, as shown in the same figure.

REFERENCE NOTES

1. Battalio, R.C.; G.P. Dwyer, Jr.; and J.H. Kagel. *Tests of some alternative theories of motivational processes underlying individual choice.* Unpublished manuscript, Economics Department, Texas A&M University.

REFERENCES

Allison, J. 1979. Demand economics and experimental psychology. *Behavioral Science 24*: 403-15.

____ 1982. Economics and operant conditioning. In P. Harzem and M.D. Zeiler, eds., *Advances in Analysis of Behaviour*, vol. 2, *Predictability, Correlation and Contiguity*. Chichester, England: Wiley.

Allison, J.; M. Miller; and M. Wozny. 1979. Conservation in behavior. *Journal of Experimental Psychology: General 108*: 4-34.

Battalio, R.C.; J.H. Kagel; and L. Green. 1979. Labor supply behavior of animal workers: Towards an experimental analysis. In V.L. Smith, ed., *Research in Experimental Economics*, vol. 1, Greenwich, Conn.: JAI Press.

Baum, W.M. 1974. On two types of deviation from the matching law: Bias and undermatching. *Journal of the Experimental Analysis of Behavior 22*: 231-42.

Catania, A.C.; T.J. Matthews; P.J. Silverman; and R. Yohalem. 1977. Yoked variable-ratio and variable-interval responding in pigeons. *Journal of the Experimental Analysis of Behavior 28*: 155-61.

Chiang, A.C. 1974. *Fundamental Methods of Mathematical Economics.* 2nd ed. New York: McGraw-Hill.

de Villiers, P.A. 1977. Choice in concurrent schedules and a quantitative formulation of the law of effect. In W.K. Honig and J.E.R. Staddon, eds., *Handbook of Operant Behavior.* Englewood Cliffs, N.J.: Prentice-Hall.

Fleshler, M., and H.S. Hoffman. 1962. A progression for generating variable-interval schedules. *Journal of the Experimental Analysis of Behavior 5*: 529-30.

Herrnstein, R.J. 1970. On the law of effect. *Journal of the Experimental Analysis of Behavior 13*: 243-66.

____. 1979. Derivatives of matching. *Psychological Review 86*: 486-95.

Herrnstein, R.J., and G.M. Heyman. 1979. Is matching compatible with reinforcement maximization on concurrent variable-interval variable-ratio? *Journal of the Experimental Analysis of Behavior 31*: 209-23.

Hogan, J.A., and T.J. Roper. 1978. A comparison of the properties of different reinforcers. In J.S. Rosenblatt, R.A. Hinde, C. Beer, and M.S. Busnel, eds., *Advances in the Study of Behavior*, vol. 8, New York: Academic Press.

Myers, D.L., and L.E. Myers. 1977. Undermatching: A reappraisal of performance on concurrent variable-interval schedules of reinforcement. *Journal of the Experimental Analysis of Behavior 27*: 203-14.

Pear, J.J. 1975. Implications of the matching law for ratio responding. *Journal of the Experimental Analysis of Behavior 23*: 139-40.

Rachlin, H.; R. Battalio; J. Kagel; and L. Green. 1981. Maximization theory in behavioral psychology. *The Behavioral and Brain Sciences 4*: 371-417.

Rachlin, H., and B. Burkhard. 1978. The temporal triangle: Response substitution in instrumental conditioning. *Psychological Review 85*: 22-48.

Rachlin, H.; L. Green; J.H. Kagel; and R.C. Battalio. 1976. Economic demand theory and psychological studies of choice. In G. Bower, ed., *The Psychology of Learning and Motivation*, vol. 10, New York: Academic Press.

Rachlin, H.; J.H. Kagel; and R.C. Battalio. 1980. Substitutability in time-allocation. *Psychological Review 87*: 355-74.

Rees, A. 1977. The labor-supply results of the experiment: A summary. In H.W. Watts and A. Ness, eds., *The New Jersey Income Maintenance Experiment*, vol. 2, New York: Academic Press.

Skinner, B.F. 1953. *Science and Human Behavior.* New York: MacMillan.

Staddon, J.E.R. 1979. Operant behavior as adaptation to constraint. *Journal of Experimental Psychology: General 108*: 48-67.

Timberlake, W. 1977. The application of the matching law to simple ratio schedules. *Journal of the Experimental Analysis of Behavior 25*: 215-17.

Walsh, V.C. 1970. *Introduction to Contemporary Microeconomics.* New York: McGraw-Hill.

Zuriff, G.E. 1970. A comparison of variable-ratio and variable-interval schedules of reinforcement. *Journal of the Experimental Analysis of Behavior 13*: 369-74.

VI MATCHING THEORY

19 MELIORATION AS BEHAVIORAL DYNAMISM

R. J. Herrnstein

The idea of maximization, deeply rooted in common sense, holds behavior to be normatively rational and adaptive. Behavior is depicted as seeking an equilibrium that maximizes something—be it total subjective utility, hedonic value, reinforcement, energy intake, or reproductive fitness—within limitations of memory and discriminative acuity as well as the limitations imposed by the environment. Each mixture of behaviors and their outcomes is viewed as a unique bundle, and the organism is supposed to select the best bundle, on whatever is the relevant dimension. Equilibrium is reached with a distribution of activities that cannot be detectably improved upon by a redistribution of choices; that is, the obtained outcomes are maximized. This is the central conception of economic theory, around which the formal structure of economic analysis has developed. Not surprisingly, an intuitively compelling idea like maximization permeates behavioral sciences besides economics, such as sociology (Homans, 1974), behavioral biology (Pyke, Pulliam, and Charnov, 1977; Schoener, 1969, 1971; and Krebs and Davies, 1978), and psychology (Rachlin, 1980; Rachlin, Battalio, Kagel, and Green, 1981).

I will argue here that this plausible, all but universally accepted, and rigorous conception of behavior should be replaced by the matching law. The argument rests primarily on data that support matching over maximization in tests where the two theories make different predictions. Secondarily, the argument will try to show that the matching theory, once grasped, is at least as compelling intui-

Preparation of this chapter was supported in part by Grant DA 02350 from the National Institute on Drug Abuse to Harvard University.

tively and as amenable to formal elaboration and evolutionary interpretation as maximization.

Like maximization, matching theory should say something about both behavioral equilibrium and change. At equilibrium, matching requires the ratio of the frequencies of any two behaviors, B_1 and B_2, to match that of their obtained reinforcements, R_1 and R_2, as follows:

$$\frac{B_1}{B_2} = \frac{R_1}{R_2} \quad . \tag{19.1}$$

This assumes that B_1 and B_2 are measured on a common scale and likewise for R_1 and R_2 (Herrnstein, 1974, 1979). The equation, for steady-state behavior and reinforcement, has been at least approximately confirmed in numerous experiments on a variety of species (Baum, 1979; de Villiers, 1977; Herrnstein, 1970).

Despite the confirming data, Eq. (19.1) leaves an explanatory vacuum into which many hypotheses have been drawn (Catania, 1973; Donahoe, 1977; Gibbon, 1977; Killeen, 1981, 1982; McDowell and Kessel, 1979; Myerson and Miezin, 1980; Nevin, 1979; Staddon and Motheral, 1978). What is missing is a matching theory of behavioral change: How does an organism come to match its distribution of choices to the obtained reinforcements? Until we know the dynamic process that yields matching at equilibrium, it will seem reasonable to argue, as indeed some do argue, that matching is fortuitous (e.g., Shimp, 1969; Silberberg, Hamilton, Ziriax, and Casey, 1978), a by-product of a more basic law of behavior. Rachlin and other maximizers have suggested that the process of maximizing sometimes yields matching at equilibrium, but not necessarily. By definition, however, maximizers believe that behavior at equilibrium always maximizes, within the particular contraints.

Is there a process that guarantees matching at equilibrium, a dynamic process that does for matching theory what maximizing does for maximization theory? The answer proposed in Herrnstein and Vaughan (1980) was that behavior shifts toward higher local rates of reinforcement. Given a repertoire with only two alternatives, for example, if B_1 earns a higher rate of reinforcement than B_2, then time allocation shifts toward B_1 in a relatively continuous manner. If some distribution of time between B_1 and B_2 earns equal local rates of reinforcement, then equilibrium has been reached. If B_1 earns more reinforcement than B_2 at all time allocations, then at equilibrium B_1 displaces B_2 altogether. This process, called melioration, differs from the process of maximization in requiring the organism to respond only to the difference between local reinforcement rates from individual behaviors. Maximization, in contrast, requires the selection of the biggest aggregation of reinforcement across behaviors.

It will be shown that quite different behavioral equilibria can result from melioration and maximization. The two hypothetical processes nevertheless resemble each other in various respects: Both are hedonistic, utilitarian theories, inasmuch as behavior is held to be driven by its psychological consequences.

Hence both can claim descent from the commonsense idea of reward and punishment and from the formal law of effect. Also, both theories involve adaptation in the sense of bettering a state of affairs, although melioration is a more limited adaptation than maximization, as will be illustrated later. Finally both require knowledge of environmental feedback functions and internal motivational dynamics before any actual behavior can be predicted. The difference between the two theories is, in short, circumscribed, though often large in predicted behavior.

In this chapter, reinforcement (or R_i) should be understood as subjective value, so that it can properly be said that maximization means obtaining the maximum aggregate reinforcement available. Maximizers may object to the term *reinforcement*, for they are much concerned with the way value interacts with consumption and also with nonlinear interactions between commodities. *Utility* is the preferred economic term to express subjective value, rather than *reinforcement*, which may seem too easily confused with the object itself. A food pellet should, in fact, not be equated with a certain magnitude of reinforcement, when its reinforcing value depends on past consumption and the mix of other commodities being taken. To avoid needless extra terminology, however, *reinforcement* will serve here as the controlling consequence of behavior in both theories. Whereas maximization centers on R_T, the total reinforcement from a distribution of choices, melioration centers on R_D, the difference between local reinforcement rates for any pair of behaviors.[a] Local reinforcement rate, in turn, is the reinforcement actually obtained from an alternative, R_i, divided by the time allocated to it, t_i. Equilibrium between t_1 and t_2 is achieved when, in the following equation, R_D is zero:

$$R_D = \frac{R_1}{t_1} - \frac{R_2}{t_2} . \qquad (19.2)$$

When $R_D > 0$, time allocation shifts toward t_1; when $R_D < 0$, time allocation shifts toward t_2. When $R_D = 0$, then

$$\frac{R_1}{t_1} = \frac{R_2}{t_2} . \qquad (19.3)$$

Equation (19.3) is the matching law for time allocation. Moreover, given, as assumed before, that responding to the two alternatives is measured on a common scale of behavior,[b] Eq. (19.3) further implies the matching law as stated in Eq. (19.1).

a. Both R_D and R_T are actually reinforcements per total available time (session duration). In the equations that follow, session duration, or $\Sigma\, t_i$, is set to 1.

b. It is assumed that frequency of responding is proportional to time spent responding. When two responses are measured on a common scale, they bear the same proportionality to time. Under these constraints, Eq. (19.3), for time matching, implies Eq. (19.1), for response matching.

Melioration should account for behavioral change if it is to fill the vacuum left by the conventional matching law for behavior at equilibrium. Even though most of the data relevant to matching are for behavior at equilibrium, tests of the process of melioration can often be inferred from steady-state responding. As for a comparison between matching and maximization, melioration implies that behavior maximizes total reinforcement, R_T, under two and only two conditions, as follows:

1. When $R_D > 0$ at all values of t_1 (see Eq. (19.2)) and R_T is at a maximum when t_2 is zero;
2. When $R_D = 0$ at the maximum value of R_T.

Under the first condition one alternative earns more than the other at all allocations; maximizing is achieved because melioration causes that alternative to become prepotent. Under the second condition melioration yields maximizing because the local reinforcement rates happen, by nature of the reinforcement feedback functions, to become equal at the maximizing time allocation across a pair of alternatives. Under any condition other than (1) or (2), behavior at equilibrium should deviate from maximizing R_T, if melioration correctly describes the dynamic process governing behavioral change. Under all conditions behavior at equilibrium should obey the matching law. Data from both maximizing conditions and several nonmaximizing conditions are presented in the next section, which is intended to illustrate melioration in relation to maximization.

A SAMPLE OF FINDINGS

Concurrent Variable-Ratio, Variable-Ratio. The simplest procedure to represent from the standpoint of melioration happens also to satisfy condition (1) and therefore to imply maximization. Moreover, conc VR VR, or an approximation to it, occurs commonly in nature. The intuitive appeal of maximization may have something to do with its appearance in environments as familiar and simple as concurrent ratio schedules. The other side of the coin is the evolutionary selection of melioration, which may have something to do with its being a process that yields maximization in common environments. The evolutionary issue is considered again later.

The essential feature of conc VR VR is that the alternative behaviors are reinforced on variable-ratio schedules, which fix a given proportionality between behavior and reinforcement.

$$\frac{1}{(\mathrm{VR})_1} = p_1 = \frac{R_1}{B_1} . \qquad (19.4)$$

The reciprocal of the VR_i value gives the probability of reinforcement p_i. The local rate of reinforcement on each of the concurrent VR schedules depends on the programmed probability of reinforcement and the local rates of responding, b_1 and b_2, which, evidence shows, are likely to be nearly equal in the range typically studied. Melioration is governed by the local reinforcement-rate difference:

$$R_D = \frac{p_1 b_1 t_1}{t_1} - \frac{p_2 b_2 t_2}{t_2} . \tag{19.5}$$

Equation (19.5) says that on conc VR VR, the time spent at each alternative, t_i, is immaterial. Assuming that $b_1 = b_2$, choice will shift toward the higher probability of reinforcement. If the probabilities are equal, all allocations are at equilibrium. Any difference in the local rates of responding simply shifts the point of equilibrium, $R_D = 0$, toward a particular unequal pair of reinforcement probabilities. All these relations are implied by setting Eq. (19.5) equal to 0.

For maximization only the numerators from Eq. (19.5) should be considered; these give the total amount of reinforcement for any distribution of time,

$$R_T = p_1 b_1 t_1 + p_2 b_2 t_2 . \tag{19.6}$$

Recall that $t_2 = 1 - t_1$, which is to say that the alternatives are exhaustive. To find the equilibrium implied by maximization, we must see how R_T varies with changes in allocation. From Eq. (19.6) and the definition of exhaustive alternatives, it follows that

$$\frac{dR_T}{dt_1} = p_1 b_1 - p_2 b_2 . \tag{19.7}$$

Total reinforcement rises as the subject shifts toward the higher product of reinforcement probability and local rate of responding; or, if local rates of responding are equal, toward the higher probability of reinforcement.

The identity of Eqs. (19.5) and (19.7) tells us that for conc VR VR, melioration and maximization are indistinguishable. This is shown graphically in Figure 19-1 (from Herrnstein and Vaughan, 1980). The left ordinate shows R_D; the right one, R_T. Except for the lined-up zeroes, the two y axes are unrelated. The two functions plot Eq. (19.5) and (19.6) against the respective ordinates. Figure 19-1 illustrates the case in which the probability of reinforcement for alternative 1 is the greater, and the local rates of responding neither change materially with time allocation nor differ from each other. Melioration predicts exclusive preference for alternative 1, since R_D remains above zero at all allocations. Maximization predicts likewise but for a different reason, which is that R_T, total reinforcement, is at its maximum at $t_1 = 1$. As both theories predict, subjects respond exclusively, or nearly so, to the better alternative when working on conc VR VR (Herrnstein and Loveland, 1975; Herrnstein and Vaughan, 1980) in which the component schedules are discriminably different. It should go without saying that exclusive preference on conc VR VR is no more evidence for maximization theory than for matching theory.

Figure 19-1. Concurrent VR VR in which alternative 1 reinforces with a higher probability than alternative 2. R_D is the local-reinforcement-rate difference between alternatives 1 and 2; R_T is the overall number of reinforcements. R_D and R_T are shown as functions of time spent responding on alternative 1, t_1, as a proportion of unit time. (From Herrnstein and Vaughan, 1980.)

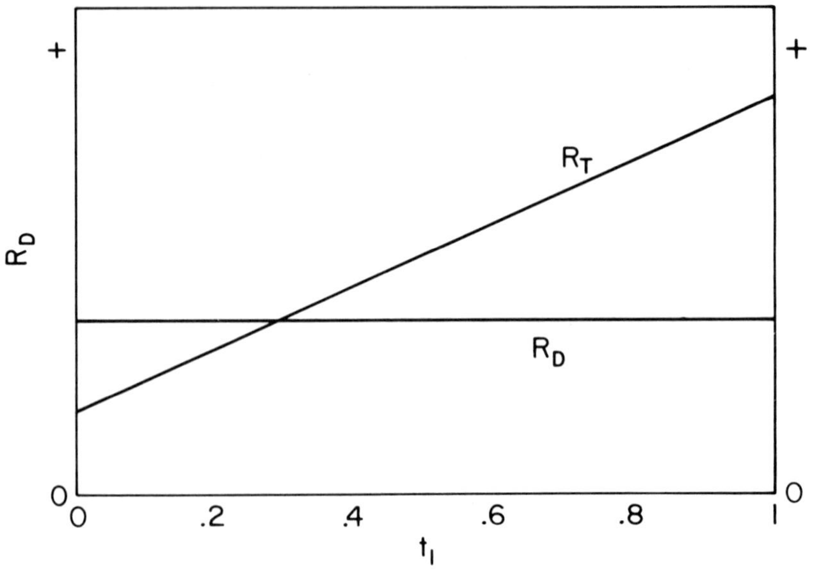

In general a comparison of matching and maximizing reduces to a comparison of the equation for R_D with that for dR_T/dt_1. Any environment that yields different expressions when the two equations are set equal to zero must result in a divergence between matching and maximizing. With conc VR VR the two expressions are identical (see Eqs. (19.5) and (19.7)), hence the convergence of matching and maximizing.

Concurrent Variable-Interval, Variable-Ratio. In conc VI VR, one alternative reinforces on a variable-ratio schedule, with its preset proportionality between responding and reinforcement. The other alternative reinforces on a variable-interval schedule, which consists of a repeating series of irregular interreinforcement intervals. In Herrnstein and Heyman, 1979, a range of VR and VI values was explored to see where behavior reached equilibrium, particularly in relation to matching and maximizing. Since the results have been reviewed in several publications besides the original (e.g., Green, Kagel, and Battalio, Chapter 18 of this volume; Herrnstein and Vaughan, 1980; Heyman, Chapter 20), the discussion here will be short.

Local rate of reinforcement behaves differently on the component schedules of conc VI VR. For VR it simply equals the product of reinforcement probability, p_2, and local rate of responding, b_2. For VI it depends (see Herrnstein and Heyman, 1979; Heyman and Luce, 1979; for details and qualifications) on the scheduled average interreinforcement time V; the average interresponse time during responding on the VI, d_1; a measure of interchangeover times between the two schedules, I; and the proportion of time spent on the VI, represented as t_1 (proportion of time on VR is $(1 - t_1)$). Designating local rates of reinforcement for VI and VR as r_1 and r_2, respectively,

$$r_1 = \frac{1}{V + d_1} + \frac{1 - t_1}{Vt_1 + I} \quad ; \tag{19.8}$$

$$r_2 = p_2 b_2 \quad . \tag{19.9}$$

Note that r_1 depends on two terms, the second of which approaches zero as time spent on the VI approaches 100 percent. Thus, on VI there is an inverse relation between time spent at it and local rate of reinforcement. The formula for melioration on conc VI VR can now be stated.

$$R_D = \frac{1}{V + d_1} + \frac{1 - t_1}{Vt_1 + I} - p_2 b_2 \quad . \tag{19.10}$$

Equilibrium is reached when R_D equals zero.

$$r_1 = r_2 \quad ;$$

$$\frac{1}{V + d_1} + \frac{1 - t_1}{Vt_1 + I} = p_2 b_2 \quad . \tag{19.11}$$

Since t_1 must fall within the $0 - 1$ interval, Eq. (19.11) may or may not be solvable for t_1, given the other parameter values. If not, then melioration predicts exclusive preference for VI or VR, depending on whether R_D is always positive or negative, respectively. In the case of exclusive preference, matching is trivially satisfied. It can easily be proved that satisying Eq. (19.11) also implies matching, but not trivially. The number (as distinguished from local rate) of reinforcements from the VI, R_1, is the product of the local rate of reinforcement and the time spent there; likewise for VR, which is R_2.

$$R_1 = t_1 r_1 \tag{19.12}$$

$$R_2 = (1 - t_1) r_2 \tag{19.13}$$

However, at equilibrium (see Eq. (19.11)), $r_1 = r_2$. Hence, from Eqs. (19.12) and (19.13), time matching is the predicted (and observed) outcome.

$$\frac{t_1}{1 - t_1} = \frac{R_1}{R_2} .$$

Now let us consider maximization, which depends on the numbers (not local rates) of reinforcement:

$$R_T = R_1 + R_2 . \tag{19.14}$$

From Eqs. (19.8), (19.9), (19.12), (19.13), and (19.14) it follows that

$$\frac{dR_T}{dt_1} = \frac{1}{V + d_1} + \frac{I - Vt_1^2 - 2It_1}{(Vt_1 + I)^2} - p_2 b_2 . \tag{19.15}$$

Equilibrium would be reached when dR_T/dt_1 equals zero;

$$\frac{1}{V + d_1} + \frac{I - Vt_1^2 - 2It_1}{(Vt_1 + I)^2} = p_2 b_2 . \tag{19.16}$$

If matching yielded maximum reinforcement, then Eq. (19.16) would be identical to the equilibrium produced by melioration (Eq. (19.11)), which it plainly is not. It can be shown that matching and maximizing can coincide only when responding to the VI stops, given realistic rates of alternation. Moreover, a comparison of Eqs. (19.11) and (19.16) indicates that matching generally requires more time spent on VI than on VR. Precisely that deviation toward spending time on the VI is the observed result (Herrnstein and Heyman, 1979).

The tendency to match on conc VI VR, as well as other features of the present analysis, are illustrated in Figure 19-2 (from Herrnstein and Heyman, 1979). Four pigeons worked on six pairs of values for conc VI VR. Steady-state responding is shown for 19 of the 24 conditions; the 5 points omitted had exclusive preference for the VI. The figure plots, in logarithms, the ratio of times on VI and VR against the reinforcements received from each, with the solid line tracing the best-fitting linear regression (of the logged variables) for the data pooled over subjects. Perfect matching is given by the dashed line. Except for a bias toward spending time on VI, which is itself consistent with the matching law (see Herrnstein and Heyman, 1979), the simple matching law accounts quite well for the data. In contrast, maximization, whose prediction is the dot-dash line, would have required substantial bias toward spending time on the VR. The optimal strategy for conc VI VR would seem to call for lots of time on the VR with occasional forays to the VI to collect a reinforcement come due, and analysis shows that intuition is right about the requirements of optimization for typical parameter values. Nevertheless, no subject displayed any such bias toward

Figure 19–2. The logged ratio of times spent on the components of conc VI VR as a function of the logged ratio of reinforcements received from them, for four pigeons on six pairs of schedule values. The variable-interval component is designated by subscript 1. Simple matching is shown by the dashed line; maximization, by the dot-dash line. The best-fitting regression line through the data is the solid line; equation and percentage variance accounted for are also given. (From Herrnstein and Heyman, 1979.)

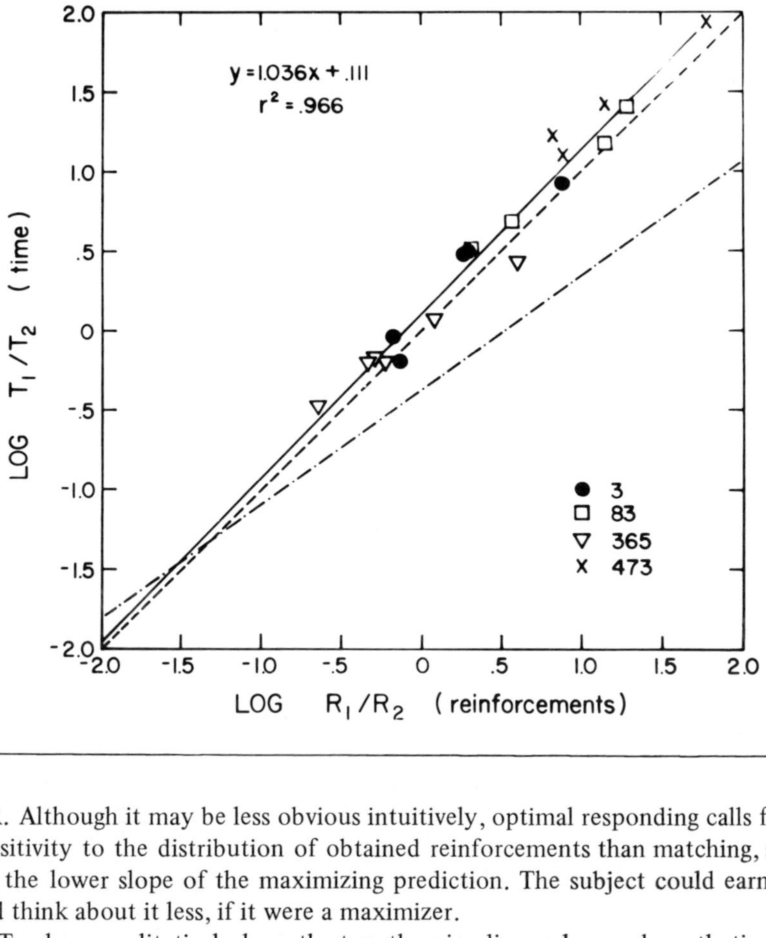

VR. Although it may be less obvious intuitively, optimal responding calls for *less* sensitivity to the distribution of obtained reinforcements than matching, shown by the lower slope of the maximizing prediction. The subject could earn more and think about it less, if it were a maximizer.

To show qualitatively how the two theories diverge here, a hypothetical conc VI VR is portrayed in Figure 19–3. The loci of Eq. (19.10) for melioration and Eq. (19.14) for maximizing are plotted against t_1, the proportion of time spent on the VI. For the values hypothesized, matching, where the local reinforcement rate difference R_D passes through zero, is for $t_1 = 0.45$; maximizing, for the same parameter values, requires $t_1 = 0.15$. The value of R_T when $R_D = 0$ is

Figure 19–3. R_D and R_T for a hypothetical conc VI VR; t_1 gives the proportion of time allocated to the VI. Crosses show where R_D intersects zero and R_T passes through its maximum. (Figure and caption from Herrnstein and Vaughan, 1980.)

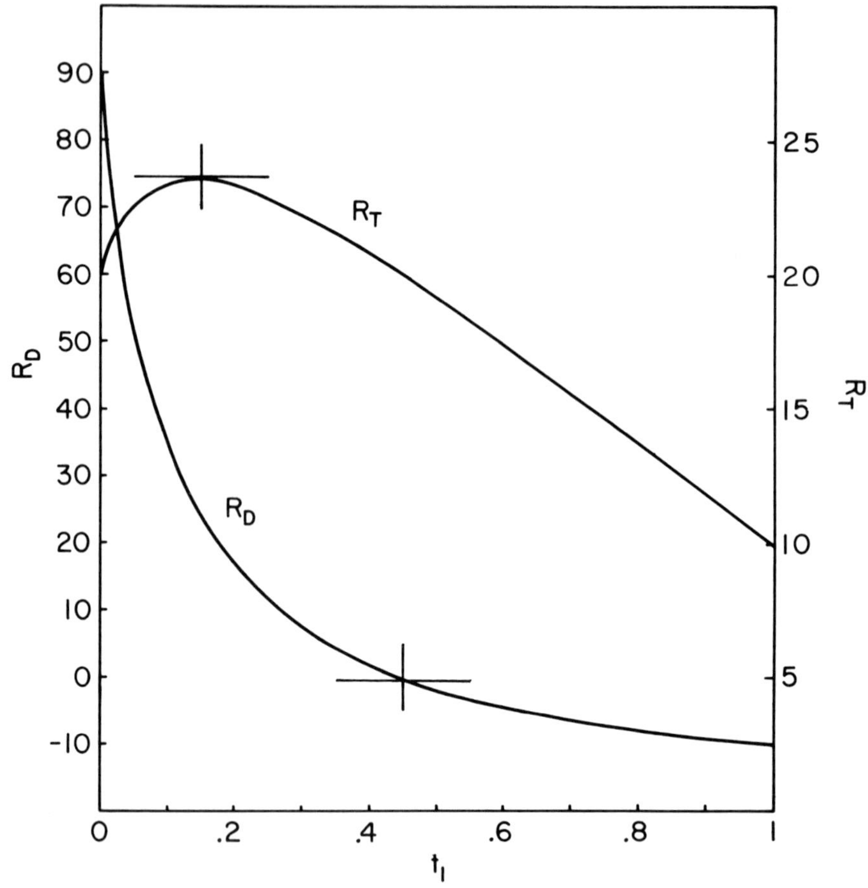

about 15 percent lower than when R_T is maximized, which is the reinforcement cost of matching. In fact, by matching, the pigeons shown in Figure 19–2 gave up food reinforcements at a rate averaging about 60 per hour, a rate known to sustain nearly asymptotic rates of responding.

In spite of the apparently devastating outcome of conc VI VR, some maximizing theorists (Rachlin, Battalio, Kagel, and Green, 1981; Green, Kagel, and Battalio, Chapter 18) have nevertheless held their ground. They point to the contribution of "leisure" as the reason subjects may sacrifice food reinforcement by overinvesting their time on VI. If a subject earns more total utility (or reinforcement, in our terms) by giving up food from the VR while replacing it with a lesser amount of food from the VI and a compensating gain from not working on VR (sparing it an aching beak?), then maximization may be saved. Just how this adds up has not been empirically demonstrated, although Green et al. (Chapter 18) believe they are on the track. But neither they nor anyone else as yet in print has even claimed to show why the leisure differential between VI and VR happens to be just the right magnitude so that subjects who appear to be matching on conc VI VR are actually maximizing. In any event, the ambiguity in conc VI VR arises from the difference in local response rates or in response topographies on the component schedules, which may indeed obscure an underlying maximizing process. The answer awaits the outcome of a more direct test of this possibility, which would be to devise a version of conc VI VR that eliminated or minimized the effect of topography or rate differences between the alternative responses. In the meantime, let us turn to two other recent experiments that confront matching and maximizing directly.

Mazur's Experiment. The procedure devised by Mazur (1981) (as well as that of Vaughan, described next) has the virtue of being so compelling intuitively that impenetrable formalizations are unnecessary. The pigeons could peck freely at either of two response keys, one green, the other red, in a standard chamber. A peck on one or the other key occasionally initiated 3 seconds of darkness. These periods of darkness were programmed by a single VI 45-second schedule, which randomly and equally often assigned each dark period to one key or the other, which makes this a Stubbs–Pliskoff (1969) procedure. Once primed, the VI timer stopped until a peck collected the assigned dark period. Hence the procedure guaranteed an exact, though random, sequence of switches in the effective peck from one alternative to the other. During a certain fraction of the 3-second dark periods, a small ration of food was delivered. For example, in the first condition every dark period was accompanied by food. In the second condition all dark periods for one alternative still yielded food, but only a random 10 percent of the dark periods for the other alternative did so. The six subsequent conditions presented further variations in the proportion of food-yielding dark periods.

Maximization and matching predict sharply different outcomes here. Consider the two conditions just described. For a maximizer the proper course of action was always to sample each alternative frequently and equally. Either key was equally likely to be primed to deliver a dark period, and the schedule only progressed as the assignments were collected, for both the first and second condition, as well as for all subsequent ones. For a matcher the tendency to equalize local reinforcement rates should have shifted preference along with the proportions of dark periods yielding food. In the first condition a matcher should have responded equally often to the two alternatives, just like a maximizer. In the second condition, however, when one alternative presented food 10 times as often as the other, a matcher should have spent over 90 percent of its time on the more lucrative alternative, even though doing so exacted a cost in food reinforcement. The cost arose because the VI timer stopped and waited at the end of each programmed interval as long as the subject was spending time on the unprimed side.

Mazur arranged four transitions for which a tendency to match would reduce overall food reinforcement (see Table 19-1 for the eight conditions), summarized in Figure 19-4. The foot of an arrow shows initial responding; the head, final responding during the decisive conditions. If pigeons were maximizers, they should have responded 50 percent to green under all conditions. Instead, in 16 out of 16 transitions (four for each pigeon), they shifted toward the matching value, which is indicated by the dashed vertical line. Fifteen of the 16 shifts caused reductions in food reinforcement, typically large ones. Unlike conc VI VR, here no problem of differential response rate or topography exists. Moreover, the maximizing strategy could hardly have been simpler: whatever happens, respond half the time to each component. Figure 19-5, on coordinates of relative responding and relative reinforcement, shows the maximizing strategy as a dashed horizontal line, which the final performances of the four pigeons invariably failed to approximate. They approximated instead a function with moderate undermatching.

Mazur does not attempt to account for the undermatching, for his point was just the departure from maximizing, but two obvious reasons suggest themselves. First, the Stubbs-Pliskoff procedure itself may produce a characteristic type of undermatching because of the nonindependence of the two components. For example, even when the nominal reinforcement probability is zero for one alternative, responding to it is differentially reinforced (with some delay) because the responding advances the sequence of intervals upon which reinforcement from the other alternative depends. Second, the unusual contingency of reinforcement—a dark period that may or may not include reinforcement—may reduce

Table 19-1. Probability of food reinforcement in a dark-key period.

	Key	
Condition	*Red*	*Green*
1	1.00	1.00
2	.10	1.00
3	1.00	1.00
4	1.00	.33
5	1.00	.00
6	1.00	.50
7	.50	1.00
8	.00	1.00

Source: Mazur, 1981.

Figure 19-4. For four pigeons, the four transitions where matching predicts a shift away from equal responding to the green and red alternatives. All eight experimental conditions are summarized in Table 19-1. Overall reinforcement rate R_T is shown as a function of the relative frequency of responding to the green alternative. Nominal matching values are shown by the vertical dashed lines. The point at the tail of each arrow is the average of the first six sessions after a transition; the point at the head of the arrow is the average of the last six sessions before the next change of conditions. (From Mazur, 1981.)

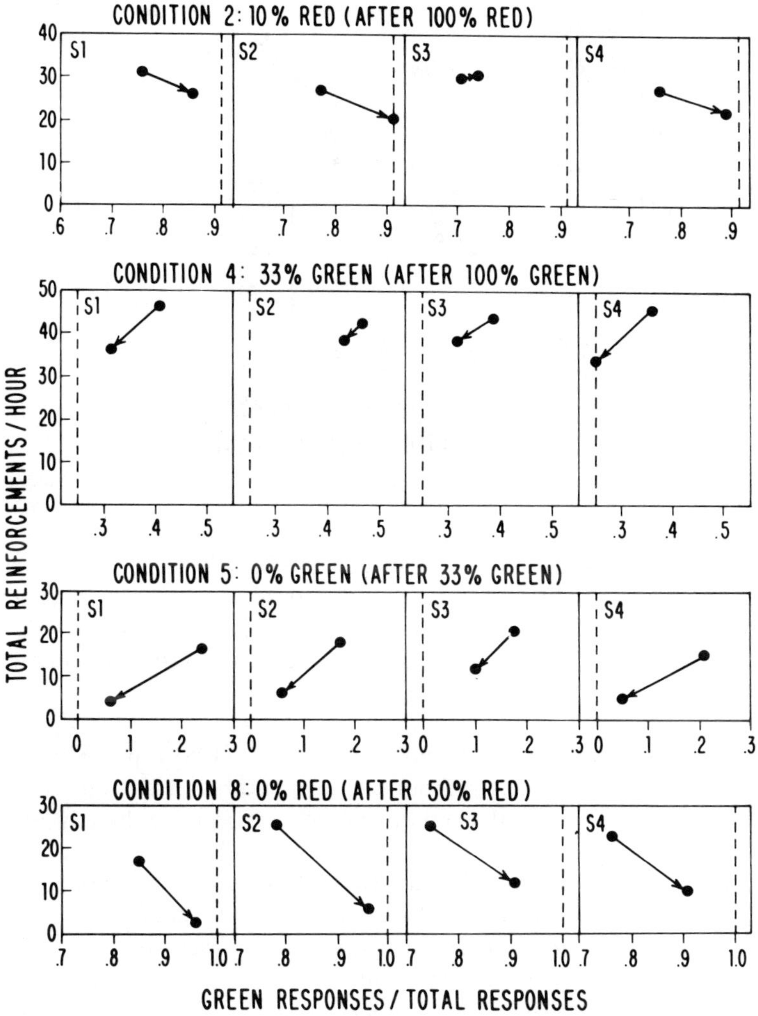

Figure 19-5. Relative responding to the green alternative as a function of relative reinforcement received therefrom, for each of four pigeons, averaging over the final six sessions of eight experimental conditions (see Table 19-1). Simple matching is the solid line; the best-fitting regression is the dashed line, whose equation is also given. (From Mazur, 1981.)

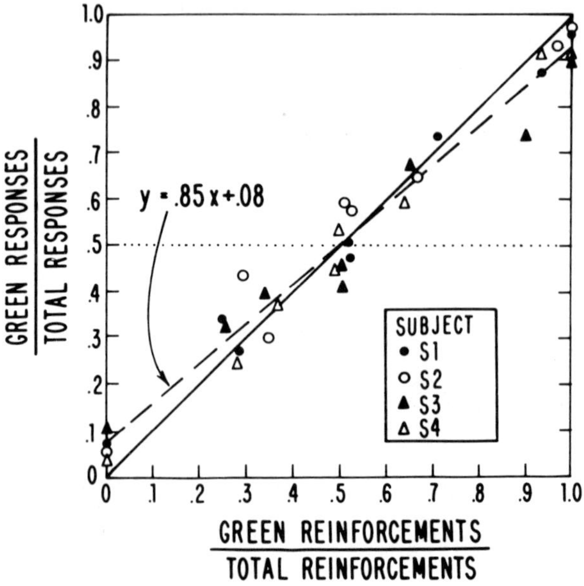

the discriminability of the two alternatives, probably the most common source of undermatching. For these reasons Mazur's procedure can hardly be conducive to matching, although from the standpoint of simplicity it is ideal for maximizing. Nevertheless, the subjects approximately matched, shifting distributions of choices at considerable costs in reinforcement.

Vaughan's Experiment. Like Mazur's, Vaughan's (1981) procedure pitted matching against maximizing without any complication from differential response topography. To test the concept of melioration more directly than any previous study, Vaughan modified a concurrent variable-interval, variable-interval schedule in two respects. First the schedule values were continuously updated in reference to the subject's behavior, as will shortly be illustrated. Second, the two VI schedules advanced only when the subject was working on them, so that reinforcement for one alternative did not grow in probability with time spent on the other alternative, as in ordinary conc VI VI. When the subject was

Figure 19–6. Programmed local rates of reinforcement on left (L) and right (R) alternatives as a function of time on right, measured as a proportion of unit time. The two experimental conditions, *a* and *b*, differ only for abscissa values between 0.25 and 0.75, as shown. (Adapted from Vaughan, 1981.)

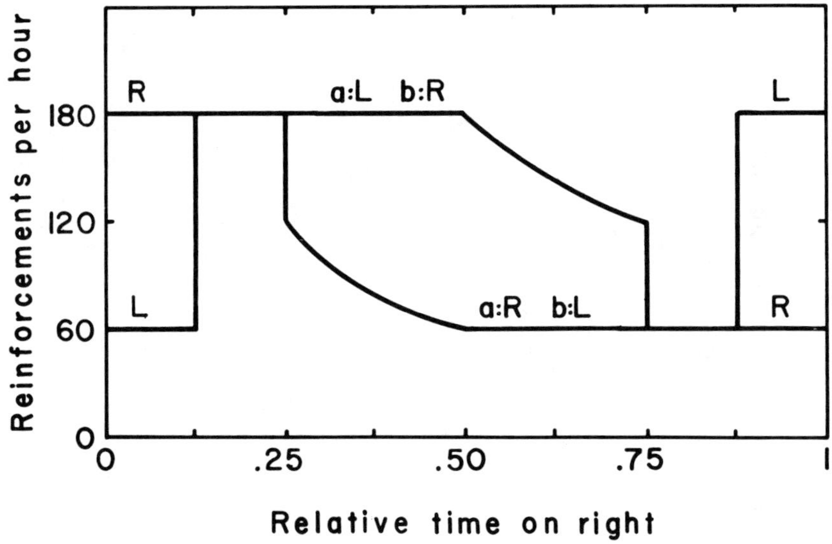

working on neither schedule, defined as any interresponse time greater than 2 seconds, neither schedule advanced.

The experiment consisted of two conditions, represented by a and b in the ensuing figures. Figure 19–6 shows the reinforcements per hour on the component VI schedules (L = left; R = right) as a function of relative time spent responding on the right. Conditions a and b are the same except when the relative time on the right falls between 0.25 and 0.75. In condition a, the left schedule then reinforces at a higher rate than the right schedule; in condition b, vice versa. Figure 19–6 shows how the subject's distribution of responding controlled the schedule values, which were updated every 4 minutes of responding. After every 4 minutes of responding the relative time on the right was automatically calculated. From that proportion the programming computer read off the appropriate values of the two component VI's to the program (as presented in Figure 19–6), which stayed in force at least until the end of the next 4-minute period of responding, when relative time on the right was again calculated.

Three pigeons ran on conditions a and b (except for minor additional adjustments in schedule values for one of them, described in Vaughan, 1981). The point of the two conditions, hence the logic of the experiment, is highlighted in

Figure 19-7. Overall reinforcements (R_T) as a function of relative time on the right alternative, for the two experimental conditions, *a* and *b*. (Adapted from Vaughan, 1981.)

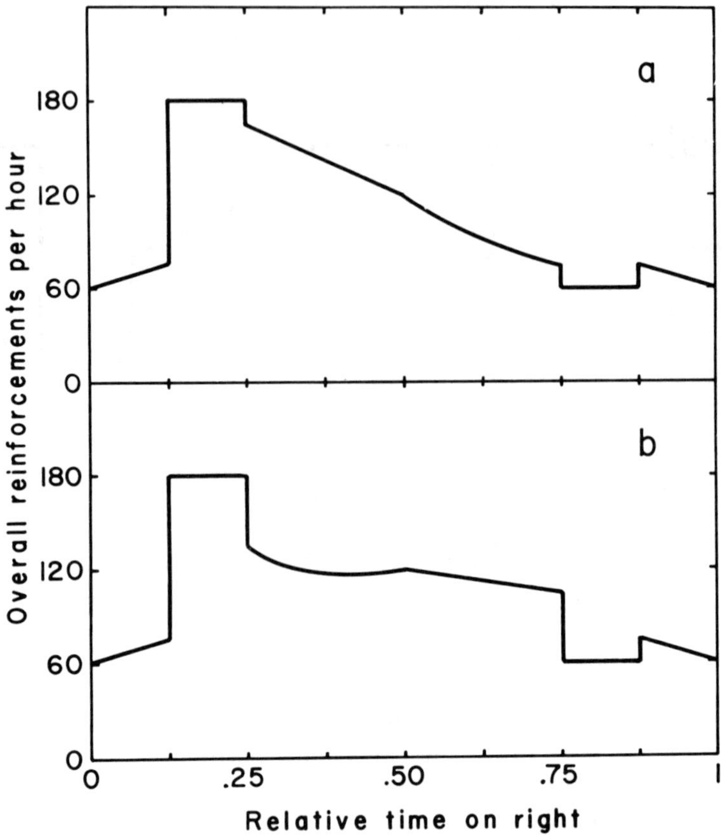

Figures 19-7 and 19-8. Figure 19-7 shows the maximization picture, namely how overall food reinforcements per hour interacted with the distribution of choices. In either condition, a subject earned the maximum, 180 reinforcements per hour, by spending 0.125-0.25 of its time responding to the right alternative. That is how a maximizer should spend its time. A sidelight that ought to be noted is that in both conditions, the minimum earnings, 60 reinforcements per hour, were given for spending 0.75-0.875 of the time on the right.

Figure 19-8 shows the melioration picture, namely the value of R_D at various distributions of choices. R_D, it should be recalled, is the local reinforcement rate difference between the two alternatives, here shown as "net reinforcements per

Figure 19-8. Local rate of reinforcement on the right minus local rate of reinforcement on the left as a function of relative time on the right alternative. Conditions *a* and *b* differ only for abscissa values between 0.25 and 0.75. (Adapted from Vaughan, 1981.)

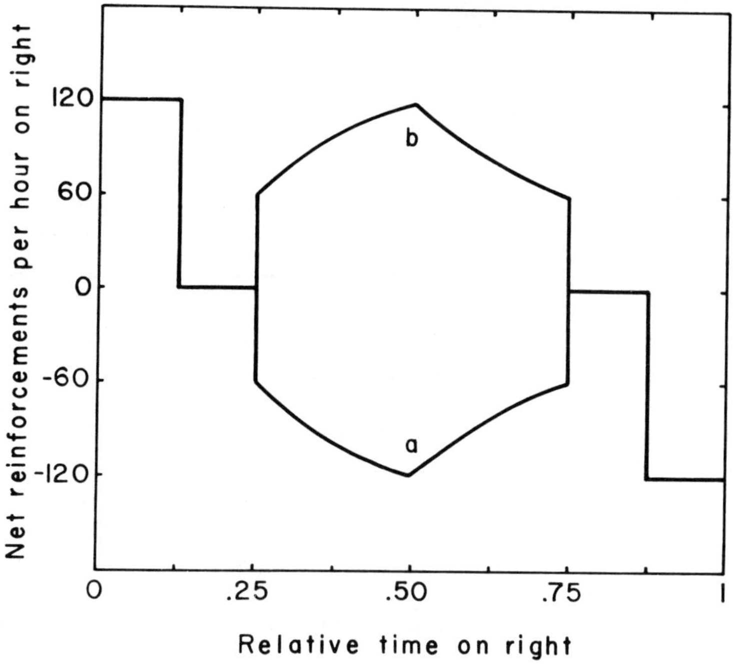

hour on right," which is obtained by subtracting the left rate of reinforcement from the right:

$$R_D = \frac{R_R}{T_R} - \frac{R_L}{T_L} \quad .$$

For condition a, the value of R_D stayed negative from 0.25 to 0.75 for time spent responding on the right; for condition b, it was always positive in that range. Otherwise the two conditions were the same. Assuming that the central range was sampled, melioration should have held choice within the interval from 0.125-0.25 during condition a and within the interval 0.75-0.875 during condition b, without regard to the accompanying changes in overall reinforcement rate.

The predictions of the two theories and the experimental results are shown in Figure 19-9 for the three subjects. In condition a both theories predict choice

Figure 19-9. Maximization ("MAX") and melioration ("MEL") predictions for conditions *a* and *b* in Vaughan's experiment (1981). The results for the three pigeons are shown on the scale for relative time spent responding on the right.

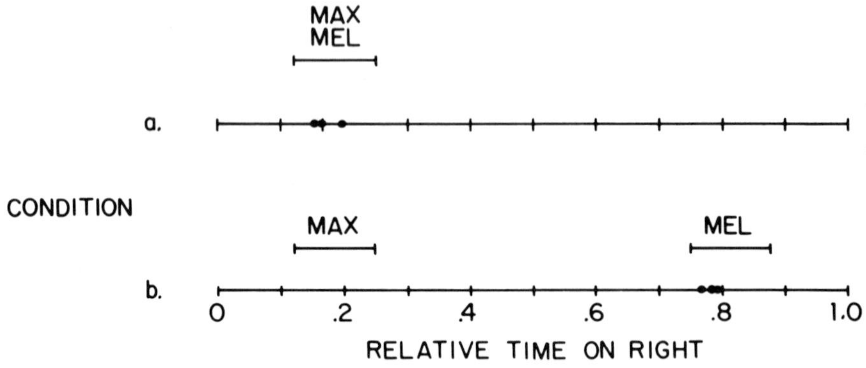

in the range from 0.125-0.25 relative time on the right. Choice by the three subjects, shown by the points on the line, fell in the predicted range. After over 50 sessions in condition b, choice had stabilized in the interval 0.75-0.875, still meliorating but no longer maximizing. By meliorating, the subjects lost 120 reinforcements per hour, a drop of 75 percent in food intake. As in Mazur's procedure the dice were loaded for maximization, since no shift in responding at all from a to b would have earned maximum food. Vaughan (1981) noted, further, that in the shift from condition a to b, choice passed through distributions of behavior that transiently increased the deviations from matching, even though final performance in both conditions closely approximated matching, as it must if melioration is in control. This incidental finding rules out a dynamic process based on minimizing deviations from matching, a theoretical possibility entertained but rejected earlier in Herrnstein and Loveland (1975).

LIMITS OF MELIORATION

Although the data make a clear case for melioration, as opposed to maximization, the theory's major limitations, empirical as well as theoretical, should be considered. The following discussion starts with the empirical questions, divided along the lines of standard psychological variables, and concludes with the inherent limits of melioration as a theory of behavioral adaptation.

Comparative. Most of the evidence for matching comes from pigeons, although there are enough data from other species, including human beings, to make matching a reasonably general principle of choice. However, the particular environments that most sharply separate maximizing from meliorating (Herrnstein and Heyman, 1979; Herrnstein and Vaughan, 1980; Mazur, 1981; Vaughan, 1981) have so far all been examined with pigeons as subjects. An argument could be made that since the equilibrium state—matching—has comparative generality, the dynamic process—melioration—must too, but this is an indirect argument. It remains conceivable that brighter creatures than pigeons do not meliorate but maximize, matching only when it involves no large cost in overall reinforcement. On the other hand, it is also conceivable that melioration can account for bright creatures as well as pigeons. Dealing seriously with comparative variation is a prime challenge to any general theory of adaptation and will therefore be reconsidered later.

Psychophysical. Melioration requires a subject to compare local rates of reinforcement, but this is "local" in a special sense. It is local with respect to response categories, not necessarily with respect to time. Pigeons, for example, compare left-key reinforcement rates to right-key reinforcement rates, but the time frame may be large or small. The rates are based, for each topography, on number of reinforcements divided by time invested in that topography, but we do not know all the parameters of the pigeon's estimation procedure. The transformation of objective local reinforcement rate into a subjective local reinforcement rate no doubt has its share of psychophysical complications (Commons, 1981, Commons, Woodford, and Ducheny, Chapter 2). Species differences in this transformation may result in different equilibrium states even though each species is meliorating and matching.

Qualitatively, a subject's recent experience must weigh more than the distant past. Otherwise, a subject who spent 60 sessions on conc VI 1 minute VI 3 minutes would need more than 60 sessions to switch preference on conc VI 3 minutes VI 1 minute. Usually, preference switches rapidly, within 3 or 4 sessions, and the switch does not get materially slower with the duration of exposure to the preceding conditions, beyond some minimal exposure. From the standpoint of melioration this indicates a relatively rapid initial updating process for local rate of reinforcement, much as called for by Killeen's averaging theory (1981) or by the Rescorla-Wagner theory (1972) of associative strength (see Vaughan, Chapter 12, for a concrete application to melioration). A subject's sensitivity to the current values of local rate of reinforcement should be reflected in the rapidity with which new equilibria are reached by melioration. One of the advantages of the melioration framework is its capacity to absorb a variety of equilibria.

The nature of the psychophysical transformation of objective rates probably explains a variety of findings that point to the importance of temporal structure above and beyond the average rates of events. It has, for example, been shown (Herrnstein, 1964) that pigeons prefer variable-interval to fixed-interval reinforcement when the objective rates are equal. Likewise, in their effects on behavior, remote reinforcers are less reinforcing (Chung and Herrnstein, 1967;

Commons, 1981; Gentry and Marr, 1980; Shull, Spear, and Bryson, 1981) and remote punishers less punishing (Hineline, 1977) than immediate ones, even after average reinforcement or punishment rate has been taken account of. Melioration theory presupposes nothing about the psychophysics of local reinforcement rate, only that behavioral equilibrium depends on what it happens to be.

Motivational. The four experiments described here were limited not only to pigeons as subjects but food as reinforcement. Choice between alternatives with the same reinforcer bypasses variations in demand elasticity, mutual substitutability, and diminishing marginal utility, complications well handled by microeconomic analysis. With only a few exceptions (e.g., Miller, 1976) matching theory has been applied to choices within single classes of reinforcers, from which some theorists (Hursh, 1978; Rachlin et al., 1981) have concluded that it cannot readily be applied across classes of reinforcers. However, as regards reinforcement interactions, there is no essential incompatibility with matching nor any essential compatibility with maximization. The difference is purely historical: Maximization theory has spelled out in great detail how motivational changes should affect choice, whereas it remains to be shown concretely how melioration and matching work in broader and biologically more relevant contexts.

Procedural. Matching was originally formulated as a description of choice, the relative frequency of a response with respect to concurrent behavior. Later it was extended to absolute frequencies of responding, which involved an interpretation of the context of reinforcement, and to behavior on multiple schedules, wherein the context of reinforcement shifts from time to time. In its present state of development, melioration has not been formulated or tested for anything but relative frequency of responding on concurrent schedules. The extension to single-response situations is in principle straightforward, however. For a single-response situation,[c] melioration presumably involves time allocation between the response in question, t_1, and all other activities t_e (such that $t_1 + t_e = t_T$, or total time) with their associated reinforcements, R_1 and R_e, respectively,

$$R_D = \frac{R_1}{t_1} - \frac{R_e}{t_e} .$$

c. The derivation assumes that the number of reinforcements R_i does not depend on time spent responding, as approximated, for example, by variable-interval or variable-time schedules. The assumption would also apply to discrete-trial paradigms in which some dimension of reinforcement other than frequency (such as amount or quality) was varied.

At equilibrium, $R_D = 0$, hence,

$$\frac{R_1}{t_1} = \frac{R_e}{t_e} \quad ,$$

from which it follows that

$$t_1 = \frac{t_T R_1}{R_1 + R_e} \quad . \tag{19.17}$$

Assuming that the frequency of the response in question, B_1, is proportional to the time spent at it, t_1, and that k measures total behavior in the units of B_1, it follows from Eq. (19.17) that

$$B_1 = \frac{kR_1}{R_1 + R_e} \quad . \tag{19.18}$$

Equation (19.18) is the familiar and widely supported hyperbolic equation (de Villiers and Herrnstein, 1976) for absolute rate of responding as a function of relative reinforcement. Melioration can therefore be generalized from concurrent to single-response procedures. It is possible that a comparable argument can be developed for multiple schedules, but no fully satisfactory formula (including the one proposed by Herrnstein, 1970) for multiple-schedule responding has yet been demonstrated. It would therefore be premature to hypothesize further about melioration in multiple schedules here.

Inherent Limits. Melioration as so far described presupposes a set of response topographies with associated rates of reinforcement. But the acquisition of a new response topography itself exemplifies melioration, as less highly reinforced movements lose time allocation to more highly reinforced alternatives. The movements that survive exposure to a new contingency of reinforcement must, by this view, have equal local rates of reinforcement. But now a question arises. Is this class of equally reinforced movements also the class of maximally reinforced movements? If it is, then we must say that the end state of the process of melioration is maximization and the gap between the two theories would have vanished. If it is not, then why not?

As a concrete example, consider the conc VI VR shown in Figure 19-3, on which the shaping of response topographies need not have stopped with pecking at each of two keys. It could, by melioration, have continued until a more inclusive unit of responding was formed, one in which the subject spent 15 percent of its time on the VI alternative, which is the overall maximization point. In Mazur's and Vaughan's procedures, too, melioration could have led to maximizing compounds of the two response alternatives. The fact that this is not observed suggests that constraints on responding impede melioration in the formation of new topographies of responding. The constraints probably arise from several sources, such as the congeniality of certain movements as compared to others (which can itself be a component of local reinforcement rate), the organism's capacity to integrate reinforcement rates over longer or shorter time periods or to categorize certain movements as operants, and reversibility in the repertoire of responses.

For the perfectly plastic behavioral repertoire, melioration yields maximization. But to achieve perfect plasticity, the organism would have to be indifferent, motivationally speaking, to the physical characteristics of its own movements, capable of abstracting any contingency of reinforcement for any of its movements, no matter how subtle, unnatural, or over how large a time frame, and the effects of its past history of reinforcement would have to be totally reversible. Every distinct reinforcing environment would require a new configuration of movements as the maximizing topography, composed of simple movements that would lose their individual identity. To pick the right configuration the subject would also need something approaching perfect stimulus control. This ideal organism, which would maximize as it meliorated, may sound a bit like the hypothetical organism of early operant theory, with its "arbitrary" stimulus and response classes. Even if no such organism exists, as apparently it does not, it at least represents a benchmark for the adaptive potentialities of melioration. As evolution produces more easily conditionable and extinguishable response topographies and the capacity to detect correlations between behavior and its consequences over increasingly large time spans, melioration approaches maximization. Different species no doubt fall on different points along this continuum. Indeed, the ability to account for a continuum of adaptation that includes maximization as an extreme case makes melioration a likely dimension of evolutionary development.

Although maximizing theorists may sometimes neglect it, their theory, too, must have inherent limits. They do not believe that every behavioral configuration is perfectly geared into its consequences, now or later, near or far. Limitations of or defects in an organism's understanding, effects of uncertainty and time discounting are among the factors that spoil an organism's progress toward perfect maximization. In some respects maximization under realistic constraints may approximate melioration, with a drift toward higher local rates of reinforcement at the cost of more remote gains. Similarly the notion of momentary maximizing (Shimp, 1969) brackets melioration from the other side, by picturing the subject as controlled by the probability of reinforcement of the next response only. In spite of the convergence of these theories, a fundamental difference remains, however. According to melioration behavioral equilibrium must always conform to matching, no matter what the constraints of response topographies or cognitive limitations. The matching law cuts across the contingent limitations on melioration. For overall or momentary maximizing no universal characteristic of equilibrium has been specified except maximization itself, which is in fact an unobserved ideal case.

WHAT IS REINFORCEMENT?

A duality for reinforcement has long been customary. New topographies are held to be shaped out of the stream of behavior by the contingency of reinforcement, as in Thorndike's early puzzle box experiments (1898) and in countless other examples of *acquisition* via the law of effect. But then, after extensive study of intermittent reinforcement (Ferster and Skinner, 1957), it became clear that the *maintenance* of behavior is a salient second function of reinforcement. The present theory puts acquisition and maintenance into a single, quantitative conception. Melioration, the displacement of less highly reinforced topographies by more highly reinforced topographies, corresponds to acquisition. Given a set of topographies, melioration proceeds until behavior is in equilibrium, which is to say, at the levels dictated by the matching law, which corresponds to the maintenance function of reinforcement. Sometimes the equilibrium state also maximizes reinforcement and sometimes it does not, depending on the characteristics of the environment and of the organism interacting with it, as outlined in preceding sections of the chapter. The present theory is, then, a more precise and focused specification of the traditional law of effect and of common sense regarding reward and punishment.

If organisms are not literal but only accidental and occasional maximizers, one may ask why not. Why has natural selection followed the path of melioration, even though it leads to nonmaximization in certain environments? Given the evolutionary pressure toward more adaptive forms, why would there not be a maximization process in store for the future? The answers to these questions have been anticipated in earlier comments. Natural selection followed the path of melioration because it is a cognitively realistic approximation to maximization in many natural environments involving choices between probabilistic alternatives (that is, ratio schedules). Taking differences between local rates of reinforcement could well be within the capacities of organisms that could not aggregate reinforcement across topographies and over long periods of time, such as is required by maximization.

As to whether there is maximization in the future, an answer is that maximization is the limiting form of melioration. As an organism meliorates over varying response topographies, larger time frames, and a more differentiated set of environments, it approaches the ideal maximizer of economic analysis. Most economic theorists realize that their theory is built on a convenient fiction, that of maximization. The present analysis suggests that it is indeed a fiction for any existing organism, but hardly convenient if it obscures a more correct view of the matter. The law of effect interpreted as the dynamism of melioration leading to the equilibrium of the matching law appears to explain when organisms maximize and when they do not.

REFERENCES

Baum, W.M. 1979. Matching, undermatching, and overmatching in studies of choice. *Journal of the Experimental Analysis of Behavior 32*: 269–81.

Catania, A.C. 1973. Self-inhibiting effects of reinforcement. *Journal of the Experimental Analysis of Behavior 19*: 517–26.

Chung, S–H., and R.J. Herrnstein. 1967. Choice and delay of reinforcement. *Journal of the Experimental Analysis of Behavior 10*: 67–74.

Commons, M.L. 1981. How reinforcement density is discriminated and scaled. In M.L. Commons and J.A. Nevin, eds., *Quantitative Analyses of Behavior*, vol. 1, Cambridge, Mass.: Ballinger, pp. 51–85.

de Villiers, P.A. 1977. Choice in concurrent schedules and a quantitative formulation of the law of effect. In W.K. Honig and J.E.R. Staddon, eds., *Handbook of Operant Behavior*, Englewood Cliffs, N.J.: Prentice–Hall.

de Villiers, P.A., and R.J. Herrnstein. 1976. Toward a law of response strength. *Psychological Bulletin 83*: 1131–53.

Donahoe, J.W. 1977. Some implications of a relational principle of reinforcement. *Journal of the Experimental Analysis of Behavior 27*: 341–50.

Ferster, C.B., and B.F. Skinner. 1957. *Schedules of Reinforcement.* New York: Appleton–Century–Crofts.

Gentry, G.D., and M.J. Marr. 1980. Choice and reinforcement delay. *Journal of the Experimental Analysis of Behavior 33*: 27–37.

Gibbon, J. 1977. Scalar expectancy theory and Weber's law in animal timing. *Psychological Review 84*: 279–325.

Herrnstein, R.J. 1964. Aperiodicity as a factor in choice. *Journal of the Experimental Analysis of Behavior 7*: 179–82.

_____. 1970. On the law of effect. *Journal of the Experimental Analysis of Behavior 13*: 243–66.

_____. 1974. Formal properties of the matching law. *Journal of the Experimental Analysis of Behavior 21*: 159–64.

_____. 1979. Derivatives of matching. *Psychological Review 86*: 486–95.

Herrnstein, R.J., and G.M. Heyman. 1979. Is matching compatible with reinforcement maximization on concurrent variable interval, variable ratio? *Journal of the Experimental Analysis of Behavior 31*: 209–23.

Herrnstein, R.J., and D.H. Loveland. 1975. Maximizing and matching on concurrent ratio schedules. *Journal of the Experimental Analysis of Behavior 24*: 107–16.

Herrnstein, R.J., and W. Vaughan, Jr. 1980. Melioration and behavioral allocation. In J.E.R. Staddon, ed., *Limits to Action*, New York: Academic Press.

Heyman, G.M., and R.D. Luce. 1979. Operant matching is not a logical consequence of maximizing reinforcement rate. *Animal Learning and Behavior 7*: 133–40.

Hineline, P.N. 1977. Negative reinforcement and avoidance. In W.K. Honig and J.E.R. Staddon, eds., *Handbook of Operant Behavior*, Englewood Cliffs, N.J.: Prentice-Hall.

Homans, G.C. 1974. *Social Behavior: Its Elementary Forms.* Rev. ed. New York: Harcourt Brace Jovanovitch.

Hursh, S.R. 1978. The economics of daily consumption controlling food- and water-reinforced responding. *Journal of the Experimental Analysis of Behavior 29*: 475-91.

Killeen, P.R. 1981. Averaging theory. In C.M. Bradshaw, E. Szabadi, and C.F. Lowe, eds., *Quantification of Steady-State Operant Behaviour.* Amsterdam: Elsevier.

_____. 1982. Incentive theory. In *Nebraska Symposium on Motivation in 1981.* Lincoln: University of Nebraska Press.

Krebs, J.R., and N.B. Davies, eds. 1978. *Behavioral Ecology: An Evolutionary Approach.* Sunderland, Mass.: Sinauer.

Mazur, J.E. 1981. Optimization theory fails to predict performance of pigeons in a two-response situation. *Science 214*: 823-25.

McDowell, J.J., and R. Kessel. 1979. A multivariate rate equation for variable-interval performance. *Journal of the Experimental Analysis of Behavior 31*: 267-83.

Miller, H.L., Jr. 1976. Matching-based hedonic scaling in the pigeon. *Journal of the Experimental Analysis of Behavior 26*: 335-47.

Myerson, J., and F.M. Miezin. 1980. The kinetics of choice: An operant systems analysis. *Psychological Review 87*: 160-74.

Nevin, J.A. 1979. Reinforcement schedules and response strength. In M.D. Zeiler and P. Harzem, eds., *Reinforcement and the Organization of Behavior.* New York: Wiley.

Pyke, G.H.; H.R. Pulliam; and E.L. Charnov. 1977. Optimal foraging: A selective review of theory and tests. *Quarterly Review of Biology 52*: 137-54.

Rachlin, H. 1980. Economics and behavioral psychology. In J.E.R. Staddon, ed., *Limits to Action*, New York: Academic Press.

Rachlin, H.; R. Battalio; J. Kagel; and L. Green. 1981. Maximization theory in behavioral psychology. *Behavioral and Brain Sciences 4*: 371-417.

Rescorla, R.A., and A.R. Wagner. 1972. A theory of Pavlovian conditioning: Variations in the effectiveness of reinforcement and nonreinforcement. In A.H. Black and W.F. Prokasy, eds., *Classical Conditioning*, vol. 2, *Current Research and Theory*, New York: Appleton-Century-Crofts.

Schoener, T.W. 1969. Optimal size and specialization in constant and fluctuating environments: An energy-time approach. *Diversity and Stability in Ecological Systems.* Brookhaven Symposia in Biology: No. 22.

_____. 1971. Theory of feeding strategies. *Annual Review of Ecology and Systematics 2*: 369-404.

Shimp, C.P. 1969. Optimal behavior in free-operant experiments. *Psychological Review 76*: 97-112.

Shull, R.L.; D.J. Spear; and A.E. Bryson. 1981. Delay or rate of food delivery as a determiner of response rate. *Journal of the Experimental Analysis of Behavior 35*: 129-43.

Silberberg, A.; B. Hamilton; J.M. Ziriax; and J. Casey. 1978. The structure of choice. *Journal of Experimental Psychology: Animal Behavior Processes 4*: 368–98.

Staddon, J.E.R., and S. Motheral. 1978. On matching and maximizing in operant choice experiments. *Psychological Review 85*: 436–44.

Stubbs, D.A., and S.S. Pliskoff. 1969. Concurrent responding with fixed relative rate of reinforcement. *Journal of the Experimental Analysis of Behavior 12*: 887–95.

Thorndike, L.L. 1898. Animal intelligence: An experimental study of the associative processes in animals. *Psychological Review Monograph Supplements 11*, No. 4 (whole, No. 8).

Vaughan, W., Jr. 1981. Melioration, matching, and maximization. *Journal of the Experimental Analysis of Behavior 36*: 141–49.

20 IS TIME ALLOCATION UNCONDITIONED BEHAVIOR?

Gene M. Heyman

In experiments that provide two reinforcement schedules (concurrent schedules), subjects switch back and forth between the two alternatives, responding for a while at each. The result that has received most attention is that the overall division of behavior between the two schedules approximates the overall division of reinforcements between the two schedules (Herrnstein, 1961, 1970). That is, $B_1/B_2 = R_1/R_2$, where B_i is either total responses or total time at schedule i, and R_i is total reinforcement at schedule i. Called *matching*, this result is often referred to as a law because of its empirical generality. For example, the available evidence shows that the matching law describes concurrent schedule performance independently of both individual and species differences.

Because of the generality of the matching law, recent theoretical accounts of instrumental behavior invariably include a derivation of the matching predictions (e.g., Killeen, 1982; Myerson and Miezin, 1980; Rachlin, Green, Kagel, and Battalio, 1976; Staddon, 1980). These derivations proceed from quite different assumptions; consequently, current theories of instrumental behavior differ quite markedly. In this chapter three types of theory are assessed according to how well they predict time allocation in concurrent schedules.

Most theories of matching can be placed in one of three groups: (1) theories that focus on the relation between responding and its overall consequences (e.g., Rachlin, 1980; Staddon, 1980); (2) theories that focus on the relation between responding and somewhat less molar consequences, such as the local reinforce-

Many thanks to R. Herrnstein and H. Rachlin for their careful reading and very helpful comments on an earlier version of the manuscript. Many thanks to J. Zuriff for his incisive comments on the interpretive content of the chapter.

ment rates (reinforcement at an alternative divided by time at the alternative, e.g., Herrnstein and Vaughan, 1980; Myerson and Miezin, 1980); and (3) theories that are compatible with the view that time allocation in concurrent schedules in an unconditioned reinforcement effect, that is, the amount of time a subject spends at a schedule is elicited (e.g., Herrnstein, 1979; Killeen, 1982; McDowell and Kessel, 1979). The first two approaches are representative of the emphasis in operant psychology on the law of effect: consequences, however measured, select an adaptive if not optimal pattern of behavior. The third approach differs somewhat from most accounts of instrumental behavior. For example, in Killeen's (in press) derivation of matching, reward elicits a change in activity level (there is no explicit or implicit feedback loop). The predictions for each approach are outlined next.

Matching is derived from overall consequences in theories that assume that reinforced behavior is governed by a maximization principle. For example, Rachlin and his colleagues (Rachlin et al., 1976) have argued that the distribution of behavior that produces the maximum overall good predicts the allocation of time spent responding in a concurrent schedule. In some accounts the overall good is the nominal reinforcement rate (e.g., Rachlin et al., 1976), and in others, the overall good is the overall reinforcement rate plus other factors, such as response cost and leisure (Rachlin, Battalio, Kagel, and Green, 1981). Whether either of these theories is correct can be determined by comparing the reinforcement rate maximizing predictions with performance in schedules that use different contingencies: If subjects maximize reinforcement rate then the predictions and performance should simply converge; alternatively, if subjects maximize some more complex response consequence, for example reinforcement less response cost, then reinforcement rate maximization solutions and performance should systematically covary. For example, if the time allocations that maximized reinforcement rate in three different procedures were shown to be 1:1, 3:1, and 9:1, then time allocation should be least biased in the first procedure and most biased in the third procedure. In effect then the test is an application of Meehl's (1950) method for determining whether a stimulus is reinforcing or not. Also note that the evaluation assumes a dynamic maximizing process: Patterns of behavior that produce larger values of the good to be maximized are more likely to persist than patterns of behavior that produce smaller values. Herrnstein and Vaughan (1980) have aptly labeled this view of the dynamics of matching "literal" maximization theory.

The independent variable in local reinforcement rate theories (Herrnstein and Vaughan, 1980; Myerson and Miezin, 1980; Staddon, 1977) is the discrepancy between the two local reinforcement rates, and the process that leads to matching is equalization of local reinforcement rates. For example, Herrnstein and Vaughan (1980) suggested that subjects switch to the alternative with the higher local reinforcement rate. Over the long run this pattern of switching will produce equal local reinforcement rates, a result equivalent to matching ($R_1/T_1 = R_2/T_2$ is equivalent to $R_1/R_2 = T_1/T_2$).

Two equalization theory predictions were tested. First, the theory predicts that subjects in concurrent schedules should match even though other distributions of behavior produce higher reinforcement rates. (In contrast, maximization theory predicts that matching will not occur if other distributions produce higher reinforcement rates.) Second, equalization theory predicts that switching should be an orderly function of changes in local reinforcement rates. For example, in concurrent procedures that use variable-interval schedules, local reinforcement rates change as a function of time since the last switch. Therefore, if subjects in experiments are compensating for local reinforcement rate differences, switching probabilities should also change as a function of time. Note that Herrnstein and Vaughan's definition of a literal theory applies here as well: The evaluation of equalization theory assumes that behavior that produces smaller differences between local reinforcement rates prevails over behavior that produces larger differences.

Elicitation models of concurrent schedule performance have not been developed in any detail. It is nevertheless possible to describe conditions that would disprove this approach. First, as with equalization theory, elicitation theory implies that concurrent schedule performance is independent of the overall reinforcement rate. Second, since elicitation is a ballistic as opposed to an adjustive process, switching should not vary dynamically with the consequences of switching. Elicitation theory can therefore be eliminated from consideration if it is shown that time allocation depends on either feedback from the overall consequences of responding or feedback from the consequences of switching.

Mathematical models provide the most practical way to describe the relation between time allocation and its consequences in concurrent schedule experiments. These models, referred to as schedule feedback functions (Baum, 1973), entail two terms. One represents the subject's performance; the other represents the schedule. For example, consider a variable-ratio schedule. The expected reinforcement rate, $E(R)$, is p/VRr, where p is the proportion of time at the schedule and r is the mean interresponse time: The expected reinforcement rate is proportional to the time spent at the schedule. In contrast, interval schedules, which are much more frequently used in concurrent procedures, are mathematically more complex. This is because changeover rate and pattern of responding affect reinforcement rate. The discussion of these issues begins with the derivation of an elementary concurrent schedule model.

INDEPENDENT INTERVAL SCHEDULES

Assume that the experiment uses a concurrent schedule with two independent timers, a continuous-response requirement, and no delay between a changeover

Figure 20-1. The sequence of events in a concurrent VT VT schedule. The rectangular excursions in the first and fourth lines indicate that a timer interval elapsed. The diagonal slashes across the second and third lines indicate that the subject just received a reinforcement. If the subject is present when the interval elapses, a reinforcer is delivered immediately, events *a* and *b*. If the subject is absent when an interval elapses, the reinforcer is held, events *c* and *d*, and not delivered until the subject returns, events *e* and *f*.

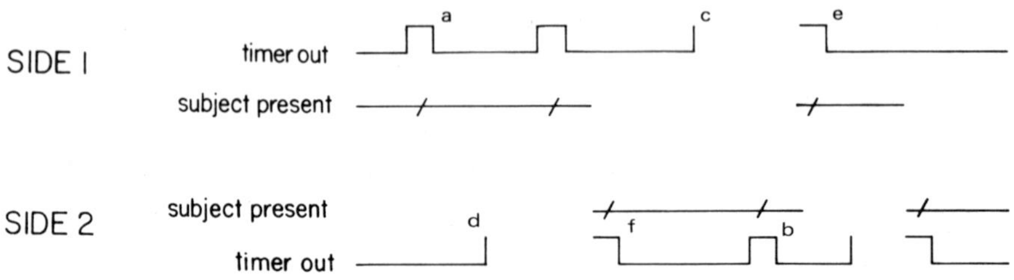

and reinforcement. Figure 20-1 shows the sequential structure of performance in this situation. Sides 1 and 2 refer to the two independently operating variable-interval timers, each distinguished by a stimulus. Lines 2 and 3, labeled "subject present," show that the subject switches back and forth between the two alternatives. This produces two different ways for reinforcement to occur. First, an interval may elapse while the subject is present. This yields an immediate reinforcement, and, simultaneously, the timer restarts with a new interval (events a and b in Figure 20-1). Second, an interval may elapse while the subject is absent (events c and d). In this case the reinforcer is held and the timer is not restarted until the subject returns. Upon returning, the reinforcement is immediately delivered and the timer is restarted (events e and f).

Second, assume an exponential distribution of intervals for each timer, with means V_1 and V_2. This approximates a commonly used schedule (Fleshler and Hoffman, 1962).

Third, assume that the temporal pattern of switching is on the continuum depicted in Figure 20-2. One endpoint is produced by a Poisson process; the other is produced by a subject that switches at precisely time t since the last switch. These two endpoints define two related continuums. First, for the Poisson process, the likelihood of the next switch does not increase as a function of time, which is to say that certainty is minimized. In contrast, for fixed-time switching, the probability of the next switch is 0 at time less than t and 1.0 at t, which is to say that certainty is maximized. Second, reinforcement rate varies with certainty of switching. For a given rate of switching, the Poisson process produces a lower reinforcement rate than any regularly increasing pattern (see Appendix 20A for a discussion of this point), whereas the fixed-time pattern produces a

Figure 20-2. Three different switching patterns in a concurrent schedule. In the first pattern the conditional probability of a switch is constant. This pattern is produced by a Poisson process. In the second pattern the conditional probability of a switch increases. This pattern was produced by a uniform distribution of interchangeover times, with each interval assigned a 10 percent relative frequency. In the third pattern the probability of a switch is 1.0 at 5 seconds. This pattern is produced by an interchangeover time distribution that takes only one value, 5 seconds.

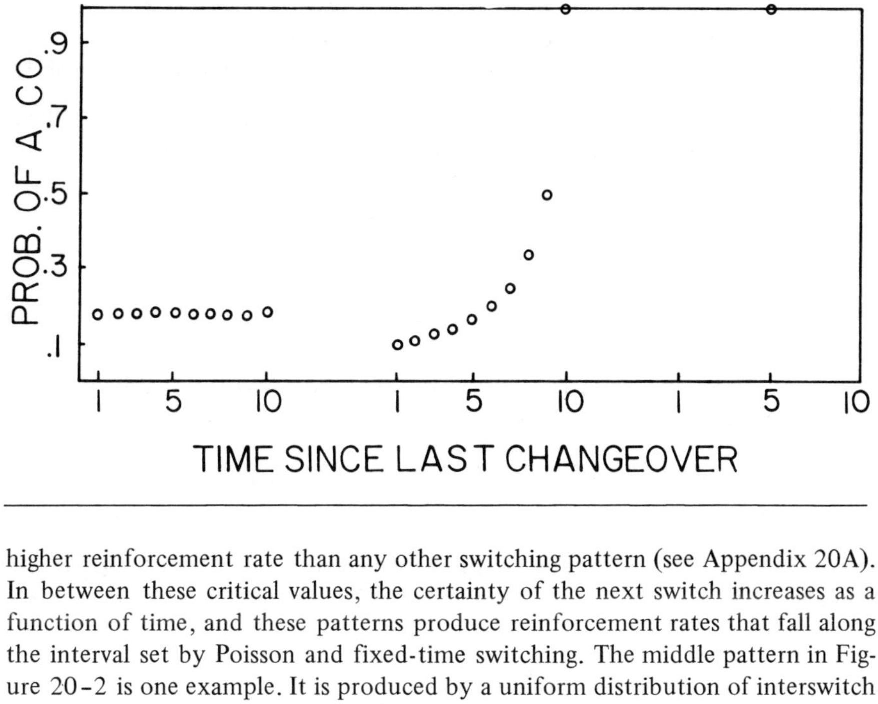

higher reinforcement rate than any other switching pattern (see Appendix 20A). In between these critical values, the certainty of the next switch increases as a function of time, and these patterns produce reinforcement rates that fall along the interval set by Poisson and fixed-time switching. The middle pattern in Figure 20-2 is one example. It is produced by a uniform distribution of interswitch times.

With the foregoing definitions it is possible to derive equations that describe the relation between time allocation and reinforcement rate in concurrent interval schedules for the range of response patterns depicted in Figure 20-2. (Note that this range is likely to include many if not most switching patterns.) The general strategy is to calculate the expected number of reinforcements per visit to an alternative and to distinguish the different ways reinforcement can occur during a visit.

First consider the case when a timer interval elapses while the subject is present. Figure 20-1 shows that these reinforcers are delivered immediately. Therefore their expected number per visit is simply $E(t_i)/V_i$, where $E(t_i)$ is the mean visit time and V_i is the mean timer interval.

Second, a timer can elapse while the subject is not present. This can occur no more than once per absence from the side (since the reinforcer is held), and the expected probability of this event depends on the mean timer interval and the mean rate and pattern of switching back. For Poisson switching, the probability that the timer interval elapsed during an absence is

$$\int_0^\infty \mu_i e^{-\mu_i t} \left(1 - e^{-t/V_j}\right) dt = \frac{1}{1 + \mu_i V_j} , \tag{20.1}$$

where $\mu_i e^{-\mu_i t}$ is the probability that the subject returned at time t, $1 - e^{-t/V_j}$ is the probability that the timer interval elapsed by time t, and $1/V_j$ is the timer rate constant. For the pattern of switching that is most different from Poisson, fixed time, the probability that a timer elapsed during an absence is $1 - e^{-T_i/V_j}$, where T_i is the fixed time the subject spends at the other alternative.

The expected number of reinforcements for the two different ways that reinforcement can occur having now been calculated, the last step is to calculate the expected time per visit. For a Poisson switching pattern this is

$$\int_0^\infty t\mu_i e^{-\mu_i t}\, dt = \frac{1}{\mu_i} . \tag{20.2}$$

For a fixed-time switching pattern, the expected time per visit is simply T_i.

Therefore the expected reinforcement rate for Poisson switching is

$$\frac{1/V_1\mu_1 + 1/(1 + \mu_2 V_1) + 1/V_2\mu_2 + 1/(1 + \mu_1 V_2)}{1/\mu_1 + 1/\mu_2} , \tag{20.3}$$

where $1/V_i$ is the programmed reinforcement rate at side i, and $1/\mu_i$ is the expected time per visit at side i. For the fixed-time switching pattern, the expected rate is

$$\frac{T_1/V_1 + 1 - e^{-T_2/V_1} + T_2/V_2 + 1 - e^{-T_1/V_2}}{T_1 + T_2} , \tag{20.4}$$

where T_i is the expected time per visit at side i.

Equations (20.3) and (20.4) may be rewritten so that they correspond to the variables usually discussed, the overall allocation of behavior, p, and reinforcement rate. In the notation for Poisson switching, the overall time proportions are $(1/\mu_1)/(1/\mu_1 + 1/\mu_2) = p$ and $(1/\mu_2)/(1/\mu_1 + 1/\mu_2) = 1 - p$. While in fixed-time switching notation, these proportions are $T_1/(T_1 + T_2) = p$ and $T_2/(T_1 + T_2) = 1 - p$. Therefore Eq. (20.3) becomes

$$E(R) = \frac{p}{V_1} + \frac{1-p}{V_1 + 1/\mu_2} + \frac{1-p}{V_2} + \frac{p}{V_2 + 1/\mu_1} , \tag{20.3a}$$

and with some algebraic manipulation, Eq. (20.4) is

$$E(R) = \frac{p}{V_1} + \frac{1-p}{V_1 + (A_1 - V_1)} + \frac{1-p}{V_2} + \frac{p}{V_2 + (A_2 - V_2)}, \tag{20.4a}$$

where

$$A_i = T_j/1 - e^{-T_j/V_i} . \tag{20.4b}$$

Put this way, the relation between responding and reinforcement rate in concurrent schedules is easily stated in the standard measures.

With probability p the subject is at side 1 when the side 1 timer sets up. This occurs at a mean rate of $1/V_1$. With probability $(1-p)$ the subject is at side 2 when the timer sets up at side 1. For Poisson switching this introduces a delay equal to the mean time away, $1/\mu_2$, so that these reinforcements occur at rate $1/(V_1 + 1/\mu_2)$. For fixed-time (optimal) switching the delay is

$$T_j/1 - e^{-T_j/V_i} - V_i . \tag{20.4c}$$

To get some idea of the effect of different patterns of switching on the delay from setup to reinforcement consider the case when a Poisson subject and fixed-time subject switch at the same rate, that is $\mu_i = 1/T_i$. For typical timer and changeover rates, the fixed time delay is approximately $T_i/2$, whereas the Poisson delay is $1/\mu_i$. This means that the Poisson delay is approximately twice as long for the same rate of switching. However, the effect on overall reinforcement rate (discussed more fully later) is quite small since the corresponding percentage difference between the two interreinforcement intervals is approximately $T_i/2V_j$, and this fraction (usually less than 1/10) only applies to reinforcers that set up at the unattended alternative.

Equations (20.3a) and (20.4a) show that the overall reinforcement rate in concurrent schedules depends on the absolute local switching rates (e.g., $1/T_i$) and on the relative magnitude of the local switching rates. These measures, however, are usually discussed somewhat differently as the overall changeover rate (e.g., $2/(T_1 + T_2)$ in fixed-time notation) and overall time allocation p.

Figures 20-3 and 20-4 show how the expected reinforcement rate changes with the usual measures, overall changeover rate, and overall time allocation in a concurrent interval schedule with a continuous response requirement (conc VT 60 seconds VT 180 seconds). The x axis in each figure is time allocation, and the y axis is expected reinforcement rate. The label I identifies the sum of the local changeover rates, and for each curve this sum is constant. This quantity has been discussed by a number of investigators (e.g., Heyman, 1977, 1979; Hunter and Davison, 1978; Myerson and Miezin, 1980; Nevin and Baum, 1980), and it seems that in situations in which the overall programmed reinforcement rates are con-

Figure 20–3. The expected reinforcement rate in a concurrent VT 60-second VT 180-second schedule as a function of time spent at the VT 60-second schedule. The switching pattern is Poisson. Each curve corresponds to a different level of the tendency to switch, I. For example, at $I = 10$ second, the local rates of switching sum to 6/minute (and the overall rate is $2p(1-p)/I$, see text).

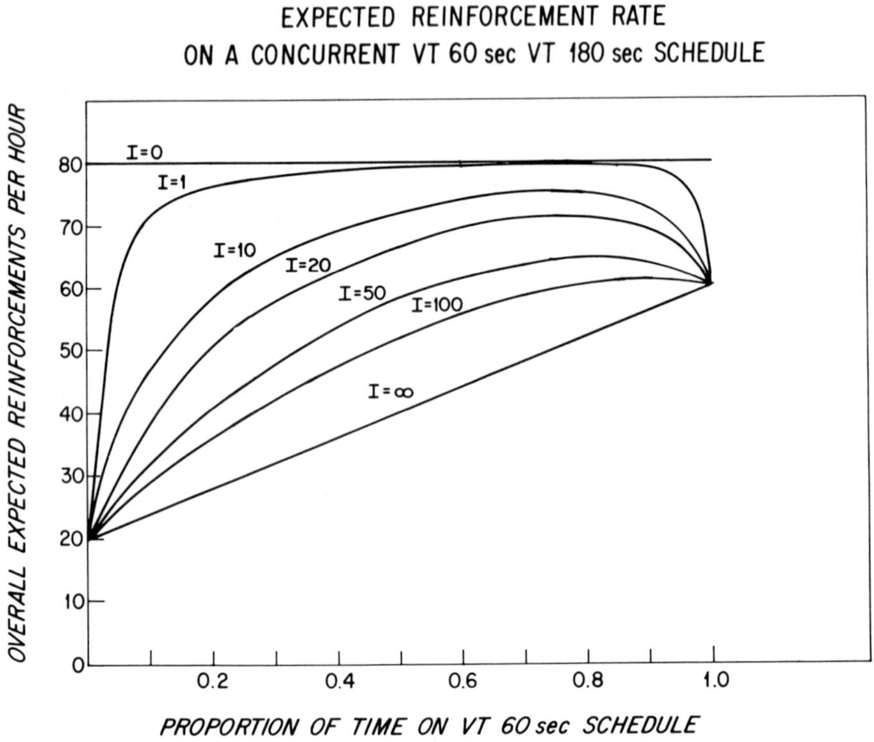

stant, local changeover rates vary reciprocally so that their sum ($1/T_1 + 1/T_2$ or $\mu_1 + \mu_2$) is approximately constant. This implies (see Appendix 20B for proof) that overall changeover rate is a function of overall time allocation according to the approximation

$$x = 2/(E(t_1) + E(t_2)) \cong 2p(1-p)/I \quad . \tag{20.5}$$

where x is overall changeover rate, I is equal to the reciprocal of the approximately constant sum of the two local changeover rates. That is, $I = 1/(\mu_1 + \mu_2)$ in the notation for Poisson switching and $I = T_1T_2/(T_1 + T_2)$ in the notation for fixed-time switching.

Figure 20-4. Expected reinforcement rate in a concurrent VT 60-second VT 180-second schedule with a fixed-time switching process. This pattern of switching is quite different than Poisson switching (see Figure 20-2), yet the consequences of the two patterns are virtually the same. This finding suggests that it is unlikely that subjects learn the optimal switching pattern in concurrent schedules.

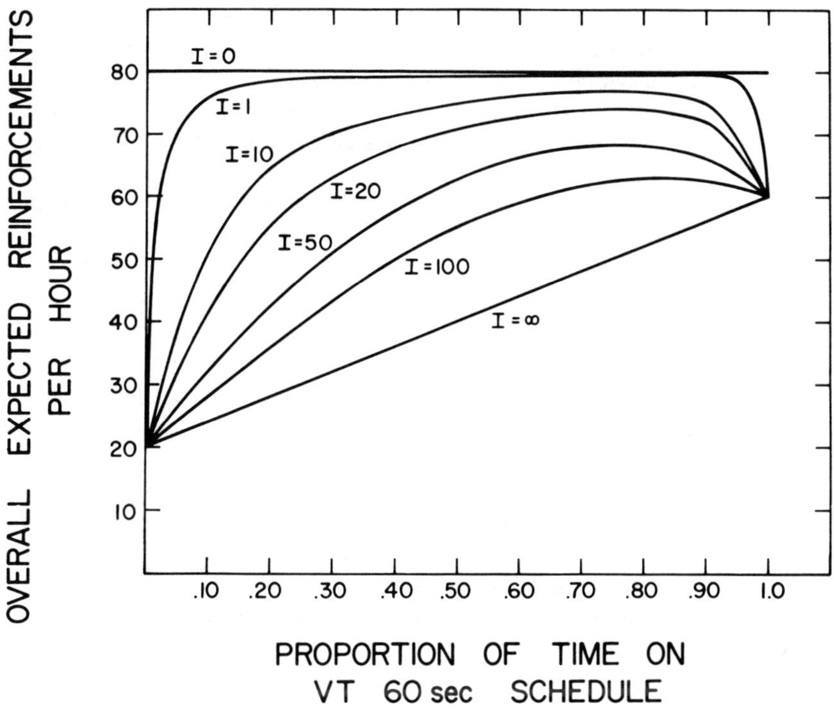

Figures 20-3 and 20-4 show (1) that for both patterns of switching, large changes in time allocation produce small changes in overall reinforcement rate; (2) different patterns of switching produce nearly the same reinforcement rate; and (3) relatedly, the relation between time allocation and overall reinforcement rate is independent of the pattern of switching. Moreover, the effect of time allocation on reinforcement is at a minimum at changeover rates that are representative of experimental subjects. For example, in a study by Brownstein and Pliskoff (1968), the median changeover rates corresponded to a value of *I* of about 6.6 seconds, and in a VT 60-second VT 180-second condition, the subjects, pigeons, shifted from about a 0.50 to 0.75 division of time. This means that nearly perfect matching was accompanied by no more than a 2 percent change in overall reinforcement rate. Similarly, given the schedule values and changeover rates, optimal and purely random (Poisson) switching could not have produced more than a 2 percent difference in reinforcement rates.

Figures 20-3 and 20-4, then, do not lend much support to theories of concurrent schedule performance that include overall reinforcement rate as an independent variable. However, these figures do establish a methodological point. Rachlin (1979) questioned whether the equation given here as (20.3a) provided a general description of concurrent interval schedule performance since it might be the case that different patterns of switching yield different reinforcement rate feedback functions. Comparison of Eq. (20.3a) and Eq. (20.4a) settles the issue. If the optimal switching pattern (fixed time) and the contingency independent pattern (Poisson) produce nearly identical expected reinforcement rates, it follows that the generality of the equations does not depend on the temporal pattern of switching. Put somewhat differently: Since pattern of switching has virtually no effect on reinforcement rate, both the Poisson and fixed-time models will accurately predict the relation between time allocation and overall reinforcement rate.

Equations (20.3a) and (20.4a) represent an elementary concurrent variable-interval schedule. In experiments there are typically additional contingencies. The most frequent additions are a discrete response requirement, such as a lever press, and a changeover penalty, called the changeover delay. The response requirement does not appear to affect the allocation of behavior (see, e.g., Brownstein and Pliskoff, 1968); the changeover delay, however, does.

The changeover delay is a changeover-initiated period, usually quite brief, during which reinforcement is withheld. This contingency greatly reduces switching rates and under certain circumstances can have a pronounced effect on how subjects allocate their behavior. The typical result is that if the subject divides its time more or less equally between the alternatives, adding a changeover delay or increasing its duration will differentiate the amounts of time the subject spends responding so that time allocation will match reinforcement allocation. Consequently, the changeover-delay duration has frequently figured in discussions of deviations from matching (e.g., Baum, 1974, 1979; de Villiers, 1977), and some authors have suggested that this contingency is a necessary condition for matching (e.g., Shimp, 1975). However, it is also the case that subjects will match in procedures that do not use the changeover delay (e.g., Bradshaw, Szabadi, and Bevan, 1976; Findley, 1958; Heyman, 1979). Possibly, then, the changeover delay amplifies the consequences of time allocation (Rachlin et al., 1976, originally suggested this) so that for some subjects, the less sensitive, the changeover delay is a necessary condition for matching, whereas for other subjects, the more acute, the changeover delay is not a necessary condition. This hypothesis can be evaluated by calculating the effects of the changeover delay on expected reinforcement rate.

For the fixed-time switching pattern it is possible to determine the effects of the changeover delay on reinforcement rate. In addition, this pattern may not be greatly unlike actual patterns of switching, because subjects learn not to switch during the delay period (Baum and Rachlin, 1969). The derivation closely follows the derivation of Eq. (20.4) and is also similar to Houston and McNamara's account (1981).

As in the previous models, it is necessary to distinguish the different ways for reinforcement to occur. First, some reinforcers are not postponed by the changeover delay. This is possible if the subject is present and the changeover delay has elapsed. The expected number of times this will happen per visit at alternative i is $(T_i - C_i)/V_i$, where C_i is the duration of the changeover-delay requirement.

Second, some reinforcers are postponed by the changeover delay. This occurs either when the reinforcement timer elapses while the subject is at the other alternative or it occurs just following a changeover during the delay period. These two possibilities are restricted to the interval $T_j + C_i$, where C_i is the delay duration at one alternative and T_j is the fixed time the subject spends at the other alternative. Therefore the probability that the reinforcer is postponed by the changeover contingency is $1 - e^{-(T_j + C_i)/V_i}$. This exhausts the different ways for reinforcement to occur. The expected reinforcement rate is, then,

$$E(R) = \frac{T_1 - C_1}{V_1(T_1 + T_2)} + \frac{1 - e^{-(T_2 + C_1)/V_1}}{T_1 + T_2} + \frac{T_2 - C_2}{V_2(T_1 + T_2)} + \frac{1 - e^{-(T_1 + C_2)/V_2}}{T_1 + T_2} . \qquad (20.6)$$

Although Eq. (20.6) is superior to Eqs. (20.3a) and (20.4a) in that it takes into account the changeover delay, it is an incomplete description of performance. Missing is the correlation between the duration of the changeover-delay requirement and the changeover rate: An increase in delay produces a decrease in rate. A number of investigators have quantified this relation (see Hunter and Davison, 1978; Stubbs, Pliskoff, and Reid, 1977); however, Eq. (20.5) gives a somewhat simpler account.

Recall that according to Eq. (20.5), changeover rate varies with time allocation and the reciprocal of the constant sum of the changeover rates, called I for inertia. The magnitude of this constant will depend, in part, on procedural factors, such as the distance between the alternatives, the delay requirement, and so on. The simplest possible structure is $I = a + b(C)$. That is, the tendency to switch, I, varies linearly with the delay requirement:

$$\frac{2}{E(T_1) + E(T_2)} = \frac{2p(1-p)}{I} = \frac{2p(1-p)}{a + b(C)} , \qquad (20.7)$$

where $E(T_i)$ is the expected time per visit to side i.

Figure 20-5. The relation between changeover delay (COD) duration and changeover rate. On the *x* axis is the changeover requirement; on the *y* axis is the changeover rate in reciprocal form (cycle duration). Rat s-1 is from an experiment conducted by Shull and Pliskoff (1967); Pigeon 10838 is from an experiment conducted by Silberberg and Schrot (1974).

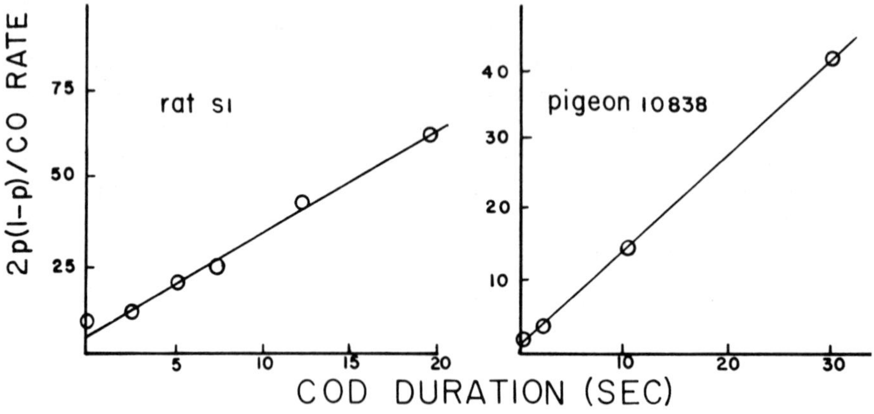

Figure (20-5) shows the fit of Eq. (20.7) to some representative data. On the *x* axis is the duration of the changeover delay requirement. On the *y* axis changeover rate is plotted in reciprocal form (the sum of the two average visit times, sometimes called cycle time). Rat s-1 is from a study by Shull and Pliskoff (1967) in which the changeover delay was varied from 0.0 to 20 seconds. Pigeon 10838 is from a study by Silberberg and Schrot (1974) in which the changeover delay was varied from 0.0 to 30 seconds. Equation (20.7) predicts changeover rate quite precisely, except for a slight tendency to underestimate the rate in the 0.0-seconds changeover-delay condition (that is, no changeover-delay requirement). Note, that Eq. (20.7) assumes a Markov model of switching (see Heyman, 1979). In contrast, momentary maximizing theories and melioration theories of switching suggest (if not imply) a feedback driven, adjustive switching process.

Figure 20-6 shows the relation between time allocation, changeover delay, and overall reinforcement rate for a conc VT 60-second VT 180-second schedule. The *y* axis is the expected overall reinforcement rate. The changeover delay was varied from 0 to 27 seconds, and changeover rate was determined from the best-fitting parameter values for *I* in a continuous response procedure in which the changeover delay was varied (Brownstein and Pliskoff, 1968).

Figure 20-6 shows that for typical changeover delays, the region circumscribed by the first and third curve, changes in time allocation have little effect on the overall reinforcement rate. For example, at a 3-second changeover delay, a shift from 0.60 to 0.75 in time spent at the VT 60-second alternative increases

Figure 20-6. The relation between time allocation and reinforcement rate in a concurrent VT 60-second VT 180-second schedule that uses a changeover delay. The changeover delay does little to change the overall consequences of time allocation, yet it frequently has a large effect on time allocation.

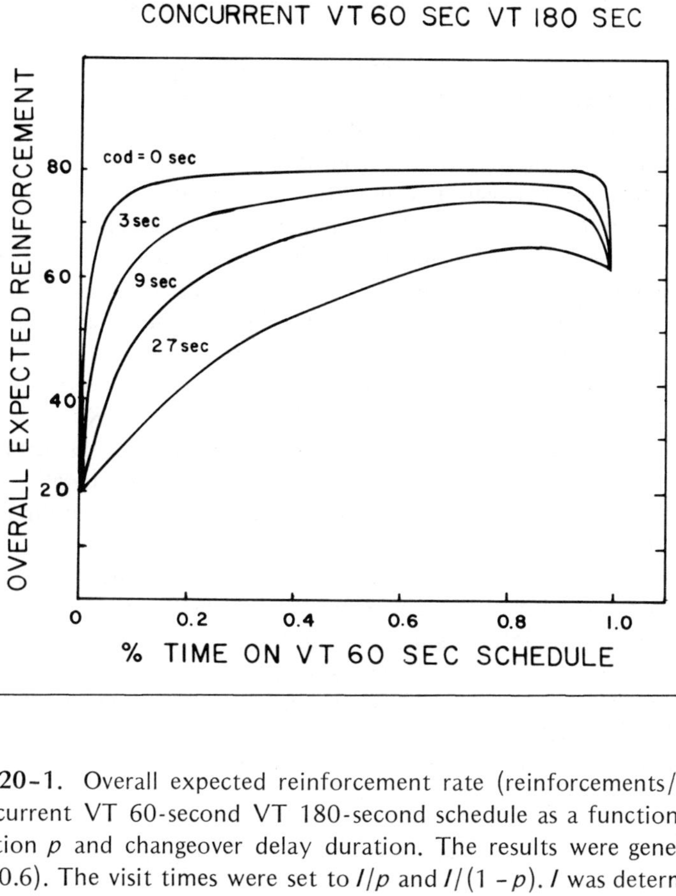

Table 20-1. Overall expected reinforcement rate (reinforcements/hour) in a concurrent VT 60-second VT 180-second schedule as a function of time allocation p and changeover delay duration. The results were generated by Eq. (20.6). The visit times were set to I/p and $I/(1-p)$. I was determined by Eq. (20.7), with the parameters $a = 1.0$ and $b = 1.1$.

	Time on VT 60-Second Schedule and Overall Reinforcements per Hour				
COD Duration	*.50*	*.60*	*.70*	*.75*	*.80*
0	77.71	78.27	78.56	78.61	78.56
3	75.90	76.20	77.27	77.46	77.57
9	69.62	71.40	72.98	73.40	73.74
27	56.60	60.10	61.66	63.07	64.80

the expected reinforcement rate by less than 2 percent. Table 20-1 provides some additional numerical values.

CONCURRENT SCHEDULE PERFORMANCE, MATCHING AND MAXIMIZING PREDICTIONS

The figures in the preceding section show how changes in the allocation of time responding affect overall reinforcement rate. In this section the allocation of time that maximizes the overall reinforcement rate is calculated. The calculations are then compared with data from three procedures, which differ in terms of the contingency between time allocation and overall reinforcement rate. That is, each procedure has a different maximizing solution. If subjects respond so as to maximize overall reinforcement rate then the predictions and observations should converge. Similarly, if subjects maximize some more complex quantity, such as reinforcement rate minus the costs of responding, then predictions and observations should vary systematically. Of course, the overall consequences of responding, however measured, need not exert any control on time allocation. For example, the acquisition models of Herrnstein (1979), Killeen (1982) and McDowell and Kessel (1979) are blind to the differences that distinguish the three procedures analyzed next, and, consequently, these models predict that subjects should come to the same asymptotic time allocations in three quite different contingencies.

The relation between time allocation and overall reinforcement rate maximization was evaluated in the following way. First, the division of time that maximizes reinforcement rate was determined. This was done by varying p in the schedule reinforcement rate feedback equations. Second, for the value of p that maximized reinforcement rate, the corresponding reinforcement rates for alternative 1 and alternative 2 were extracted. Third, the maximizing solutions were plotted as a function of their corresponding reinforcement rates according to the linear model (Baum, 1974):

$$\log T_1/T_2 \text{ max} = \log a + b \log R_1/R_2 \text{ max} \quad , \qquad (20.8)$$

where T_i is the optimal amount of time spent at alternative i and R_1 plus R_2 sum to the maximum reinforcement rate. Fourth, the best-fitting parameter values (Eq. (20.8)) for the maximizing solutions were compared to the best-fitting parameter values (Eq. (20.8)) for the typical subject. In other words, Eq. (20.8) measured whether subjects in experiments maximized overall reinforcement rate.

The three different procedures that are analyzed in this way are concurrent variable-interval with two independent timers, concurrent variable-interval with interdependent timers, and a concurrent procedure in which one alternative is a variable-interval schedule and the other is a variable-ratio schedule (conc VI VR). In addition, for the independent-interval procedure, there is an analysis of the effects of the changeover delay and response requirement on the maximizing solutions.

Maximizing in Concurrent VI VI

Figures 20-7 and 20-8 show the maximizing policies for experiments with a continuous-response requirement and no changeover delay. Figure 20-7 is for fixed-time switching (Eq. (20.4a)), and Figure 20-8 is for Poisson switching (EQ. (20.3a)). The schedule values, V_1 and V_2, were varied so as to maintain a constant programmed reinforcement rate of 80/hour, and I was set at two different values. The smaller one, 4.3 seconds, corresponds to the parameters for

Figure 20-7. A comparison of typical performance and reinforcement rate maximization predictions in concurrent VT VT schedules. The maximization predictions were generated by Eq. (20.4a) for fixed-time switching. The description of the typical subject is based on Baum's (1979) review. The shallower slope corresponds to a lower changeover rate.

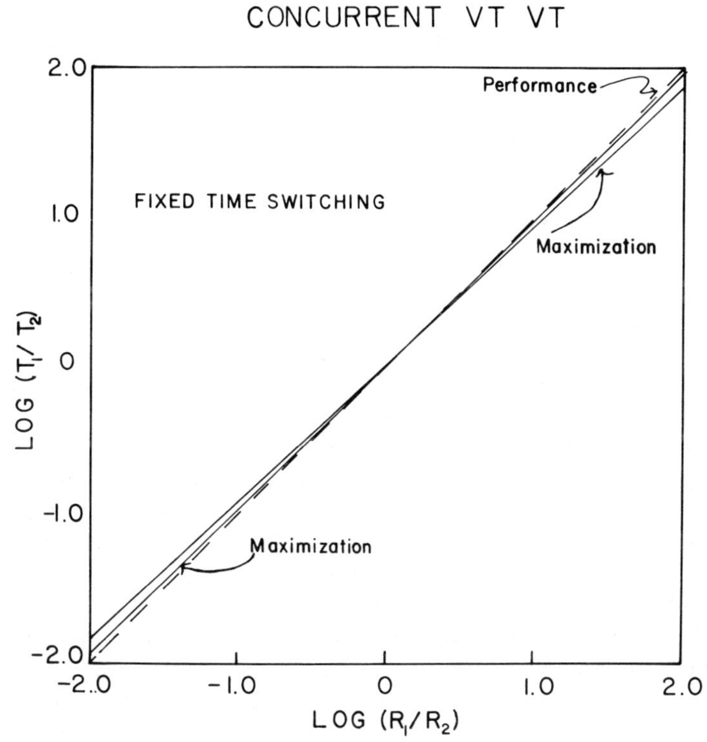

Figure 20–8. A comparison of typical performance and reinforcement rate maximization predictions (Poisson switching) in concurrent VT VT schedules. The maximization predictions were generated by Eq. (20.3a) (Poisson switching). Note that the maximization solutions are virtually independent of pattern of switching. Put somewhat differently, large differences in response pattern in concurrent schedules yield only small differences in outcome.

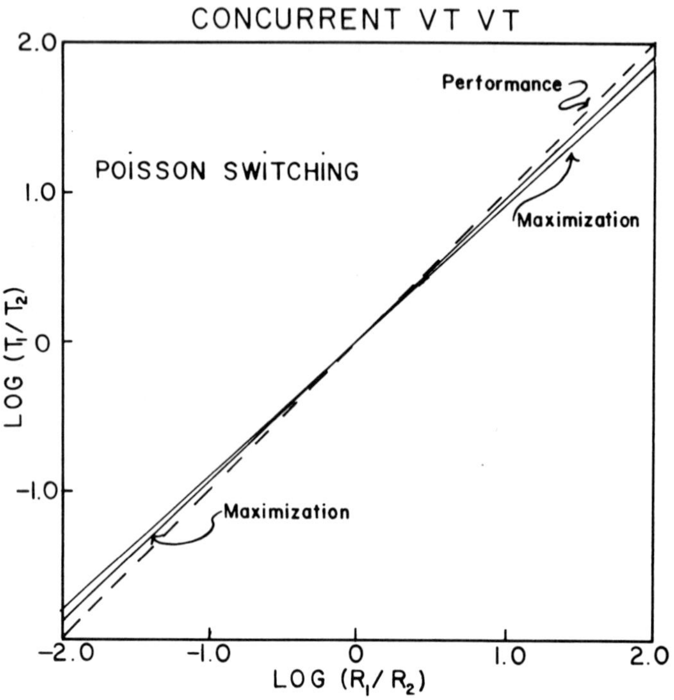

I in an experiment with pigeons (Brownstein and Pliskoff, 1968), and it sets changeover rate at approximately 7/minute for $p = 1/2$. The larger value, $I = 34$ seconds, gives a changeover rate of about 1/minute at $p = 1/2$. The continuous diagonals are the best-fitting straight lines to the ratios of time spent responding that maximized reinforcement rate, and the broken diagonals are the best-fitting straight lines to the ratios of time spent responding for the typical subject.

The parameters for the typical subject are based on reviews by Baum (1974, 1979) and de Villiers (1977). According to those authors, fitting Eq. (20.8) to the available time-allocation data yields a median slope of approximately 1.0 and a median intercept, for experiments with symmetrical response requirements, of approximately 0.0. In other words, time allocation in concurrent schedules does not systematically vary from simple matching (slope of 1.0 and intercept of 0.0).

Equation (20.8) reveals that the maximizing solutions closely approximated performance in concurrent interval schedules. However, the linear model also reveals that the discrepancies, albeit small, were systematic. First, the slopes for the maximizing solutions fell slightly short of 1.0, with lower slopes for lower changeover rates. For fixed-time switching, the slopes were 0.93 and 0.98; for Poisson switching, the slopes were 0.91 and 0.96. Second, the relation between log T_1/T_2 max and log R_1/R_2 max was not precisely linear. Rather, there was the suggestion of a sigmoidal function. In contrast, the empirical deviations from Equation (20.8) are symmetrical (see Baum, 1979).

Figure 20-9 shows the maximizing predictions for procedures that use a changeover delay. As pointed out previously, the changeover delay has the effect of shifting time allocation from indifference, 1/1, to matching, R_1/R_2, so that from the point of view of maximization theory, the changeover delay should have a sizable effect on the maximizing solutions. Comparison of Figure 20-9 with Figures 20-7 and 20-8, however, show that the changeover delay has a rather modest effect of the relation between time allocation and reinforcement allocation. For example, without the changeover delay, the slopes range from about 0.91 to 0.98, and with the changeover delay, the slopes range from about 1.1 to 1.2.

Figure 20-10 displays the maximizing solutions for concurrent schedules that have a changeover delay and a response requirement. In this type of procedure each alternative has a manipulandum and the subject switches back and forth between them. Reinforcement occurs either after a changeover or after an interresponse time on one of the manipulandums. Formally, the analysis of interresponse times is the same as the previous analysis of interchangeover times (Eqs. (20.3), (20.4) and (20.6)). For example, an exponential interresponse time distribution extends the expected interreinforcement interval by the mean interresponse time (see Herrnstein and Heyman, 1979; Nevin and Baum, 1980).

Figure 20-10 shows that the slopes fit to the maximizing solutions vary as a function of local response rate. At high local response rates (120/minute), the slope is greater than 1.0, and at low local response rates (20/minute) the slope is lower than 1.0. Since the lines fit to the maximizing solutions cross in this way, there may exist some combination of local response rates and changeover values for which matching and maximizing are identical. However, the evidence presented in this chapter indicates that the similarity of matching and maximizing in independent schedules is fortuitous.

Maximizing in Interdependent VI VI

Figures 20-7 to 20-10 show the maximizing solutions for concurrent schedules with independent timers. Next schedules are analyzed in which the alternatives are not independent. The purpose of this modification is to ensure that the subject receives the programmed relative reinforcement rate (see, e.g., Stubbs and Pliskoff, 1969, and subsequent discussion).

Figure 20–9. Performance and maximization theory predictions in procedures that use a changeover delay. The changeover delay can have a quite sizable effect on time allocation, yet it has a small effect on maximization policy. This is consistent with the general finding that large differences in behavior produce virtually the same outcomes in concurrent schedules. Note that increasing the changeover delay duration does not yield steeper slopes because of the way that the delay interacts with obtained reinforcement proportions (see Herrnstein, 1970).

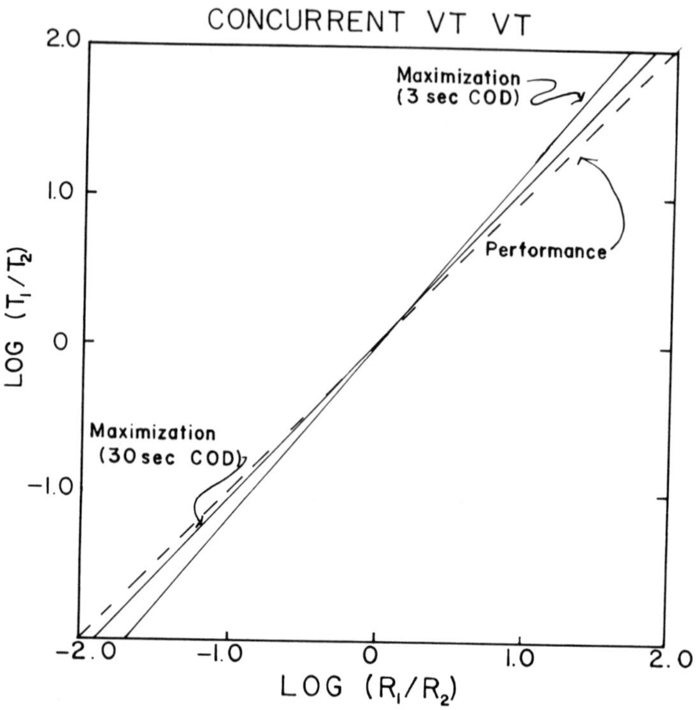

An interdependent procedure can be arranged in two nearly equivalent ways (Fantino, Squires, Delbruck, and Peterson, 1972; Stubbs and Pliskoff, 1969). The model presented (Eq. (20.9)) is based on Fantino et al.'s (1972) method. These researchers used two variable-interval timers. When one timer set up a reinforcer, the other was halted and not restarted until the reinforcer was collected. Thus, if the unattended timer was primed, the attended timer stopped running also. This method ensures that reinforcements are obtained in the order that they are made available, which, in turn, implies that the obtained reinforcement proportions are equal to the arranged reinforcement proportions. The method arranged by Stubbs and Pliskoff (1969) results in a nearly identical con-

Figure 20–10. Performance and maximization theory in concurrent schedules that use a changeover delay and a response requirement. The tendency to switch, *I*, was calculated according to Eq. (20.7), with the parameters a = 1.0 and b = 1.1 (Shull and Pliskoff, 1967). The maximization predictions vary with local response rate and for some combinations of changeover delay duration and interresponse time duration, it appears logically possible for performance and maximization to converge.

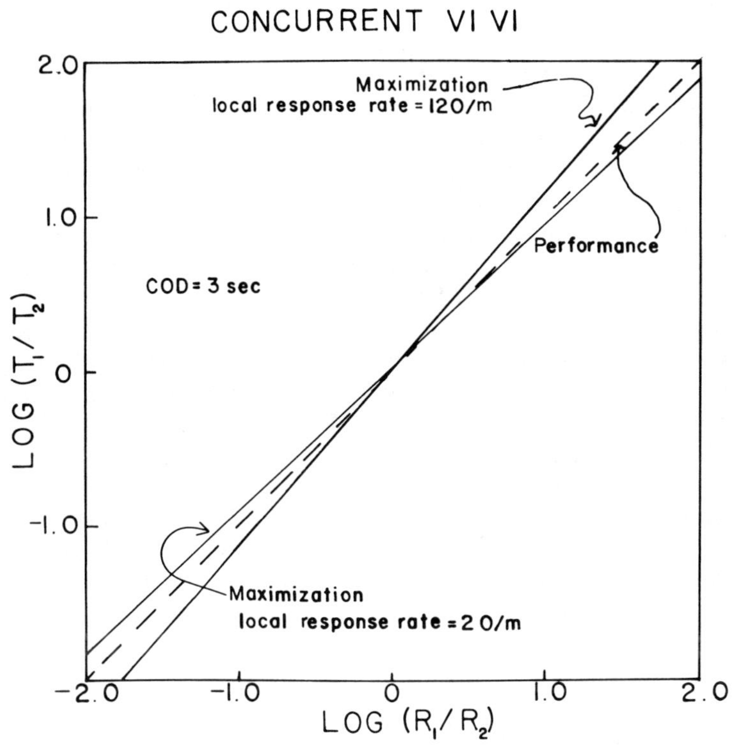

tingency: (1) Reinforcement availability at the unattended alternative prevents reinforcement at the attended alternative; (2) reinforcements are obtained in the arranged order; (3) and obtained reinforcement proportions equal arranged reinforcement proportions. The only difference is that Stubbs and Pliskoff used a single timer so that a reinforcement rather than a changeover resets the expected time to the next reinforcement (see Staddon, Hinson, and Kram, 1981, for a fuller discussion of this point). However, this difference has little effect on the relation between time allocation and reinforcement rate, so that Eq. (20.9) quite accurately predicts the expected reinforcement rate for both methods of arranging interdependent schedules (see, e.g., Heyman and Luce, 1979).

Appendix 20C provides a derivation of the equation for expected reinforcement rate in an interdependent procedure. The result is as follows (also see Heyman and Luce, 1979):

$$E(R) = \frac{p}{V_1} - \frac{pI}{V_1(I+(1-p)V_2)} + \frac{1-p}{V_1 + 1/\mu_2} + \frac{1-p}{V_2} - \frac{(1-p)I}{V_2(I+pV_1)} + \frac{p}{V_2 + 1/\mu_1} . \quad (20.9)$$

The second and fifth terms give the expected rate of loss at an attended but not running timer. For example, while the subject is at side 1, the probability that the timer there has come to a halt increases according to the expression given by Eq. (20.1). Once it stops, it remains so until the next switch, and this event has the expectation given by Eq. (20.2). The result is included in Eq. (20.9). Note that this model differs quite markedly from the equations for independent timer procedures. For example, if the subject never switches in an interdependent procedure, reinforcement rate decreases to zero.

Figure 20-11 shows the relation between typical performance and reinforcement rate maximization in interdependent schedules. The maximization solutions reflect the structure of the contingency. That is, since not switching reduces reinforcement rate at both sides, the subject must not spend too much time at the richer alternative. Thus the optimal policy is to allocate time about midway between 1/1 (indifference) and V_1/V_2 (perfect matching). Subjects in experiments do not allocate time according to this contingency, however. Rather, as in independent procedures, time allocation approximates reinforcement allocation. For example, in the most exhaustive review to date, Baum (1979) concluded that time allocation was the same in independent schedules and schedules that maintained a fixed relative reinforcement rate.

Maximizing in Concurrent VI VR

The third procedure in this comparison uses a ratio schedule on one alternative and an interval schedule on the other alternative (conc VI VR). On the ratio schedule, reinforcement rate is proportional to time spent responding so that changes in time allocation produce large changes in reinforcement rate. In contrast, Figures 20-3 and 20-6 showed that on interval schedules changes in time allocation have little effect on reinforcement rate. Accordingly, it is not unreasonable to expect subjects to allocate time rather differently when a ratio schedule controls reinforcement.

The equation for expected reinforcement rate on the ratio schedule is $E(R) = (1-p)/\mathrm{VR}r_2$, where $1-p$ is the proportion of time spent responding, VR is the mean ratio requirement, and r_2 is the mean local interresponse time. Rein-

Figure 20-11. Performance and maximization theory in interdependent interval schedules. The characterization of performance is based on Baum's (1979) review. Maximization theory predictions diverge widely from how subjects actually behave.

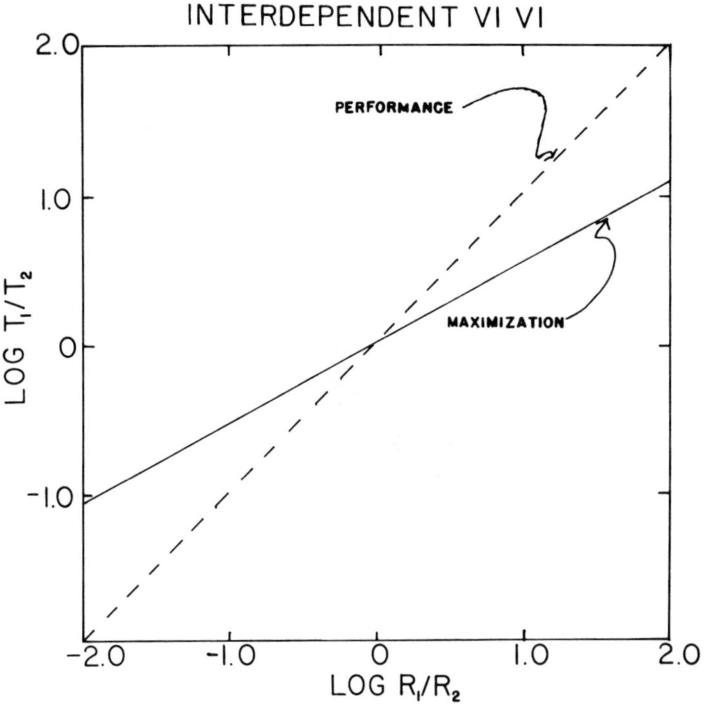

forcement rate on the interval schedule is as previously described in Eqs. (20.3) and (20.4). Therefore, the expected overall reinforcement rate for a conc VI VR is

$$E(R) = p/(\mathrm{VI} + r_1) + (1 - p)/(\mathrm{VI} + 1/\mu_2) + (1 - p)/\mathrm{VR}r_2 \quad , \qquad (20.10)$$

where r_1 is the local interresponse time on the VI alternative, r_2 is the local interresponse time on the VR alternative, and the other symbols are as defined previously.

The first two expressions give the reinforcement rate on the VI alternative. With probability p the subject receives rate $1/(\mathrm{VI} + r_1)$, the mean programmed interval plus the mean local interresponse time, and with probability $1 - p$, the subject receives rate $1/(\mathrm{VI} + 1/\mu_2)$, the mean programmed interval plus the mean time per visit on the ratio alternative. Note that these two rates are not

Figure 20-12. Performance and maximization theory predictions in a concurrent VI VR experiment (Herrnstein and Heyman, 1979). The open symbols show the time-allocation measures for individual subjects. The filled symbols show reinforcement rate maximization predictions.

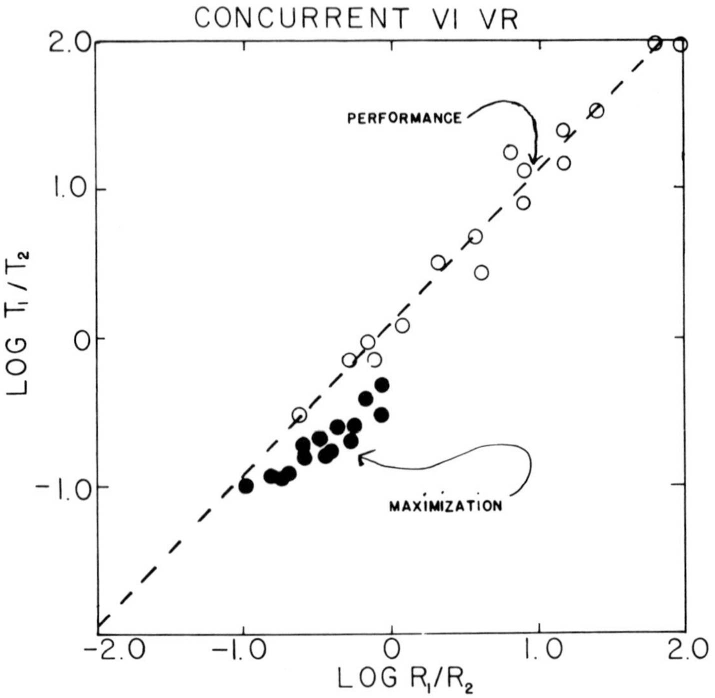

very different from each other for typical schedule and behavior rates. The third expression is the rate on the ratio alternative. In contrast to the interval schedule, reinforcement rate is zero if the subject is absent, and this occurs with probability p.

Figure 20-12 compares the maximizing solutions with performance in a concurrent VI VR experiment (Herrnstein and Heyman, 1979). The maximizing solutions were derived from Eq. (20.9) (filled symbols) and the subjects were pigeons responding for grain (open symbols). The coordinates are in terms of the ratio VI/VR so that positive values on the y axis indicate that more time was spent on the VI alternative and negative values indicate that more time was spent on the VR alternative.

The maximizing predictions invariably fall into the lower left quadrant of the figure, with a slope of 0.72 for the best-fitting line. This means that in order to maximize overall reinforcement rate, the subjects should have spent most of their time on the VR alternative and from time to time briefly sampled the VI alternative to check if a reinforcer had set up there.

In contrast to these predictions, the subjects spent most of their time on the VI alternative and did so in a way that was consistent with performance in concurrent-interval schedules. The slope of the best-fitting line is 1.04, which is about the value found in independent and interdependent interval schedules. This means that the allocation of time spent responding did not respect the differences in reinforcement contingencies. Moreover, the costs of contingency indifference were rather great: On the average, the subjects would have received about 60 more reinforcements an hour by maximizing (see Herrnstein and Heyman, 1979).

SUMMARY AND OTHER APPROACHES

Figures 20-7 to 20-12 show that in three quite different procedures subjects responded so that the ratios of time spent at each alternative were equal to the ratios of reinforcement obtained at each alternative; that is, $T_1/T_2 = R_1/R_2$ (correcting, if necessary, for bias). In contrast, there was no systematic relation between the maximizing policies and time allocation. This absence of covariation with the maximizing solutions implies that subjects in concurrent schedules do not maximize value, where value includes the overall reinforcement rate. Put somewhat differently, these results say that in concurrent schedules time allocation is not controlled by its overall consequences.

The conclusion that subjects in operant reinforcement schedules are not controlled by the molar consequences of their behavior is at odds with recent economic (Rachlin et al., 1981) and biologically (Staddon, 1980) oriented theories of instrumental (adaptive) behavior. Some support for these theories comes from Staddon and Motheral's (1978) mathematical model of the concurrent VI VI contingency. Since this model does not agree with the mathematics presented in this chapter, some comments are necessary.

Staddon and Motheral claim that the expected rate of reinforcement in a concurrent schedule with independent timers is

$$E(R) = 1/(V_1 + 1/x) + 1/(V_2 + 1/y) \quad , \tag{20.11}$$

where x and y are the overall response rates (responses at a side divided by the entire session time, in contrast to the local response rates, discussed in the previous sections of this chapter, which are responses at a side divided by the portion of time spent at a side). Staddon and Motheral's equation says that the expected time between reinforcements is the mean timer interval V_i plus the mean overall interresponse time. But this cannot possibly be correct since in concurrent schedules reinforcement occurs after a local interresponse time or an interchangeover time (there are no other possibilities), and the overall response rates have no necessary connection with either of these measures. For example, it is quite possible for two subjects to produce equal overall response rates but obtain quite different reinforcement rates, because one switches frequently and the other switches infrequently.

Staddon and Motheral (1979) anticipated this criticism and write that Eq. (20.11) implies a changeover rate and local response rate. They write that changeover rate is

$$C = 2xy/(x+y) \tag{20.11a}$$

and that local response rate is

$$2xy/C \quad , \tag{20.11b}$$

where C is the overall changeover rate (Eq. (20.11a)). These definitions are incorrect, however. In other words, according to Staddon and Motheral changeover rate is a function of response rate. However, one of the basic features of concurrent schedule performance is that switching rates and response rates are independent measures. For example, switching rate varies with time allocation, whereas response rates remain approximately constant with changes in time allocation (e.g., Stubbs and Pliskoff, 1969).

EQUALIZATION AND MOMENTARY MAXIMIZING THEORIES

Equalization theories of matching (e.g., Herrnstein and Vaughan, 1980; Myerson and Miezin, 1980; Staddon, 1977) predict that concurrent schedule performance is independent of molar consequences and that switching will vary in an orderly way with local reinforcement rates. Figures 20-7 to 20-12 confirmed independence. Next the predictions for switching are tested.

In a concurrent schedule that uses interval schedules, the local rates of reinforcement rise and fall with switching. A switch produces an increase, and, then, there is a decline until the next switch. This means that the longer the subject stays at one side, the relatively greater the local rate of reinforcement at the other side. Therefore, if equalization theory describes concurrent schedule performance, the probability of a changeover should increase as a function of time since the last changeover. Figure 20-13 tests whether this prediction holds.

The figure shows the temporal pattern of switching in three different concurrent-schedule experiments. On the x axis is the number of responses or amount of time (pigeon B 64) since the last switch; the filled circles are the conditional probabilities of switching from the richer alternative; and the open circles are the conditional probabilities of switching from the leaner schedule.

Pigeons 209 and 241 were in an experiment that used an interdependent schedule (Heyman, 1979), and the temporal pattern for pigeon 209 is representative of the main findings of the study. The switching probabilities were station-

Figure 20-13. Switching in three different concurrent schedule experiments. The *x* axis is the number of responses since the last changeover or time since the last changeover (pigeon B 64). The *y* axis is the conditional probability of a switch. The concurrent schedule reinforcement contingency requires that the probability of reinforcement and local rate of reinforcement increase at the unattended side. The figure shows that subjects did not switch in accordance with these consequences.

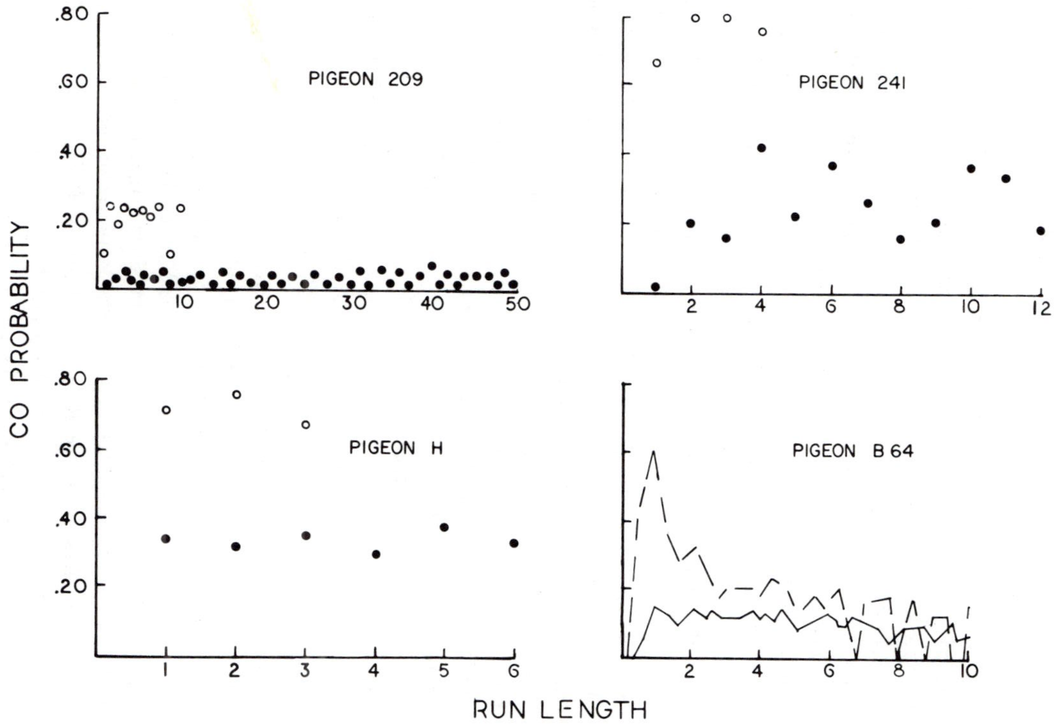

ary (chi-square test) and independent of run length. This means that switching did not depend on changes in local reinforcement rates. The data from pigeon 241 are characteristic of the minority of instances in which switching probabilities were not stationary (3 of 24 cases). The probabilities rose and fell in an alternating pattern. This corresponds with previous reports (e.g., Blough, 1966) of response bursts, but it is incompatible with the changes in local reinforcement rate described in equalization theory.

Pigeon H was in a discrete-trials experiment (Herrnstein, unpublished, described in de Villiers, 1977). Reinforcement was arranged by a concurrent VI 135-second VI 270-second schedule (independent), the intertrial intervals were 22 seconds, and the relative probability of reinforcement for a switch, counting from the last changeover, rose from about 0.50 to about 0.90. Nevertheless, the subjects were no more likely to switch after 6 trials than after 1 trial.

Pigeons B 64, like pigeons 209 and 241, was in an experiment that used an interdependent procedure (Silberberg, Hamilton, Casey, and Ziriax, 1978). The x axis is time, not responses; the broken line plots the probabilities of switching from the leaner schedule; and the solid line plots the conditional probabilities for the richer schedule. Again, the pattern predicted by local reinforcement rate control does not emerge. Instead, switching was either independent of time (from the richer schedule) or showed a peak and then a gradual decline with time (from the poorer schedule).

Figure 20-13 also evaluates a somewhat similar theory of matching, momentary maximizing. In this account (Shimp, 1969), the local reinforcement probabilities that accompany switching are the independent variables, and the psychological principle is that subjects switch to the schedule that is most likely to provide the next reinforcer. For example, in a concurrent variable-time schedule (no response requirement, e.g., Baum and Rachlin, 1969; Brownstein and Pliskoff, 1968) the subject should switch when the probability of reinforcement at the unattended alternative, $1 - e^{-t_i / V_j}$, is greater than the probability of reinforcement at the attended alternative, $1/V_i$, where t is time since the last changeover. For other procedures the calculations are slightly different, but for any concurrent schedule, momentary maximizing predicts that the subject should switch at some target (optimal) time. Assuming a normally distributed error about the target time, it can be shown that the resulting distribution of interchangeover times will yield a regularly increasing pattern when plotted in the coordinates of Figure 20-13 (see Thomas, 1971, for a proof). Figure 20-13, however, shows that in three different concurrent-schedule experiments, the momentary maximizing pattern does not occur.

The data presented in Figure 20-13 indicate that switching is independent of local reinforcement rates and local reinforcement contingencies. In contrast, Shimp (1969) and more recently Silberberg et al. (1978) claim that momentary maximizing theory predicts switching. These authors, however, do not plot their data in terms of the relevant reinforcement probabilities. Rather, they use a cumulative-run length measure that bears no obvious relation with the reinforcement contingencies. And, in fact, when the data are replotted in a way that corresponds to how reinforcement is obtained, the momentary maximizing predictions are absent, as in Figure 20-13 (see Nevin, 1979).

CONCLUSION

The basic tenet of operant psychology is that behavior is controlled by its consequences. Accordingly theories of matching have typically assumed that some sort of behavioral consequence determines time allocation in concurrent schedules. Figures 20-7 to 20-12, however, showed that time allocation is independent of its overall consequences, and Figure 20-13 showed that the dynamics of time allocation, switching, is independent of its immediate consequences. Together, these data suggest the conclusions that how much time an organism spends at a reinforcement schedule is an unconditioned effect and that the elicitation processes conform to the matching equation. Finally, since the matching relation was initially formulated in terms of overall response rates (Herrnstein, 1961) something needs to be said about this way of measuring behavioral allocation.

Overall response rate depends on the amount of time spent at the schedule, and it also depends on the topography of the emitted response. For example, a subject in a concurrent schedule experiment may spend exactly the same amount of time at each alternative but respond at very different overall rates because at one side the lever takes 1 second to depress and at the other side the lever takes 5 seconds to depress. In turn, these two dimensions of response rate depend on reinforcement contingencies. Time spent responding, as described previously, is a function of variables having to do with shear reinforcement frequencies, independent of the feedback relation between a response and reinforcement. In contrast, response topography, as is shown next, appears to depend on the feedback relation independently of reinforcement frequency. This is a very different account of response rate than given by cost/benefit analyses (e.g., Baum, 1973; Rachlin, 1980), and this paper concludes with an outline of some of the relevant evidence.

In experiments in which variable-interval and variable-ratio schedules maintain equal reinforcement rates, subjects respond approximately twice as fast for the variable-ratio reinforcement (Herrnstein and Heyman, 1979). Importantly, the local response rates ratio was invariant; it did not depend on absolute reinforcement rates or absolute response rates. In contrast, variable-interval and variable-ratio reinforcers maintained virtually identical amounts of time spent responding (see Figure 20-12 and corresponding discussion). That is, time allocation depended only on the frequencies of reinforcement. These facts are compatible with the view that a reinforcement schedule simultaneously elicits and shapes behavior: Reward impels activity, independent of the contingency, and the contingency shapes activity into an effective response, independent of activity level.

APPENDIX 20A

The pattern of switching affects the probability of reinforcement for a switch but has no affect on reinforcement obtained between switches. Therefore the pattern that produces the highest return on switches will produce the highest overall reinforcement rate. Equations (20A.1)–(20A.3) give the probabilities for the patterns shown in Figure 20-2. For an exponential distribution of interchangeover times, the expected probability of reinforcement for a switch is

$$1/(1 + V_j \mu_i) \quad . \tag{20A.1}$$

For a uniform distribution of interchangeover times, the expected probability of reinforcement for a switch is

$$\frac{1 - e^{-T_1/V_j} + 1 - e^{-T_2/V_j} \ldots + 1 - e^{-T_n/V_j}}{n} \quad , \tag{20A.2}$$

where the average time between switches is $(T_n/2) + 1/2$ and $(T_n/2) + (1/2) = 1/\mu_i$ (Eq. (20A.1)) $= T_i$ (Eq. (20A.3)). For a point-mass distribution of interchangeover times (fixed-time switching) the expected probability of reinforcement for a switch is equal to:

$$1 - e^{-T_i/V_j} \quad . \tag{20A.3}$$

The relative magnitude of Eqs. (20A.1) − (20A.3) take the order (20A. 3) > (20A.2) > (20A.1). Therefore Poisson switching produces the lowest reinforcement rate and fixed-time switching produces the highest reinforcement rate. In addition Houston and McNamara (1981) have shown that fixed-time switching is the optimal pattern in that no other pattern of switching can produce a higher reinforcement rate. In contrast, Poisson switching produces a lower reinforcement rate than any regularly increasing pattern.

APPENDIX 20B

Overall changeover rate may be written $2/(E(t_1) + E(t_2))$, where $E(t_i)$ is the expected time per visit to side i. Note that the sum of the two local changeover rates is approximately constant (Heyman, 1979; Hunter and Davison, 1978; Nevin and Baum, 1980; Myerson and Miezin, 1980). Let I equal the reciprocal of this sum: $I = 1/(\mu_1 + \mu_2)$. Next, note that if $p = \mu_2/(\mu_1 + \mu_2)$ then $1/\mu_1 = I/(1-p)$ and $1/\mu_2 = I/p$. That is, the average time at side 1 is $I/(1-p)$ and the average time at side 2 is I/p. Therefore changeover rate can be rewritten as

$$\frac{2}{E(t_1) + E(t_2)} = \frac{2}{\frac{I}{p} + \frac{I}{1-p}} = \frac{2p(1-p)}{I} \quad . \tag{20B.1}$$

APPENDIX 20C

Assume an interdependent concurrent VT VT schedule with exponentially distributed intervals and a Poisson switching process. If the interval at the attended side elapses, the reinforcement is delivered immediately and the timer at the unattended side is not halted. In contrast, if the timer at the unattended side sets up, both timers are halted. The conditional probability of this occurring was calculated previously (see Eq. (20.1)). The result is $1/(1 + V_j \mu_i)$. Therefore the expected reinforcement rate at side 1 is

$$\frac{p}{V_1} + \frac{1-p}{V_1 + 1/\mu_2} - \frac{p}{V_1(1 + V_2 \mu_1)} \quad . \tag{20C.1}$$

For side 2 the expected rate is

$$\frac{1-p}{V_2} + \frac{p}{V_2 + 1/\mu_1} - \frac{1-p}{V_2(1 + V_1 \mu_2)} \quad . \tag{20C.2}$$

Note that reinforcement rate goes to zero if the subject does not switch (for example, $p = 0$ and $\mu_2 = 0$).

REFERENCES

Baum, W.M. 1973. The correlation based law of effect. *Journal of the Experimental Analysis of Behavior 20*: 137–53.

_____. 1974. On two types of deviation from the matching law: Bias and undermatching. *Journal of the Experimental Analysis of Behavior 22*: 231–42.

_____. 1979. Matching, undermatching, and overmatching in studies of choice. *Journal of the Experimental Analysis of Behavior 32*: 269–81.

Baum, W.M., and H.C. Rachlin. 1969. Choice as time allocation. *Journal of the Experimental Analysis of Behavior 12*: 861–74.

Blough, D.S. 1966. The reinforcement of least-frequent interresponse times. *Journal of the Experimental Analysis of Behavior 9*: 581–91.

Bradshaw, C.M.; E. Szabadi; and P. Bevan. 1976. Behavior of humans in variable-interval schedules of reinforcement. *Journal of the Experimental Analysis of Behavior 26*: 135–41.

Brownstein, A.J., and S.S. Pliskoff. 1968. Some effects of relative reinforcement and changeover delay in response-independent concurrent schedules of reinforcement. *Journal of the Experimental Analysis of Behavior 11*: 683–88.

de Villiers, P.A. 1977. Choice in concurrent schedules and a quantitative formulation of the law of effect. In W.K. Honig and J.E.R. Staddon, eds., *Handbook of Operant Behavior*, Englewood Cliffs, N.J.: Prentice-Hall.

Fantino, E.; N. Squires; N. Delbruck; and C. Peterson. 1972. Choice behavior and the accessibility of the reinforcer. *Journal of the Experimental Analysis of Behavior 18*: 35–43.

Findley, J.D. 1958. Preference and switching under concurrent scheduling. *Journal of the Experimental Analysis of Behavior 1*: 123–44.

Fleshler, M., and H.S. Hoffman. 1962. A progression for generating variable-interval schedules. *Journal of the Experimental Analysis of Behavior 5*: 529–30.

Herrnstein, R.J. 1961. Relative and absolute strength of response as a function of frequency of reinforcement. *Journal of the Experimental Analysis of Behavior 4*: 267–72.

_____. 1970. On the law of effect. *Journal of the Experimental Analysis of Behavior 13*: 243–66.

_____. 1979. Derivatives of matching. *Psychological Review 86*: 486–95.

Herrnstein, R.J., and G.M. Heyman. 1979. Is matching compatible with reinforcement maximization on concurrent variable interval, variable ratio? *Journal of the Experimental Analysis of Behavior 31*: 209–23.

Herrnstein, R.J., and W. Vaughan. 1980. Melioration and behavioral allocation. In J.E.R. Staddon, ed., *Limits to Action*, New York: Academic Press, pp. 143–76.

Heyman, G.M. 1977. Matching and maximizing in concurrent schedules. Unpublished doctoral dissertation. Harvard University.

_____. 1979. A Markov model description of changeover probabilities on concurrent variable-interval schedules. *Journal of the Experimental Analysis of Behavior 31*: 41–51.

Heyman, G.M., and R.D. Luce. 1979. Operant matching is not a logical consequence of maximizing reinforcement rate. *Animal Learning and Behavior 7*: 133–40.

Houston, A.I., and J. McNamara. 1981. How to maximize reward on two variable-interval paradigms. *Journal of the Experimental Analysis of Behavior 35*: 367–96.

Hunter, I.W., and M.C. Davison. 1978. Response rate and changeover performance on concurrent variable-interval schedules. *Journal of the Experimental Analysis of Behavior 29*: 535–56.

Killeen, P.R. 1982. Incentive theory. In *Nebraska Symposium on Motivation 1981*. Lincoln: University of Nebraska Press.

McDowell, J.J., and R. Kessel. 1979. A multivariate rate equation for variable-interval performance. *Journal of the Experimental Analysis of Behavior 31*: 267–83.

Meehl, P.E. 1950. On the circularity of the law of effect. *Psychological Bulletin 47*: 52–75.

Myerson, J., and F.M. Miezin. 1980. The kinetics of choice: An operant systems analysis. *Psychological Review 87*: 160–74.

Nevin, J.A. 1979. Overall matching versus momentary maximizing: Nevin (1969) revisited. *Journal of Experimental Psychology: Animal Behavior Processes 5*: 300–306.

Nevin, J.A., and W.M. Baum. 1980. Feedback functions for variable-interval reinforcement. *Journal of the Experimental Analysis of Behavior 34*: 207–217.

Rachlin, H. 1979. Comment on Heyman and Luce: "Operant matching is not a logical consequence of maximizing reinforcement rate." *Animal Learning and Behavior 7*: 267–68.

____. 1980. Economics and behavioral psychology. In J.E.R. Staddon, ed., *Limits to Action*, New York: Academic Press, pp. 205–36.

Rachlin, H.; L. Green; J.H. Kagel; and R.C. Battalio. 1976. Economic demand theory and psychological studies of choice. In G. Bower, ed., *The Psychology of Learning and Motivation*, vol. 10, New York: Academic Press, pp. 129–54.

Rachlin, H.; R. Battalio; J. Kagel; and L. Green. 1981. Maximization theory in behavioral psychology. *The Behavioral and Brain Sciences 4*: 371–417.

Shimp, C.P. 1969. Optimal behavior in free-operant experiments. *Psychological Review 76*: 97–112.

____. 1975. Perspectives on the behavioral unit: Choice behavior in animals. In W.K. Estes, ed., *Handbook of Learning and Cognitive Processes*, vol. 2, Hillsdale, N.J.: Lawrence Erlbaum, pp. 225–68.

Shull, R.L., and S.S. Pliskoff. 1967. Changeover delay and concurrent performances: Some effects on relative performance measures. *Journal of the Experimental Analysis of Behavior 10*: 517–27.

Silberberg, A.; B. Hamilton; J.M. Ziriax; and J. Casey. 1978. The structure of choice. *Journal of Experimental Psychology: Animal Behavior Processes 4*: 368–98.

Silberberg, A., and J. Schrot. 1974. A yoked-chamber comparison of concurrent and multiple schedules: The relationship between component duration and responding. *Journal of the Experimental Analysis of Behavior 22*: 21–30.

Staddon, J.E.R. 1977. On Herrnstein's equation and related forms. *Journal of the Experimental Analysis of Behavior 28*: 163–70.

____. 1980. Optimality analyses of operant behavior and their relation to optimal foraging. In J.E.R. Staddon, ed., *Limits to Action*, New York: Academic Press, pp. 101–41.

Staddon, J.E.R.; J.M. Hinson; and R. Kram. 1981. Optimal choice. *Journal of the Experimental Analysis of Behavior 35*: 397–414.

Staddon, J.E.R., and S. Motheral. 1978. On matching and maximizing in operant choice experiments. *Psychological Review 85*: 436–44.

____. 1979. Response independence, matching, and maximizing: A reply to Heyman. *Psychological Review 86*: 501–505.

Stubbs, D.A., and S.S. Pliskoff. 1969. Concurrent responding with fixed relative rate of reinforcement. *Journal of the Experimental Analysis of Behavior 12*: 887–95.

Stubbs, D.A.; S.S. Pliskoff; and H.M. Reid. 1977. Concurrent schedules: A quantitative relation between changeover behavior and its consequences. *Journal of the Experimental Analysis of Behavior 27*: 85–96.

Thomas, E.A.C. 1971. Sufficient conditions for monotone hazard rate. An application to latency probability curves. *Journal of Mathematical Psychology 8*: 303–32.

VII BEYOND MATCHING OR MAXIMIZING

21 A UNIFIED PRINCIPLE OF REINFORCEMENT
Some Implications for Matching

John W. Donahoe, Michael A. Crowley, Will J. Millard, and Kenneth A. Stickney

The transcendent task of psychology is, arguably, the formulation of an adequate principle of reinforcement. Such a principle would stand in the same relation to problems of ontogeny as does the principle of natural selection to those of phylogeny. That is, a well-formulated principle of reinforcement should illuminate individual development and provide a context congenial to the assimilation of those subordinate principles required by an exhaustive analysis of the behavior of the individual organism.

The history of science reveals that the initial formulations of even fundamental principles are often functional and descriptive rather than reductive in character. The principle of natural selection is now buttressed by analyses at the levels of genetics and molecular biology, but it was not always so. Initially, the fundamental principle of phylogeny was a purely functional statement expressing the relation between the occurrence of certain structural and behavioral characteristics and the likelihood of leaving surviving offspring. Darwin was ignorant of the biological mechanisms that mediated the functional statement of natural selection; he had kept his appointment with Newton at Westminster Abbey some eighteen years prior to the rediscovery of Mendel's work. Following the formulation of a functionally adequate principle of reinforcement, we may anticipate supportive analyses at the neurophysiological and neurochemical levels. But, for now, the search for the fundamental principle of ontogeny is probably best

This research was supported by grant MH 28610 from the National Institute of Mental Health and by the Faculty Research Fund of the University of Massachusetts, Amherst. Appreciation is expressed to Andrea Bugen, Laine Donnelly, and Robert Koolkin for experimental assistance.

guided by work at the behavioral level. (See Donahoe and Wessells, 1980 for a more extended discussion of these issues.)

Psychology has sought a comprehensive reinforcement principle using the two procedures of classical and instrumental (operant) conditioning. The current best understanding of a principle of reinforcement comes from the theoretical analysis of the classical procedure provided by Rescorla and Wagner (1972) and of the operant procedure provided by the interrelated proposals of Herrnstein (1970) and Premack (1965, 1971). The present chapter briefly characterizes the essential features of the theoretical analyses of these two procedures and then proposes an approach to a unified formulation of a reinforcement principle that draws upon both. Some implications of this unified formulation are then explored for the analysis of matching. A major conclusion of the approach is that a precise statement of a reinforcement principle awaits the collection of observations that often go unmonitored in current conditioning procedures.

TOWARD A UNIFIED PRINCIPLE OF REINFORCEMENT

Modern analyses of conditioning both acknowledge and respect the procedural distinction between the classical and instrumental paradigms. Procedurally, the reinforcer is contingent upon the presentation of a stimulus in the former case and upon the occurrence of a response in the latter. However, unanimity does not exist regarding whether fundamentally different behavioral processes are involved in the two paradigms (cf. Bindra, 1976; Guthrie, 1935; Hull, 1943; Konorski and Miller, 1937; Skinner, 1938; Trapold and Overmier, 1972).

Perhaps the most general description of conditioning is provided by the language of situation-transition (Baum, 1973). In these terms, the classical procedure is defined as a transition from an environment containing the conditioned stimulus to one containing the unconditioned stimulus and the instrumental procedure as a transition from an environment permitting the instrumental response to one containing the reinforcing stimulus. When viewed as situation-transition, the central problem for a theoretical analysis of conditioning becomes one of discovering those components of the transition that are essential for behavioral change—most especially those components that are common to both the classical and instrumental procedures. This section describes a search for such components.

Theoretical Analysis of the Classical Procedure

In the classical procedure the experimenter manipulates the spatiotemporal relation between environmental stimuli and the reinforcer but does not systematically vary the relation between behavior and the reinforcer. As a consequence of the transition from the conditioned stimulus (CS) to the reinforcer (the unconditioned stimulus, or US), the CS comes to control a response (the conditioned response, or CR) that often resembles the reinforcer-elicited response (the unconditioned response, or UR).

Whereas earlier formulations (e.g., Estes, 1950; Hull, 1943) attempted to account for the behavioral changes produced by the classical procedure in terms of the contiguous transition from the CS to the US, the seminal work of Kamin (1968, 1969) showed that other events play a critical role in conditioning. Specifically, Kamin discovered that the acquisition of control of behavior by a putative CS could not be understood independently of the control exerted by other contemporaneous stimuli. If other stimuli already controlled the CR, a new stimulus placed in an otherwise favorable temporal relation with the reinforcer would not acquire control over the CR; that is, control of the CR by the nominal CS was "blocked."

Building upon Kamin's findings, and supplemented with many of their own, Rescorla and Wagner (1972) proposed a linear-operator model of the change in associative value V that was postulated to underlie the observed change in behavior. The model is summarized by Eq. (21.1), in which ΔV represents the trial-to-trial change in associative strength, λ represents the asymptotic associative strength for a given US, V_t represents the total associative strength supported by all prevailing stimuli and θ is a coefficient whose value depends on various parameters of the conditioning situation (e.g., CS intensity).

$$\Delta V = \theta(\lambda - V_t) \quad . \tag{21.1}$$

The Rescorla-Wagner model is a relational principle of reinforcement in that the change in association value is a function of the asymptotic associative strength relative to the total associative strength of all other stimuli, only one of which is the putative CS. Experimental tests of the model have yielded many successful ordinal predictions (e.g., Kamin and Gaioni, 1974; Kremer, 1978).

To facilitate further developments it is convenient to use an alternative expression that depicts the total change in behavior as a function of steady-state probabilities of responses rather than trial-to-trial changes in association value.[a] In Eq. (21.2), the total change in the probability of the CR given the occurrence of the CS ($\Delta p_{CR/CS}$) is proportional to a fraction (θ) of the discrepancy be-

a. Equation (21.2) and the other equations that follow may be characterized as steady-state equations since they do not describe trial-to-trial changes in behavior but the total change from one state prior to the institution of the experimental contingencies to another state after those contingencies have been applied for an indefinitely long period. The equations do not represent a commitment to a particular view of the acquisition process (e.g., Estes, 1950, 1959). At this juncture, steady-state models are adequate for many purposes and the adoption of a particular view of the acquisition process would seem premature (cf. Prokasy and Gormezano, 1979).

tween the probability[b] of the UR given the US ($p_{UR/US}$) and the initial probability (operant level) of the CR given both the CS and whatever other stimuli (S_o) are also present at the outset of conditioning (i.e., $p_{CR/S_t} = p_{CR/CS} + p_{CR/S_o}$).

$$\Delta p_{CR/CS} = \theta(p_{UR/US} - p_{CR/S_t}) \quad . \tag{21.2}$$

It is a task for an experimental analysis to determine which of the events that covary with the discrepancy defined by Eq. (21.2) are crucial for conditioning. A detailed examination of this issue is beyond the intent of the present chapter, but two points may be made: (1) Changes in responses specific to the CS (the orienting response, or OR) do not appear to contribute directly to the discrepancy, and (2) changes specific to US–UR events are sufficient to bring about conditioning in many instances.

The conclusion that the OR does not contribute fundamentally to the discrepancy may be drawn from the work of Holland (1977). Using the Kamin two-phase blocking design, rats were presented a tone followed by food. The tone (CS) initially elicited a head-jerk response (OR) whereas food elicited approach to the feeder and eating. After a number of tone-food transitions, both head-jerking and approaching were controlled by the tone. In the second phase of the experiment, a tone-light simultaneous compound CS was paired with food. Even though other work had shown that the light elicited a different OR of rearing, the conditioning of approach behavior to the light was blocked. As shown by test trials when the light was presented alone, approaching the feeder did not occur. Blocking of acquisition to the light—in spite of the fact that the light and tone elicited different ORs—indicates that a discrepancy in the OR is not sufficient for conditioning. Thus, although the OR may contribute importantly to the ultimate form of the CR (e.g., Timberlake and Grant, 1975), it does not appear to contribute to the fundamental discrepancy underlying conditioning.

The contribution of US–UR events to the discrepancy is well documented, however. As Kamin (1969) has shown, increases in US intensity (and concomitant but unmeasured changes in the UR) in the second phase of the blocking design produce conditioning to the new component of the compound CS. Also, if the physical properties of the US are held constant, but the affectors and effectors activated by the US are changed by moving the site of application of the US, blocking is eliminated (Stickney and Donahoe, Note 1). Although more

b. Since a central consideration in the selection of a US is that it controls a measureable UR (Gormezano and Kehoe, 1975), it might be assumed that $p_{UR/US}$ is always near unity in a classical procedure. This conception is valid only if $p_{UR/US}$ is defined by the relative frequency of the UR given the occurrence of the US. If the probability measure is defined otherwise, such as the area under the UR curve relative to the maximum area found with an intense US, then $p_{UR/US}$ may vary over a considerable range. The area under the response curve is sensitive to CS–US manipulations (e.g., Smith, 1968) and area measures have been found to provide orderly indices of conditioning (Martin and Levey, 1969).

experimental work is required, it is clear that a transition in US-related events is often sufficient to produce conditioning. Some effective elicitors of behavior (such as eye blinking elicited by an auditory click) do not function as USs, however, and future experimental analysis must identify the manner in which these elicitors differ from those that are capable of serving as reinforcers (cf. Donahoe and Wessells, 1980:114-15 and Frey, Maisiak, and Dague, 1976).

To summarize, the situation-transition responsible for conditioning in the classical procedure involves a discrepancy in US-UR events. At the outset of conditioning, the CS controls behavior that is discrepant with respect to the responses elicited by the US. At the conclusion of conditioning, the CS controls US-elicited responses and the discrepancy in behavior has been reduced; the behavior during the CS now more closely matches the behavior during the US.

Theoretical Analysis of the Instrumental Procedure

In the simplest instrumental—or operant—procedure, the experimenter manipulates the spatiotemporal relation of behavior to the reinforcer while the relation of stimuli to the reinforcer is not systematically varied. (All three classes of events—stimulus, response, and reinforcer—may, of course, be manipulated in a discriminated operant procedure.) The most often measured result of the instrumental procedure is that some aspect of the environment (the discriminated stimulus, or S) acquires control of the instrumental response R.

Among the efforts to formulate a principle of reinforcement based upon findings with the instrumental procedure, the interrelated proposals of Premack (1959, 1965) and Herrnstein (1961, 1970) are the most general and highly developed. Premack (1971) stated that an instrumental response will change in probability if the opportunity to engage in a response having a different initial probability is made contingent upon the occurrence of the instrumental response. Thus a transition between responses having different probabilities of occurrence defines instrumental conditioning according to this formulation. As an example, the instrumental response of bar-pressing (R) on the presence of the sight of the bar (S) increases when the reinforcing stimulus of food (S^R), which elicits the ingestion of food (R_e), is made contingent on bar pressing. By measuring the amounts of time spent bar pressing and eating during a baseline period when the stimuli that control these responses are both available, the initial probabilities of responses may be determined.

Using symbols appropriate to the events in the instrumental procedure, the content of the foregoing may be represented by the following equation (Donahoe, 1977):

$$\Delta p_{R/S} = \theta(p_{R_e/S^R} - p_{R/S_t}) \ . \qquad (21.3)$$

In this steady-state equation, $\Delta p_{R/S}$ is the total change in the probability of R given S, θ is an empirical coefficient,[c] p_{R_e/S^R} is the probability of the response elicited by the reinforcer given the reinforcing stimulus,[d] and (p_{R/S_t}) is the initial probability of R given the total stimulus environment (S_t) (that is, $p_{R/S_t} = p_{R/S} + p_{R/S_o}$).

The reinforcement principle that emerges from Premack's analysis of the instrumental procedure is also relational in character. A given event will function as a reinforcer depending upon the probability of the contingent response (the response elicited by the reinforcer) relative to the initial probability of the instrumental response in that environment. The effect of a putative reinforcer on the instrumental response depends, therefore, on the stimulus control of the instrumental response that predates the experimental contingencies. As an example of the relational character of the analysis, control of an instrumental response by a new stimulus may be blocked if that response is already controlled by another stimulus (e.g., vom Saal and Jenkins, 1970; cf. Donahoe, 1977: 346–47). To summarize Premack's analysis, a change in behavior is produced by the instrumental procedure when a transition occurs between discrepant baseline probabilities of the responses controlled by S and S^R.

The matching law proposed by Herrnstein (1970) accommodates the truth that every instrumental procedure contains more than one reinforcing stimulus (and its elicited response) and more than one instrumental response (and its controlling stimulus). Strictly speaking, Premack's analysis confines itself to only these events—a convenient fiction, but a fiction nevertheless. Herrnstein's analysis explicitly acknowledges the existence of other responses R_o that are correlated with the occurrence of other reinforcers r_o. For purposes of experimental analysis, some of these other responses may be monitored and reinforcers scheduled contingent upon their occurrence, in which case the procedure is described as a concurrent schedule (Ferster and Skinner, 1957).

c. The symbol θ represents a weighting factor associated with the discrepancy postulated to underlie conditioning. To facilitate exposition the same symbol is used throughout the chapter although the specific antecedents of which θ is a function may vary with the procedure.

d. In commonly employed operant procedures, such as key pecking for grain by pigeons, the reinforcing stimulus controls a readily observed elicited response R_e such as ingestion or swallowing. However, not every R_e is easily measured on the behavioral level in every environmental context. Some reinforcer-elicited responses are subtly expressed in some environments (cf. Valenstein, Cox, and Kakolewski, 1969: 256, 263) and, by various interventions such as curariform compounds, may be aborted altogether on the behavioral level. Limitations of this sort are viewed as technical and not fundamental to the analysis. In still other operant paradigms (e.g., negative reinforcement), further problems are encountered since the experimenter manipulates only the stimuli prior to the instrumental response and the postresponse stimuli m, which function as reinforcers and presumably elicit the R_{em}, are not well specified (cf. Donahoe and Wessells, 1980: 208).

When concurrent schedules are instituted, the relative frequency of responding to an alternative has been found to match the relative frequency of reinforcement for that alternative. The relation between responding R and reinforcement r in concurrent schedules is known as the matching law (Herrnstein, 1970) and may be stated as

$$R/(R + \sum_{o=1}^{n} R_o) = r/(r + \sum_{o=1}^{n} r_o) \quad , \qquad (21.4)$$

where there are n other responses ($o = 1, 2, \ldots, n$), each with its corresponding reinforcer. The matching law has received extensive experimental support (cf. de Villiers, 1977) and may be obtained from Eq. (21.3) under certain conditions (Donahoe, 1977: 342-44).

A Unified Theoretical Analysis

As interpreted here, Eqs. (21.2) and (21.3) represent the reinforcement principles that have arisen from experimental work with the classical and instrumental procedures, respectively. For convenience, these equations are repeated.

$$\text{Classical:} \quad \Delta p_{CR/CS} = \theta(p_{UR/US} - p_{CR/CS_t})$$

$$\text{Instrumental:} \quad \Delta p_{R/S} = \theta(p_{R_e/S^R} - p_{R/S_t}) \; .$$

The equations are similar in form (both are relational in the sense described earlier), but they differ in the terms exemplifying that common form. That is, they differ in the type of transition postulated to underlie behavioral change. The differences between the two interpretations are now examined and a proposal is advanced for their reconciliation.

Comparison of the Two Principles of Reinforcement. Leaving aside for the moment the different responses by which behavioral change is measured in the two procedures, consider the nature of the terms defining the critical discrepancies within the parentheses. Although the notation differs, the left-hand terms refer to the same class of event, namely, the probability of the response elicited by the reinforcer given the occurrence of the reinforcing stimulus. As a concrete example, salivation is elicited by food in the mouth whether food follows the presentation of a stimulus in the classical procedure or the occurrence of a response in the instrumental procedure. Thus except for historical reasons, the left-hand terms of the two discrepancies—$p_{UR/US}$ and p_{R_e/S^R}—may be symbolized as $p_{UR/US}$ in both procedures.

Consider now the right-hand terms within the parentheses, because therein resides the fundamental difference between the two formulations. In the classical procedure the discrepancy represents a comparison of the levels of responding within the same response system, that elicited by the reinforcer. Returning to the example of food as a reinforcer, both the UR and the CR are salivary responses. The two responses in the classical procedure differ primarily in the stimuli that control responding within a common salivary system. Because the discrepancy is based upon differences within the same response system, it is hereafter termed a *homogeneous* discrepancy. In the instrumental procedure, however, the right-hand term of the discrepancy refers to responding within a different response system, that of the instrumental response. As an example, the different response systems might be salivating and bar pressing. This latter discrepancy is termed *heterogeneous* because it involves a comparison between different response systems, that elicited by the reinforcer and that of the instrumental response.

If the distinction between the reinforcement principles describing behavioral change with the classical and instrumental procedures resides in a presumed difference in the nature of the discrepancy, it behooves us to examine the reasons that have led to that distinction. Does the evidence require the assumption that a homogeneous discrepancy underlies one procedure whereas a heterogeneous discrepancy underlies the other?

A complete presentation of relevant information is beyond the present treatment, but two points may be made here. First, the assumption that a heterogeneous discrepancy is crucial to conditioning with the instrumental procedure is gratuitous on logical grounds. At the outset of a typical instrumental procedure, the occurrence of a heterogeneous discrepancy is confounded with the simultaneous presence of a homogeneous discrepancy. As an illustration, at the outset of the instrumental conditioning of bar pressing for food, bar pressing not only occurs at a lower probability than food-elicited salivation but conditioned salivation also occurs at a lower probability. Thus typical instrumental procedures contain a homogeneous as well as a heterogeneous discrepancy, although the former discrepancy is rarely apparent since reinforcer-elicited responses are seldom monitored in instrumental procedures (cf. Kimble, 1961).

Second, empirical grounds also exist for denying a fundamental role to a heterogeneous discrepancy in the genesis of conditioning with the instrumental procedure. Considerable evidence, due primarily to the work of Allison and Timberlake, indicates that the reinforcer-elicited response need not be more probable than the instrumental response in order for the latter to increase in probability when an instrumental contingency is instituted (Allison, 1978; Timberlake and Allison, 1974; Timberlake and Wozny, 1979; but see Dunham, 1977; Premack, 1971). In short, neither logical nor experimental analysis compels the belief that a heterogeneous discrepancy underlies conditioning with the instrumental procedure.

Synthesis of the Two Reinforcement Principles. Since heterogeneous and homogeneous discrepancies are confounded in the typical instrumental procedure, parsimony encourages the exploration of the possibility that a homogeneous discrepancy underlies conditioning with both the classical and instrumental procedures. Consider the following alternative formulation of a relational principle of reinforcement for the instrumental procedure where $p'_{R/S}$ is the asymptotic probability of the instrumental response and the other terms are as defined earlier:

$$\Delta p_{R/S} = \theta(p_{\mathrm{UR/US}} - p_{\mathrm{CR}/S_t})$$

$$p'_{R/S} - p_{R/S} = \theta(p_{\mathrm{UR/US}} - p_{\mathrm{CR}/S_t})$$

$$p'_{R/S} = p_{R/S} + \theta(p_{\mathrm{UR/US}} - p_{\mathrm{CR}/S_t}) \ . \tag{21.5}$$

Equation (21.5) states that the asymptotic probability of the instrumental response is equal to the operant level of the instrumental response plus some fraction of the homogeneous discrepancy within the response system activated by the reinforcer. This equation represents the unified principle of reinforcement for the instrumental procedure.

What is the view of instrumental conditioning that arises from this reformulation of the reinforcement principle? As described by Eq. (21.5), instrumental conditioning is purely a procedural arrangement whereby the instrumental response is made to occur in a specified spatiotemporal relationship to a homogeneous discrepancy. As the result of the procedural arrangement, the environment acquires control over behavior arising from two principal sources—the instrumental response and the reinforcer-elicited response. From this perspective the instrumental procedure promotes the conditioning of the instrumental response because the learning mechanism fails to distinguish fully between reinforcer-elicited responses that are inherent constituents of the homogeneous discrepancy and the instrumental response that is contiguous because of procedural constraints only. Behaviors of both origins potentially come under the control of the environment and, therefore, the net behavioral outcome of the instrumental procedure depends upon interactions between instrumental and reinforcer-elicited responses (see Donahoe and Wessells, 1980:159–62).

Parallel reasoning applied to the classical procedure indicates that the environment should acquire control over not only reinforcer-elicited responses but also responses of whatever other origins may reliably occur in a favorable relation to a homogeneous discrepancy. Since the classical procedure does not specify the relation of the reinforcer to behavior, considerable variability may be anticipated in the control of responses other than those elicited by the reinforcer, although some components of the orienting response to the CS are prime candidates (cf. Holland, 1977; 1980; Rescorla, 1978; Skinner, 1948). As with the instrumental procedure, the behavior that ultimately comes under environmental control will be determined by interactions between responses of diverse origins (see Donahoe and Wessells, 1980:111–12).

The common conditioning rule proposed for the behavioral changes produced by the classical and instrumental procedures may be summarized as follows: The environment acquires control over whatever responses occur in a favorable spatiotemporal relation to a homogeneous discrepancy (a discrepancy confined to the responses elicited by the reinforcer). This formulation is hereafter referred to as the *unified principle of reinforcement.* It describes a process, analogeous to that of adaptation via natural selection, through which a discrepancy is reduced between the initial behavior in an environment and the elicited behavior that occurs because of the introduction of a reinforcer into that environment.

If a homogeneous discrepancy does not exist when the instrumental contingency is imposed, the unified reinforcement principle predicts that the occurrence of an otherwise effective reinforcing stimulus after the instrumental response will *not* produce conditioning. The following experiment, conducted for another purpose by Konorski and Miller (1936; cited in Konorski, 1967: 362), illustrates this prediction. A dog was first conditioned to salivate by pairing food with a CS. Following this classical procedure an instrumental procedure was initiated in which the CS was presented and now the leg of the dog was passively flexed prior to the presentation of the food. Although passive leg flexion followed by food was sufficient to bring leg movement under the control of the CS in naive animals, the flexion-food sequence was insufficient for this animal for which the environment had previously been paired with food. From the perspective of a unified reinforcement principle, the prior resolution of the homogeneous discrepancy (the dog salivated to the CS from the onset of the instrumental procedure) had eliminated the capacity of food to function as a reinforcer in that environment. Although considerable indirect evidence is now consistent with this finding (e.g., Mackintosh and Dickinson, 1979; Pearce and Dickinson, 1978; St. Claire-Smith, 1979; Williams, 1975; Zanich and Fowler, 1978), more work is sorely needed in which the reinforcer-elicited as well as the instrumental response is directly monitored in the instrumental procedure. The deactivation of a reinforcer in an instrumental procedure by prior response-independent presentations of the reinforcing stimulus may also prove helpful in understanding some instances of the phenomenon of learned helplessness (Maier and Seligman, 1976) and learned laziness (Engberg, Hansen, Welker, and Thomas, 1972).

Some Enduring Issues

A unified reinforcement principle seeks to identify the common conditions that are essential to produce changes in the environmental control of behavior in the classical and operant procedures. An adequately formulated reinforcement principle, in addition to addressing this central issue, must ultimately confront a number of closely interrelated matters. Among these are schedules of reinforcement, biological constraints on learning, and the analysis of acquisition. Before examining some implications of the unified reinforcement principle for the treatment of matching, each of these three areas is characterized and its relation to the present formulation briefly examined.

Schedules of Reinforcement. In the operant procedure a schedule of reinforcement is a constraint imposed upon the relation that may exist among certain events, notably between the instrumental response and the reinforcer. The most general analysis of the changes in total operant responding produced by schedule constraints is provided by the proposals of Rachlin and Burkhard (1978) and Staddon (1979). The treatment may be characterized thus: An organism is viewed as allocating its behavioral resources among n alternative responses. Under baseline conditions when no explicit schedule constraints are imposed, the organism allocates its resources among the n alternatives in a manner that is assumed to be optimal in some sense. The baseline point B represents this allocation of resources and, if allocation is measured by the time t apportioned to each response, then B has coordinates $(t_1, t_2, \ldots, t_n)$. (See Figure 21-1 for a graphic representation with three alternative behaviors.) When a constraint is imposed in the form of a schedule of reinforcement, the organism reallocates its time among the n alternatives. For example, a schedule might constrain the relation between bar pressing and eating such that t_1' seconds of bar pressing is required for t_2' seconds of eating. The constrained point C represents the new allocation of resources under the schedule of reinforcement. According to this approach to schedules of reinforcement, C has coordinates $(t_1', t_2', \ldots, t_n')$ such that the deviation of C from B is minimal subject to the satisfaction of the schedule constraint (see Figure 21-1). Since B is an optimal allocation of resources and C departs minimally from B given the schedule constraint, then C represents an optimal solution to the problem imposed by the schedule of reinforcement.

The unified reinforcement principle has a thrust that is not inconsistent with the analysis of schedules of reinforcement, but with a caveat. The foregoing analysis asserts that C is chosen such that the distance $\overline{BC}$ is minimized subject to the schedule constraint whereas the unified reinforcement principle states that asymptotic instrumental responding is achieved when the discrepancy between conditioned ($p_{CR/CS}$) and reinforcer-elicited responding ($p_{UR/US}$) is minimized. The caveat is that the baseline point B may itself be thought of as a deviation from some ideal point I at which the discrepancy between conditioned and elicited responding is simultaneously zero for all responses. Since even the baseline environment imposes some constraints, B represents the outcome of these implicit constraints. For example, it is impossible to allocate time to approaching food without simultaneously decreasing the time available for some other activity such as grooming. Thus, a baseline environment in which food and water occupy adjacent locations might support different levels of eating and drinking than one in which these stimuli were physically removed from each other. Accordingly, the baseling point B may be viewed as a minimum deviation from a hypothetical ideal point I at which all discrepancies are simultaneously zero.

Figure 21-1. A space displaying the time allocated among three hypothetical behaviors. Point B, with coordinates t_1, t_2, t_3, represents the baseline conditions in which no explicit constraints are placed on the relation among the behaviors. Point C, with coordinates t'_1, t'_2, t'_3, represents a constrained condition in which a schedule of reinforcement specifies the relation between behaviors 1 and 2. The directed path between points B and C represents the change in time allocation as the organism adapts to the schedule constraint.

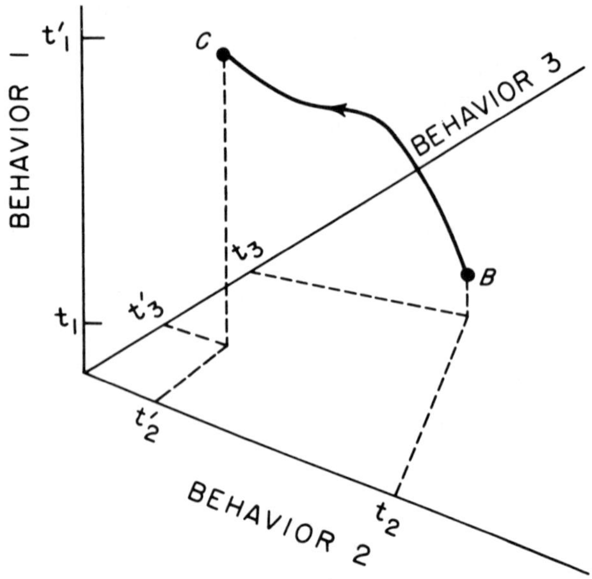

From this perspective, the contingency point C is more precisely described as a minimum deviation (subject to the schedule constraint) from I rather than B. The minimum-distance relation is not transitive and, hence, if C departs minimally from B and B departs minimally from I, it does not follow that C departs minimally from I. With well-chosen baseline conditions, B should not differ systematically from I and the theoretical analysis of schedules of reinforcement should be relatively little affected by this consideration (but see Allison, Miller, and Wozny, 1979: experiment 3 for a possible counterexample).

Biological Constraints. A well-conceived principle of reinforcement must not only accommodate the ontogenic constraints imposed by schedules of reinforcement, but it must also prove congenial to the phylogenic constraints arising from evolution—so-called biological constraints on learning (e.g., Bolles, 1970; Breland and Breland, 1961; Rozin and Kalet, 1971; Seligman and Hager, 1972; Shettleworth, 1972).

Whereas biological constraints are sometimes regarded as competing or even conflicting with a principle of reinforcement, the present formulation regards at least some of these constraints as implied by the reinforcement principle. Since the behavior that comes under environmental control potentially includes the OR as well as reinforcer-elicited responses in the classical procedure and reinforcer-elicited as well as the instrumental response in the operant procedure, stimulus-reinforcer and response-reinforcer interactions are anticipated by the unified principle of reinforcement. These phenomena are not supplementary to and independent of the reinforcement principle but are inherent in its statement. The ultimate behavioral outcome of a particular conditioning procedure requires an analysis of the interactions among the various responses that are evoked by the particular stimuli used to implement the procedure, but a common framework for their analysis is provided by the reinforcement principle.

Certain general classes of complexities may be anticipated in both the analysis of schedules of reinforcement and biological constraints. First, behaviors differ from one another in the degree and origins of their interdependence. Some interrelations are mechanically imposed by the prevailing environment; for example, a pigeon cannot simultaneously peck two response keys. Others are biologically mediated by the ancestral environment; for example, many animals drink immediately after eating (postprandial drinking). These interdependencies complicate the prediction of both the reallocation of responding produced by the schedule of reinforcement and the final outcome of conditioning procedures (see the discussions of response substitutability in Rachlin and Burkhard, 1978: 40–42 and Rachlin, Kagel, and Battalio, 1980).

As a specific illustration of response interactions in an instrumental procedure, an omission contingency is relatively ineffective when reduction in salivation is the operant and food is the reinforcer (Sheffield, 1965) but is quite effective if water is the reinforcer (Miller and Carmona, 1967). From the present perspective, this biological constraint arises from the fact that food elicits salivation, which then competes with the instrumental response of "dry mouth," whereas water does not elicit salivation. The interrelation between instrumental and food-elicited pecking in pigeons probably plays a similarly important role in autoshaping and negative automaintenance (Brown and Jenkins, 1968; Williams and Williams, 1969).

A second complication in the analysis of both biological and schedule constraints is that the behavioral changes wrought by conditioning procedures are not restricted to the particular responses that occur within that procedure. The potential behavioral change appears to include a class of responses of which the measured CR is only one member. As an example, consider the reinforcing stimulus of electrical stimulation of the brain (ESB). ESB of constant physical properties delivered to the same brain site will elicit eating, drinking, or a variety of other responses dependent upon which environmental stimuli are present when ESB is given (Valenstein, Cox, and Kakolewski, 1965; see Antelmen and Cagguila, 1978 for studies of similar import using peripheral stimulation). It may be anticipated that a stimulus paired with ESB in a classical procedure will also acquire control of behavior that is similarly sensitive to the context of environmental stimulation. Because of the sensitivity of the form of the unconditioned response to the prevailing stimuli, the interaction of the instrumental response with reinforcer-elicited responses may be expected to vary with the particular features of the training environment. In summary a class of reinforcer-elicited responses is conditioned rather than merely the particular responses elicited by the reinforcer in the specific conditioning situation (cf. Timberlake and Grant, 1975) and the expression of the biological constraint associated with the reinforcer-elicited response may vary somewhat from one situation to the next. A thoroughgoing experimental analysis of response interactions in various environmental contexts is required if the net outcome of conditioning is to be understood fully.[e]

Acquisition. In addition to schedules of reinforcement and biological constraints on learning, a third problem that must be faced with a well-formulated reinforcement principle is the analysis of acquisition. In terms of concepts introduced in the discussion of schedules of reinforcement, the analysis of acquisition seeks to describe the path taken as behavioral resources are reallocated when the organism moves from the baseline point B to the contingency point C (see Figure 21-1).

The unified reinforcement principle as presently described is a steady-state model and its relation to a model of acquisition remains to be developed. (Because of surface similarities with other formulations, Eq. (21.3) has sometimes been misinterpreted as a commitment to a linear-operator model of acquisition; but see Donahoe, 1977: 347). As a steady-state model, the unified reinforcement principle attempts to specify the conditions necessary for learning with the classical and instrumental procedures. Those conditions are a discrepancy between the unconditioned and conditioned activity in the response system activated by the reinforcer, a discrepancy that is measured in a well-chosen behavioral preparation by a difference in the strengths of the UR and CR.

Since a discrepancy measured by differences in unconditioned and conditioned responding is postulated to underlie learning, the unified reinforcement principle is congenial to acquisition models that are sensitive to response as well as to reinforcer events. That is, the effect of a conditioning trial on subsequent

e. Not mentioned in this treatment is another source of response interactions—those arising from homeostatic mechanisms that oppose the continued expression of reinforcer-elicited responses after repeated presentations of the reinforcing stimulus. As an example, a US consisting of the injection of insulin produces a decrease in blood sugar level through causing the conversion by the liver of glucose into glycogen. With continued CS–US pairings, however, the CR becomes an *increase* in blood glucose. This opposing response occurs, in part, through the conditioned release of glucagon, causing the conversion of glycogen back into glucose. Opposing response interactions of this type are most often studied under the rubric of opponent-process theory (Solomon, 1980; Solomon and Corbit, 1974).

performance is potentially a function of the occurrence of the CR as well as the US. Unlike traditional law-of-effect formulations, trials on which a CR occurs before the US (CR–US trials) may not foster behavioral change since the discrepancy between the UR and CR is minimal on such trials. Instead, other things being equal, trials on which a discrepancy occurs ($\overline{\text{CR}}$–US trials) should provide the impetus for the transition from the baseline to the contingency point. Unfortunately, experimental evidence is once again inadequate to evaluate the issue fully. Scattered findings do indicate, however, that the effect of a constant reinforcer differs as a function of the occurrence or nonoccurrence of a CR on the previous trial (e.g. Prokasy and Gormezano, 1979). Such findings are inconsistent with simple linear models of the acquisition process unless supplemented by other principles.

The potential importance of the molecular sequence of events to the analysis of acquisition has disquieting implications for the ultimate utility of the molar approach to the analysis of schedules of reinforcement described earlier. The total amount of time allocated to a particular behavior provides no representation of the sequences of responses upon which that total is based (cf. Rachlin, Kagel, and Battalio, 1980). To the extent that the sequence of responses is crucial to the production of behavioral change, then to that extent is the molar analysis insensitive to the acquisition processes that underlie the transition from the baseline to the contingency point. Two organisms, each allocating their total behavioral resources identically so as to occupy the same baseline point, may nevertheless respond differently to the same schedule of reinforcement because the underlying sequence of responses is different. For example, assume that during baseline conditions two rats allocate identical total times to wheel running. If an instrumental contingency is then instituted such that *t* seconds of running is required prior to eating and if one rat emits longer bursts of running than the other, then the first rat may encounter food before the second. Thus, the two rats would require different paths to describe the transition from the baseline to the contingency point.

The notion that performance can be understood independently of the sequence of events that constitutes the total antecedent conditions is a form of the independence-of-path assumption in mathematical learning theory. There is a pressing need to interrelate steady-state and acquisition theories (Staddon, 1977), but it should be noted that several recent theoretical analyses of acquisition during operant conditioning are path-dependent (e.g., Blough, 1975; Herrnstein, 1979); that is, they are sensitive to the historical sequence of events and not merely to molar characterizations of response strength.

APPLICATIONS RELATED TO MATCHING

As already noted, the matching law explicitly acknowledges the multiplicity of responses and reinforcers in every situation and states that the relative frequency of these responses matches their relative frequency of reinforcement. In terms of the unified principle of reinforcement described by Eq. (21.5), the asymtotic probability of one instrumental response ($R1$) relative to that of a second instrumental response ($R2$) is given by Eq. (21.6). (The subscripts denoting the different stimuli controlling the various responses are omitted to simplify the notation.)

$$\frac{p'_{R1}}{p'_{R2}} = \frac{p_{R1} + \theta\,(p_{\mathrm{UR1}} - p_{\mathrm{CR1}})}{p_{R2} + \theta\,(p_{\mathrm{UR2}} - p_{\mathrm{CR2}})} \,. \tag{21.6}$$

Under the assumptions that the operant levels of the two instrumental responses are zero (that is, $p_{R1} = p_{R2} = 0$) and that the conditioning of reinforcer-elicited responses is initially zero ($p_{\mathrm{CR1}} = p_{\mathrm{R2}} = 0$), Eq. (21.6) reduces to Eq. (21.7).

$$\frac{p'_{R1}}{p'_{R2}} = \frac{p_{\mathrm{UR1}}}{p_{\mathrm{UR2}}} \,. \tag{21.7}$$

By means of arguments parallel to those presented in the development of the matching law from Premack's principle (Donahoe, 1977: 342–44), Eq. (21.7) may be shown to be algebraically equivalent to Eq. (21.4), the matching law. Thus under conditions that may reasonably to assumed to exist in most concurrent procedures, the unified reinforcement principle is also consistent with the matching law.

In the remainder of this section, we shall explore some implications of the relation between the unified reinforcement principle and matching. The following issues are of specific concern: (1) the effect of the operant level of the instrumental response on the empirical matching function and (2) a reevaluation of some findings that have been interpreted as inconsistent with matching.

Operant Level of the Instrumental Response

The matching law developed from the unified principle of reinforcement predicts undermatching when the operant levels of the instrumental responses are nonzero (cf. Donahoe, 1977: 343; Timberlake, Chapter 23 of this volume). By undermatching is meant that the proportionality between relative response and relative reinforcer frequency is less than expected from a strict matching relation (Baum, 1974). This can be seen most readily by expressing the ratio of the probabilities of the instrumental responses for two alternatives in terms of Eq. (21.6). When this is done and only the initial probability of CRs during the two alternatives is assumed to be zero (i.e., $p_{\mathrm{CR1}} = p_{\mathrm{CR2}} = 0$), Eq. (21.8) results.

$$\frac{p'_{R1}}{p'_{R2}} = \frac{p_{R1} + \theta\,(p_{UR1})}{p_{R2} + \theta\,(p_{UR2})} \,. \tag{21.8}$$

As demonstrated elsewhere (Donahoe, 1977), Eq. (21.8) may be expressed in terms of event frequencies in Eq. (21.9).

$$\frac{R_1}{R_2} = \frac{r_b + \theta r_1}{r_b + \theta r_2} \,. \tag{21.9}$$

R_1 and R_2 are the frequencies of the two instrumental responses, r_1 and r_2 are the corresponding response-contingent reinforcement frequencies with the weighting coefficient θ, and r_b represents the experimenter-independent events that maintain the baseline (operant) level of the instrumental responses. Since r_1/r_2 must be more extreme than $(r_b + \theta r_1)/(r_b + \theta r_2)$ when the ratio of response-contingent reinforcers departs from unity, undermatching should occur with $r_b > 0$.

It should be noted that Herrnstein's matching law, independent of any presumed relation to the unified principle of reinforcement, may also accommodate undermatching when the operant levels of the instrumental responses are nonzero. According to the matching law, nonzero operant levels result from the presence in the situation of other reinforcers r_o for the instrumental response (Herrnstein, 1970). Thus undermatching with respect to the experimenter-controlled reinforcers of r_1 and r_2 is matching with respect to the total reinforcers available in the situation (that is, $r_1 + r_o$ and $r_2 + r_o$). The construct r_b in the present formulation is intended to reflect not only other reinforcers but also preexisting control of the instrumental response that arises from extraexperimental ontogenic and phylogenic sources (cf. Skinner, 1977b:1011).

A Demonstration Experiment. To illustrate the undermatching that occurs with nonzero operant levels, running in an activity wheel was selected as the instrumental response. It is well known that wheel-running occurs in food-deprived rats in the absence of experimenter-controlled reinforcers.

Six hooded rats, kept at 80 percent of their free-feeding weights, were individually placed in modified activity wheels in which retractable response levers and food cups were mounted on the stationary sidewalls. For the three lead animals a two-component concurrent schedule was arranged in which depressing the lever changed the schedule component (Findley, 1958). During one component, running was reinforced according to a variable-interval 20-second (VI-20) schedule of reinforcement in the presence of an 85-decibel white noise. During the second component, an extinction (EXT) schedule was operative in the presence of an 85-decibel, 1,000-Hertz tone. (Prior work with other animals had shown that the noise and tone stimuli did not have differential unconditioned effects on the rate of wheel-running.) A 45-milligram food pellet functioned as the reinforcer. For the other three animals the presentations of the auditory stimuli and food were yoked to the behavior of the corresponding lead animals and were thus independent of the behavior of the yoked animals. Yoking provided a control for the possible unconditioned eliciting effects of the food on lever pressing and running (Ferster and Skinner, 1957).

Prior to the institution of the concurrent schedule for the lead animals, all subjects had received training consisting of extensive exposure to multiple schedules of alternating components of VI-20 and EXT with the same discriminative stimuli (white noise and tone). Thus wheel running was stabilized prior to beginning concurrent training. When the bar was first extended at the outset of the concurrent schedule, insufficient lever presses occurred during five daily sessions to ensure frequent contact with the new contingency. Accordingly, all lead and yoked subjects were given three sessions in a standard operant chamber in which lever pressing was continuously reinforced with food (60 pellets/session). When returned to the modified running wheels, lever pressing generalized sufficiently to produce many exposures to the concurrent contingency. For the lead animals lever pressing changed the schedule component but with a 2-second changeover delay that prevented the production of a number of changes in schedule components by a single burst of lever pressing or the immediate occurrence of food after a lever press. To ensure that running came into contact with both schedules, the components alternated independently of responding if more than 2 minutes elapsed without a lever press (cf. Winton and Beale, 1971). For the yoked animals lever pressing had no programmed consequences. A session consisted of 30 alternating presentations of each component.

The findings with respect to the changeover response of lever pressing were as follows. For the lead animals, lever pressing was maintained at high levels for two rats but dropped to a lower level for the third. Subjects H1 and H3 terminated 68 percent and 84 percent of the schedule components with a changeover lever press during the last 10 days of training. However, lead subject H5 terminated only 23 percent of the components during the same period. For all lead animals most changeovers occurred during the EXT component, but substantial responding persisted during the VI-20 component as well. The percentage of VI-20 and EXT components, respectively, that were terminated by a changeover lever press were 37 and 99 percent for H1, 69 and 98 percent for H3, and 20 and 26 percent for H5. For the three yoked animals lever pressing quickly declined to near zero levels and remained there for the duration of training. Thus the changeover contingency, present only for the lead animals, was necessary for the maintenance of lever pressing.

The results obtained with the instrumental response of wheel-running indicated that undermatching occurred by both relative time and relative rate measures. For the relative time measure, the time allocated to the VI-20 component was expressed as a proportion of the total duration of the session (time during the changeover delay was excluded from the calculation). Figure 21-2 displays the relative time measures as a function of sessions for the two lead animals (H1 and H3), whose frequencies of changeover responses were sufficient to make this measure meaningful. The relative time allocated to the VI-20 component (shown in Figure 21-2) was 0.86 and 0.84 for H1 and H3, respectively, during the last 10 sessions of training. The relative rate of responding during the VI-20 component is also shown in Figure 21-2 and averaged 0.66 and 0.59 for H1 and H3 during the final 10 sessions. Although both the relative time and rate measures were somewhat variable, there was no systematic trend in the slopes of either measure over the last 10 days ($0.08 > p < 0.50$; Hollander and Wolfe, 1973: 200-204) and the 0.95 confidence interval for the mean in no case included 1.0, the point of exclusive preference for the VI-20 component. For all lead animals, including H5, the contingency between running and food produced higher rates during the VI-20 than the EXT component when compared to that found with the corresponding yoked partner. Thus differential rates of running could not be attributed to the presentation of food alone.

The foregoing experiment demonstrates that preference is not exclusive when all experimenter-controlled reinforcers are confined to one alternative if an instrumental response is used whose operant level is nonzero. Undermatching is also often observed in concurrent schedules employing pigeons with key pecking as the instrumental response (Baum, 1979). Whether the present observations are relevant to such undermatching is problematic, although increases in key pecking in the absence of a contingency between key pecking and reinforcement have been reported after exposure to food alone (e.g., Downing and Neuringer, 1976).

Violations of the Empirical Matching Law

Although instrumental responses having nonzero operant levels may produce departures from the empirical matching law, departures have most often been reported with temporally extended operants such as the time between successive occurrences of a response (an interresponse time, or IRT). In this section, findings obtained with IRTs are briefly summarized and an interpretation is offered that is consistent with a unified principle of reinforcement.

In a programmatic series of investigations conducted by Shimp and his students, concurrent schedules were studied in which the operant was defined as an IRT between successive key pecks by pigeons. Behaviors associated with criterion IRTs from t_1 to $(t_1 + \Delta t)$ seconds or from t_2 to $(t_2 + \Delta t)$ seconds served as instrumental responses and were maintained by VI schedules of food presentation. Relative preference for the shorter of the two IRTs was assessed by plotting the relative frequency of that IRT as a function of the relative reciprocal of its duration (that is, its relative harmonic mean).

Figure 21–2. The relative amount of time allocation and the relative rate of wheel running during the reinforced component of a concurrent VI 20-second, EXT schedule. Data are shown for subjects H1 and H3. The parentheses contain the baseline rates of wheel running (revolutions/minute) during sessions prior to the introduction of food reinforcement.

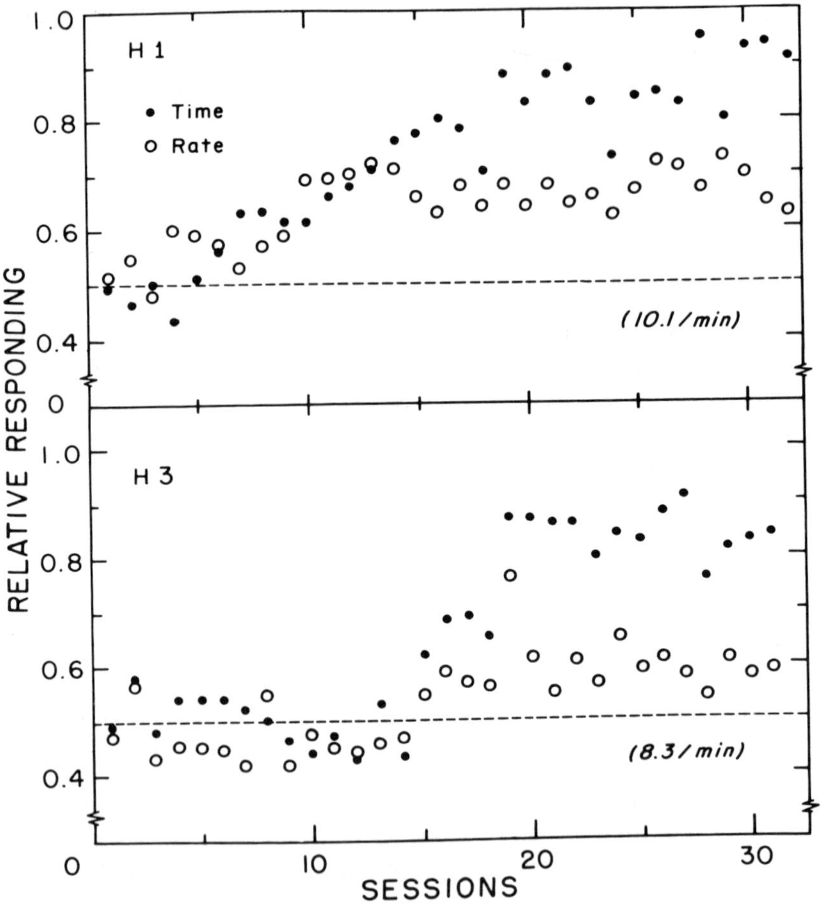

Using this methodology a number of studies found matching with temporally extended responses (e.g., Shimp, 1969b; Moffitt and Shimp, 1971), but later parametric work discovered boundary conditions on the matching relation. Specifically, when the absolute durations of the criterion IRTs were varied with the rate of reinforcement maintained above 0.5 per minute, matching was sometimes violated even though the relative reciprocal of the criterion IRT was held constant (Hawkes and Shimp, 1974). When the shorter criterion was less than 0.4-0.7 seconds undermatching occurred; when the shorter IRT criterion was greater than 5.0-6.0 seconds overmatching occurred. (Overmatching signifies that more shorter criterion IRTs were observed than predicted if matching to the relative reciprocal had occurred.) In other studies, when the rate of reinforcement was varied with the absolute duration of the criterion IRTs held constant, matching was also sometimes violated (Shimp, 1970). When the rate of reinforcement fell below approximately 0.3 per minute, the relative frequency of the shorter IRT was less than the matching value using criterion IRTs of 1.43-2.43 and 3.43-4.43 second. In summary, matching was obtained with temporally extended operants only within a limited range of criterion IRTs and reinforcement rates.

Matching is not a fundamental principle from the perspective of a unified principle of reinforcement; it is an important derivative relation found only under certain conditions. Accordingly, departures from matching are not necessarily inconsistent with a unified reinforcement principle and we may inquire how the foregoing departures may be accommodated.

First, the choice of metric for assessing matching with temporally extended operants may not have been appropriate (cf. Anderson, 1970, 1978; Donahoe, 1977). The foregoing studies used the relative reciprocal of the shorter IRT as the index of matching, a measure that assumes that the harmonic mean provides a proper metric. If behavioral time were a power function of clock time (Catania, 1970), and the geometric mean were therefore selected as the metric, then the complement of the relative logarithm would provide an appropriate index of matching. Indeed, the complement of the relative logarithm is empirically indistinguishable from the relative reciprocal in many of these experiments and, in some studies, more closely approximates the trend of the observations (Hawkes and Shimp, 1974).

Most centrally from the perspective of a unified principle of reinforcement, the analysis of experiments with temporally extended operants must incorporate the potential effects of reinforcer-elicited responses on the outcome of conditioning. The preceding experiments used a discriminated IRT procedure in which the keylight was illuminated differentially during the periods that met the temporal criteria for the two reinforced IRTs. These added stimuli were intended to hasten and amplify the effects of the IRT criteria on key pecking. Although control studies indicated that the discriminated IRT procedure yielded results comparable to those obtained without such added stimuli (Moffitt and Shimp, 1971), the control conditions were not otherwise identical to those employed in studies reporting departures from matching. The control experiments employed criteria for the shorter IRT that ranged only from 1.5-2.5 seconds to 2.0-3.0 seconds. This range does not include the very brief (e.g., 0.4-0.7 seconds) or relatively long (e.g., 5.0-6.0 seconds) criteria for the shorter IRTs that produced departures from matching. Furthermore, the control experiments employed reinforcement rates that typically exceeded 0.6 per minute—well above the value of 0.3 per minute below which matching was not found with the discriminated IRT procedure.[f]

Since the control experiments do not force the conclusion that the discriminated IRT procedure is behaviorally neutral, let us consider what possible effects the added stimuli might have upon responding. According to a unified principle of reinforcement, the conditioning rule for the instrumental procedure is that the stimuli preceding the operant acquire control over responses of two origins; those constituting the operant and those elicited by the reinforcer. In the discriminated IRT procedure, the stimuli are the wavelengths transilluminating the key during the IRT criterion and the responses are the instrumental responses of key pecking and the food-elicited response of pecking. What is suggested here is that manipulating the duration of the criterion IRT and the rate of reinforcement simultaneously varied the degree to which food-elicited pecking came under the control of the added stimuli on the key.

It is well known that a stimulus-food sequence may result in the acquisition of key pecking in the absence of any response-reinforcer contingency whatso-

f. The interpretation of these studies is further complicated by the method whereby reinforcers were scheduled. Following the delivery of a reinforcer, the nominal duration of the next interreinforcement interval was first selected and then the IRT criterion that would produce the reinforcer was specified. The same distribution of interreinforcement intervals was sampled for both IRT criteria. The consequence of this method is that the obtained length of the next interreinforcement interval is longer for the longer IRT criterion; that is, the mean of the obtained distribution of interreinforcement intervals is a direct function of the length of the IRT criterion. Where the two IRT criteria differ appreciably in duration, the difference in the obtained distribution of their interreinforcement intervals may be substantial. If the criterion IRTs also differ in their baseline probabilities of occurrence, then the correlation between the durations of the IRT criteria and the means of their respective interreinforcement distributions may be increased still further. Such a correlation is particularly troublesome for a molecular analysis of responding since it permits the time from the previous reinforcer to become a potential source of discriminative stimuli for the criterion IRT. The partial confounding of the duration of the IRT criterion with the length of the preceding interreinforcement interval may be reduced by an alternative method in which the duration of the next criterion IRT is first specified and a separate interreinforcement distribution is then sampled for each criterion IRT. By an appropriate choice of interreinforcement distributions for each IRT, the obtained distributions of interreinforcement intervals may be made nearly identical for different IRT criteria, and control over the local rate of reinforcement may be improved.

ever. This is the phenomenon of autoshaping (Brown and Jenkins, 1968; Hearst and Jenkins, 1974). Moreover, the degree of autoshaping is known to vary with such factors as the duration of the stimulus and the length of the intertrial interval (e.g., Gibbon, Baldock, Locurto, Gold, and Terrace, 1977; Perkins, Beavers, Hancock, Hemmendinger, and Ricci, 1975; Terrace, Gibbon, Farrell, and Baldock, 1975). Therefore the possibility is raised that variations in the duration of the criterion IRT and in the rate of reinforcement produced departures from matching because these manipulations simultaneously altered stimulus-reinforcer relationships that modulate autoshaping (cf. Schwartz and Gamzu, 1977).

CONCLUDING COMMENTS

An approach to the formulation of a comprehensive principle of reinforcement has been described. This formulation, termed the unified principle of reinforcement, proposes a common conditioning rule for the classical and instrumental procedures: The environment acquires control over whatever behavior occurs in a favorable temporal relation to a change in the level of responding *within* the response system activated by the reinforcing stimulus (that is, a homogeneous discrepancy). In the classical procedure the conditioned behavior consists, to a first approximation, of the responses elicited by the reinforcer but may also include CS-elicited responses (for example, orienting responses). In the instrumental procedure the conditioned behavior consists, to a first approximation, of the responses upon which the homogeneous discrepancy is contingent but also includes those reinforcer-elicited responses that are conditionable in the classical procedures.

The treatment of reinforcement given here is at most an approach to an adequate principle and not a completed theoretical analysis. One of its chief features is the identification of two types of information about which present knowledge is insufficient: (1) the occurrence of reinforcer-elicited responses during the instrumental procedure and (2) the interaction of responses of diverse origins (CS-elicited, reinforcer-elicited, and instrumental responses) during conditioning. Direct information of the first type is essential if we are to determine whether a homogeneous discrepancy underlies conditioning in both the classical and instrumental procedures. Knowledge of the second type is required if we are to understand the responses that ultimately come under environmental control as the result of conditioning.

The relation of the present approach to alternative formulations of a reinforcement principle is a complex one. In agreement with a radical behavioral perspective, instrumental conditioning is purely a procedural arrangement; that is, superstitious in the most fundamentsl sense (Skinner, 1948, 1977a). The environment acquires control of the operant, not because the operant "causes" the reinforcer but because the instrumental procedure ensures that the operant stands in a specified temporal relation to a homogeneous discrepancy. The approach is in disagreement, however, with the assumption that rate of responding provides a straightforward measure of probability of response (Skinner, 1950). The observed rate of a response is a function of the probability of that response *relative* to the probability of all other responses in the prevailing environment, a view more congenial with earlier notions such as the behavior ratio (Tolman, 1938) and recent statements such as the matching law (Herrnstein, 1970). Unlike the matching law—and other molar principles such as maximizing overall reinforcement frequency (Rachlin, Green, Kagel, and Battalio, 1976)—molar formulations are not seen as fundamental but as derivative from the moment-to-moment (molecular) interrelationships among stimuli, responses, and reinforcers (Heyman, 1979: 500; Shimp, 1969a, 1975; Silberberg, Hamilton, Ziriax, and Casey, 1978; Staddon and Motheral, 1979: 501). In agreement with Premack's analysis (1959) of reinforcement, the reinforcing function is viewed as a relative and not an absolute property of stimulation and as not transituational in character. Whether a given stimulus will function as a reinforcer is dependent upon the existence of a homogeneous discrepancy in that particular environment. Contrary to Premack's analysis, however, the difference in the probabilities of the contingent (reinforcer-elicited) and noncontingent (instrumental) responses is not held to be the origin of these characteristics of reinforcers. As already noted, it is a difference in level of responding within the response system activated by the reinforcer that is assumed to underlie conditioning. By focusing on the response system of the reinforcer, the present analysis is consistent with theoretical work that stresses the central role of response deprivation (Allison, Miller, and Wozny, 1979; Rachlin and Burkhard, 1978; Staddon, 1979; Timberlake and Wozny, 1979; Timberlake and Allison, 1974; cf. Allison, 1978). Deprivation has the effect of increasing the probability of the response elicited by the reinforcer and, therefore, of increasing the homogeneous discrepancy thought to subserve conditioning in the unified principle of reinforcement.

REFERENCE NOTES

1. Donahoe, J.W., and K.A. Stickney. Unified relational principle of reinforcement: Further implications. Paper presented at the meeting of the Psychonomic Society, San Antonio, Texas, November, 1978.

REFERENCES

Allison, J. 1978. Beyond the relational principle of reinforcement. *Journal of the Experimental Analysis of Behavior 29*: 557–60.

Allison, J.; M. Miller; and M. Wozny. 1979. Conservation in behavior. *Journal of Experimental Psychology: General 108*: 4–34.

Anderson, N.H. 1970. Functional measurement and psychophysical judgment. *Psychological Review 77*: 153–70.

_____. 1978. Measurement of motivation and incentive. *Behavioral Research Methods and Instrumentation 10*: 360–75.

Antelman, S.M., and A.R. Cagguila. 1977. Nonrepinephrine-dopamine interactions and behavior. *Science 195*: 646–53.

Baum, W.M. 1973. The correlation-based law of effect. *Journal of the Experimental Analysis of Behavior 20*: 137–54.

_____. 1974. On two types of deviation from the matching law: Bias and undermatching. *Journal of the Experimental Analysis of Behavior 22*: 231–42.

_____. 1979. Matching, undermatching, and overmatching in studies of choice. *Journal of the Experimental Analysis of Behavior 32*: 269–81.

Bindra, D. 1976. *A Theory of Intelligent Behavior.* New York: Wiley–Interscience.

Blough, D.S. 1975. Steady state data and a quantitative model of operant generalization and discrimination. *Journal of Experimental Psychology: Animal Behavior Processes 104*: 3–21.

Bolles, R.C. 1970. Species-specific defense reactions and avoidance learning. *Psychological Review 77*: 32–48.

Breland, K., and M. Breland. 1961. The misbehavior of organisms. *American Psychologist 16*: 661–64.

Brown, P.L., and H.M. Jenkins. 1968. Autoshaping of the pigeon's key-peck. *Journal of the Experimental Analysis of Behavior 11*: 1–8.

Catania, A.C. 1970. Reinforcement schedules and psychophysical judgments: A study of some temporal properties of behavior. In W.N. Schoenfeld, ed., *The Theory of Reinforcement Schedules*, New York: Appleton–Century–Crofts.

de Villiers, P.A. 1977. Choice of concurrent schedules and a quantitative formulation of the law of effect. In W.K. Honig and J.E.R. Staddon, eds., *Handbook of Operant Behavior*, Englewood Cliffs, N.J.: Prentice–Hall.

Donahoe, J.W. 1977. Some implications of a relational principle of reinforcement. *Journal of the Experimental Analysis of Behavior 27*: 341–50.

Donahoe, J.W., and M.G. Wessells. 1980. *Learning, Language, and Memory.* New York: Harper & Row.

Downing, K., and A. Neuringer. 1976. Autoshaping as a function of prior food presentation. *Journal of the Experimental Analysis of Behavior 26*: 463–69.

Dunham, P.J. 1977. The nature of reinforcing stimuli. In W.K. Honig and J.E.R. Staddon, eds., *Handbook of Operant Behavior*, Englewood Cliffs, N.J.: Prentice–Hall.

Engberg, L.A.; G.A. Hansen; R.L. Welker; and D.R. Thomas. 1972. Acquisition of keypecking via autoshaping as a function pf prior experience. *Science 178*: 1002–1004.

Estes, W.K. 1950. Toward a statistical theory of learning. *Psychological Review 57*: 94–107.

_____. 1959. Component pattern models with Markovian interpretations. In R.R. Bush and W.K. Estes, eds., *Studies in Mathematical Learning Theory*, Stanford, Calif.: Stanford University Press.

Ferster, C.B., and B.F. Skinner. 1957. *Schedules of Reinforcement.* New York: Appleton-Century-Crofts.

Findley, J.D. 1958. Preference and switching under concurrent schedules. *Journal of the Experimental Analysis of Behavior 1*: 123-44.

Frey, P.W.; R. Maisiak; and G. Dague. 1976. Unconditioned stimulus characteristics in rabbit eyelid conditioning. *Journal of Experimental Psychology: Animal Behavior Processes 2*: 175-90.

Gibbon, J.; M.D. Baldock; C. Locurto; L. Gold; and H.S. Terrace. 1977. Trial and intertrial durations in autoshaping. *Journal of Experimental Psychology: Animal Behavior Processes 3*: 264-84.

Gormezano, I., and E.J. Kehoe. 1975. Classical conditioning: Some methodological and conceptual issues. In W.K. Estes, ed., *Handbook of Learning and Cognitive Processes*, vol. 2, Hillsdale, N.J.: Lawrence Erlbaum, pp. 143-79.

Guthrie, E.R. 1935. *The Psychology of Learning.* New York: Harper.

Hawkes, L., and C.P. Shimp. 1974. Choice between response rates. *Journal of the Experimental Analysis of Behavior 21*: 109-15.

Hearst, E., and H.M. Jenkins. 1974. *Sign-Tracking: The Stimulus-Reinforcer Relation and Directed Action.* Austin, Texas: The Psychonomic Society.

Herrnstein, R.J. 1961. Relative and absolute strength of response as a function of frequency of reinforcement. *Journal of the Experimental Analysis of Behavior 4*: 267-72.

____. 1970. On the law of effect. *Journal of the Experimental Analysis of Behavior 13*: 243-66.

____. 1979. Derivatives of matching. *Psychological Review 86*: 486-95.

Heyman, G.M. 1979. Matching and maximizing in concurrent schedules. *Psychological Review 86*: 496-500.

Holland, P.C. 1977. Conditioned stimulus as a determinant of the form of the Pavlovian conditioned response. *Journal of Experimental Psychology: Animal Behavior Processes 3*: 77-104.

____. 1980. CS-US interval as a determinant of the form of Pavlovian appetitive conditioned responses. *Journal of Experimental Psychology: Animal Behavior Processes 6*: 155-74.

Hollander, M., and D.A. Wolfe. 1973. *Nonparametric Statistical Methods.* New York: Wiley.

Hull, C.L. 1943. *Principles of Behavior.* New York: Appleton-Century-Crofts.

Kamin, L.J. 1968. "Attention-like" processes in classical conditioning. In M.R. Jones, ed., *Miami Symposium on the Prediction of Behavior: Aversive Stimulation.* Miami, Fla.: Miami University Press.

____. 1969. Predictability, surprise, attention, and conditioning. In B.A. Campbell and R.M. Church, eds., *Punishment and Aversive Behavior*, New York: Appleton-Century-Crofts.

Kamin, L.J., and S.J. Gaioni. 1974. Compound conditioned emotional response conditoning with differentially salient elements in rats. *Journal of Comparative and Physiological Psychology 87*: 591-97.

Kimble, G.A. 1961. *Hilgard and Marquis' Conditioning and Learning.* Englewood Cliffs, N.J.: Prentice-Hall.

Konorski, J. 1967. *Integrative Activity of the Brain.* Chicago: University of Chicago Press.

Konorski, J., and S. Miller. 1937. On two types of conditioned reflex. *Journal of General Psychology 16*: 264–72.

Kremer, E.F. 1978. The Rescorla–Wagner model: Losses in associative strength in compound conditioned stimuli. *Journal of Experimental Psychology: Animal Behavior Processes 4*: 22–36.

Mackintosh, N.J., and A. Dickinson. 1979. Instrumental (type II) conditioning. In A. Dickinson and R.A. Boakes eds., *Mechanisms of Learning and Motivation: Memorial Volume for Jerzy Konorski*, Hillsdale, N.J.: Lawrence Erlbaum, 143–69.

Maier, S.F., and M.E.P. Seligman. 1976. Learned helplessness: Theory and evidence. *Journal of Experimental Psychology: General 105*: 3–46.

Martin, I., and A.B. Levey. 1969. *The Genesis of the Classical Conditioned Response.* London: Pergamon Press.

Miller, N.E., and A. Carmona. 1967. Modification of a visceral response, salivation in thirsty dogs, by instrumental conditioning with water reward. *Journal of Comparative and Physiological Psychology 63*: 1–6.

Moffitt, M., and C.P. Shimp. 1971. Two-key concurrent paced variable-interval schedules of reinforcement. *Journal of the Experimental Analysis of Behavior 16*: 39–49.

Pearce, J.M., and A. Dickinson. 1978. Overshadowing the instrumental conditioning of a lever-press response by a more valid predictor of the reinforcer. *Journal of Experimental Psychology 4*: 356–67.

Perkins, C.C.; W.O. Beavers; R.A. Hancock; D. Hemmendinger; and J.A. Ricci. 1975. Some variables affecting rate of keypecking during response-independent procedures (autoshaping). *Journal of the Experimental Analysis of Behavior 24*: 59–72.

Premack, D. 1959. Toward empirical behavior laws. I. Positive reinforcement. *Psychological Review 66*: 219–33.

_____. 1965. Reinforcement theory. In D. Levine, ed., *Nebraska Symposium on Motivation*, Lincoln: University of Nebraska Press.

_____. 1971. Catching up with common sense or two sides of a generalization: Reinforcement and punishment. In R. Glaser, ed., *The Nature of Reinforcement*, New York: Academic Press.

Prokasy, W.F., and I. Gormezano. 1979. The effects of US omission in classical aversive and appetitive conditioning of rabbits. *Animal Learning and Behavior 7*: 80–88.

Rachlin, H., and B. Burkhard. 1978. The temporal triangle: Response substitution in instrumental conditioning. *Psychological Review 85*: 22–47.

Rachlin, H.; L. Green; J.H. Kagel; and R.C. Battalio. 1976. Economic demand theory and psychophysical studies of choice. In G. Bower, ed., *The Psychology of Learning and Motivation*, vol. 10, New York: Academic Press.

Rachlin, H.; J.H. Kagel; and R.C. Battalio. 1980. Substitutability in time allocation. *Psychological Review 87*: 355–74.

Rescorla, R.A. 1978. Some implications of a cognitive perspective on Pavlovian conditioning. In S.H. Hulse, H. Fowler, and W.K. Honig, eds., *Cognitive Processes in Animal Behavior*, Hillsdale, N.J.: Lawrence Erlbaum.

Rescorla, R.A., and A.R. Wagner. 1972. A theory of Pavlovian conditioning: Variations in the effectiveness of reinforcement and nonreinforcement. In

A.H. Black and W.F. Prokasy, eds., *Classical Conditioning*, vol. 2, *Current Research and Theory*. New York: Appleton-Century-Crofts.

Rozin, P., and J.W. Kalat. 1971. Specific hungers and poison avoidance as adaptive specializations of learning. *Psychological Review 78*: 459-86.

Schwartz, B., and E. Gamzu. 1977. Pavlovian control of operant behavior: An analysis of autoshaping and its implications for operant conditioning. In W.K. Honig and J.E.R. Staddon, eds., *Handbook of Operant Behavior*, Englewood Cliffs, N.J.: Prentice-Hall.

Seligman, M.E.P., and J.L., Hager, eds. 1972. *Biological Boundaries of Learning*. Englewood Cliffs, N.J.: Prentice-Hall.

Sheffield, F.D. 1965. Relation between classical conditioning and instrumental learning. In W.F. Prokasy, ed., *Classical Conditioning: A Symposium*, Englewood Cliffs, N.J.: Prentice-Hall.

Shettleworth, S.J. 1972. Constraints on learning. In D.S. Lehrman, R.A. Hinde, and E. Shaw, eds., *Advances in the Study of Behavior*, New York: Academic Press.

Shimp, C.P. 1969a. Optimal behavior in free-operant experiments. *Psychological Review 76*: 97-112.

_____. 1969b. The concurrent reinforcement of two interresponse times: The relative frequency of an interresponse time equals its relative harmonic length. *Journal of the Experimental Analysis of Behavior 12*: 402-11.

_____. 1970. The concurrent reinforcement of two interresponse times. *Journal of the Experimental Analysis of Behavior 13*: 1-8.

_____. 1975. Choice behavior in animals. In W.K. Estes, ed., *Handbook of Learning and Cognitive Processes*, vol. 2, *Conditioning and Behavior Theory*, Hillsdale, N.J.: Lawrence Erlbaum.

Silberberg, A.; B. Hamilton; J.M. Ziriax; and J. Casey. 1978. The structure of choice. *Journal of Experimental Psychology: Animal Behavior Processes 4*: 368-98.

Skinner, B.F. 1938. *The Behavior of Organisms*. New York: Appleton-Century-Crofts.

_____. 1948. Superstition in the pigeon. *Journal of Experimental Psychology 38*: 168-72.

_____. 1950. Are theories of learning necessary? *Psychological Review 57*: 193-216.

_____. 1977a. The force of coincidence. In B.C. Etzel, J.M. LeBlanc, and D.M. Baer, eds., *New Developments in Behavioral Psychology*, Hillsdale, N.J.: Lawrence Erlbaum, pp. 3-6.

_____. 1977b. Herrnstein and the evolution of behaviorism. *American Psychologist 33*: 1006-12.

Smith, M.C. 1968. CS-US interval and US intensity in classical conditioning of the rabbit's nictitating membrane response. *Journal of Comparative and Physiological Psychology 66*: 679-87.

Solomon, R.L. 1980. The opponent-process theory of acquired motivation. *American Psychologist 35*: 691-712.

Solomon, R.L., and J.D. Corbit. 1974. An opponent-process theory of motivation. I. Temporal dynamics of affect. *Psychological Review 81*: 119-45.

St. Claire-Smith, R. 1979. The overshadowing and blocking of punishment. *Quarterly Journal of Experimental Psychology 4*: 51-61.

Staddon, J.E.R. 1977. On Herrnstein's equation and related forms. *Journal of the Experimental Analysis of Behavior 28*: 163-70.

_____. 1979. Operant behavior as adaptation to constraint. *Journal of Experimental Psychology: General 108*: 48-67.

Staddon, J.E.R., and S. Motheral. 1979. Response independence, matching, and maximizing: A reply to Heyman. *Psychological Review 86*: 501-505.

Terrace, H.S.; J. Gibbon; L. Farrell; and M.D. Baldock. 1975. Temporal factors influencing the acquisition and maintenance of an autoshaped key-peck. *Animal Learning and Behavior 3*: 53-62.

Timberlake, W., and J. Allison. 1974. Response-deprivation: An empirical approach to instrumental performance. *Psychological Review 81*: 146-64.

Timberlake, W., and D.L. Grant. 1975. Autoshaping in rats to the presentation of another rat predicting food. *Science 190*: 690-92.

Timberlake, W., and M. Wozny. 1979. Reversibility of reinforcement between eating and running by schedule changes: A comparison of hypotheses and models. *Animal Learning and Behavior 7*: 461-69.

Tolman, E.C. 1938. The determiners of behavior at a choice point. *Psychological Review 45*: 1-41.

Trapold, M.A., and J.B. Overmier. 1972. The second learning process in instrumental learning. In A.H. Black and W.F. Prokasy, eds., *Classical Conditioning*, vo. 2, *Current Research and Theory*, New York: Appleton-Century-Crofts.

Valenstein, E.S.; V.C. Cox; J.W. Kakolewski. 1969. The hypothalamus and motivated behavior. In J.T. Tapp, ed., *Reinforcement and Behavior*, New York: Academic Press, pp. 242-85.

vom Saal, W., and H.M. Jenkins. 1970. Blocking the development of stimulus control. *Learning and Motivation 1*: 52-64.

Williams, B.A. 1975. The blocking of reinforcement control. *Journal of the Experimental Analysis of Behavior 24*: 215-25.

Williams, D.R., and H. Williams. 1969. Auto-maintenance in the pigeon: Sustained pecking despite contingent nonreinforcement. *Journal of the Experimental Analysis of Behavior 12*: 511-20.

Winton, A.S.W., and I.L.Beale. 1971. Peak shift in concurrent schedules. *Journal of the Experimental Analysis of Behavior 15*: 73-82.

Zanich, M.L., and H. Fowler. 1978. Transfer from Pavlovian appetitive to instrumental conditioning: Signalling versus discrepancy interpretations. *Journal of Experimental Psychology: Animal Behavior Processes 4*: 37-49.

22 CONSTRAINTS ON PERFORMANCE IN TWO ELEMENTARY PARADIGMS

James Allison

Two familiar paradigms are analyzed in terms of their experimental constraints on behavior under various schedules. Response rate, amount of responding per unit of time, is the focal measure of performance throughout. In the *controlled-time* paradigm the experimenter holds session duration constant across sessions and the subject varies rate by varying the amount of responding. In the *controlled-amount* paradigm the experimenter holds the amount of responding constant and the subject varies rate by varying the time to complete that amount. Theories of performance impose further constraints by specifying which outcomes will be observed among those allowed by the purely experimental constraints. Several theories are examined in light of pertinent data. They include labor supply theory, conservation theory, and theories proposing that performance under the schedule represents a minimum deviation from performance in a baseline condition without constraint by the schedule.

Each of the two paradigms takes its name from the way the session ends. In the controlled-time paradigm the experimenter ends each session a fixed time after starting; in controlled-amount each session ends after a fixed amount of responding (Allison and Timberlake, 1975). Should each session run for exactly one hour, regardless of the number of lever presses emitted under some fixed-ratio schedule, the procedure would fall into the controlled-time category. But were each session ended upon delivery of the fiftieth food pellet, the procedure would fall into the controlled-amount category, as would the runway trial that ends as soon as the rat has run to the goal box and eaten the food therein. The focal measure of performance in both paradigms is the average rate of responding over the session as a whole, amount of responding per unit of time. The subject varies rate by varying the amount of responding in the controlled-time session and the time to complete the response requirement in the controlled-amount session.

Each paradigm sees wide use in its pure form, and sometimes appears in combination with the other. I intend to compare the two in terms of their experimental constraints on performance, along with some theoretical constraints drawn from psychology and economics.

As a key example of experimental constraints, consider a hypothetical controlled-time setting where the rat runs in a wheel for water. In each contingency session, 1,200 seconds long, the schedule requires *I* seconds of running for access to water and *C* seconds of drinking to free the wheel for the next bout of running. Figure 22-1 plots total time spent drinking against total time spent running and defines the possible results in terms of five vectors that represent five different schedules (see Staddon, 1979: fig. 1, for a similar representation). The

Figure 22-1. Hypothetical example of experimental constraints in the controlled-time paradigm. Time spent drinking (response *c*) as a function of time spent running (response *i*) on five schedules, and in baseline without constraint by a schedule (open circle). The slope of each vector is *C*/*I* (drink time/run time, as specified by the schedule). Broken lines demarcate facilitation and suppression relative to baseline rates.

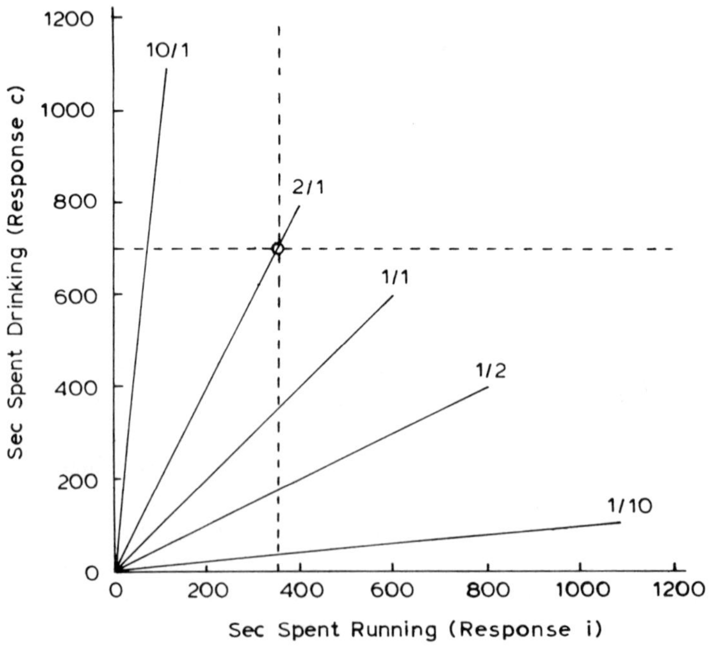

slope of each vector is C/I, the drink requirement divided by the run requirement. If the rat fails to run, the data point falls at the origin (0, 0). Each time the rat completes the scheduled sequence of response requirements, $I + C$, the point moves up the vector toward a high rate of responding. Given no overrunning of either requirement and no fractional completion of $I + C$, the point will lie on the vector. Otherwise, the point may lie a little above or below the a priori vector with slope C/I. Letting N signify the number of times the rat completes the sequence, the rat would total roughly NI seconds of running, and NC seconds of drinking. In a T-second session, the average rates for the session as a whole would come to NI/T, and NC/T.

The length of the vector reflects the maximum rates of responding, which depend in our example on I, C, and the duration of the session. If the rat spent the whole session running and drinking on the 2/1 schedule, the maximum amounts would be 800 seconds of drinking, 400 of running. The vector could be shortened by taking into account any inability to run and drink without pause.

Several theorists have proposed models for performance under this sort of schedule (Allison, 1976; Allison, Miller, and Wozny, 1979; Heth and Warren, 1978; Mazur, 1975; Premack, 1971; Rachlin and Burkhard, 1978; Staddon, 1979; Timberlake and Allison, 1974), each attributing special importance to a paired baseline condition where the subject may perform each of the two responses without constraint by the schedule. In the present example, the wheel would be free to rotate and the tube freely available throughout the paired baseline session, also 1,200 seconds long. The open circle in Figure 22-1 shows time spent running and drinking in the paired baseline session, and the broken lines through the basepoint divide the space into four regions often used in classifying performance under the schedule. The upper right quadrant denotes facilitation of both responses relative to basepoint; the lower left, suppression of both. The two remaining quadrants denote facilitation of one response, suppression of the other. Each of the five schedules could produce some of these effects, but none could produce all.

Theory imposes further constraints by specifying which of the possible outcomes will occur. As a primary example, the left panel of Figure 22-2 shows constraints from labor supply theory, plotting total wages per day against two complementary variables on the horizontal axis, total hours of work per day and the complement of work, total hours of leisure per day. The example assumes 15 hours each day for work plus leisure. The five vectors represent five different wage rates, dollars/hour, each analogous to the ratio C/I in a fixed-ratio schedule. (Vector lengths are arbitrary.) Each point in the space represents a particular combination of daily work hours, daily leisure hours, and daily dollars earned (cf. Chapter 18 by Green, Kagel, and Battalio).

The theoretical constraints come from the nonintersecting convex curves called "indifference curves." All points on a curve are equally attractive; thus, much leisure with little income is equivalent to less leisure with more income.

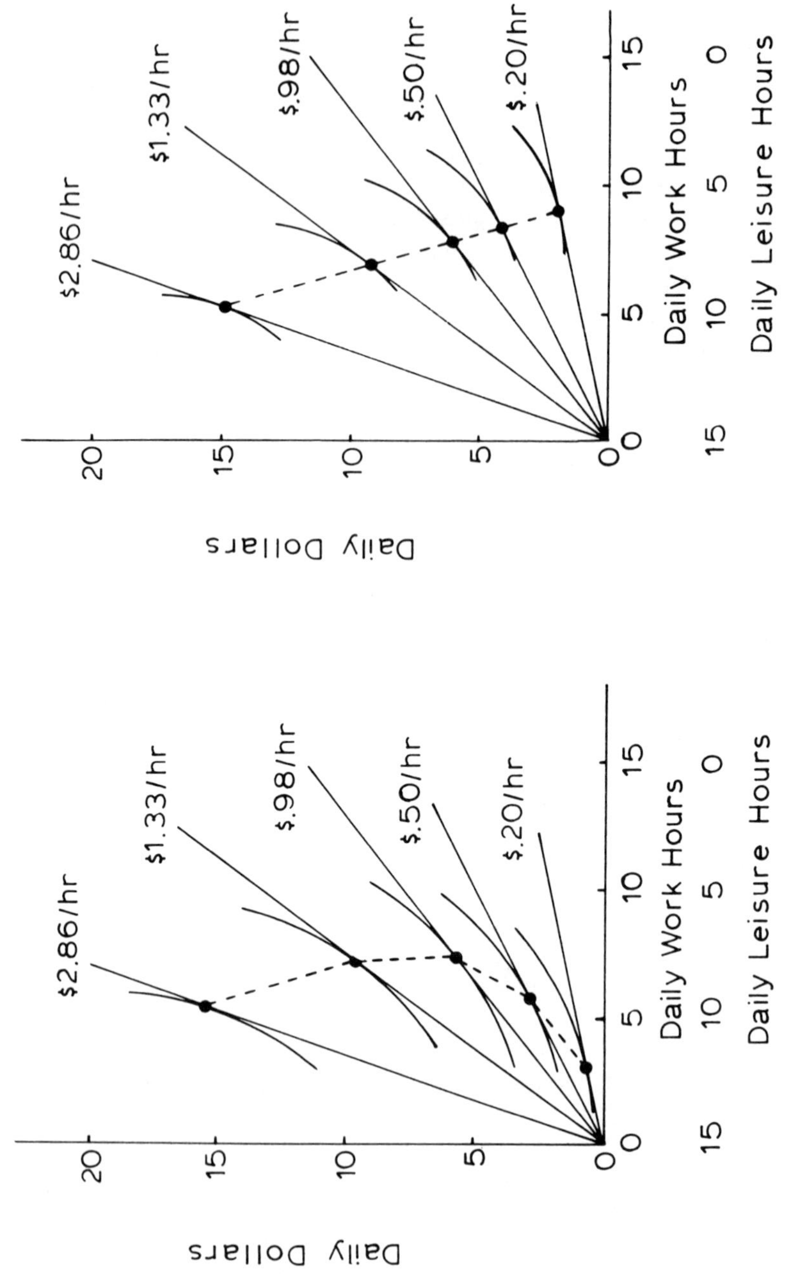

Figure 22-2. Theoretical labor supply curves. Daily income as a function of daily hours of work or leisure. The slope of each vector is the wage rate, dollars/hour. Convex curves are indifference curves. The broken lines connecting the points of equilibrium represent the theoretical labor supply curves.

However, the worker prefers any higher curve to any lower curve, because every point on the higher one offers more utility, or psychological satisfaction, than any point on the lower one. As proof, look at the top two indifference curves. At 10 hours of leisure the upper curve offers more daily dollars than the lower, so the upper point offers more attraction than the lower. The indifference property therefore implies that the worker prefers each point on the upper curve to any point on the lower curve.

Behavioral predictions follow from the same logic. Under the constraint of a particular wage rate, the worker will maximize utility by moving up the wage rate vector to the highest indifference curve attainable on that vector. This unique point of equilibrium is the point at which the vector is tangent to an indifference curve.

In the left panel, the broken lines joining the equilibrium points comprise the well-known backward bending labor supply curve of neoclassical economic theory (Awh, 1976: 96; Samuelson, 1976: 580). Starting at the lowest wage rate, \$0.20/hr, the curve implies that a modest increase in the wage rate, say to \$0.50/hr, will induce the worker to work longer, giving up some daily leisure in exchange for more daily dollars. But if the wage rate is fairly high to begin with, a further increase will induce the worker to work less, taking more daily leisure and more daily dollars.

In theory the curve changes slope from positive to negative because of a shifting balance between two opposing factors: the substitution effect and the income effect. As the wage rate rises, work time becomes more valuable and leisure time more costly in terms of lost opportunity to work. The worker would therefore substitute more work for leisure than before. But the additional income gained would enable the worker to buy more consumer goods than before, including the good known as leisuretime activities. This income effect therefore opposes the substitution effect and leads the worker to do less work than if governed by the substitution effect alone. If the substitution effect exceeds the income effect, the labor supply curve will have positive slope. But if the income effect exceeds the substitution effect, the curve will have negative slope. The theory assumes a relatively large substitution effect at a relatively low wage rate because an additional dollar should have more value for the poor than for the rich. Relative to that of leisure, the marginal utility of money increases as wealth decreases.

The backward bending labor supply curve implies that the greatest supply of labor will come at an intermediate wage rate. Analogously, to get the highest rate of lever pressing from the rat, the fixed-ratio requirement I must be kept large, but not too large, and the size of the food pellet C small, but not too small. As the wage rate ratio C/I is decreased, the risk of an extinction-like drop in instrumental responding is higher. Increasing C/I entails a higher risk that the rat will eat its fill on a small amount of labor. In terms of theoretical predictions, many will recognize some similarities between this economic theory and psychological theory from Rachlin and Burkhard (1978: fig. 17) and Staddon (1979), who derive a function much like the backward bending labor supply curve from rather different theoretical premises.

In evaluating this class of theories perhaps we can ask whether economic studies of the labor supply market reveal a backward bend in the empirical labor supply curve. Economists who specialize in labor supply claim that the answer is no: Studies of the labor supply market, both experimental and nonexperimental, reveal no backward bend; the curve has negative slope throughout (Watts and Rees, 1977:12). In other words, the economic evidence suggests that higher wage rates go hand in hand with less labor, as measured by such indices as hours worked by individual workers and rate of participation in the work force. Of course the theoretical curve could be endowed with negative slope throughout; the right-hand panel of Figure 22-2 shows that this could be done simply by drawing the indifference curves a little differently than in the left-hand panel.

The curve on the right imposes theoretical constraints similar in form to those imposed by the earliest version of the conservation model for this kind of schedule (Allison, 1976; Allison et al., 1979; Shapiro and Allison, 1978). The model assumes a tradeoff between the two responses controlled by the schedule, such that an increase in one produces a linear decrease in the other (for a conservation model consistent with the backward bending labor supply curve, see Allison, 1981b). Figure 22-3 applies the analysis to an experiment (Allison et al., 1979) much like our hypothetical example on running and drinking. The six thirsty rats could hold a lever or drink from a water tube during paired baseline sessions with tube and lever freely available throughout. They performed the same two responses in contingency sessions under nine different schedules requiring *I* seconds of lever holding for each access to water and *C* seconds of drinking for the next access to the lever, all sessions 60 minutes long. In accordance with the model, drinking decreased linearly as lever holding increased; higher wage rates were associated with less work, more leisure, and more income in the form of drinking time. The predicted effects include the facilitation of lever holding and the suppression of drinking relative to the open-circle basepoint in Figure 22-3. The model would predict both of these effects for any schedule vector passing beneath the basepoint; because the theoretical line descends through the basepoint, performance under each schedule should lie in the lower right-hand quadrant.

Figure 22-4 applies the same analysis to an unusual variation on the conventional fixed-ratio schedule (Collier, Hirsch, and Hamlin, 1972). The schedule requirements ranged from 1 lever press for food up to 5,120 presses, but the rat determined the size of the ensuing meal, because the food bin stayed open until a 10-minute pause in eating. The rats continued to respond as the response requirement rose to levels that would normally discourage continued responding, so the descending linear function predicted by the model fit the data reasonably well. Note too that the results of Figure 22-4 seem to contradict the notion that

Figure 22-3. Total drinking time as a function of total lever-holding time in baseline (open circle) and on nine cumulative time schedules (closed circles). Each schedule is represented by a vector with slope *C/I* (drink time/lever time, as specified by the schedule). The 20/10 and 40/20 schedules are represented by the same vector, as are the 20/20 and 30/30 schedules. (Based on Allison et al., 1979.)

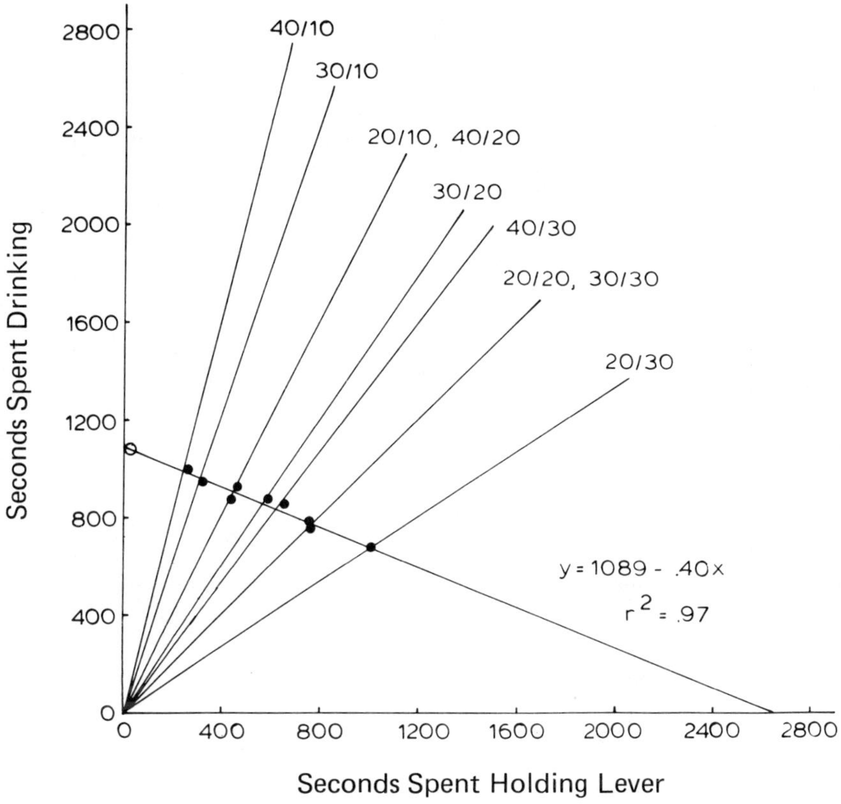

the rat performs on schedule so as to minimize its deviation from basepoint performance free of schedule constraints (Heth and Warren, 1978; Rachlin and Burkhard, 1978; Staddon, 1979). For example, if the meals taken on FR-160 had been as large as those taken on FR-1,280, the rats could have approached the basepoint more closely than they actually did on FR-160 (Allison, 1981a).

Controlled-time experiments with variable-interval schedules are surprisingly scarce. The experiment of Figure 22-5, recently completed in my laboratory and reported elsewhere (Allison, 1980; Allison, Note 1), used VI 7-second and VI 14-second schedules in 60-minute sessions. I selected these relatively short

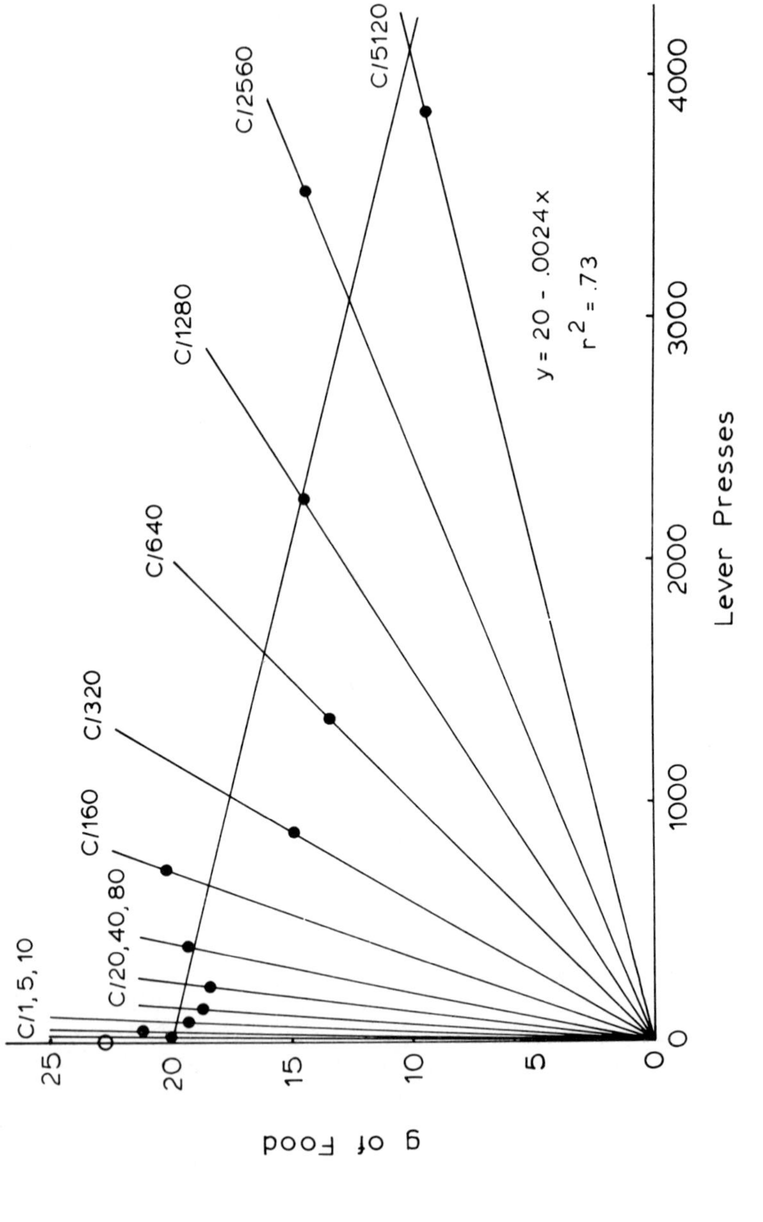

Figure 22-4. Total food consumed as a function of total lever presses in baseline (open circle) and on 12 fixed-ratio schedules (closed circles). Each schedule is represented by a vector with slope *C/I* (mean size of the meal/lever presses required for access to the food bin). (Based on Collier et al., 1972.)

Figure 22-5. Total licks as a function of total lever presses in baseline (open circles) and on variable-interval schedules (closed symbols). Each schedule is represented by a vector whose slope is the number of licks/press as specified by the schedule, 50/1 or 25/1. Each 25/1 vector has two terminal points; the lower point approximates the maximum number of licks attainable on the 14-second schedule, the upper point the maximum attainable on the 7-second schedule.

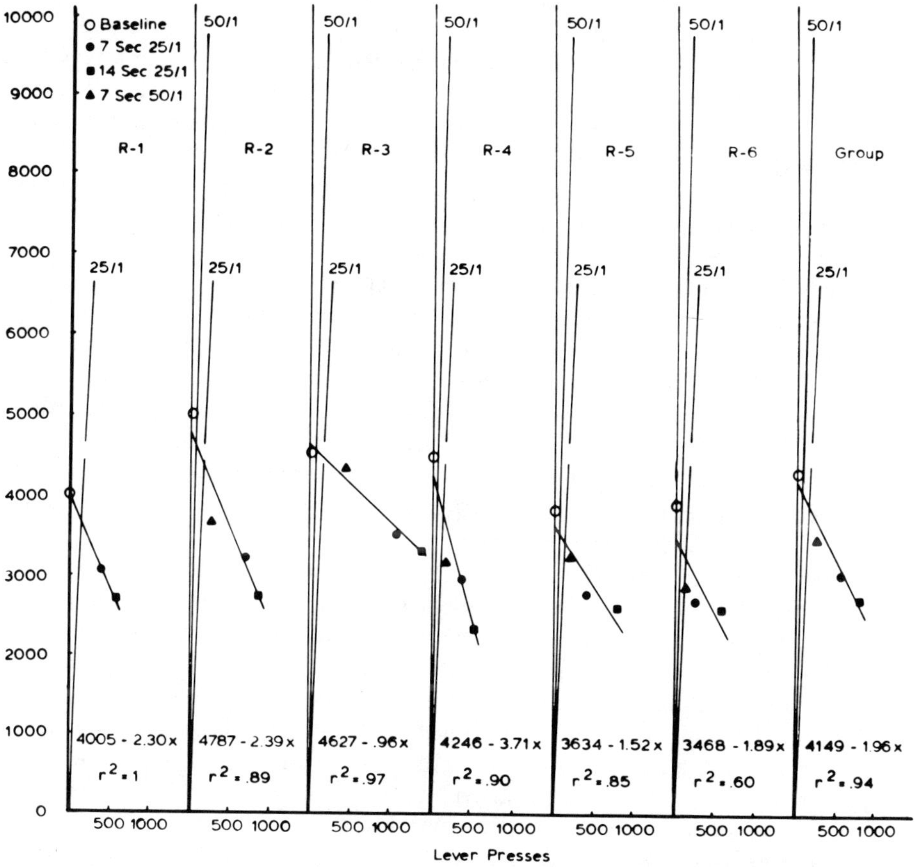

intervals so as to provide nonartifactual tests of the suppression of drinking predicted by the model (if the interval is too long, the schedule may provide so few setups that the rat could not possibly equal or exceed its basepoint number of licks). The six rats got all of their daily water by pressing a lever once for 25 licks on VI 7-second or VI 14-second schedule; five of the six were tested on a third VI 7-second schedule, one press for 50 licks. Each of the 25/1 vectors in Figure 22–5 has two terminal points; the higher one represents the VI 7-second schedule, the lower VI 14-seconds. Relative to the open-circle basepoints, each of the six showed a nonartifactual suppression of licking on VI 7-second 25/1, and each of the five on VI 14-second 50/1. Nonartifactual tests on VI 14-second 25/1 were possible for three of the six (R-1, R-5, and R-6), each of which showed the expected suppression of licking.

The grouped data on the right show that licks decreased linearly as presses increased. Note that none of the rats approached the steepest vector attainable on schedule, because each performed a great many unscheduled lever presses in the time before the next setup was due. As a result, they deviated widely from their basepoints. Like those of Collier et al. (1972), such results may be difficult to reconcile with the minimum deviation models. Perhaps they would have approached the basepoint more closely if told when the next setup was due (Segal, 1962).

Like the variable-interval schedule, the concurrent fixed-ratio schedule allows the subject some control over the slope of the vector actually traced through the course of the session. It does so by offering a choice between two alternative components, such as 60 licks for 2 lever presses at one station, and 60 for 8 at another station nearby. The slopes of the possible vectors would range from $60/2 = 30$ to $60/8 = 7.5$, each extreme attainable by responding exclusively on either component, never sampling the other. The possible vectors include a large number with intermediate slopes, each attainable by responding sometimes on one component, sometimes the other.

The conservation model for such a schedule (Shapiro and Allison, 1978) predicts again that total licks will decrease linearly with an increase in total presses. It does not predict how the rat will favor the two components, but other models do. For example, if the basepoint lies above the steeper of the two schedule vectors, the rat can minimize its deviation from the basepoint by responding exclusively on the component with the steeper vector, ignoring the other component.

Many such experiments have found exclusive or near-exclusive responding (e.g., Herrnstein, 1958; Herrnstein and Loveland, 1975; Striefel, 1972) but in others rats sampled both components extensively (Lea and Roper, 1977; Shapiro and Allison, 1978). As a result, they failed to approach the basepoint as closely as they might have. An example is shown in Figure 22–6, based on data from Shapiro and Allison (1978). The figure plots, for individual rats, total licks at a water tube against total lever presses in baseline (open circles) and on schedule (closed circles). Although each of the eight basepoints lay above the steepest

Figure 22-6. Total licks as a function of total lever presses in baseline (open circles) and on concurrent fixed-ratio schedules (closed circles). Each component of the schedule is represented by a vector whose slope is the number of licks/press as specified by the schedule. Proportions refer to the proportion of times the rat sampled a particular component. The tail of each closed circle points to the vector the rat would have traced by sampling five times the component with the shallower vector. (Based on Shapiro and Allison, 1978.)

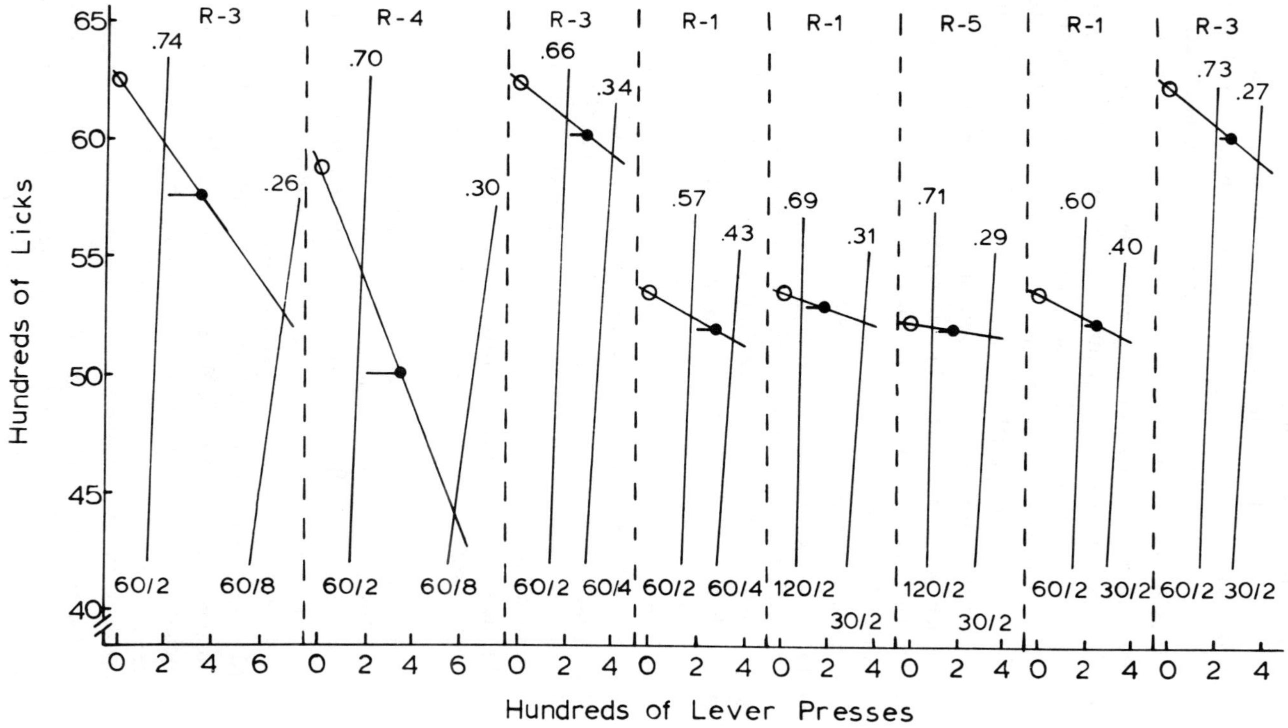

vector attainable, in each case the rat chose to trace a vector less steep than it could have traced, sampling each component liberally throughout the course of the session. Note that in all cases the schedule facilitated lever pressing and suppressed licking, as the conservation model predicts. (The length of each component vector is arbitrary, but none of the suppression effects shown in Figure 22-6 was an artifactual ceiling effect.)

Why did these rats sample so freely? The literature suggests that indeed we should expect a good deal of sampling, given absolute response requirements as small as these (Shapiro and Allison, 1978). Necessary searching was probably not the cause. Each closed circle in Figure 22-6 has a tail pointing to the vector the rat would have traced by sampling each component five times, otherwise responding exclusively on the component with the steeper vector. Note that each such vector is steeper, thus nearer the basepoint, than the one actually traced. These rats simply did more foraging than needed.

Lea and Roper (1977) minimized the forage factor by keeping each component at the same station from session to session. Figure 22-7 shows their results, plotting total food pellets against total lever presses. The four pairs of components pitted 1 against 8 lever presses per pellet, 8 against 16, 8 against 11, and 6 against 8. Each vector has arbitrary length. The authors reported no basepoints, but each basepoint surely would have fallen above the steepest vector attainable under the schedule. Each of the four tests revealed some sampling of each component, and three revealed a considerable deviation from the steepest vector attainable. Figure 22-8 combines the four schedules and shows a fairly decent linear fit. These two experiments suggest that the rat might minimize its deviation if the basepoint lay *between* the two component vectors, but I know of no such experiment.

In the second paradigm, controlled-amount, the experimenter ends each trial or session on completion of a fixed amount of responding; the subject varies rate by varying the latency of completion. Its characteristic experimental constraints can be understood through a simple but general hypothetical example with 1 lever press and 100 licks required on each trial. With lever pressing instrumental, the rat might press the lever 10 seconds after the trial began, start licking 5 seconds after the water tube appeared, and continue at 5 licks/second. The completion times would come to 10 seconds and 35 seconds, the average session rates 1/10 = 0.1 presses/second and 100/35 = 2.9 licks/second.

In this particular example the experimenter ends each trial as soon as the rat completes the scheduled sequence, I lever presses followed by C licks, exactly once. In other words, the experimenter sets $N = 1$. But for any predetermined value of N, the two response totals on each trial are NI and NC. To specify the response rates, t_I is defined as the time the rat takes to complete NI in the contingency trial and t_C as the time to complete NC. Then the two average rates for the contingency trial come to NI/t_I, and NC/t_C.

Figure 22-7. Total 45-milligram food pellets as a function of total lever presses on four concurrent fixed-ratio schedules. Each component of the schedule is represented by a vector whose slope is the number of pellets/press as specified by the schedule. (Based on Lea and Roper, 1977.)

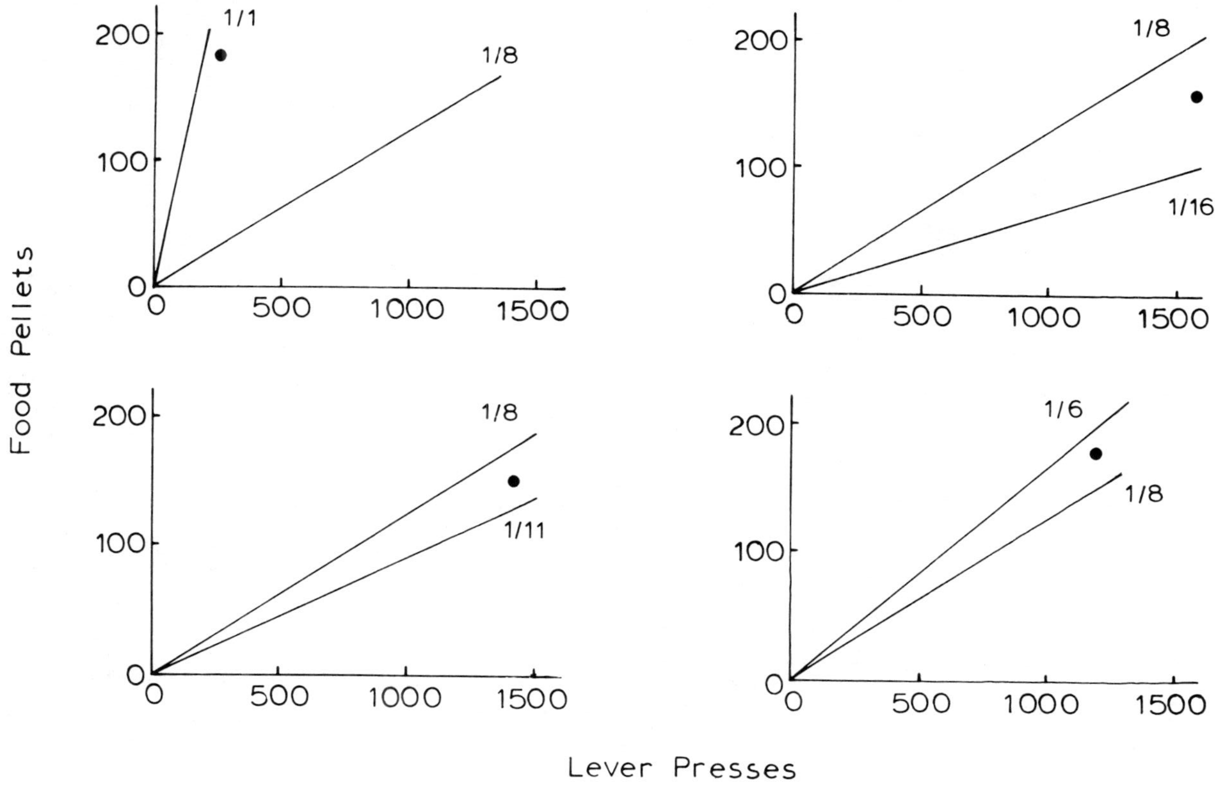

Figure 22-8. Total 45-milligram food pellets as a function of total lever presses on four concurrent fixed-ratio schedules. Reading from left to right, the requirements were 1 press against 8 presses per pellet, 6 against 8, 11 against 8, and 16 against 8. (Based on Lea and Roper, 1977.)

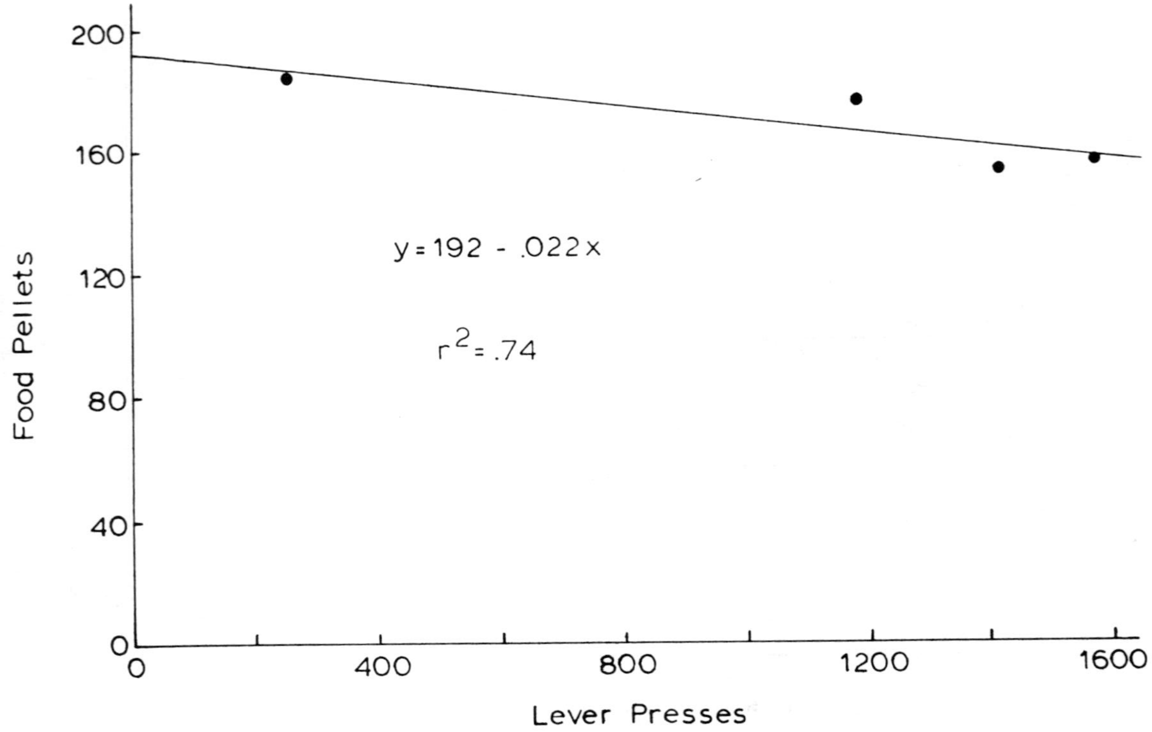

Figure 22-9. Hypothetical example of experimental and theoretical constraints in the controlled-amount paradigm. Rate of licking (response c) as a function of rate of lever pressing (response i). The upward vector with slope NC/t_I represents the experimental constraints imposed by the schedule. Baseline rates are represented by the open circle. Broken lines demarcate facilitation and suppression relative to the baseline rates. Possible outcomes with response i instrumental lie to the right of the vector; possible outcomes with response c instrumental lie to the left of the vector. The line of theoretical constraint imposed by the conservation model descends through the basepoint (the slope shown is arbitrary).

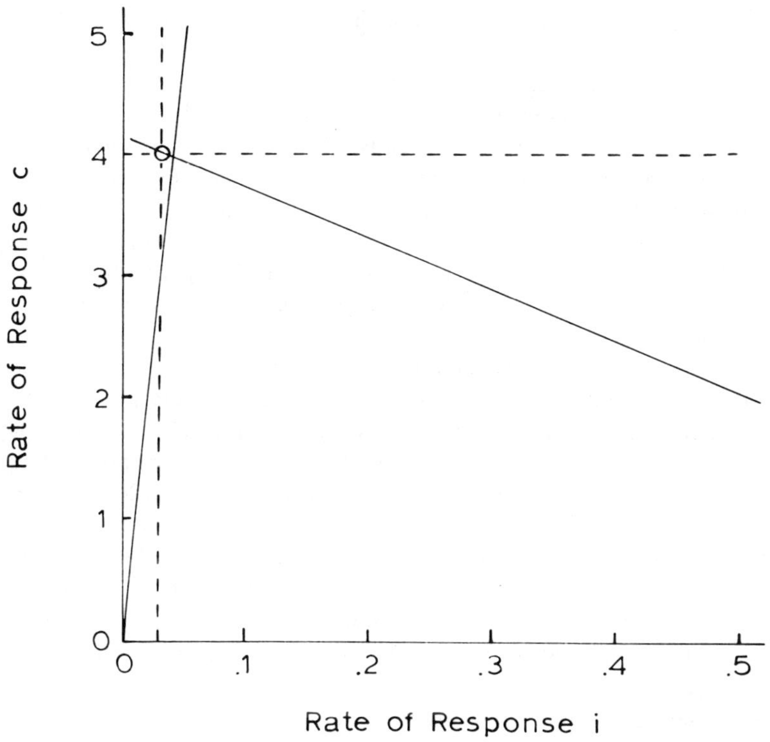

Figure 22-9 shows the experimental constraints, plotting the average rate of licking against the average rate of lever pressing, with an upward vector dividing the space into two sets of mutually exclusive possibilities. With lever pressing instrumental and licking contingent, the possible outcomes lie to the right of the upward vector. The slope of the vector is C/I, the number of licks (100) divided by the number of lever presses (1) specified by the schedule requirements. The length of the vector is arbitrary but would depend in practice on the highest response rate the subject could manage. To understand the slope, recall that if lever pressing is instrumental, the rat cannot start its contingent licking until it has pressed the lever once—at time t_I. At the absolute (impossible) best, the rat would both start and finish licking at time t_I. The maximum rate of licking, NC/t_C, is therefore NC/t_I. For a particular rate of instrumental lever pressing, NI/t_I, the maximum rate of contingent licking is therefore NC/t_I. It follows that the slope of the vector with lever pressing instrumental is $(NC/t_I)/(NI/t_I)$, or C/I—the same as in the controlled-time paradigm.

What if licking were instrumental and lever pressing contingent? In that case, for a particular rate of instrumental licking, NC/t_C, the maximum rate of contingent lever pressing would be NI/t_C. The slope of the vector would then be $(NI/t_C)/(NC/t_C)$, or I/C. Because I/C is the inverse of C/I, the space to the left of the upward vector in Figure 22-9 defines the possible outcomes with licking instrumental and lever pressing contingent.

The open-circle basepoint in Figure 22-9 shows hypothetical rates if the rat completed the two response requirements without constraint on the order of their completion. At the start of each paired baseline trial, lever and tube appear simultaneously. The tube retreats at the hundredth lick, the lever at the first press, regardless of order. As in the contingency trial, the completion times are measured from the beginning of the baseline trial. The paired baseline differs from the contingency procedure by placing no constraint on the relative latency of completion.

The broken lines identify facilitation and suppression effects. As in Figure 22-1, the upper right quadrant denotes facilitation of both responses, the lower left suppression of both. In both paradigms the experimental possibilities depend on whether the schedule vector passes above or below the basepoint. Here they depend additionally on which response is instrumental. With lever pressing instrumental, the schedule can suppress both responses, facilitate both, or facilitate lever pressing while suppressing licking. It cannot facilitate licking while suppressing lever pressing, nor can it produce the null effect. But if licking is instrumental, anything could happen: suppression of both responses, facilitation of both, either facilitated with the other suppressed, and no effect on one or both responses.

The conservation model for this fixed-ratio schedule has a distinctive notation but is formally identical to the corresponding model for the controlled-time schedule (see Appendix 22A). Its theoretical line of constraint would have negative slope and would pass through the basepoint, as illustrated in Figure 22-9. With lever pressing instrumental, the model would therefore predict facilitation of lever pressing, and suppression of licking. In no other way could the rat reach the theoretical line. But with licking instrumental, the rat could reach the theoretical line in three distinct ways: The schedule could facilitate either response while suppressing the other, or have no effect on either, but should not suppress or facilitate both responses.

Figure 22-10. Rate of 25 water licks as a function of the rate of 1 lever press. Possible outcomes with lever pressing instrumental lie to the right of the vector; possible outcomes with licking instrumental lie to the left of the vector. The open circle shows baseline rates, the closed circles rates under the schedules.

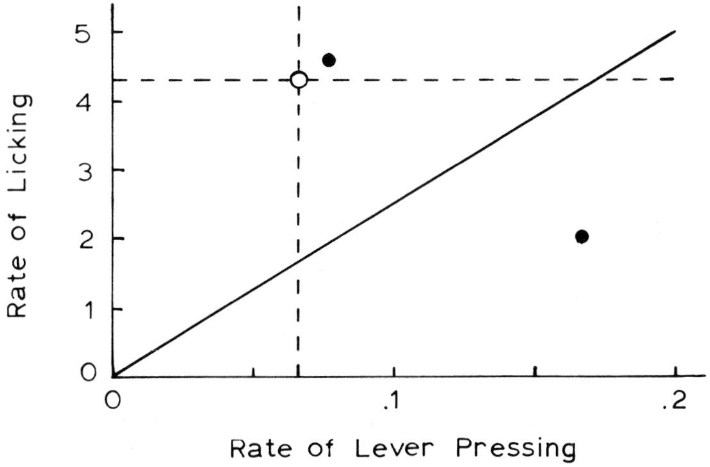

I applied this analysis to an experiment with three thirsty rats, thoroughly practiced in an earlier experiment with the same two responses. On four trials each day, 60 seconds apart, they pressed the lever once and made 25 licks at a water tube. During the single day of baseline trials, each rat chose to begin each trial by licking and end by pressing the lever. Next came two days of training, with lever pressing instrumental on two of the four daily trials, licking instrumental on the other two, counterbalanced in ABBA order on day 1, BAAB on day 2. Figure 22-10 shows the results. In accordance with the model, when lever pressing was instrumental the schedule facilitated lever pressing and suppressed licking in all three rats. When licking was instrumental the schedule had no consistent effect on either response. That too conformed to the model, because the three theoretical possibilities included the null effect.

The hundreds of published reports on controlled-amount experiments rarely report the completion times incorporated in the model. One of the few is analyzed in the two upper panels of Figure 22-11, based on data from Allison and Timberlake (1975). Four groups of thirsty rats made 10 or 100 contingent licks at a metal spout filled with saccharin solution or water, and one instrumental lick at an identical dry spout. The left panel plots the rate of licking saccharin against the rate of licking the dry spout. The steeper vector applies to the 100-lick group (circles), the other to the 10-lick group (squares). Baseline rates

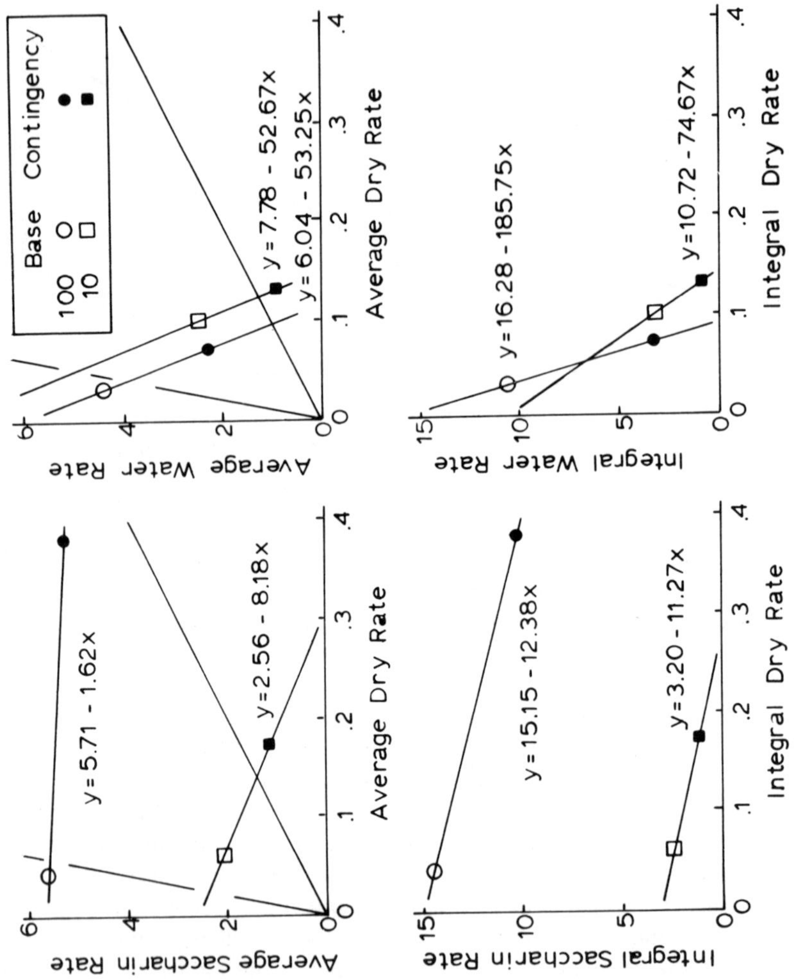

Figure 22-11. Rate of 10 (squares) or 100 (circles) saccharin licks (*left panels*) or water licks (*right panels*) as a function of the rate of licking the dry spout. Baseline rates are shown as open symbols. The steeper vector is for the 100-lick condition, the shallower for the 10-lick condition. (Based on Allison and Timberlake, 1975.)

appear as open symbols, rates under the schedule as closed symbols. In both groups the schedule facilitated instrumental, and suppressed contingent licking. The upper right panel shows the same pattern of results for the two water groups—all eight effects predicted by the model, because each basepoint lay above the corresponding schedule vector.

Granted their rate of instrumental responding, the 100-lick group on the right strayed pretty far from the baseline rate of water licking. Further analysis revealed two reasons for their large deviation. First, after gaining access to the contingent spout, they started licking about 5 seconds later than the other groups—a significant difference. Second, after starting they licked at a slightly lower rate than their own rate in baseline (4.3 versus 5.1 licks/second), whereas the other three groups licked at the same rate in contingency and baseline alike—another significant difference.

The lower panels test an alternative conservation model against data from the same experiment. The alternative model depends on the latency of each lick and not just the latency of the final lick. Analysis of the 29 rats showed that the first model fit 79 percent, the second 86 percent.

These analyses go some way toward integrating a variety of experimental procedures whose common ground is otherwise not so apparent. Behavior theory has long been divided along lines of apparatus and procedure, with one body of theory for free responding in operant conditioning boxes, mostly under the controlled-time paradigm and another body of theory for discrete-trials experiments with straight runways, mostly under the controlled-amount paradigm. In conjunction with the vector analysis of experimental constraints, the theories described here seem to cut across those lines rather easily, yet manage to address the data in some detail. All of the psychological theories bear the mark of Premack (1965) in that all attend to performance under the schedule relative to performance free of schedule constraints; they differ in matters of detail, and they differ on the exact character of the governing principle. I hope this chapter helps set the stage for further exploration of their limits of applicability.

APPENDIX 22A

The conservation model for the controlled-time schedules of Figures 22-1 and 22-2 is approximated by the equation

$$N(kI + C) = kO_i + O_c \quad , \qquad (22A.1)$$

where I refers to the amount of response i; C to the amount of response c specified by the schedule requirements. The dependent variable N is the number of times the subject completes the required sequence $I + C$. On the right-hand side, O_i refers to the amount of response i under the paired baseline condition, O_c to the amount of response c. The k on both sides is a positive dimensional constant, the number of measured response c units equal to one measured unit of response i in the theoretical rate of exchange between the two responses. Each side specifies, in units of response c, the total amount of some unspecified dimension, physical (e.g., energy expenditure) or psychological (e.g., utility), attributable to the two behaviors represented in the equation.

The predicted linear relation can be derived by solving Eq. (22A.1) for the total amount of response c performed under the schedule. The result is

$$NC = (kO_i + O_c) - k(NI) \ , \tag{22A.2}$$

the equation for a line with intercept $kO_i + O_c$ and slope $-k$. Equation (22A.2) predicts that response c will decrease linearly as response i increases. If the model is applied to the experiment of Figure 22-4, I refers to the number of lever presses required for access to the food bin, C to the average size of the ensuing meal in g.

The model's implications for facilitation and suppression depend on the fact that the line defined by Eq. (22A.2) descends through the basepoint. We can see that it does by noting that when $NI = O_i$, the equation implies that $NC = O_c$. Next we must ask what schedule vector would permit such a result, $(NC/NI) = (O_c/O_i)$. The slope of this schedule vector must be $(C/I) = (O_c/O_i)$. It follows that if the schedule vector passes beneath the basepoint, $(C/I) < (O_c/O_i)$, it must intersect the theoretical line at the lower right of the basepoint (facilitation of response i, suppression of response c). If the schedule vector passes above the basepoint, $(C/I) > (O_c/O_i)$, it must intersect the theoretical line at the upper left of the basepoint (facilitation of response c, suppression of response i).

The conventional variable-interval schedule offers abundant opportunity for unscheduled responding in the form of instrumental responding in the time before the next setup is due. For the controlled-time experiment of Figure 22-5, the applicable model would be

$$N(kI + C) + kU_i = kO_i + O_c \ , \tag{22A.3}$$

where U_i is the total number of lever presses in the contingency session that would be classified as unscheduled lever presses. Solving Eq. (22A.3) for the total number of licks performed under the variable-interval schedule,

$$NC = (kO_i + O_c) - k(NI + U_i) \ . \tag{22A.4}$$

This equation predicts that response c will decrease linearly as response i increases, whether response i be scheduled (NI) or unscheduled (U_i).

Another notable feature of the variable-interval schedule is that the schedule requirements I and C place at best only an upper limit on the slope of the schedule vector. Assuming that unscheduled occurrences of response c are negigible, the slope of the vector as measured at the end of the session will be $NC/(NI + U_i)$, the total amount of response c divided by the total amount of response i. It follows that the slope will be C/I only if there is no unscheduled instrumental responding ($U_i = 0$). Otherwise, the slope can only be less than C/I. Because C/I was already less than O_c/O_i, the model predicted the effects seen in Figure 22-5: facilitation of lever pressing and suppression of licking. Note that three of the 17 suppression effects in Figure 22-5 were artifactual, because the schedule vector ended below the baseline level of licking (on VI 14-second, R-2, R-3 and R-4). In those three cases the rat could not have attained its baseline level of licking even if it had taken the maximum number of setups allowed by the schedule.

The model for the concurrent fixed-ratio schedules of Figures 22-6 to 22-8 is

$$N_1(kI_1 + C_1) + N_2(kI_2 + C_2) = kO_i + O_c \quad , \qquad (22A.5)$$

where the subscripts 1 and 2 signify the two components of the schedule. Solving for the total amount of response c under the schedule,

$$N_1 C_1 + N_2 C_2 = (kO_i + O_c) - k(N_1 I_1 + N_2 I_2) \quad . \qquad (22A.6)$$

This equation predicts that response c will decrease linearly as response i increases. Note that Eqs. (22A.5) and (22A.6) can be satisfied whether $N_1 = 0$ (exclusive responding on component 2), or N_1 and N_2 both exceed zero (nonexclusive responding).

The example of Figure 22-9 is a simple fixed-ratio schedule in the controlled-amount paradigm, with $I = 1$ lever press, $C = 100$ licks, and N fixed experimentally at 1. The average rate of response i in the contingency trial is NI/t_I, the total amount of response i divided by the time to complete that total as measured from the start of the trial. Similarly, the average rate of response c is NC/t_C, the total divided by the completion time as measured from the start of the contingency trial. The conservation model for such a schedule is

$$(kNI/t_I) + (NC/t_C) = (kNI/t_i) + (NC/t_c) \quad , \qquad (22A.7)$$

where t_i and t_c are the completion times as measured from the start of the paired baseline trial. This model is formally identical to the corresponding model for the controlled-time paradigm, $N(kI + C) = kO_i + O_c$ (Eq. (22A.1)), where N is the dependent variable that determines the average rate of each response. The formal identity of the two models can be seen by recalling that in the controlled-time paradigm the duration of the session is constant and may therefore be set at unity. If the four durations in Eq. (22A.7) are each replaced by 1, and NI and NC on the right-hand side are replaced by O_i and O_c, Eq. (22A.7) becomes Eq. (22A.1).

Both models predict a linear relation between the rates of the two responses in contingency. Solving Eq. (22A.7) for the rate of response c, the result is

$$NC/t_C = (kNI/t_i + NC/t_c) - k(NI/t_I) \quad . \qquad (22A.8)$$

Like Eq. (22A.2), the line defined by Eq. (22A.8) must descend through the basepoint; when $NI/t_I = NI/t_i, NC/t_C = NC/t_c$.

The slope of the schedule vector is C/I, as explained in the text. If the schedule vector passes beneath the basepoint, it must intersect the theoretical line at the lower right of the basepoint. With response i instrumental the model would then predict facilitation of response i and suppression of response c. But with response c instrumental the model would allow any of three outcomes: facilitation of either response and suppression of the other and the null effect (no facilitation or suppression of either response).

Unlike the controlled-time paradigm there is no schedule for which the controlled-amount model would predict the null effect unequivocally. Although the theoretical line must pass through the basepoint, no schedule vector can pass through the basepoint. To prove this conclusion note that the baseline rate of response c, relative to that of response i, is $(NC/t_c)/(NI/t_i)$ or $(C/I)(t_i/t_c)$. The schedule vector, with slope C/I, cannot pass through the basepoint unless $(C/I) = (C/I)(t_i/t_c)$, which can happen if, and only if, $t_i = t_c$. But t_i cannot equal t_c, because the subject cannot complete the two response requirements simultaneously. Thus the schedule vector must pass at least a little above or below the basepoint and cannot pass through the basepoint.

Equation (22A.7) focuses on the final unit of each response requirement; there, the rate depends on the latency of completing the requirement as a whole. It does not depend directly on the latency of initiating the response, except for the special case in which the requirement is exactly one unit of the response and $N = 1$, wherein the first unit is also the last. The alternative to this completion model incorporates the possibility that the subject might adjust not only the rate of the final unit but also the rate of any other unit of the response requirement. The alternative model can be expressed as

$$k \sum_{j=1}^{m} (1/t_{Ij}) + \sum_{j=1}^{n} (1/t_{Cj}) = k \sum_{j=1}^{m} (1/t_{ij}) + \sum_{j=1}^{n} (1/t_{cj}) \quad , \qquad (22A.9)$$

where m refers to the number of units in the response i requirement, n to the number in the response c requirement. Each parenthetical term signifies the rate of the jth unit of the response and is simply the reciprocal latency of that unit. Each side of the equation specifies an integral rate across all units of each response. The model proposes that the subject conserves this integral rate as between the contingency trial on the left and the baseline trial on the right. Unlike the completion model of Eq. (22A.7), this integration model would allow the subject to complete the requirement "late" by starting the requirement "early." Similar compensations could be made by varying the latencies of other response units besides the first, but of course the reciprocal transformation would assign less weight to an adjustment of unit j than unit $j - 1$.

In the integration model each of the rates measured experimentally is a harmonic sum. Because the two experiments of Figure 22-11 recorded start times as well as finish times, the integration model could be tested against the data from those experiments with the help of one plausible assumption. Thirsty rats will make 10 or 100 water or saccharin licks at a fairly uniform rate from the first lick to the last. It was therefore reasonable to approximate the necessary harmonic sums by assuming that the 10 or 100 licks followed an arithmetic progression of latencies starting with the latency of the first lick, a seconds, and ending with the latency of the final lick, f seconds. From elementary algebra,

$$f = a + (n - 1)d \quad ,$$

where n is the total number of licks, 10 or 100, and d is the common difference in the arithmetic progression of latencies. It follows that the necessary harmonic sums are approximated by

$$1/a + 1/(a + d) + \ldots + 1/[a + (n - 1)d] \quad .$$

Analysis of individuals showed that 86 percent of the rats satisfied the integration model of Eq. (22A.9): 25/29, as compared with the 23 that satisfied the completion model of Eq. (22A.7). The two additional rats included one that completed both requirements sooner in contingency than in baseline. By the measures incorporated in the completion model, this result would have been classified as facilitation of both responses, in violation of that model. The other rat's results would have been classified as facilitation of the instrumental response and no effect on the contingent response, also a violation of the completion model. If its terms are rearranged and partitioned appropriately, it can easily be seen how the integration model can handle results that would violate the completion model:

$$k \sum_{j=1}^{m-1} (1/t_{Ij} - 1/t_{ij}) + \sum_{j=1}^{n-1} (1/t_{Cj} - 1/t_{cj}) +$$
$$k(1/t_{Im} - 1/t_{im}) + (1/t_{Cn} - 1/t_{cn}) = 0 \quad . \qquad (22A.10)$$

Reading from left to right, the third and fourth terms compare the rate of completion in contingency with the rate of completion in baseline. It is these four rates on which the completion model depends. The first and second terms make the same comparison with respect to all units but the final unit of each response: $m - 1$ units of the response i requirement, and $n - 1$ units of the response c requirement. Because the first two terms are experimentally independent of the rest, the equation can still be satisfied even if the third and fourth terms have values that would violate the completion model: both positive (facilitation of both responses), both negative (suppression of both), or either zero and the other positive or negative.

Further work with the controlled-amount paradigm is needed for a more decisive comparison of the completion and integration models than the experiments of Figure 22-11. When the instrumental requirement is just one unit of response i, as in those experiments, the first term in Eq. (22A.10) is zero. The two models are then distinguished by only one term, namely the second. A more decisive test must therefore use an instrumental requirement that specifies several units of response i, such as five lever presses for access to water. A more precise test of the integration model must record the latency of each response unit instead of depending on the approximation method.

Statistical analysis of k supported the same conclusions, whether k was calculated from Eq. (22A.7) or Eq. (22A.9): The numerical value of k did not vary significantly with the number of saccharin or water licks but was significantly larger with water than with saccharin. In terms of the model, the pattern of results suggests that the underlying dimension cannot be the sweetness of saccharin relative to water or the dry spout but might be the frictional resistance of saccharin relative to the others. In these two experiments, k refers to the amount of the dimension entailed by a lick at the dry spout relative to a lick at a spout filled with saccharin or water. Because k was generally greater than 1 and less with saccharin than with water, the pattern suggests that a lick at the dry spout entailed more of the dimension than a saccharin lick, which in turn entailed more than water. The underlying dimension could hardly be sweetness, unless a dry spout is sweeter than saccharin. Energy expenditure might be a more likely candidate, inasmuch as a dry spout probably offers more frictional resistance than 0.4 percent saccharin, which in turn offers more fluid friction than water (Allison, 1971:411).

REFERENCE NOTE

1. Allison, J. *Demand economics and conservation in ratio and interval schedules.* Paper presented at the meeting of the Psychonomic Society, San Antonio, Texas, November 1978.

REFERENCES

Allison, J. 1971. Microbehavioral features of nutritive and nonnutritive drinking in rats. *Journal of Comparative and Physiological Psychology 76*: 408-17.

____. 1976. Contrast, induction, facilitation, suppression, and conservation. *Journal of the Experimental Analysis of Behavior 25*: 185-98.

_____ . 1980. Conservation, matching, and the variable-interval schedule. *Animal Learning and Behavior 8*: 185–92.

_____ . 1981a. Paired baseline performance as a behavioral ideal. *Journal of the Experimental Analysis of Behavior 35*: 355–66.

_____ . 1981b. Economics and operant conditioning. In P. Harzem and M. Zeiler, eds., *Advances in Analysis of Behaviour*, vol. 2, *Predictability, Correlation and Contiguity*, Chichester, England: Wiley.

Allison, J.; M. Miller; and M. Wozny. 1979. Conservation in behavior. *Journal of Experimental Psychology: General 108*: 4–34.

Allison, J., and W. Timberlake. 1975. Response deprivation and instrumental performance in the controlled-amount paradigm. *Learning and Motivation 6*: 122–42.

Awh, R.Y. 1976. *Microeconomics: Theory and Applications.* New York: Wiley.

Collier, G.H.; E. Hirsch; and P.H. Hamlin. 1972. The ecological determinants of reinforcement in the rat. *Physiology and Behavior 9*: 705–16.

Herrnstein, R.J. 1958. Some factors influencing behavior in a two-response situation. *Transactions of the New York Academy of Sciences 21*: 35–45.

Herrnstein, R.J., and D.H. Loveland. 1975. Maximizing and matching on concurrent ratio schedules. *Journal of the Experimental Analysis of Behavior 24*:107–16.

Heth, C.D., and A.G. Warren. 1978. Response deprivation and response satiation as determinants of instrumental performance: Some data and theory. *Animal Learning and Behavior 6*: 294–300.

Lea, S.E.G., and T.J. Roper. 1977. Demand for food on fixed-ratio schedules as a function of the quality of concurrently available reinforcement. *Journal of the Experimental Analysis of Behavior 27*: 371–80.

Mazur, J.E. 1975. The matching law and qualifications related to Premack's principle. *Journal of Experimental Psychology: Animal Behavior Processes 4*: 374–86.

Premack, D. 1965. Reinforcement theory. In D. Levine, ed., *Nebraska Symposium on Motivation*, vol. 13, Lincoln: University of Nebraska Press.

_____ . 1971. Catching up with common sense or two sides of a generalization: Reinforcement and punishment. In R. Glaser, ed., *The Nature of Reinforcement*, New York: Academic Press.

Rachlin, H., and B. Burkhard. 1978. The temporal triangle: Response substitution in instrumental conditioning. *Psychological Review 85*: 22–47.

Samuelson, P.A. 1976. *Economics.* 10th ed. New York: McGraw-Hill.

Segal, E.F. 1962. Exteroceptive control of fixed-interval responding. *Journal of the Experimental Analysis of Behavior 5*: 49–57.

Shapiro, N., and J. Allison. 1978. Conservation, choice, and the concurrent fixed-ratio schedule. *Journal of the Experimental Analysis of Behavior 29*: 211–23.

Staddon, J.E.R. 1979. Operant behavior as adaptation to constraint. *Journal of Experimental Psychology: General 108*: 48–67.

Striefel, S. 1972. Timeout and concurrent fixed-ratio schedules with human subjects. *Journal of the Experimental Analysis of Behavior 17*: 213–19.

Timberlake, W., and J. Allison. 1974. Response deprivation: An empirical approach to instrumental performance. *Psychological Review 81*:146–64.

Watts, H.W., and A. Rees, eds. 1977. *The New Jersey Income Maintenance Experiment*, vol. 2, *Labor-Supply Responses.* New York: Academic Press.

23 THE EMPEROR'S CLOTHES Assumptions of the Matching Theory

William Timberlake

The matching account has stood for several years as the emperor of a large domain of steady-state responding on operant schedules. The original conceptual clothing of this account was the plain empirical generalization that organisms match responses to obtained reinforcements on concurrent variable-interval schedules (Herrnstein, 1961). Subsequently Herrnstein (1970, 1974, 1979) and others[a] have attempted to clothe the matching account in the elegance of a general theory that extends its potential application to all behavior. The purpose of this chapter is to determine if the assumptions underlying this matching theory will support such a radical expansion of the matching domain.

This chapter will focus on two aspects of the matching theory: first, the clarity, consistency, and reasonableness of its assumptions; and second, the relation of these assumptions to problems of survival that an organism typically encounters in its selection environment (such as regulation of intake of important commodities, distribution of time and search effort among "patches" of these commodities, and the selection of diet among alternative sources of a commodity—Krebs, 1978). If the matching theory accounts for all behavior, it must provide solutions to such typical survival problems as well as account for regularities that occur in the laboratory.

Preparation of this manuscript was partly supported by Grant BNS79-15117 from the National Science Foundation. I thank Michael L. Commons, Richard Ellis, Richard J. Herrnstein, Alexandra Logue, Philip Podsakoff, and Mark Wozny for their comments.

a. Other theories have attempted to explain much the same data as the matching account (e.g., Catania, 1963; Staddon and Motheral, 1978), but they have not been yet as explicitly ambitious and influential as the matching account. Though some remarks may apply to them, I will not consider them directly.

Before evaluating the assumptions underlying the matching theory, I will review briefly the history, application, and accuracy of the matching account. The theory has passed through three phases, beginning as the original empirical generalization, or simple matching law, of Herrnstein (1961), becoming the generalized matching law (Baum, 1974b), and finally the matching theory (Herrnstein, 1970, 1974).

THE SIMPLE MATCHING LAW AND THE GENERALIZED MATCHING LAW

The simple matching law was first advanced by Herrnstein (1961) as a reliable description of the key pecking of pigeons on concurrent variable-interval schedules of food reinforcement. One phrasing of the matching law is that subjects match the ratio of behavior to the ratio of obtained reinforcers.

$$\frac{P_1}{P_2} = \frac{R_1}{R_2}, \tag{23.1}$$

where P_1 and P_2 refer to the frequency or rate of responding on two schedules, and R_1 and R_2 are the obtained frequency or rate of reinforcement on two schedules. This simple expression has proved impressively accurate in its description of a large amount of data (cf. de Villiers, 1977). Baum (1974b) further increased its accuracy by including the relation within a power function, producing the generalized matching law:

$$\frac{P_1}{P_2} = k\left(\frac{R_1}{R_2}\right)^{a}, \tag{23.2}$$

where k is related to bias for one response alternative and a is related to deviations from matching.

Despite their ability to fit data, both forms of the matching law have several deficiencies as general laws of behavior. First, Eq. (23.2) borders on the tautalogous (Rachlin, 1971; Killeen, 1972). The multiplicative constant and power were added with little formal justification, (though see Rachlin, Chapter 16 of this volume). Baum (1974b) developed the generalized matching law by assuming that the simple matching law was true but that transformations of the data were needed to account for the effect of variables such as the type of reinforcement and schedule, the level of deprivation, and individual differences in preference and discrimination. However, unless the values of a and k in Eq. (23.2) can be related systematically to measurable independent variables, they function primarily as free parameters to increase the accuracy of fit to data.

Accuracy of prediction is an important quality of a law, but the history of science shows that improving the empirical accuracy of an equation, in the absence of accompanying conceptual development, rarely advances understanding of a phenomenon. To take an extreme case, Ptolemy developed an amazingly accurate description of a planetary motion as a series of epicycles, but his work in fitting data with equations did little to increase general understanding of planetary motion (Kuhn, 1957).

Second, the simple matching law, Eq. (23.1), is limited in its range of application. Precise matching of response rate to rate of obtained reinforcement occurs primarily under concurrent variable-interval schedules, with symmetric delays of 1.5–20 seconds between switching schedules and receiving a reinforcer (though zero delay has been effective in some conditons—Baum, Note 1), and with the same responses, reinforcements, and deprivation states for each schedule. The reasons for such limitations usually are not explained, and they appear unrelated to the basic concepts of the matching law. The generalized matching law, Eq. (23.2), has considerably more flexibility in its application, but its limitations are also conceptually unclear.

Third, the matching law is fundamentally a correlational law, though with causal overtones. The most basic conceptual statement of the matching law is that for a particular set of concurrent schedules, the ratio of an operant response to some transformation of its obtained reinforcements is unknown but constant. The cause of this correlation is loosely attributed to the strengthening ability of the reinforcements (Herrnstein, 1970). However, the causal overtones are of little use in predicting whether an organism will obtain any or all of the available reinforcements, how it will arrange its choices among alternative responses, or the precise ratio of responses to obtained reinforcements. Thus, the matching law makes little contact with traditional independent variables such as drive, and frequency, magnitude, or delay of reinforcement, nor with the survival problems outlined in the introduction.

For example, given two sources of food with different fixed handling times, similar nutrient values, and limited foraging time, most survival theories predict exclusive preference for the food with the least handling time (e.g., Krebs, 1978; Staddon, 1979). The matching law predicts only that the animal will show an exclusive preference for one or the other (Herrnstein and Loveland, 1975). Further, this prediction is made only by using a form of Eq. (23.1) that cannot be falsified if the subject does not respond on one alternative. From the latter approach it is only a small step to argue that animals match when asleep in the chamber, or, for that matter, when dead. My preference is to eliminate this problem (and the possibility of matching on concurrent ratio schedules) by representing the simple matching law only as $P_1/R_1 = P_2/R_2$.

In short, the matching law is fundamentally a claim that a correlation exists between responses and some transformation of the reinforcements obtained on concurrent schedules. It makes the empirically derived prediction that the ratio of responses to obtained reinforcers, though of unknown value, will be constant for two or three concurrent variable-interval schedules, and perhaps for some other combinations of schedules as well (de Villiers, 1977).

THE MATCHING THEORY

In an attempt to produce a principle of greater scope, Herrnstein (1970, 1974) generalized the matching law by introducing three additional theoretical terms: k, the total amount of behavior; R_e, the reward for background behavior in the situation; and P_e, the performance of the background responses. By assuming that k and R_e are constant, Herrnstein (1970) combined these terms with Eq. (23.1) to produce the hypothesis that the absolute rate or amount of instrumental responding was a function of relative reinforcement rate:

$$P_1 = \frac{k \cdot R_1}{\sum_{i=1}^{n} R_i + R_e}, \qquad (23.3)$$

where all terms are defined above, and P_e is contained in k. This equation and its underlying assumptions constitute the matching theory.[b]

The matching theory has been fit to data from a wide variety of schedules: concurrent interval, concurrent ratio, concurrent interval and ratio, multiple, simple interval, simple ratio, concurrent delay, and discrete trials. In most of these cases, if k and R_e are assumed constant, the matching theory predicts absolute levels of responding with a high degree of accuracy (de Villiers, 1977; de Villiers and Herrnstein, 1976; Pear, 1975; though see Timberlake, 1977). However, as in the case of the matching law, there are reasons to question whether accuracy of prediction is the only, or even the best way to evaluate the matching theory.

First, there has been little concern with the evaluation of goodness of fit between Eq. (23.3) and data, other than the size of r^2. As pointed out by Anderson (1978) and his colleagues, for any monotonic function there exist transformations of the data that produce very high r^2 values whether or not the function is correct. Anderson (1978) suggested several techniques for assessing functions more carefully (e.g., Farley and Fantino, 1978).

Second, several researchers have derived or simulated Eq. (23.3) beginning with a variety of different theoretical assumptions (Catania, 1963; Herrnstein, Chapter 19; Killeen, 1979; Rachlin, Chapter 16; Staddon and Motheral, 1978). Thus, the accuracy of Eq. (23.3) cannot be presumed to be a strong test of the accuracy of the matching theory, since the equation may be true for any of several different reasons.

Third, though k and R_e are theoretical terms, they are frequently used in tautalogous fashion as fitting parameters, free to vary without restriction. Only a few experiments have tested their presumed constancy or functional variation across rewards, deprivation states, or schedules (e.g., McSweeney, 1975; de Villiers, Note 2). Even in these tests, because k and R_e are fit simultaneously from

b. Herrnstein (1979) derived Eq. (23.3) differently but with the same basic assumptions.

the same data, they can be estimated in a variety of ways yielding rather different results. Further, the flexibility provided by these two parameters in fitting data may be too great to be encompassed by the underlying theory. There has been much discussion of the difficulties of including operant and discrete trial's results within the same model (e.g., Platt, 1971), but Eq. (23.3) can fit data from both paradigms (de Villiers and Herrnstein, 1976). It seems potentially odd that a theory that does not predict local (momentary) effects in operant experiments can readily deal with local incentive effects in a discrete-trials paradigm.

Last, the matching theory still maintains elements of a correlational approach because it is most often applied to obtained rather than programmed reinforcements. (See Commons, Woodford and Ducheny, Chapter 2, for an exception). The relation of obtained reinforcement to independent variables such as deprivation, and the pattern, amount, and frequency of presentation of reinforcement are not well worked out within the theory. Work by Baum (1973) and Prelec and Herrnstein (1978) has introduced the concept of a feedback function to relate obtained reinforcements to reinforcements programmed by the experimenter. Though a step in the right direction, parts of most feedback functions must be fit from data, and the theory still does not make good contact with most of the aforementioned independent variables and survival problems.

In short, though the development of the matching theory appeared to increase the variety of data and schedules that could be fit, there was little improvement in the ability of the matching account to predict solutions to survival problems or to explain limits of application in terms of the theory. It is clear that Eq. (23.3) performs impressively in fitting a wide variety of examples of steady-state responding (though see Timberlake, 1977), and perhaps data on acquisition as well (Herrnstein, 1979). However, for the reasons given, a good fit to data as indicated by r^2 values is not sufficient evidence to accept the internal coherence or theoretical importance of the matching theory. Thus it appears there is a critical need to evaluate the matching theory by examining its underlying assumptions.

Assumptions of the Matching Theory

Let us consider three assumptions of the matching theory:

1. Frequency or rate of a response is proportional to the frequency or rate of reinforcement obtained for that response relative to the total reinforcement obtained in the situation, irrespective of the nature of the schedules, responses, or reinforcements.
2. All behaviors, including background behaviors not associated with any obvious reinforcer, occur because they are reinforced. Put another way, there exists no unreinforced behavior.
3. The total behavior in a situation is constant despite changes in instigating conditions such as deprivation, food quality, or amount of reinforcement.

Assumption 1: Responding is directly related to relative obtained reinforcement. Herrnstein (1970, 1974) clearly intended the matching of behavior to relative obtained reinforcement to be a general law governing both learned and unlearned behavior. With reference to learned behavior, Herrnstein (1974) suggested that matching "appears to be a general framework to replace the traditional informal (or nonformal) vocabulary of the law of effect." As for unlearned behavior, Herrnstein makes no conceptual distinction between behavior that is reinforced on an experimenter imposed schedule and behavior that is "intrinsically" reinforced on its own schedule. Both scheduled and intrinsic reinforcements enter Eq. (23.3) in the same way, a point I will discuss further in the next section.

A significant problem with the application of the matching theory is that Eq. (23.3) cannot be used unless a way can be found to make the different reinforcements commensurate without making the theory tautalogical (Rachlin, 1971; Killeen, 1972). An obvious solution is to convert the different reinforcements to a single scale by using a value transform h appropriate to each. For example, applying appropriate value transforms for a simple interval schedule yields

$$P_1 = \frac{k \cdot h_1 R_1}{h_1 R_1 + h_e R_e} \qquad (23.4)$$

where h_1 is the value transform for the food reinforcements R_1 and h_e is the value transform for the background (intrinsic) reinforcements R_e. Values of R_1 cannot be used to calculate P_1 without knowing the value of h_1, but k, R_e, h_1, and h_e are all unknowns that must be fit from the same data. Fortunately the number of unknowns can be reduced to two by dividing top and bottom of the right side of Eq. (23.4) by h_1, the value of the transform for R_1, and treating $h_e/h_1\ R_e$ as one unknown.

$$P_1 = \frac{k \cdot R_1}{R_1 + \frac{h_e}{h_1} R_e} \quad . \qquad (23.5)$$

Equation (23.5) represents the procedure followed in applying the matching theory. Researchers enter R_1 directly in terms of frequency or seconds of obtained reinforcement. Only the value of R_e is modified by the value transforms. Fitting the two unknowns in Eq. (23.5) from the same data has its own procedural problems (e.g., McSweeney, 1975), but having two unknowns is far better than having three or four.

The transformation of R_e shown in Eq. (23.5) allowed Herrnstein to introduce the notion of a modulus (unit size) of measurement to account for deprivation effects in schedule responding. According to Herrnstein (1974), as the value of food reinforcement is increased by deprivation, the modulus for measuring

R_e expands, thereby decreasing the relative value of R_e and its contribution to total reward. On the other hand, as the value of food reinforcement is decreased, the modulus for measuring R_e shrinks, thereby increasing the relative value of R_e and its contribution to total reinforcement. The mechanism for this effect is the change in the ratio of h_e to h_1 as deprivation changes the value of h_1. As h_1 increases with deprivation, the importance of R_e decreases. As h_1 decreases with satiation, the importance of R_e increases.

It is unfortunate that the precise nature of these functional relations is not clear. By a simple extension of the foregoing argument, the value of h_1 might also be expected to vary with amount or frequency of reinforcement. If so, the contribution of R_e could be expected to change over a session and to be greater for a fatter schedule (VI 30 seconds) than for a leaner schedule (VI 240 seconds). Herrnstein (1979) indicated that little evidence exists for such changes, but few experiments have explored a wide range of reinforcement frequency and amount (Timberlake and Peden, Note 3).

The procedure contained in Eq. (23.5) for making different types of reinforcement commensurate works equally well on concurrent variable-interval schedules with similar reinforcements. Simply measure both reinforcement frequencies directly and estimate the transformed value of R_e and k. Unfortunately, this procedure does not work when the second reinforcement on a concurrent schedule is different than the first. Adding a second type of reinforcement, such as water, requires another value transform h_2 to convert it to the same measurement scale as R_1 and R_e. In a concurrent interval schedule involving food and water, key pecking for food (P_1) and for water (P_2) can be written as

$$P_1 = \frac{k \cdot R_1}{R_1 + \frac{h_2}{h_1} R_2 + \frac{h_e}{h_1} R_e} \; ; \qquad (23.6)$$

$$P_2 = \frac{k \cdot \frac{h_2}{h_1} R_2}{R_1 + \frac{h_2}{h_1} R_2 + \frac{h_e}{h_1} R_e} \; . \qquad (23.7)$$

It can be seen that the frequency of food reinforcement can be entered directly in the equations, but the frequency of water reinforcement cannot be entered unless the transformation h_2/h_1 is known. It should always be possible to set up equations for different schedule intervals or different types of reward, and eventually solve for h_2/h_1; but until this value is known one cannot reasonably fit k and $(h_e/h_1)\,R_e$ from the data using the obtained frequency of R_2.

An analogous application problem is raised by the problem of making different operant responses commensurate. Dimensional analysis of Eq. (23.3) requires that all instrumental behaviors be measured in the same units. Herrnstein

(1974, 1979) further constrained the measurement scale by requiring that total responding always sum to a constant. Thus, if a concurrent schedule for food reinforcement involved key pecking (P_1) and treadle pressing (P_2), it would be necessary to transform both expressions.

$$q_1 P_1 = \frac{k \cdot R_1}{R_1 + R_2 + \frac{h_e}{h_1} R_e} , \qquad (23.8)$$

$$q_2 P_2 = \frac{k \cdot R_2}{R_1 + R_2 + \frac{h_e}{h_1} R_e} . \qquad (23.9)$$

Since Herrnstein (1974) has indicated that "the value of k depends entirely on which response is taken as P_1," the more appropriate expression for these equations might be

$$P_1 = \frac{\frac{k}{q_1} \cdot R_1}{R_1 + R_2 + \frac{h_e}{h_1} R_e} ; \qquad (23.10)$$

$$P_2 = \frac{\frac{k}{q_2} \cdot R_2}{R_1 + R_2 + \frac{h_e}{h_1} R_e} . \qquad (23.11)$$

The latter expressions raise no problems in measuring P_1 or P_2, but do create the problem of how to fit the k/q ratio, which now varies with the response measured.

It should be apparent that combining different responses and different reinforcers produces large difficulties in applying the matching theory. Not only is there the problem of an extra reinforcement transform, but there is also a response transform. It might be argued that most of these problems could be resolved by measuring both responses and reinforcements in units of time, thus reducing the above equations back to Eq. (23.3). Whatever the empirical problems with this procedure, it violates the third assumption of the matching theory that total responding remains constant. If all responding is measured in time, as the frequency of reinforcement is increased, the time available for total responding (key pecking and background responding) must be decreased. If one solves this problem by including in total responding the time spent eating, the matching theory would be a tautology. Total responding that takes up the entire session is a function only of the session time.

The problem of value transformations for responses and reinforcers also makes it impossible for the ratio of responding to equal the ratio of obtained reinforcements in Eq. (23.1), because

$$\frac{P_1}{P_2} = \frac{\frac{h_1}{q_1} R_1}{\frac{h_2}{q_2} R_2} \quad . \tag{23.12}$$

The generalized matching law, Eq. (23.2), might fit such data by collapsing the four value transformations into one or two variables, but it is not clear how such a prediction would deal with changes in the transforms, as might occur if the deprivation level of one reinforcer were increased.[c]

Two other features of the assumption that response rate is proportional to obtained relative reinforcement rate merit comment. The first feature concerns the precision of matching data. There are many behavioral strategies that an organism could use to match behavior to relative obtained reinforcements. For example, an organism could match by failing to respond, by spending equal time on each schedule and pecking less frequently on the fat schedule, or responding at a constant rate throughout the session but switching schedules once during the session at the appropriate time.

Instead of showing multiple strategies the subjects seem restricted to a few. They tend to match both responses and time to relative obtained reinforcement (de Villiers, 1977) and to switch schedules with a characteristic distribution of times (Heyman, 1979, and Chapter 20). Further, the subject tends to pick up the majority of the reinforcements in a given situation. Thus it seems that characterizing a subject's behavior only as matching is an incomplete account of the regularity of its behavior (Myerson and Miezin, 1980; Pear, Rector, and Legris, Chapter 1). The explanation of these regularities seems extremely important to an understanding of the matching phenomenon.

The last problematic feature of the basic matching assumption concerns how the organism constitutes the total pool of reinforcement active at any one time. Herrnstein (1970) argued that in a particular component of a multiple schedule the subject decreased the weight assigned to reinforcers in the other component. By similar logic the organism should weight reinforcements differently depending on whether they are interspersed irregularly through the session (as in a variable-interval schedule) or concentrated near the beginning (as often happens on low fixed-ratio schedules). In a similar vein, how does the animal value the reinforcements it received yesterday, 30 minutes ago, or the food it receives in its home cage after the session? Work by Killeen (1975, 1979) suggests that

c. It should be noted that the problem of value transformations for responses and reinforcers does not preclude the possibility of referential hedonic scaling using the matching law (Miller, 1976). It is still reasonable to predict the ratio of responding for two different reinforcers from prior measures of responding for each reinforcer paired with a third common reinforcer.

pigeons give little weight to rewards separated temporally by more than 10-20 minutes. Other research of this sort seems critical to discover how reinforcement influences behavior.

In short, the assumption that responding is directly related to relative obtained reinforcement is at best an incomplete statement of the variables controlling behavior. Insufficient attention has been paid to the importance of the value transformations necessary to deal with different responses and reinforcers. It may well be that the value transformation problem hides a larger problem of interactions and divisions between types of responses and reinforcements. Further, the rules for specifying the total reinforcement pool are not worked out, and the regularities of matching are not adequately dealt with.

Assumption 2: All Behavior is Reinforced. This assumption stems from the long hedonist-utilitarian tradition of Digby (1644), Bentham (1789), Spencer (1870), and others. In this tradition animals are assumed always to do what feels good and to avoid or stop what feels bad. If we replace the concept of hedonism with that of reinforcement, we are forced to conclude that whatever an organism is doing is reinforcing, otherwise it would be doing something else. Following this tradition Herrnstein (1970) postulated the existence of various unspecified background reinforcements in the experimental situation, which included, for example, hedonic affect from "the creature's own body and physiological processes" (Herrnstein, 1974:160). When the organism was not key pecking or eating, it was presumed to be under the control of these background reinforcers.

The concept of background reinforcement allowed the matching theory to treat simple schedules within the same framework of choice used for concurrent interval schedules, and, in addition, provided a parameter for fitting the data (R_e). However, these advantages had their cost. The concept of background reinforcement is not clear. Despite the fact that Eq. (23.3) treats background reinforcements as formally equivalent to "foreground" (experimenter designated) reinforcements, background reinforcements are treated differently in practice.

Herrnstein (1970, 1974) assumed that R_e is constant, but that its transformed value ($h_e/h_1 \ R_e$ in Eq. (23.5)) changes as a function of the value of the other reinforcements. However, there is no compelling theoretical justification that the number of obtained background reinforcements should be constant. The number of reinforcers obtained from programmed sources is not presumed constant. Even if the same number of background reinforcements were always available, the animal should be free to pass up or delay obtaining at least some of these reinforcements, just as it does with food reinforcements.

Second, background reinforcements are apparently presented on schedules, but what is the nature of these schedules? The answer to this question should determine how R_e is used in equations. Prelec and Herrnstein (1978) had to assume that R_e was not presented on a ratio schedule in order to make their derivation (or was presented on a ratio schedule equal to the VR schedule), but how did they know? It seems reasonable that different background reinforcements might be presented on different schedules: for example, exploration on variable-interval schedules, itches on variable-time schedules, heartbeats and digestion on fixed-time schedules, and so forth. It is possible to make qualitative predictions by a simplifying assumption (e.g., Prelec and Herrnstein, 1978), but the precise prediction of absolute and relative responding requires knowledge of the nature of these schedules and how they interact with each other and with the programmed schedules.

A most significant problem with the consistency of the matching theory is raised by applying the logic of intrinsic reinforcement to an organism's behavior in a paired baseline and then using these equations to predict responding on simple schedules. If there are background reinforcements for responses such as scratching, there should be background reinforcements for responses such as key pecking and eating (Herrnstein, 1979). Following an unpublished paper of Herrnstein (Note 4), a matching account can be written for key pecking and eating when both are freely and simultaneously available in paired baseline (leaving aside the complexities of response and reinforcement transformations):

$$P_{KP} = \frac{k \cdot R_{KP}}{R_{KP} + R_E + R_e} \quad ; \tag{23.13}$$

$$P_E = \frac{k \cdot R_E}{R_{KP} + R_E + R_e} \quad . \tag{23.14}$$

where *KP* refers to key pecking, *E* refers to eating, and the other terms were defined in Eq. (23.3). When a variable-interval schedule is imposed on this baseline, the equation for key pecking must change because key pecking is now based both on background reinforcements (R_{KP}) and the now contingent reinforcement from subsequent eating (some portion of R_E). Thus, key pecking and eating under a simple interval schedule can be written as

$$P_{KP} = \frac{k(R_{KP} + sR_E)}{R_{KP} + R_E + R_e} \quad , \tag{23.15}$$

$$P_E = \frac{k(1-s)R_E}{R_{KP} + R_E + R_e} \quad , \tag{23.16}$$

where *s* refers to the proportion of the reinforcements for eating that are used to support key pecking under the schedule. The portion of eating reinforcement affecting key pecking may be modified by yet another parameter to account for the relative value of the reinforcement. It is difficult to imagine how to deal with the complexity of an equation for concurrent-interval performance that involved both response and reinforcement transforms and intrinsic reinforcement for all behaviors.

The problem of dealing with the background reinforcements of key pecking and eating does not affect the application of the simple matching law, Eq. (23.1), and the generalized matching law, Eq. (23.2), to concurrent schedules with the same responses and reinforcers. The presence of such background reinforcements does raise practical or conceptual problems with both forms of the matching law when different responses and reinforcers are used.

There is a striking resemblance between the issue of background reinforcements for key pecking and eating, and the concerns of recent equilibrium theories of learned performance (e.g., Allison, 1976; Rachlin and Burkhard, 1978; Staddon, 1979; Timberlake, 1980; Timberlake and Allison, 1974). In essence, an equilibrium theory attempts to account for how instigating forces that underlie paired baseline behavior are expressed under a schedule. It would appear that the application of the matching theory to baseline situations would produce approachment with the language and concepts of such equilibrium theories.

Mazur's (1975 1977) theory is a step in this direction (Herrnstein, 1977). However, the predictions of Mazur's value-averaging theory have been applied only to ratio schedules, and here they appear to contradict the predictions of the matching theory. According to Pear's (1975) derivation, the matching theory (Eq. (23.3)) predicts that responding on simple ratio schedules will decrease with increased ratio requirement and show no change with variation in amount of reward. Mazur (1975) predicts that responding will typically increase with increased ratio requirement and decrease with increased reward magnitude. A number of lines of evidence support Mazur's predictions at a qualitative level but suggest that his predictions are quantitatively inaccurate (Allison, Miller, and Wozny, 1979; Timberlake and Wozny, 1979; Wozny, 1979).

In short, if it is appropriate to treat background reinforcement the same as scheduled reinforcement, the appropriate application of the matching theory (and the matching law) seems limited to concurrent variable-interval schedules with the same zero operant level response and the same type of reinforcers.

Assumption 3: Total Amount of Behavior is Constant. Herrnstein (1974, 1979) designated total behavior as k, defined as the total of all responses "in units commensurate with the observed response," and as "the amount of behavior that the observed response would display if there were no sources of reinforcement other than the one associated with the observed response" (Herrnstein, 1974:163). The assumption of constancy of k increases the predictive power of the matching theory, but the assumption seems intuitively wrong, as well as potentially inconsistent with other assumptions.

By definition the effective value of k varies with the instrumental behavior that is measured. Further, unless I misunderstand, there is nothing in the other assumptions of the matching theory to prevent k from varying as a function of the number of reinforcements (McSweeney and Dericco, 1976), or the level of deprivation (de Villiers, Note 2; McSweeney, 1975). Certainly, there is considerable evidence that the activity of a deprived organism markedly increases in the presence of food related stimuli (Bolles, 1967), or intermittent food (Reberg, Innis, Mann, and Eizenga, 1978). On the other hand, exploratory behavior markedly declines in the presence of stimuli eliciting fear (Montgomery and Monkman, 1955) and with repeated exposure to a stimulus situation (Berlyne, 1960; Timberlake, 1969).

Thus, the claim that an organism maintains a constant amount of behavior across all conditions of deprivation, density of reinforcement-related cues, schedule types, and responses seems unwarranted unless it were somehow possible to argue that all forms of overt activity and inactivity are weighted the same on some underlying common dimension, such as time. This solution is not appealing because using the time measure makes the assumption of constant total behavior a tautology. Total behavior is constant if and only if the session length is constant. Further, since the effective value of k presumably varies with characreristics of the instrumental response of interest (Herrnstein, 1974), the potential values of k in a situation involving two instrumental responses, such as key pecking and treadle pressing, cannot both equal the same session time.

Given that total behavior is not always constant, a practical problem is the necessity of defining the limits within which invariance may hold. Is total behavior invariant across elements of a multiple schedule, across a 1-hour session, across 24-hour periods, across seasons, or across years? Momentary phenomena such as predatory chase and prey flight, and seasonal phenomena such as migration and hibernation suggest that total behavior varies with the demands of survival and reproduction. These considerations appear to provide better predictions of the variation and persistence of behavior than an assumption of constant total behavior tied to an ad hoc inference of reinforcement.

THE CURRENT STATUS OF MATCHING

More than 20 years of data support the simple matching law. These data show that pigeons, and several other species, rather consistently match relative responding to some transformation of relative obtained reinforcements on concurrent interval schedules. There has been some question of how precisely animals match (e.g., Myers and Myers, 1978), but the generalized matching law has handled most of these objections at the level of fitting the results. Within its range of application, the matching law appears as accurate a description of data as any law of chemistry or physics that predicts the behavior of individual units. Herrnstein's (1961, 1970) original insight, that an organism's behavior is controlled by one source of reinforcement in the context of other sources, seems as reasonable today as when he first wrote it.

Most researchers view the matching theory as a reasonable extension of the matching law, which markedly increased the ability of the matching approach to fit data. However, as indicated in the previous pages, the conceptual status of the matching theory is problematic. Its theoretical assumptions do not appear to be entirely reasonable, do not lead to a complete account of the regularities of matching, and effectively add only a little flexibility to the predictions of the original matching law. Application of the matching theory to different reinforcers and different instrumental behaviors is confounded by problems of response and reinforcement value transforms. The concept of background reinforcement seems more convenient than coherent, and very little is known about the nature of contact between background reinforcement and the organism's behavior. Further, intrinsic reinforcements for the instrumental and contingent responses are rarely considered (Herrnstein, 1979). For example, if the intrinsic reinforcements for key pecking and eating are considered, the matching law becomes equivocal and the matching theory incomplete.

The relations of the matching theory to important independent variables such as deprivation, amount of reinforcement, and contingency relations in the schedule (Gibbon, Berryman, and Thompson, 1974) are not well worked out. The assumptions of constant total behavior and the irrelevance of total reinforcement seem questionable. Rules for calculating the total pool of reinforcement and the boundary conditions and measurement of total behavior are rarely considered.

Perhaps most important, the ability of Eq. (23.3) to fit data has appeared to inhibit rather than encourage an experimental analysis of the determinants of matching, and the precise forms of its behavioral expression. Mathematical formulations, especially those with complex form and free parameters, have a life of their own. They may clarify theories and lead to new insights, but they also may be used to produce predictions unrelated to the concepts from which they were developed, or to ignore differences that bear on further development of a theory.

The matching and maximizing debate (Herrnstein and Loveland, 1975; Heyman, 1979, Chapter 20; Heyman and Luce, 1979; Nevin, Chapter 6; Shimp, 1969, 1975, Chapter 4; Silberberg and Ziriax, Chapter 5), though sometimes bordering on acrimony, at least called attention to potential mechanisms underlying matching. The animal is clearly doing more than just matching and less than perfect maximizing. What we call the phenomenon is considerably less important than what we find out about it.

My suspicion is that the notions underlying matching came from classical psychophysical procedures in which subjects adjust one stimulus to equal their perception of another (Baum, 1974b; Commons, Woodford, and Ducheny, Chapter 2). Such a source would explain the correlational emphasis of matching and why there has been so little concern with the relation of matching to survival problems. In the study of behavior empirical consistency is not its own end, but an important beginning. At the least, it seems time to move away from proof by cumulated examples, and toward a synthesis of psychophysics and survival. Investigators must relate variation in matching to variation in independent variables rather than simply observe correlations.

THE FUTURE OF MATCHING RESEARCH

It is evident that a phenomenon of such empirical consistency as matching in standard concurrent schedules will continue to interest scientists. It seems wise, therefore, to consider what steps might be taken to study it. I think it is useful to assume that species' typical regularities in behavior, such as simple matching (Eq. (23.1)), are likely to be based on phylogenetic solutions to recurring problems in the organism's selection environment. Within the constraints imposed by the gene pool of the species, such phylogenetic solutions should tend toward behavior that is optimal with respect to some naturally occurring circumstance. The argument is straightforward: for any group of organisms the genes of those whose behavior is more efficient in promoting survival and, ultimately, reproduction will be differentially represented in the next generation. To the degree that such efficient behaviors are heritable, the species will tend toward optimal behavior in its selection environment.

This is not to argue that organisms will respond efficiently with respect to some arbitrary standard imposed by the experimenter. A particular experimental setting may not contain the variables controlling behavior in the organism's niche, or it may provide only enough similarity to "trick" the organism into providing an unnecessary or inappropriate phylogenetic solution, as in the case of autoshaping in pigeons (Hearst and Jenkins, 1974), or the misbehavior of chickens, raccoons, and pigs (Breland and Breland, 1961). Further, an organism may not achieve the level of efficiency demanded by the experimenter because of limitations in its genetic possibilities or developmental path.

From this viewpoint it is not surprising that pigeons do not precisely maximize their intake in the matching situation (Heyman, Chapter 20; Heyman and Luce, 1979). That the birds consistently approach maximum intake despite the arbitrary nature of the circumstances suggests there may be an important relation between matching and natural foraging mechanisms (Krebs, 1978).

Beginning with this orientation one step in the further analysis of matching would be to examine the niche that has been created in the laboratory. The laboratory niche is quite complex from the standpoint of discovering an optimal behavioral solution that minimizes the cost per benefit. Each peck, or at least each relatively brief group of pecks, directed at either key, has some probability of producing a reinforcement. This reinforcement probability increases with lengthened interpeck interval, but the cumulative probability of payoff grows at different rates for any one schedule depending on which interval is up, and at different average rates for the two schedules. Therefore, no strategy involving fixed stay or switch times is clearly best. What is clear is that it is possible to earn most of the available reinforcements by alternating schedules in some probabilistic fashion while pecking at some minimal rate (Heyman, 1979, and Chapter 20).

In contrast to its complexity for optimal distribution of instrumental behavior, the laboratory niche is very simple from the point of view of locating and differentiating types of reinforcement. Only one type of reinforcement is usually available, and it is always present as the same location. The two "patches" are side by side, and, once the food is delivered, the animal need not search for food in space, or in time. The animal is almost an observer that will get food if it opens its mouth after appropriate variable intervals. From this viewpoint it is most interesting that Baum (Note 1) reports that placing a partition between the keys, in place of a changeover delay, disrupted matching. The birds would stay on the side with the fatter schedule and occasionally run around and pick up the reinforcement on the other side. See also Baum (1982).

A second step would be to analyze behavior in the real world to see what variables are important in determining distribution of time and effort. Baum's (1974a) work with wild pigeons in a Skinner box is a partial step in this direction, but it includes the peculiar limitations of the Skinner box and schedule mentioned earlier. There is no search or identification, the patches are very close together, and the reinforcer is clearly signaled. Observations like Kamil's (1978) on honey creepers, and Goss-Custard's (1976) on redshanks seem a more appropriate starting point.

A third step would be to use information from the subject's real-world niche to modify our laboratory procedures and determine the sorts of variables manipulated. Kamil (1978; and with Peters and Lindstrom, Chapter 8), Olton (1978, and Chapter 9, Lea, (1979, and Chapter 7), and Collier, Hirsch, and Kanarek (1977) have all attempted to bring aspects of real world feeding behavior into the laboratory.

A last step in the analysis of matching would be to attempt to force the organism to operate in an inefficient manner within the laboratory niche. If the mechanisms underlying matching are indeed phylogenetic in origin, matching should be under the relatively automatic control of a particular set of external and internal conditions. The efficiency of behavior related to this mechanism should depend on the relation of the experimental circumstances to the selection environment. Thus it should be possible to make the organism behave inefficiently by providing stimulus conditions that engage the underlying mechanism, but in the absence of the result characteristic of the animal's niche (Breland and Breland, 1961; Timberlake, Wahl, and King, 1982). Vaughan'work (Chapter 12) appears to be a step in this direction. Discovering conditions that disrupt matching should be very helpful in discovering the underlying mechanism(s) that control the phenomenon.

To return to the emperor's clothes, it seems they are not the elegant finery of a general theory, but still the work clothes of empirical consistency. So far, the expansion of the matching account has been based more on the excitement of conquering new fields of data, than on subduing and governing its own theoretical domain. The importance of the matching account cannot be denied. But those concerned with the emperor's clothes might be of better service by attending more to the quality of the fabric and less to the elegance and precision of the tailoring.

REFERENCE NOTES

1. Baum, W.M. General communication at the Harvard Symposium on Quantitative Analyses of Behavior: Matching and Maximizing Accounts, 1979.
2. de Villiers, P.A. *The matching law and aversive control of behavior.* Final Report to National Science Foundation, 1977.
3. Timberlake, W., and B. Peden. *Key pecking for food in pigeons: Regulation or reinforcement?* Paper presented at Meetings of the Midwestern Association for Behavior Analysis, Chicago, Illinois, May, 1977.
4. Herrnstein, R.J. *Premackian relativity and the matching law.* Unpublished manuscript, 1977.

REFERENCES

Allison, J. 1976. Contrast, induction, facilitation, suppression, and conservation. *Journal of the Experimental Analysis of Behavior 25*: 185–99.

Allison, J.; M. Miller; and M. Wozny. 1979. Conservation in behavior. *Journal of Experimental Psychology: General 108*: 4–34.

Anderson, N.H. 1978. Measurement of motivation and incentive. *Behavior Research Methods & Instrumentation 10*: 360–75.

Baum, W.M. 1973. The correlation-based law of effect. *Journal of the Experimental Analysis of Behavior 20*: 137–53.

_____. 1974a. Choice in free-ranging wild pigeons. *Science 185*: 78–79.

_____. 1974b. On two types of deviation from the matching law: Bias and undermatching. *Journal of the Experimental Analysis of Behavior 22*: 231–42.

_____. 1982. Choice, changeover, and travel. *Journal of the Experimental Analysis of Behavior 38*: 35–50.

Baum, W.M., and H.C. Rachlin. 1969. Choice as time allocation. *Journal of the Experimental Analysis of Behavior 12*: 861–74.

Bentham, J. 1789. *An Introduction to the Principles of Morals and Legislation.* London: Payne.

Berlyne, D.E. 1960. *Conflict, Arousal, and Curiosity.* New York: McGraw-Hill.

Bolles, R.C. 1967. *Theory of Motivation.* New York: Harper & Row.

Breland, K., and M. Breland. 1961. The misbehavior of organisms. *American Psychologist 16*: 681–84.

Catania, A.C. 1963. Concurrent performances, reinforcement, interaction and response independence. *Journal of the Experimental Analysis of Behavior 6*: 253–63.

Collier, G.; E. Hirsch; and R. Kanarek. 1977. The operant revisited. In W.K. Honig and J.E.R. Staddon, eds., *Handbook of Operant Behavior*, Englewood Cliffs, N.J.: Prentice-Hall.

de Villiers, P.A. 1977. Choice in concurrent schedules and a quantitative formulation of the law of effect. In W.K. Honig and J.E.R. Staddon, eds., *Handbook of Operant Behavior*, Englewood Cliffs, N.J.: Prentice-Hall.

de Villiers, P.A., and R.J. Herrnstein. 1976. Toward a law of response strength. *Psychological Bulletin 83*: 1131–53.

Digby, K. 1644. *Two treastises. In the one of which, the nature of bodies; in the other, the nature of mans soule; is looked into.* Paris.

Farley, J., and E. Fantino. 1978. The symmetrical law of effect and the matching relation in choice behavior. *Journal of the Experimental Analysis of Behavior 29*: 37–60.

Gibbon, J.; R. Berryman; and R.L. Thompson. 1974. Contingency spaces and measures in classical and instrumental conditioning. *Journal of the Experimental Analysis of Behavior 21*: 585–605.

Goss-Custard, J.D. 1976. Variation in the disperson of redshank *Tringa totamus* on their winter feeding grounds. *Ibis 118*: 257–63.

Hearst, E., and H.M. Jenkins. 1974. *Sign Tracking: The Stimulus-Reinforcer Relation and Directed Action.* Austin, Texas: The Psychonomic Society.

Herrnstein, R.J. 1961. Relative and absolute strength of response as a function of frequency of reinforcement. *Journal of the Experimental Analysis of Behavior 4*: 267–72.

____. 1970. On the law of effect. *Journal of the Experimental Analysis of Behavior 13*;243–66.

____. 1974. Formal properties of the matching law. *Journal of the Experimental Analysis of Behavior 21*: 159–64.

____. 1977. The evolution of behaviorism. *American Psychologist 32*: 593–603.

____. 1979. Derivatives of matching. *Psychological Review 86*: 486–95.

Herrnstein, R.J., and D.H. Loveland. 1975. Maximizing and matching on concurrent ratio schedules. *Journal of the Experimental Analysis of Behavior 24*: 107–16.

Heyman, G.M. 1979. A Markov model description of changeover probabilities on concurrent variable-interval schedules. *Journal of the Experimental Analysis of Behavior 31*: 41–52.

Heyman, G.M, and R.D. Luce. 1979. Operant matching is not a logical consequence of maximizing reinforcement rate. *Animal Learning and Behavior 7*: 133–40.

Kamil, A.C. 1978. Systematic foraging by a nectar-feeding bird, the amakiki (Loxops virens). *Journal of Comparative and Physiological Psychology 92*: 388–96.

Killeen, P. 1972. The matching law. *Journal of the Experimental Analysis of Behavior 17*: 489–95.

____. 1975. On the temporal control of behavior. *Psychological Review 82*: 89-115.

____. 1979. The genesis, modulation, and extinction of arousal. In M.D. Zeiler and P. Harzem, eds., *Reinforcement and the Organization of Behavior*, New York: John Wiley.

Krebs, J.R. 1978. Optimal foraging: Decision rules for predators. In J.R. Krebs and N.B. Davies, eds., *Behavioral Ecology*, London: Blackwell.

Kuhn, T.S. 1957. *The Copernican Revolution.* Cambridge, Mass.: Harvard University Press.

Lea, S.E.G. 1979. Foraging and reinforcement schedules in the pigeon: Optimal and non-optimal aspects of choice. *Animal Behavior 27*: 875-86.

Mazur, J.E. 1975. The matching law and quantifications related to Premack's principle. *Journal of Experimental Psychology: Animal Behavior Processes 104*: 374-86.

____. 1977. Quantitative studies of reinforcement relativity. *Journal of the Experimental Analysis of Behavior 27*: 137-50.

McSweeney, F.K. 1975. Concurrent schedule responding as a function of body weight. *Animal Learning and Behavior 3*: 264-70.

McSweeney, F.K., and D.A. Dericco. 1979. Rates of responding in the pigeon generated by simple and complex schedules which provide the same rates of reinforcement. *Animal Learning and Behavior 4*: 379-85.

Miller, H.L., Jr. 1976. Matching-based hedonic scaling in the pigeon. *Journal of the Experimental Analysis of Behavior 26*: 335-47.

Montgomery, K.C., and J.A. Monkman. 1955. The relation between fear and exploratory behavior. *Journal of Comparative and Physiological Psychology 48*: 132-36.

Myers, D.L., and L.E. Myers. 1977. Undermatching: A reappraisal of performance on concurrent variable-interval schedules of reinforcement. *Journal of the Experimental Analysis of Behavior 27*: 203-14.

Myerson, J., and F.M. Miezin. 1980. The kinetics of choice: An operant systems analysis. *Psychological Review 87*: 160-74.

Olton, D.S. 1978. Characteristics of spatial memory. In S.H. Hulse, H. Fowler, W.K. Honig, eds., *Cognitive Aspects of Animal Behavior*, Hillsdale, N.J.: Lawrence Erlbaum.

Pear, J.J. 1975. Implications of the matching law for ratio responding. *Journal of the Experimental Analysis of Behavior 23*: 139-40.

Platt, J.R. 1971. Discrete trials and free behavior situations. In H.H. Kindler and J.T. Spence, eds., *Essays in Neobehaviorism: A Memorial Volume to Kenneth W. Spence*, New York: Appleton-Century-Crofts.

Prelec, D., and R.J. Herrnstein. 1978. Feedback functions for reinforcement: A paradigmatic experiment. *Animal Learning and Behavior 6*: 181-86.

Rachlin, H. 1971. On the tautology of the matching law. *Journal of the Experimental Analysis of Behavior 15*: 249-51.

Rachlin, H., and B. Burkhard. 1978. The temporal triangle: Response substitution in instrumental conditioning. *Psychological Review 85*: 22-45.

Reberg, D.; N.K. Innis; B. Mann; and C. Eizenga. 1978. "Superstitious" behavior resulting from periodic response-independent presentations of food or water. *Animal Behavior 26*: 507-19.

Shimp, C.P. 1969. Optimal behavior in free-operant experiments. *Psychological Review 76*: 97–112.

_____. 1975. Perspectives on the behavioral unit: Choice behavior in animals. In W.K. Estes, ed., *Handbook of Learning and Cognitive Processes*, vol. 2., Hillsdale, N.J.: Lawrence Erlbaum.

Spencer, H. 1870. *Principles of Psychology*. 2nd ed. London: Williams and Morgate.

Staddon, J.E.R. 1979. Operant behavior as adaptation to constraint. *Journal of Experimental Psychology: General 108*: 48–67.

Staddon, J.E.R., and S. Motheral. 1978. On matching and maximizing in operant choice experiments. *Psychological Review 85*: 436–44.

Timberlake, W. 1969. *Continuous coding of general activity in the rat during repeated exposure to a constant environment and to stimulus change*. Unpublished doctoral dissertation, University of Michigan.

_____. 1977. The application of the matching law to simple ratio schedules. *Journal of the Experimental Analysis of Behavior 25*: 215–17.

_____. 1980. An equilibrium theory of learned performance. In G.H. Bower, ed., *Psychology of Learning and Motivation*, vol. 14. New York: Academic Press.

Timberlake, W., and J. Allison. 1974. Response deprivation: An empirical approach to instrumental performance. *Psychological Review 81*: 146–64.

Timberlake, W., and M. Wozny. 1979. Reversibility of reinforcement between eating and running by schedule changes: A comparison of hypotheses and models. *Animal Learning and Behavior 7*: 461–69.

Timberlake, W.; G. Wahl; and D. King. 1982. Stimulus and response contingencies in the misbehavior of rats. *Journal of Experimental Psychology: Animal Behavior Processes 8*: 62–85.

Wozny, M. 1979. *Models and microeconomics of performance under fixed ratio schedules*. Unpublished doctoral dissertation, Indiana University.

NAME INDEX

Ainslie, G., 84
Alferink, L.A., 97
Allison, J., 79, 282, 376, 395, 411, 412, 423, 500, 504, 516, 523, 525, 528, 529, 532, 534, 539, 546, 560
Anderson, J.R., 127
Anderson, N.H., 513, 552
Anger, D., 115, 116, 268
Antelman, S.M., 505
Arnett, F., 297
Arnold, S., 190
Atkinson, J.W., 82, 256
Augerinos, G., 206
Autor, S.M., 181, 328
Awh, R.Y., 527
Azrin, N.H., 328, 332

Baddley, A.D., 207
Baldock, M.D., 515
Basmann, R.L., 365
Battalio, R.C., 25, 47, 154, 227, 263, 323, 347, 363, 365, 370, 371, 375, 396, 400, 401, 418, 423, 433, 438, 442, 443, 459, 460, 505, 507, 516, 525
Bättig, K., 210
Baum, W.M., 4, 47, 112, 116, 133, 136, 154, 161, 179, 181, 227, 228, 230, 231, 239, 243, 252, 253, 263, 268, 282, 305, 322, 323, 337, 341, 348, 350, 356, 365, 368, 369, 413, 434, 461, 465, 468, 469, 472, 474, 475, 484–86, 494, 508, 511, 550, 551, 553, 562, 564, 565
Bauman, R.A., 328
Beale, I.L., 174, 510
Beatty, W.W., 206, 222
Beavers, W.O., 515
Becker, G.S., 353, 355
Becker, J.T., 207
Benson, L., 376
Bentham, J., 558
Berg, B., 206
Berlyne, D.E., 561
Berryman, R., 562
Bevan, P., 332, 341, 468
Bindra, D., 494
Birch, D., 82, 256
Bitterman, M.E., 191
Black, A.H., 206
Black, R.W., 178
Blackman, D.E., 258
Blough, D.S., 7, 484, 507
Blough, P.M., 7
Boice, R., 220
Bolles, R.C., 327, 504, 561
Boren, M.C.P., 101
Borotsky, S., 206
Bouzas, A., 253
Bower, G.H., 127
Bradshaw, C.M., 332, 341, 342, 468
Breland, K., 504, 563, 564
Breland, M., 504, 563, 564
Brown, P.L., 505, 515
Brownstein, A.J., 328, 376, 467, 468, 470, 474

Bryson, A.E., 452
Buck, S.L., 258, 287, 289, 293
Burkhard, B., 245, 282, 350fn, 370, 375, 376, 379, 395, 401, 412, 414–15, 423, 503, 505, 516, 525, 528, 529, 560

Cabral, B.D., 75
Cagguila, A.R., 505
Calhoun, J.B., 220
Carman, J.B., 330, 331
Carmona, A., 505
Casey, J., 114, 131, 134-37, 139, 140, 143, 149, 155, 434, 484, 516
Catania, A.C., 7, 22, 59, 112, 118, 145, 254, 282, 319, 368, 419, 434, 513, 549fn, 552
Cerella, J., 198
Charnov, E.L., 170, 172, 174, 190, 194, 195, 208, 218, 227, 235, 433
Cheney, C.D., 90
Chiang, A.C., 425
Chung, S-H., 44, 84, 451
Church, R.M., 306, 307, 311, 340–41
Cole, B.K., 28
Collier, G.H., 170, 174, 181, 197, 227, 528, 532, 564
Collison, C., 206
Commons, M.L., 7, 26, 28, 41, 48, 55, 59, 62, 67, 72, 75, 451, 452
Corbit, J.D., 506fn
Cowie, R.J., 195
Cox, V.C., 498fn, 505
Crossman, E.K., 80, 81, 97
Cumming, W.W., 28
Curio, E., 200

Dague, G., 497
D'Amato, M.R., 207, 222
Dashiel, J.F., 210
Davies, N.B., 173, 200, 227, 433
Davison, M.C., 80, 181, 357, 367, 368, 465, 469, 486
Delbruck, N., 476
Deluty, M.Z., 306, 307, 309, 311, 332-34, 337, 340–41
Dember, W.N., 210
Dericco, D.A., 561
DeVilliers, P., 113, 132, 133, 153, 155, 161, 281, 305, 310, 319, 323, 332–35, 337, 338, 340, 341, 353, 355–57, 361, 366, 370, 371, 423, 434, 453, 468, 474, 484, 499, 550, 551, 553, 557, 561, 565
DeWeese, J., 101
Dickinson, A., 502
Digby, K., 558
Dinsmoor, J.A., 327, 330
Donahoe, J.W., 434, 494, 496-98, 501, 506, 508, 509, 513, 516
Douglas, R.J., 210
Dow, S., 184
Downing, K., 511
Driscoll, P., 210
Ducheny, J.R., 7, 26fn, 75, 451
Dunham, P.J., 500
Dwyer, G.P., Jr., 418

Edmon, E.L., 245
Egan, J.P., 62, 73
Eizenga, C., 561
Engberg, L.A., 502
Erichsen, J.T., 172
Estes, W.K., 26, 59, 74, 113, 127, 327, 332, 495
Evans, S.H., 126

Fallon, D., 314
Falls, J.B., 208, 209
Fantino, E., 96, 181, 298, 310, 333, 476, 552
Farley, J., 310, 332-34, 336-41, 552
Farrell, L., 515
Ferster, C.B., 80, 81, 101, 174, 350, 351, 455, 498, 510
Findley, J.D., 81, 174, 276, 314, 468, 509
Fisher, J., 183
Fitzpatrick, J.W., 195-97
Fleshler, M., 99, 352, 405, 462
Fowler, H., 210, 502
Frey, P.W., 497

Gage, F.H., 126, 210, 219, 220–22
Gaioni, S.J., 495
Gamzu, E., 245, 252, 281, 286, 287, 297, 515
Gandelman, R., 368
Gause, G.F., 170
Gay, R., 290
Gentry, G.D., 452
Gibb, J.A., 194, 208, 218
Gibbon, J., 434, 515, 562
Gilbert, T.F., 4
Gill, F.B., 192, 200
Glander, K.E., 200
Gofseyeff, M., 80
Gold, L., 515
Goldstein, D.A., 153
Gollub, L.R., 253
Gormezano, L., 495fn, 496fn, 507
Goss-Custard, J.D., 173, 192, 200, 564
Graham, C.L.D., 106
Grant, D.L., 496, 506
Gray, L., 190, 221

Green, D.M., 41, 62
Green, L., 47, 154, 227, 263, 323, 347, 363, 365, 370, 371, 396, 400, 401, 423, 433, 438, 442, 443, 459, 460, 516, 525
Green, S.L., 275
Gustavson, C.R., 368
Guthrie, E.R., 494

Hager, J.L., 504
Hale, J.M., 127
Hall, D.J., 172
Hamblin, R.L., 368
Hamilton, B.E., 114, 134-37, 139, 140, 143, 149, 155, 256, 287, 289, 434, 484, 516
Hamilton, W.J., 207
Hamlin, P.H., 528
Hancock, R.A., 515
Handelman, G.E., 207
Hanson, N.R., 112
Hanson, S.J., 227
Harzem, P., 101, 106
Hassell, M.P., 227, 228
Hawkes, L., 114, 116, 161-63, 513
Hearst, E., 28, 281, 515, 563
Hefferline, R.F., 22
Heinrich, B., 190, 200
Hemmendinger, D., 515
Herrnstein, R.J., 4, 7, 22, 26fn, 44, 76, 82, 84, 112, 116, 131-34, 136, 153-55, 161, 164, 192, 198, 220, 222, 227, 230, 231, 243, 244, 246, 252, 253, 258-59, 264, 274, 281, 282, 297, 299, 305, 319, 327, 331-33, 338, 340, 341-42, 347, 348, 350, 352, 358, 361, 366, 370, 371, 395, 412, 413, 419, 420, 422, 423, 434, 437, 438, 439, 440, 450, 451, 453, 460, 461, 472, 475, 480-82, 484, 485, 494, 497-99, 507, 509, 516, 532, 549, 550, 551-62, 565
Heth, C.D., 525, 529
Heyman, G.M., 25, 113, 136, 149, 155, 158, 161, 164, 184, 231, 360, 370, 419, 420, 422, 438, 439, 440, 451, 465, 468, 470, 475, 478, 480-82, 485, 486, 516, 557, 562-64
Hinde, R.A., 182, 183
Hineline, P.N., 452
Hinson, J.M., 227, 245, 247, 252, 256, 283, 297, 477
Hirsch, E., 528, 564
Hitch, G., 207
Hoffman, H.S., 99, 352, 405, 462
Hogan, J.A., 185, 407, 411
Holland, P.C., 496, 501
Hollander, M., 511
Hollard, V., 368
Holling, C.S., 190, 228
Holz, W.C., 332
Homans, G.C., 433
Honig, W.K., 174, 206, 207, 219
Houston, A.I., 135, 227, 469, 486
Hughes, R.N., 192
Hull, C.L., 84, 132, 494, 495
Humphrey, N.K., 185
Hunter, I.W., 465, 469, 486
Hursh, S.R., 369, 452
Hutchinson, G.E., 228

Inman, O.P., 90
Innis, N.K., 561
Itani, J., 183
Ivlev, V.A., 172

Jenkins, H.M., 174, 281, 498, 505, 515, 563
Johnson, C.T., 210, 220-22

Kacelnik, A., 74, 184, 191, 201
Kagel, J.H., 25, 47, 154, 227, 263, 323, 347, 363, 365, 370, 371, 375, 396, 400, 401, 418, 423, 433, 438, 442, 443, 459, 460, 505, 507, 516, 525
Kakolewski, J.W., 498fn, 505
Kalat, J.W., 504
Kamil, A.C., 74, 192, 194, 196, 198, 200, 208, 209, 564
Kamin, L.J., 495, 496
Kanarek, R., 564
Kaplan, M., 90, 94
Kaufman, L., 174, 181
Keehn, J.D., 81
Keenan, B., 22
Kehoe, E.J., 496fn
Keller, K., 286
Kessel, R., 434, 460, 472
Killeen, P.R., 227, 256, 282, 290, 296, 297, 314, 434, 451, 459, 460, 472, 550, 552, 554, 557-58
Kintsch, W., 101
Klemm, W.R., 365
Konorski, J., 494, 502
Kram, R., 227, 477
Krebs, J.R., 74, 170, 172, 174, 184, 190, 191, 194, 201, 208, 218, 227, 433, 549, 551, 563
Kremer, E.F., 495
Kuhn, T.S., 112, 551

Lander, D., 174
Landford, A., 376
Lea, S.E.G., 74, 170, 174-77, 179, 181, 197, 532, 534, 564

Leigland, S., 357
Levey, A.B., 496fn
Lobb, B., 181
Locurto, C., 515
Logan, C.A., 181
Logan, F.A., 84, 174, 350
Loveland, D.H., 134, 153, 192, 252, 269, 274, 319, 361, 437, 450, 532, 551, 562
Lowe, C.F., 101, 106
Luce, R.D., 82, 136, 158, 360, 388, 439, 475, 478, 562, 563
Lyons, J., 295

MacArthur, R.H., 169, 190
MacCrimmon, K.R., 371
McDowell, J.J., 434, 460, 472
Mackintosh, N.J., 178, 502
McLeod, A., 115, 116
McMillan, A. St. C., 330, 331
McNally, K., 297
McNamara, J., 135, 227, 469, 486
McSweeney, F.K., 244, 322, 352, 552, 554, 561
Maier, N.R.F., 214
Maier, S.F., 502
Maisiak, R., 497
Maki, W.S., 206
Malagodi, E.F., 101
Malone, J., 288, 293, 297
Malott, R.W., 102
Mandell, C., 55
Mann, B., 561
Mansfield, R.J.W., 330
Marr, M.J., 452
Martin, I., 496fn
Matthews, L.R., 369
Mazur, J.E., 40, 51, 99, 100, 376. 443-46, 450, 451, 453, 525, 560
Meehl, P.E., 460
Menlove, R.L., 115, 143, 149, 253
Menzel, E.W., 183
Meunzinger, K.T., 179
Miezin, F.M., 434, 459, 460, 465, 482, 486, 557
Millenson, J.R., 330, 331, 332
Miller, H.L., Jr., 269, 368, 369, 452, 557fn
Miller, M., 282, 395, 412, 504, 516, 525, 560
Miller, N.E., 505
Miller, S., 494, 502
Mintz, D.E., 22, 80
Moffitt, M., 513, 514
Monkman, J.A., 561
Montgomery, K.C., 561
Moore, B.R., 174
Morse, W.H., 112
Motheral, S., 154, 245, 247, 282, 323, 359, 369, 434, 481-82, 516, 549fn, 552
Mourer, D.J., 80
Mowner, O.H., 327
Muir, D., 174
Murton, R.K., 200
Muth, J.F., 51
Myers, D.L., 181, 305, 414, 561
Myers, L.E., 181, 305, 414, 561
Myerson, J., 434, 459, 460, 465, 482, 486, 557

Nägeli, H.H., 210
Neimark, E.D., 26
Neuringer, A., 511
Newman, P., 350fn, 354
Nevin, J.A., 26fn, 41, 42, 113, 114, 115, 117, 133, 134-37, 147-49, 155, 156, 159, 161, 164, 252, 256, 282, 287, 434, 465, 475, 486, 562
Norman, W.D., 322
Notterman, J.M., 22

Oaten, A., 192, 235
Olton, D.S., 126, 183, 206, 207, 211, 218, 219, 220, 222, 564
Over, R.F., 80
Overmier, J.B., 494

Panksepp, J., 368
Pareto, V., 78
Partridge, L., 200
Pear, J.J., 4, 6, 413, 552, 557, 560
Pearce, J.M., 502
Peck, J.R., 26fn, 75
Peden, B.F., 555
Perkins, C.C., 515
Peterson, C., 476
Pianka, E.R., 169, 190
Pietrevicz, A.T., 192, 196, 198
Platt, J.R., 7, 126, 127, 553
Pliskoff, S.S., 155, 198, 253, 254, 290, 322, 334, 352, 443-44, 467-70, 474-77
Poirier, F.E., 183
Powell, R.W., 85-92, 96, 97, 99, 101
Prelec, D., 341, 342, 553, 558, 559
Premack, D., 376-77, 384, 494, 497, 500, 516, 525, 541
Priddle-Higson, P.J., 101, 106
Prokasy, W.F., 495fn, 507
Ptolemy, 551
Pulliam, H.R., 170, 190, 194, 198, 227, 433, 467
Pyke, G.H., 170, 190, 194, 200, 227, 433

Rachlin, H.C., 4, 25, 26fn, 47, 112, 113, 114, 132, 133, 154, 171, 227, 245, 256, 263, 281-82, 286, 287, 289, 292, 300, 323, 327-30, 347, 348fn, 350, 356, 359, 360, 362, 363, 365, 369-71, 375,

379, 395, 396, 400, 401, 412, 414-15, 423, 433, 434, 442, 452, 460, 469, 481, 484, 485, 503, 505, 507, 516, 525, 528, 529, 550, 552, 554, 560
Rapport, D.J., 172
Reberg, D., 561
Rector, B.L., 4, 6
Rector, M., 557
Rees, A., 429, 528
Reichman, O.J., 200
Reid, H.M., 469
Renner, K.E., 331
Rescorla, R.A., 264, 277, 451, 494, 495, 501
Reynolds, G.S., 7, 22, 115, 116, 118, 253
Ricci, J.A., 515
Richards, J.J., 258
Rider, D.P., 319
Rilling, M., 287
Roberts, W.A., 206
Roper, T.J., 185, 407, 411, 532, 534
Rothstein, B., 258, 287, 289
Rovee-Collier, C.K., 170, 174, 197, 227
Rowe, D., 297
Royama, T., 190, 200
Rozin, P., 504
Ryan, J.C., 174, 194, 208, 218

Sahlins, M., 170
St. Claire-Smith, R., 502
Samuelson, R.J., 206, 527
Sargent, T.D., 192, 194, 196, 200, 209
Schaeffer, R.W., 376
Schild, M.E., 314
Schlatter, J., 210
Schlosberg, P., 206, 211
Schoener, T.W., 194, 198, 433
Schoenfeld, W.N., 28
Schrot, J., 470
Schuster, C., 80
Schuster, R., 328-30
Schwartz, B., 90, 127, 163, 245, 252, 281, 286, 287, 297, 515
Segal, E.F., 532
Seligman, M.E.P., 182, 502, 504
Seraganian, P., 174
Shapiro, N., 528, 532, 534
Shavalia, R.T., 206, 222
Sheffield, F.D., 505
Sherman, J.A., 98
Shettleworth, S.J., 256, 287, 504
Shimp, C.P., 7, 112-15, 117, 118, 120, 122, 125-27, 131, 133-36, 153-56, 160-63, 220, 222, 257, 282, 284, 323, 352, 434, 454, 468, 484, 511, 513, 514, 516, 562
Shull, R.L., 198, 253, 322, 328, 452, 470
Sidman, M., 101, 102
Siegel, S., 153, 273
Silberberg, A., 115, 126, 131, 134-37, 139, 140, 143, 149, 155, 156, 161, 184, 256, 287, 289, 300, 434, 470, 484, 516, 562
Silverman, P.J., 25
Silverman, L.T., 81, 97
Simmelhag, V.L., 246, 254, 283, 285, 370
Sitgler, G.J., 355
Skinner, B.F., 7, 22, 80, 81, 101, 112, 113, 132, 174, 350, 351, 395, 455, 494, 498, 501, 509, 510, 515, 516
Small, W.S., 182
Smith, J.N.M., 191, 200, 209
Smith, J.P., 227
Smith, M.C., 496fn
Solomon, R.L., 327, 506fn
Spealman, R.D., 286, 287
Spear, D.J., 452
Spencer, H., 558
Squires, N., 181, 476
Staddon, J.E.R., 22, 115, 116, 125, 154, 161, 227, 245, 246, 247, 252, 254, 256, 259-60, 282, 283, 285, 296, 323, 359, 369, 370, 376, 395, 401, 412, 415, 423, 426, 434, 459, 460, 477, 481-82, 503, 507, 516, 524, 525, 528, 529, 549fn, 551, 560
Stahl, E.P., 214
Stahl, J.M., 214
Stebbins, W.C., 101, 102
Stelmach, G.E., 127
Stevenson-Hinde, J., 182
Stickney, K.A., 496, 516
Striefel, S., 532
Stubbs, D.A., 118, 155, 334, 352, 443-44, 469, 475-77
Suzuki, S., 206
Sweatman, H.P.A., 191, 209
Swets, J.A., 41, 62
Szabadi, E., 332, 341, 468

Tardiff, R.R., 221
Taylor, P.J., 74, 184, 191, 201
Temple, W., 181, 369
Terrace, H.S., 515
Thomas, D.R., 295, 502
Thomas, E.A.C., 484
Thomas, J.R., 98
Thompson, D.M., 314
Thompson, R.L., 562
Thorndike, E.L., 327, 455
Timberlake, W., 376, 413, 414, 496, 500, 506, 508, 516, 523, 525, 539, 552, 553, 555, 560, 561, 564, 565
Tinbergen, L., 190, 191, 192, 200
Toda, M., 371
Todorov, J.C., 257, 282, 284
Tolman, E.C., 516
Trapold, M.A., 494

Trevett, A.J., 357, 367
Trowill, J.A., 368
Tversky, A., 82, 377-78

Uster, H.J., 210

Valenstein, E.S., 498fn, 505
Vaughan, W., Jr., 264, 272, 274, 348, 352, 434, 437-39, 443, 446, 447, 450, 451, 453, 460, 461, 482, 564
Vieth, H., 287
vom Saal, W., 498

Wagner, A.R., 264, 277, 451, 494, 495
Walker, J.A., 206, 210, 219, 220-22
Walsh, V.C., 429
Ware, D.M., 172
Warren, A.G., 525, 529
Watt, K.E.F., 207
Watts, H.W., 528
Webbe, F.M., 101
Webber, M.I., 172
Weisman, R.G., 376
Weissman, A., 28
Welker, R.L., 502
Werner, E.E., 172
Werz, W.A., 206
Wessells, M.G., 494, 497, 498fn, 501
Wheatley, K.L., 257, 282, 284
White, K.G., 253
Whitham, T.G., 209
Wickelgren, W.A., 51
Wilkie, D.M., 22
Williams, B.A., 258, 287, 289-94, 296, 299, 300, 502
Williams, D.R., 505
Williams, H., 505
Williams, J.L., 328
Williams, R.J., 357, 367
Wilton, R., 290
Winograd, E., 90, 91, 94
Winton, A.S.W., 510
Wolf, L.L., 192, 200
Wolfe, D.A., 511
Woodford, M., 7, 75, 451
Wozny, M., 282, 395, 412, 500, 504, 516, 525, 560

Zach, R., 208, 209
Zanich, M.L., 502
Ziriax, J.M., 114, 131, 134-37, 139, 140, 143, 149, 155, 156, 184, 300, 434, 484, 516, 562
Zoladek, L., 206
Zuriff, G.E., 419

SUBJECT INDEX

Abortive response chains, 330-31
Accuracy of prediction (goodness of fit), 550-51, 552-53, 562
Acquisition, 75, 455, 502, 506-07; of data, 75
Additivity of reinforcers, 41, 48, 55
Additivity theory, 286-90, 296, 299, 300
Adenosine triphosphate, 183
Adjunctive behavior, 376
Advance key, 174
Adjustment procedure, 46
See also Titration
Aggregate: date, 25-28; functions, 48, 74; level of observation, 27; molar theory, 47; reinforcement value, 27; reinforcers, 36, 41, 48
Algorithm, 263
Amount of reinforcement, 36, 37, 42, 47, 55, 59, 62
Analysis: hierarchical, 127; local, 162
Analysis, levels of, 113; macro, 31 def, 75; micro, 34 def; molar, 31 def, 75, 111-16, 126, 127, 131-33, 139, 154 def, 507; molecular, 31 def, 75, 131, 133-36, 139, 150, 154 def, 161, 164, 507
Anti-matching, 369
Arcsine transformation, 49, 50
Association theory, 112
Assumptions of the matching theory, 549-50, 553-61, 562. *See also* Matching; Matching law
Autoshaping, 505-06, 514-15, 563
Averaging theory, 451
Aversive stimuli, 327-29, 340

Bayes' law, 60
Behavior: adjunctive, 376; all-or-none, 173, 184; choice, 131-33; commensurate measures of, 555-56, 560-61; instrumental, 169-70; local organization, 111-17, 119-20, 123, 127-28; maintenance of, 455; natural, 169-70, 172, 174; optimal, 170, 182, 184, 563-64; plasticity of, 454; schedule-induced, 22; structure of, 8, 9, 17, 22-23; total, 553-54, 556-57, 560-62; unlearned, 554, 557-58; wave, 8, 11, 17, 22, 23
Behavioral change, 434, 436
Behavioral contrast, 283-86, 287-88, 290, 297
Behavioral ecology, 169, 171, 173, 176, 183
Behavioral equilibrium, 433-40, 451, 452-56
Bias, 41, 43-44, 46, 49, 59-60, 67, 161, 162, 230, 232, 356, 360, 365, 369-70
Biological constraints, 502, 504-06
Blocking: of reinforcement, 502; of stimulus control, 495, 499
Blue jays, 189, 192-94, 196, 198-99
Boyle's law, 248
Brain stimulation, reinforcing, 368
Budget, 353-55, 362-67, 369

Catocala moths, 192-94, 196, 198, 200
Causation, 182

Central place foragers, 207, 219. *See also* Foraging
Changeover, 267, 268
Changeover delay, 551, 564; in punishment experiments, 306, 322; in reinforcement experiments, 319, 322, 323
Changeover key, 174
Children, 377, 384-88
Choice, 136, 139, 149, 347-71; allocation, 134, 135, 139, 143, 149; alternatives, 133, 143, 145, 147; asymmetrical, 350; axioms, 353; behavior, 131-33; ethological interpretation, 74; microstructure, 135, 139, 143fn, 149; period, 28-30, 62, 76; probability, 42-44, 74; probability correct, 43
Classical conditioning: compared with operant (instrumental), 494, 497-500; of response classes, 505-06
Cognitive limitations, 454
Commodity bundles, 397, 400, 424
Commodity space, 353-55
Comparison: key defined, 37; schedule, 27, 56; sequence of reinforcers, 27; substimulus, 29, 37, 40, 44, 46
Competition, 245-47, 249, 254-55, 258-59
Competitive exclusion, 170
Completion model, 544-46
Component duration, 283-84, 287-90, 291-92, 297
Concurrent interdependent procedure, 40
Conditioning: classical. *See* Classical conditioning. evolution of, 170, 184-85; as laboratory artifact, 173, 182, 185; operant (instrumental), 494, 497-500
Conservation, 523, 528, 532-34, 538, 541, 543
Conservation model, 395, 412-13, 423
Constant elasticity of substitution (CES), 418-20
Consumption, 174
Contingency, 375-77, 379
Contingent behavior, duration of, 81, 85, 91, 95, 101, 104
Contingent shock, 328, 329
Contrast, 170, 171 def, 173, 177-78, 181, 184, 245 def, 252, 254, 258; local, 256-57
Controlled-amount paradigm, 523-24, 534, 541, 543-44, 546
Correct rejection, 30
Correlational law, 551, 553, 562
Cost-benefit analysis, 229
Cross-situational test of maximizing (molar) theory, 472-78
Cumulative record, 113
Cycle (of a *T*-schedule), 28-30, 36-37; base length, 30; four-cycle case, 31, 34-37, 43, 50-52, 60, 67, 68, 70; length, 30, 55; number of, 68-69; six-cycle case, 34-35, 43, 50-52, 68, 70

Daily income, 525, 527. *See also* Income; Wage rate
Decision rule, 42 def, 43-45
Decision theory, 171
Decrementing functions, 41, 48
Delay of punishment gradient, 331
Demand theory, 354-55, 364fn, 371
Density of prey. *See* Prey
Depletion, 236, 238; of food supply, 215-18
Deprivation, 550-51, 553-55, 561-62
Dimensional analysis, 555
Diminishing returns, 353, 359, 366
Diminishing substitution, 353, 366
Discrete response, 348, 360
Discrete trial non-correction choice procedure, 26
Discrete trial test procedure, 219-20, 222, 541, 553
Discriminability, 25, 55, 76; imperfect, 72, 74; index of, 62; measure of, 68; of reinforcement, 64, 67, 68; of substimuli, 51, 75
Discrimination, 26, 27 def; data, 38, 41, 43-44, 51, 56, 57, 74; four-cycle case, 28, 36, 37, 43-44, 48-50, 54, 57; imperfect, 28, 42, 67, 76; procedure, 27, 44, 46, 48, 50, 57, 76; six-cycle case, 35, 43-44, 50, 54; task, 35
Drifting response rate, 5, 7

Eating, 528-29, 534
Ecology, 169, 173, 176, 189, 190-92, 194, 197-201
Economic theory, 74, 353-55, 433, 452, 455
Economics, 347-71, 524, 527-28
Elasticity, 452
Elemental data, 46, 55
Encounter frequency, 171-72, 176-77
Energy: maximization, 172; net worth of prey, 171-72; rate of intake, 171, 176
Environmental niche, 169, 183
Equilibrium points, 527
Equilibrium theories, 560
Ethological theories, 74
Evolution, 182, 184-85, 436, 454-55; of conditioning principles, 170, 184-85; of human intelligence, 185
Evolutionarily conservative traits, 183, 185
Excitatory tendency, 22
Exclusive preference, 362, 437, 439
Exclusive responding, 532, 534
Expansion paths, 415-18
Expected payoff, 26, 44, 59, 61, 69

Expected time: until punishment, 310, 311, 313, 323; until reinforcement, 323
Exponential decay function, 51
Exponential waiting-time distribution of punishment, 306
Extinction, 30

Facilitation, 525, 538, 539, 541–46 *passim*
False alarm, 30
Feedback function, 229–31, 233, 237–38, 243, 329, 341–42, 350–53, 359–61, 434, 461 def
Feedback system, 229
Field studies, 172–73, 181
Flycatcher, 169, 173
Following-schedule effect, 290–300
Foraging, 74, 169–76, 181–85, 189–92, 194–95, 197–98, 200–201, 227–28, 231–33, 239, 347, 534; central place, 207, 219; optimal, 169–71, 173, 174, 176, 184, 190, 194 def, 195–96, 233; rate of, 231–33, 235–37
Forgetting, 27, 28, 59, 67, 74, 75; curves, 57; hyperbolic function of, 68, 69, 70; in laboratory tests, 210–11, 221–22; model, 54, 67; in natural habitats, 205, 207–10, 220–21
Free behavior paradigm, 174
Free food, 401–02, 403, 406–07, 410–12, 415–17, 423
Free operant concurrent choice procedure, 25
Free parameter, 550, 559, 562
Free responding, 541
Functional response theory, 228

Gene-pool, 173
General process learning theory, 182, 184
Giving up time, 196, 197, 199, 235
Glucose, 183
Great tit, 172, 184

Handling time, 171–72, 175–76, 183
Hedonist tradition, 558
Heuristic, 263 def, 264, 279
Hit, 30
Homothetic contours, 360–61
Human intelligence, 185
Human learning, 185
Hyperbolic transformation, 51

Incentive, 350
Income, 353, 361, 365; -compensated wage change, 400–403, 412, 415; effect, 527. *See also* Daily income; Wage rate
Independent variables, 551, 553, 562–63
Indifference, 40, 44, 46, 56, 76, 243
Indifference contour (curve), 353–55, 357, 358, 361, 364fn, 366–67, 378–80, 382–84, 388–91, 397 def, 398–401, 403–04, 406, 412, 414–15, 418, 420–21, 525–28
Inertia, as determinant of behavior, 82–83, 85–97, 102
Inhibitory tendency, 22
Instrumental behavior, duration of, 81, 84, 85, 87, 91, 95, 101
Integration model, 544–46
Interchangeover times, 114–16, 126, 139 def, 140–49, 439
Interim activities, 246–47, 255, 257, 370. *See also* Leisure; Nonpecking activities
Interresponse time, 8, 11, 16, 17, 22, 116, 122, 125–27, 161–63, 439, 447

k, 552–53, 555–56, 560–61
k parameter, 332, 337–40
Key-pecking rate, 40

Labor, 525, 527
Labor supply, 371, 395, 396, 400, 401, 403–04, 412, 414, 415, 419, 423, 523, 525, 527–28; bitonic function, 403, 411–15, 423
Laboratory artifact, 173, 182, 185
Laboratory niche, 563–64
Law of demand, 411
Law of effect, 435, 455–56, 460, 485
Learning: constraints on, 182; general process theory, 182–83, 184–85; human, 185; probability, 190–91; of song, 183; of wild animals, 183
Leisure, 370, 395–98, 401, 412, 418, 419–20, 422–24, 442–43, 525–27. *See also* Interim activities; Nonpecking activities
Levels of analysis. *See* Analysis
Lever pressing, 523, 527–29, 532–34, 537–39, 542–43, 546
Licking (drinking), 524, 525, 528–29, 532, 537–43, 545, 546
Likelihood ratio criterion, 63
Local contrast, 256, 258
Local organization of behavior, 111–17, 119–20, 123, 127–28
Local response rates, 86 def, 101–02, 287–90
Logit transformation, 49 def, 50, 61, 62, 73
Lost-shift search strategy, 221
Luce's axiom, 388

Macro level results, 75. *See also* Analysis, levels of
Macro level stimuli, 28 def, 31
Mammals, 173
Marginal rate of subsitution, 379 *See also* Substitution

Marginal utility, 247, 452
Marginal value theorum, 194-95
Matching, 25 def, 132 def, 133-37, 139, 149, 154-55, 157, 159, 161, 163-64, 178, 220, 243, 246, 251-53, 258, 263, 268-69, 274, 331, 333-34, 337-39, 341, 433-36, 438-46, 450-52, 454-56, 459 def, 460, 461, 467; apparent, 28, 72; biased, 232, 246; curve, 44; equation, 61, 67; expected payoff, 26, 61, 69, 72; law. *See separate listing.* limits, 551, 553, 556, 560, 562; local, 70, 72, 73; molar level, 44, 46, 47, 59, 61-62, 69, 74; molecular level, 44, 46, 47, 59; in multiple schedules, 282-83, 290, 295, 300; obtained aggregate, 75; obtained reinforcement, 43-44, 47, 75; probability of response, 74; programmed payoff, 26, 44, 47, 61, 74; relative expected payoff, 59, 61; strategies, 557, 562-63; theory, 69, 549-50, 553-61, 562, 549, 552-55, 558-60, 562; time-allocation. *See separate listing.* traditional, 75; unconditioned, 485
Matching law, 4, 7, 26, 42, 47, 59, 61, 72, 76, 126, 132-33, 154, 161, 164, 227 def, 228, 248-49, 347-71, 395, 412-14, 422-26; generalized, 174, 181, 184, 549-52, 557, 560; for punishment, 306-07, 311, 313, 322-23; for reinforcement, 305, 317, 319, 322-23; for reinforcement and punishment, 309, 310; simple, 549-51, 560-62
Mathematical model of concurrent variable interval-variable interval schedules, 464-66, 473-75
Mathematical model of concurrent variable interval-variable interval schedules with a changeover delay, 471
Mathematical model of interdependent concurrent variable interval-variable ratio schedules, 478-81
Mathematical model of concurrent variable interval-variable ratio schedules, 480
Mathematical model of changeover rate, 465-68
Mathematical model of changeover rate with a changeover delay, 470
Maximizing, 25, 26fn, 26-27 def, 42, 47, 134-36, 140, 220, 263, 264, 268, 269, 272, 274-75, 277, 347-71, 375, 383, 390-91, 412, 419, 422, 424, 433 def, 433-38, 440-48, 450, 452, 454, 460-61 def, 470, 472-81, 562, 563; choice payoff, 75; with errors, 135, 139, 149; expected payoff, 63, 69, 74; full-information aggregate, 27; function, discontinuous, 44; law, 42; local, 27-28, 70, 72, 73, 75, 155; model, 26, 28, 72, 74; molar level, 43-44, 47, 154, 159; molecular level, 44, 47, 59, 74, 134-36, 140, 149, 460; momentary, 134, 136, 137, 153-59, 160-63, 454-55; obtained aggregate, 75; obtained reinforcement, 26, 44, 75, 76; payoff, 44; programmed reinforcement, 26, 47; of reinforcement, 314, 317-19, 322, 323, simple, 62; theory, 69, 75; utility, 62, 72
Maximum, global, 360
Maximum, local, 263
Mazes, 210-18, 220, 222; radial arm, 206, 207, 211-14; three table, 214-19, 222
Mean response rate, 112, 126, 128
Melioration, 184, 264, 434 def, 434-37, 439-40, 441, 446, 448-51, 453, 455-56, 470; comparative limits, 450-51; inherent limits, 450-51; motivational limits, 452; procedural limits, 452-53; psychophysical limits, 451-52
Memory, 26, 28, 42, 205-07, 215; working, 205-09, 222. *See also* Forgetting
Microeconomics, 353-54, 375
Micro level stimuli, 28 def, 34 def
Migration, 183, 184
Minimizing of punishment, 311-13, 322, 324
Minimum deviation, 523, 529, 532, 534
Minimum-distance model, 247, 395, 412, 415 def, 416-18, 423, 426-28
Misbehavior, 563-64
Miss, 30
Models of reinforcement value, 47-48; additive, 48; four-cycle case, 51-52; exponential, 50, 51, 54-55, 57; hyperbolic, 50, 54-55, 57; linear, 48, 50-51, 54-55, 57; molecular level, 32; multivariate, 50-51, 57; preference experiment, 48, 75; six-cycle case, 51-52
Molar aggregation, 154, 161, 163
Molar analysis. *See* Analysis, levels of
Molar level, 32, 43, 44, 48, 61, 76; data, 43, 62, 69, 70, 72; function, 48; measure of density, 44; results, 75, 154, 155, 161; stimuli, 31, 32, 41
Molar theory, 347-50; cross-situational test of, 472-78
Molecular aggregation, 154
Molecular analysis. *See* Analysis, levels of
Molecular level, 34, 43, 44, 48; data, 32, 43, 62, 70, 72

Natural selection, 169, 170, 182, 455
Need of predator, 228, 232, 235-36, 238
Negative law of effect, 327-28, 331, 333, 340

Noise, 42, 67, 74; additive model, 67, 69, 70, 75; additive process, 75; linear model, 67, 74; process, 75
Noncontingent shock, 328-29, 332-33, 340
Nondepletion of food supply, 215-18
Nonexclusive responding, 532, 534, 543
Nonpecking activities, 5. *See also* Interim activities; Leisure
Nonterminal troughs, 11 def, 16, 17, 22-23

Obtained reinforcement, 47, 75, 76. *See also* Value
OMNI model, 136-39
Operant, 7, 8, 17, 23; arbitrary, 173-74; chamber, 174, 182; conditioning box, 541; conditioning experiments, 25, 348, 350, 352, 356, 360, 371
Opponent-process theory, 506fn
Opportunity cost, 172, 184
Optimal strategy, 227, 236
Optimality, 190, 194, 247
Optimization, 227-28, 236-39, 323
Optimizing ideal observer, 43
Orienting response, 496
Overmatching, 181, 333-34, 337-38, 368

Paired baseline, 525, 529, 532-34, 538-39, 541-44, 559-60
Patches, 189, 190-92, 194-96, 198, 231-33, 235, 238-39; management of, 228, 232, 233, 235-37, 239
Pattern of reinforcement, 37 def, 38, 43, 47
P_e, 552
Perseveration, 155, 156, 159
Phylogenetic solution, 563-64
Pigeon, 174-81, 184
Ping-pong ball effect, 114-17, 127
Postdiscrimination phase, 249
Postreinforcement pauses, 80, 85, 86, 88-92, 95, 97, 101-02, 107
Power function, 550
Predation, 227-28, 231, 232, 235-36, 239
Predator population, 228, 238
Predator-prey system, 173, 189, 192, 200
Prediscrimination phase, 249
Preference: absolute, 184; data, 38, 43, 44, 47-48, 52, 54, 56, 57, 74-75; procedure, 27 def, 35-41, 46-48, 57, 75; for proportion, 367; stochastic, 184
Premack's probability hypothesis, 377, 497-98
Prey, 74 def, 169, 171-74, 176, 177, 183, 184; density, 74, 172-73, 177, 183, 185, 228, 231, 233; energy worth of, 171, 183; population, 228, 231-33, 236-37; predator-prey system, 173, 189, 192, 200
Price, 353
Priming, in mixed fixed ratio schedules, 81 def, 97, 106
Probability: of behavior, 25; learning, 190-91; of reinforcement, 25, 28, 48, 59; of responding. *See* Bias
Probe trial, 30
Probit transformation, 49, 62, 73
Procurement, 174
Programmed payoff. *See* Expected payoff
Programmed reinforcement. *See* Expected payoff
Propensity, 4 def, 5-8
Punchboard, 170
Punishment: avoidance theory, 327-29, 330-31; delayed, 452; competing response theory, 327, 331, 332, 338, 340; frequency, 328-29, 332, 337-38, 340-41; gradient, 331; intensity, 329, 330, 334, 340; subtractive model, 333-42
Punishment schedules: variable interval, 332-34, 338, 340; variable time, 306, 311; variable ratio, 341-42

Rat, 174, 183
Ratio size effects, 83, 85-92, 94-95, 99-101, 103-05
Rationality, 72
R_e, 552-55, 558, 559
Redshank, 169, 173
R_e parameter, 332, 333, 337-40
Regression technique, 177
Reinforcement, 433-36, 440, 442, 444-46, 451, 452, 455-56; adventitious, 5, 22; amount of, 36-37, 42, 47, 55, 59, 62; context of, 452; delay, 180-81, 444, 452; density, 31, 42, 44; frequency, 32-34, 47; immediacy of, 180; local probability, 155, 163; local rate, 265-66, 268, 434-37, 439, 441, 443, 448-49, 451-52, 454, 455. local relative probability, 160; negative, 327; obtained rate, 47, 75-76. *See also* Value. opportunity, 31, 34, 35; pattern of, 37, 38, 43, 47; probability of, 25, 28, 35, 43, 48, 59, 76, 155-57, 164, 189, 191, 198; rate, 171, 181, 183, 229-31, 233, 235, 237, 264-65, 268-69, 271-75, 277; samples from *T*-schedules, 36; sequence. *See separate listing.* shaping effect of, 111-12, 115, 116 def, 117, 120, 126-27; signaled, 283-86; strengthening effect of, 111-12, 116 def, 126; unified principle of, 500-02, 515-16; utility of, 37, 59, 61; value models. *See* Models of reinforcement value
Reinforcement, background (intrinsic), 554, 558-61, 562; commensurate measures of, 554-55; magnitude, 553, 555,

560; programmed (available), 553, 558; strengthening, 551, total, 553, 555-58, 562
Reinforcement schedules, 25, 27, 74, 227, 231, 375, 502-04; concurrent, 36, 40, 42, 113-15, 126-28, 132, 133, 136, 140-46, 171, 173, 174, 178, 181, 244, 247-49, 269, 274, 277, 281-82, 297-99, 300, 350, 357, 371, 509-11; concurrent chain, 27, 35, 36, 40, 75, 178, 181, 328, 331-32, 333-34, 338, 340-42; concurrent fixed ratio, 532, 543; concurrent fixed ratio 1-extinction, 42-43, 76; concurrent interval, 350, 366-69, 370, 549-52, 555-56, 558-61; concurrent ratio, 350, 361-66, 369, 370, 551-52; concurrent ratio-interval, 369-71; concurrent variable-interval, 153-55, 157-58, 160-61, 163-64, 243, 248, 258, 352, 366, 368-69; concurrent variable interval-fixed interval, 251-52; concurrent variable interval-variable interval, 446, 451; concurrent variable interval-variable ratio, 25, 29, 76, 395, 419-23, 438-43, 444, 453, 461, 472; concurrent variable ratio-variable ratio, 75, 436-38; fixed interval, 174-76, 178-79; fixed ratio, 30, 80-95, 101, 174, 257, 525, 529; fixed variable interval, 253, 356; interdependent, 274, 475-78; interval, 381-82, 386; lean, 27-28, 30-32, 34-35, 63, 74, 255-56; lean variable ratio, 76; linear, 269-70; mixed, 42; mixed fixed ratio, 80, 95-98, 106; modified, 28; multiple, 42, 171, 173-74, 243-44, 247-49, 251, 255, 257-60, 281-300, 452; multiple fixed ratio, 80 def, 81, 106-07; multiple variable interval-variable interval, 253, 256; probabilistic, 31-32, 37, 40; random interval, 31, 313-14, 317, 319; random ratio, 95-96, 99-101; ratio, 231-32, 381-82, 386, 395, 398, 400-03, 405-07, 410-12, 413-15, 419, 422-24; rich, 27-28, 30-32, 34, 35, 41, 62, 74, 255-56; rich variable ratio, 76; sample, 30, 74; sample utility, 74; simple, 554, 559, 560; simple fixed ratio, 523, 527-28, 538; simple variable interval, 529-32, 542-43; synthetic concurrent, 116-17, 127; *T*, four-cycle, 60; variable interval, 5-8, 22, 25, 31, 35-37, 112-13, 118-19, 126-28, 229, 230, 232, 246, 258, 265, 268-69, 271-72, 305, 328, 330, 332-34, 340, 351, 352, 356, 422, 423, 438-41, 443, 451, 529-32, 542-43; variable ratio, 76, 98-101, 258, 271-72, 274, 351-52, 436, 438-41, 443; variable time, 352, 405-07, 411-12, 419, 423
Reinforcement sequence, 27, 31, 36; value of, 26fn, 28
Reinforcer magnitude, effects, 83, 90-91, 107
Reinforcer value, 82-85, 90, 101-06
Relative: pattern duration, 115, 122-26; payoff probability, 43, 59; reinforcement rate, 42-43, 46; response rate, 26, 40, 42, 46; value, 40
Replenishment, 228, 235-37, 239
Reproductive fitness, 433
Rescorla-Wagner mode, 264, 275-76, 451
Response: allocation in reinforcement, 305, 309, 314-15, 317-22, 324, 357, 485; chains, 329-31; competition, 282-83; 285; deprivation, 499-500; discrete, 348, 360; distribution, 189, 191-92; effort, 83, 87, 91-95; free, 541; instrumental, 256, 396-97, 400, 412-13, 415, 424; local rate, 86 def, 101-02, 287-90; mean rate, 112, 126, 128; nonexclusive, 532, 534, 543; orienting, 496; pattern, 37; probabilistic, 42, 47, 67, 69. *See also* Bias. rate, 171, 181, 229-31; rate ratio, 179-80; sequences, 120-26; species-typical, 173; strength, 75, 249, 274; substitutability, 505; suppression, 327-28, 332-34; topography, 443-44, 446, 453-56; unscheduled, 532, 542-43; value, 376
Runway, 180, 523, 541

Sample (from a schedule), 48, 55, 57, 74, 75; perceived richness, 42, 74; *T*-schedule, 28-31, 34-36
Satiation, 102-06
Satiation-deprivation, 256, 258
Schedule-induced behavior, 22
Schedule vector, 524-27, 528-29, 532-34, 537-38, 541-44
Search strategies, 205-18, 221-22
Sensitivity, imperfect, 75
Sequential statistics, 133-35, 137-49
Selection, natural, 169, 170, 182
Selective advantage, 182, 184
Shift-depletion, 215, 219, 221-22
Shift-nondepletion, 215, 219, 221-22
Signal, 41, 67, 74
Signal detection analysis, 59, 75
Signal detection theory, 41, 62, 73
Species-specific defense reactions, 327
Speed, 4 def, 5-8
Stable performance, 30
Staddon & Motheral model, 481-82
Standard key, 37 def
Standard sequence of reinforcement, 27

Stay-depletion, 215, 219, 221–22
Stay-nondepletion, 215, 219, 221–22
Stay-shift, 208, 219, 220, 221
Stimulus, 28–30, 31 def, 32, 34 def, 41, 44, 48; period, 28–30, 76
Structure of behavior, 8, 9, 17, 22–23, 127–28
Stubbs-Pliskoff procedure, 40, 443, 444
Subpatch, 74
Substimulus, 29–31, 34–36, 38, 40–42, 46, 48, 62, 66; density, 44, 47–49, 51, 53, 54, 70, 72, 75; standard, 29, 37 def, 40, 48
Substitution, substitutability, 357–61, 365–69, 375–76, 377–80, 384, 388–92, 395, 401–03, 407, 414–15, 452, 505
Substitution effect, 403–04, 407, 410, 411, 412, 423, 527
Successive discrimination procedure, 194
Suppression, 525, 528, 532–34, 538–39, 541–44, 546
Survival problems, 549, 553, 561, 563
Switching, 153, 155–56, 160–61; rate 158–59

Tape-moves procedure: involving punishment, 313, 322, 323–24; involving reinforcement, 319, 322, 324
Tape-stops procedure: involving punishment, 313, 323–24; involving reinforcement, 305, 324
Taste, 368
Tautology, 360, 550, 552, 554, 556
Temporal triangle, 395, 412, 414
Time: constraints, 350, 353, 355; integration of, 347–50; until punishment, 310, 311, 313, 323; until reinforcement, 323
Time-allocation matching, 115–16, 245, 252–54, 347–50, 440, 556, 561; involving punishment, 306, 309–11, 313; involving reinforcement, 305, 309, 311, 314–15, 318–22, 324
Titration, 38, 40 def, 46, 58, 75
Topography, 4, 22, 23
Transformations, 41, 46, 48–50, 55, 59, 73; arcsine, 49, 50, 73; logit, 49, 50, 61–62, 73; probit, 49, 62, 73; *z*, 49, 70, 72
Travel time, 193 def, 194–201, 235
Trial, 27, 29, 30, 34, 35, 42, 76; probe, 30, 35
Trial and error, 179

Undermatching, 44, 47, 75, 76, 243 def, 244, 247–48, 251–53, 258, 270, 333, 337, 339, 341, 365, 367–68, 444, 446; of obtained payoff (reinforcement), 26, 47, 75; of response probability, 75
Utility, 375, 433, 434 def, 443, 452, 527, 542; effective, 75; expected, 26fn, function, 26fn, 55, 56; marginal, 247, 452; maximizing, 62, 72; of reinforcement, 37, 59, 61

Value, 353–55, 369, 433, 434; effective, 27 def, 28, 38, 41, 46, 49, 53, 55–56; exponentially decreasing, 51; function, 378, 380–81, 383, 390–92; models of, 50–55, molar, 47; mean perceived, 59, perceived, 27–28, 38, 41, 46, 48, 49, 54, 55, 57; predicted, 49; of reinforcement, 26fn, 27–28, 40, 50, 67, 73; of standard substimulus, 38, 44, 46, 48, 49, 51, 54, 56, 61; total, 47; transformed, 48, 49, 70, 554–58, 562

Wage rate, 395, 400–02, 404, 411, 412, 527–28. *See also* Income
Wheel running, 524–25
Win-shift, 205; in natural habitats, 208–10; in laboratory tasks, 210–19
Win-stay, 205, 220; in natural habitats, 208–10; in laboratory tasks, 211–19
Within-ratio pauses, 81 def, 86–88, 95, 102, 106
Work. *See* Response, instrumental

Zero-bias experiment, 62, 66

ABOUT THE EDITORS

Michael Lamport Commons was born October 1939 in Hollywood, California. He did his undergraduate work at the University of California at Berkeley and then at Los Angeles, where in 1965 he received a B.A. in mathematics and psychology. In 1967 he received his M.A. and in 1973 his Ph.D. in psychology from Columbia University. Before coming to Harvard University in 1977, where he is now a research associate, he was an assistant professor at Northern Michigan University. He co-edited *Quantitative Analyses of Behavior, Volume I: Discriminative Properties of Reinforcement Schedules.* His research is on the perception of utility and how it develops in infrahumans and humans across the lifespan, including the subareas of perception and knowledge of causal relations and of value.

Richard J. Herrnstein is Edgar Pierce Professor of Psychology at Harvard University, where he has been teaching since 1958. He obtained his A.B. in 1952 from the College of the City of New York and his Ph.D. in 1955 from Harvard University. He is past editor of *Psychological Bulletin*, has edited and authored several books, and has contributed to numerous periodicals. Among his long-standing interests is the quantitative analysis of behavior, especially with regard to the effects of reward and punishment.

Howard Rachlin received his B.E. in mechanical engineering from Cooper Union in 1956, his M.A. in psychology from The New School for Social Research in 1962, and his Ph.D. in psychology from Harvard University in 1965.

He taught at Harvard from 1965 until 1969. He is Professor of Psychology at the State University of New York at Stony Brook where he has been teaching since 1969. His interests include self-control, the interaction of economics and psychology, the matching law, the psychology of eating and drinking, and the philosophy of psychology.

LIST OF CONTRIBUTORS

James Allison
Department of Psychology
Indiana University

Raymond Battalio
Department of Economics
Texas A&M University

William M. Baum
Department of Psychology
University of New Hampshire

Barbara Burkhard
Department of Psychology
State University of New York
at Stony Brook

Michael L. Commons
Department of Psychology and
Social Relations
Harvard University

Michael Crowley
Department of Psychology
University of Massachusetts at Amherst
Now at:
Computing Center
Mount Holyoke College

Marvin Z. Deluty
Department of Physiology
University of Pennsylvania School
of Medicine
Now at:
Behavior Therapy Unit
Department of Psychiatry
Temple University Medical School

Peter A. de Villiers
Department of Psychology and
Social Relations
Harvard University
Now at:
Department of Psychology
Smith College

John W. Donahoe
Department of Psychology
University of Massachusetts at Amherst

John R. Ducheny
Department of Psychology
Boston University
Now at:
Department of Behavioral Sciences
University of Chicago

Leonard Green
Department of Psychology
Washington University

Richard J. Herrnstein
Department of Psychology
and Social Relations
Harvard University

Gene M. Heyman
Department of Psychology
and Social Relations
Harvard University
Now at:
Department of Pharmacological
and Physiological Sciences
University of Chicago

John H. Kagel
Department of Economics
Texas A&M University

Alan C. Kamil
Departments of Psychology and Zoology
University of Massachusetts at Amherst

Stephen E.G. Lea
Department of Psychology
University of Exeter

Joseph A. Legris
Department of Psychology
University of Manitoba

Frederick J. Lindstrom
Department of Zoology
University of Massachusetts at Amherst

Will J. Millard
Department of Psychology
University of Massachusetts at Amherst
Now at:
The Rockefeller University

James E. Mazur
Department of Psychology
Yale University
Now at:
Department of Psychology
and Social Relations
Harvard University

John A. Nevin
Department of Psychology
University of New Hampshire

David S. Olton
Department of Psychology
Johns Hopkins University

Joseph J. Pear
Department of Psychology
University of Manitoba

Jerrilynn Peters
Department of Psychology
Rutgers University, New Brunswick

Howard Rachlin
Department of Psychology
State University of New York
at Stony Brook

Brian L. Rector
Department of Psychology
University of Manitoba
Now at:
Valley View Centre
Moosejaw, Saskatchewan

Charles P. Shimp
Department of Psychology
University of Utah

Alan Silberberg
Department of Psychology
American University

John E.R. Staddon
Department of Psychology
Duke University

Kenneth A. Stickney
Department of Psychology
University of Massachusetts at Amherst

William Timberlake
Department of Psychology
Indiana University

William Vaughan, Jr.
Department of Psychology
and Social Relations
Harvard University

Ben A. Williams
Department of Psychology
University of California, San Diego

Michael Woodford
Department of Economics
Massachusetts Institute of Technology

John M. Ziriax
Department of Psychology
American University
Now at:
Health Effects Research Laboratory
Neurotoxicology Division
Environmental Protection Agency
Research Triangle Park